Introduction to PHYSIOLOGY

VOLUME 1

Introduction to PHYSIOLOGY

VOLUME 1
BASIC MECHANISMS

PART 1

HUGH DAVSON
Physiology Department, University College
London, England

M. B. SEGAL
Sherrington School of Physiology, St. Thomas's Hospital
London, England

1975

ACADEMIC PRESS
and
GRUNE & STRATTON

U.K. edition published and distributed by
ACADEMIC PRESS INC. (LONDON) LTD.
24/28 Oval Road
London NW1

United States Edition published and distributed by
GRUNE & STRATTON INC.
111 Fifth Avenue
New York, New York 10003

Second printing 1976

Library of Congress Catalog Card Number: 75 5668
ISBN (Academic Press): 0 12 206801 7
ISBN (Grune & Stratton): 0 8089 0896 0

Printed in Great Britain by
The Whitefriars Press Ltd., London and Tonbridge

PREFACE

We can think of two ways of composing an Introduction to Physiology. First we may take a large standard text and sieve the material contained in it to free it of as much experimental material, argument and other "extraneous matter" to reduce its bulk to about one third of the original. Alternatively one may compose something entirely new, expounding as simply as feasible the basic scientific principles governing the functioning of the animal. The former method has, we think, been employed before, and the result has been a *synopsis* of, rather than an *introduction* to, physiology. By memorizing it almost word-for-word the medical student probably passes muster at an undiscriminating examination, and completes his medical education with a very poor understanding of the basic principles of medicine.

Pursuing the latter method we found that the first draft was embarrassingly large, and in reviewing what had been written with a view to shortening the book, it became clear that any serious surgery would destroy its character since it was, in effect, rather more than an Introduction containing—to use a musical term—a great deal of "development" too. Rather than abandon the project of producing an Introduction that was both short and adequate, we carried out a different kind of surgery, namely the division into several volumes. Volumes 1 and 2, which we now present, are an introduction to the basic mechanisms whereby the animal absorbs, distributes and transforms its energy-giving materials; and whereby the energy thus made available is utilized in such fundamental activities as muscular contraction, the transmission of messages by both nerves and hormones, the defence mechanisms and in reproduction.

The difficulties in understanding physiology arise in the fundamental principles governing the activities of the animal's parts, such as the flow of fluids, the conduction of the nervous impulse, the elimination of secretions from a cell or epithelium and so on. If the student has a firm grasp of these principles, the way is clear for the understanding of the rest of physiology, which consists in the analysis of control mechanisms. The remaining volumes are designed to enable the student to take up where the first two left off; thus Volume 3 is devoted to visceral control mechanisms and may be regarded as the "development" of the themes introduced mainly in Volume 1. Very arbitrarily the control of somatic motor activity and of reproduction have been put together to make Volume 4; this is only because their inclusion in Volume 3 would have made it too large for convenience. Volume 5 deals with sensory mechanisms and higher integrative processes, involving the cerebral cortex.

A few words on the way the volumes have been written. The present two volumes, being concerned largely with fundamentals, require little or no documentation, so that we have contented ourselves mostly with general references to reviews and texts at the end of each volume. This does not mean that the information has been culled only from these sources, and it is rare if we have quoted work that we have not read in the original. In the remaining volumes the subject matter has been treated in greater experimental depth, so that a more elaborate documentation, comparable with that found in Starling's *Principles of Human Physiology*, has been employed.

To conclude, we think that a study of the completed work will provide the student of physiology, taking this as part of a larger course, such as in medicine or dentistry, with knowledge of the subject sufficient for his requirements; for the student intending to make physiology his career the book will, we trust, be a proper "Introduction".

HUGH DAVSON
M. B. SEGAL

October 1974

CONTENTS

ACKNOWLEDGMENTS

We should like to record our indebtedness to those authors and publishers who gave us permission to reproduce illustrations, also to Jane Barnett for secretarial assistance, and to Moyra Harding for assistance with illustrations.

CHAPTER 1

Structure and Function

Introduction

Gross Anatomy

Physiology is the science of the functioning of living organisms; in order to discover how these living organisms work, i.e. the physical and chemical laws governing their behaviour, we must ultimately dissect them so as to reveal the relations of their parts.

With the naked eye to observe, and the scalpel to dissect, we can discover a great deal, and this "gross anatomical" approach has been followed exhaustively in man and other large animals, so that the subject of gross anatomy is essentially a completed story. In this way we have discovered the way in which blood is distributed to the body, the manner in which the bones are articulated and their muscles attached, and so on. Thus with only this gross anatomical knowledge the physiologist, by virtue of the ingenious design of his experiments on the living animal, has been able to attempt explanations for a great many of the phenomena of life in higher organisms, e.g. the mechanism of breathing, the intimate connection of breathing with the supply of blood to the lungs and the rest of the body, the mode of intake and digestion of food, the mechanics of the contraction of muscle and the relation of energy consumption to this process, and many features of the control mechanisms exerted through the nervous system.

Microscopy

The study of anatomy on a microscopical level—called *histology* and *cytology*—is necessary if the physiologist is to be permitted to interpret and design his experiments with more meaning; thus, with only a knowledge of gross anatomy we can derive an enormous amount of information regarding the circulation of the blood, its changes with exercise, with climbing to high altitudes, and so on. However, the manner in which the circulation fulfils its functions (namely, absorbing oxygen from the air and delivering it to the tissues of the body; carrying absorbed materials from the intestine to the liver and depositing these materials in the organs) can only be more fully elucidated by a knowledge of the structure of the smallest blood vessels—the capillaries, which are fine tubes in the region of 10 μm diameter or less that form the connecting links between the arterial and venous systems. These tubes are too small to be seen with the unaided eye, and it required the development of the microscope and its application to the tissues of living organisms to demonstrate the structures of these vessels which, besides representing the connecting links between the arterial and venous systems postulated by Harvey, represent the locus in the blood vascular system at which materials are able to escape from, or enter, the blood. The microscope can reveal many details in the structure of these small vessels, and their relation to the tissues with which they come into contact. With this histological and cytological information the physiologist may carry out functional experiments designed both to show how materials can escape from these fine tubes and to elucidate the mechanisms of control of these transport phenomena.

Electron Microscopy

The ordinary light microscope does not permit the physiologist to see the complete details of the structures he is concerned with. To keep to our example, his physiological experiments tell him that the fine tubes or capillaries probably have holes in them that allow molecules of a certain size to pass through, whilst larger ones are retained either completely or partially, i.e. the experiments indicate that the tubes are behaving like sieves or filters, thereby exerting some control over the types of molecule that can pass into or out of the blood and the rates at which the permeating molecules pass to and fro. The holes, deduced theoretically, would have diameters of about 80 Ångstrom units (80 Å = 8 nm). Now the limit of resolution of the conventional light microscope is about 0·2 μm, and since 1 μm is 10^4 Ångstrom units, the resolution is some 2000 Ångstom units, so that it would not be possible to see the holes in the capillaries. Within the past twenty years, however, the sciences of anatomy and histology have been fortified by the development of the electron microscope, which has revealed details in structure that are quite unresolvable in the light microscope; its theoretical limit of resolving power is of the order of a few Ångstrom units, and as techniques of preparation of the specimens have improved, structural details of this order of magnitude have become visible. In the capillaries, the postulated holes have, indeed, been resolved, but of course the electron microscope is used to examine dried and chemically treated tissue, so that the holes revealed by this instrument may not be of the same size in the living material.

Structure and Function

The electron microscope has certainly revolutionized the anatomist's way of life, and has helped the physiologist to a great extent, often confirming deductions as to structure that were made entirely as a result of functional experiments, just as the light microscope confirmed Harvey's hypothesis of the connection between the arteries and veins. For example, the muscle of heart was recognized to have very special properties, in that the individual fibres of which it was composed were in some way connected with each other, in marked contrast to the muscles of the skeleton which behaved as though the fibres were quite separate. The resolution of the light microscope did not permit the anatomist to say categorically whether the cardiac muscle fibres were, indeed, connected or fused together, and it was not until the electron microscope was applied to the problem that the existence of localized regions of fusion between adjacent fibres was confirmed. This is an example of the usefulness of the anatomical studies in *confirming* a deduction from physiological experiments. It would be easy to cite other cases where, instead, the morphological discovery suggested a theory of function. Thus we may, once again, choose muscle.

Shortening of Muscle

The special feature of muscle is its ability to shorten, and thus perform mechanical work; this is illustrated in Fig. 1.1, where we see

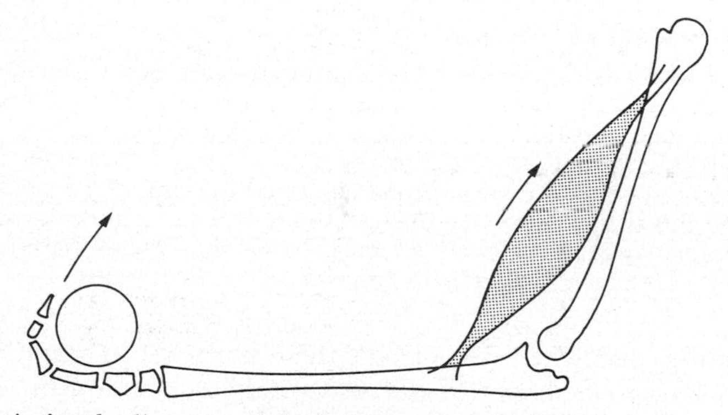

Fig. 1.1. A simple diagram of the arm showing the biceps muscle shortening to lift an object.

a muscle of the arm; it has its origin on the humerus and is inserted or attached to the bone of the forearm. By shortening, it causes the arm to bend, or flex, and mechanical work will be done if the bending of the arm causes a weight to be lifted.

Folding of Protein Chain. If we examine this muscle in the light microscope we see that it is composed of bundles of fibres running

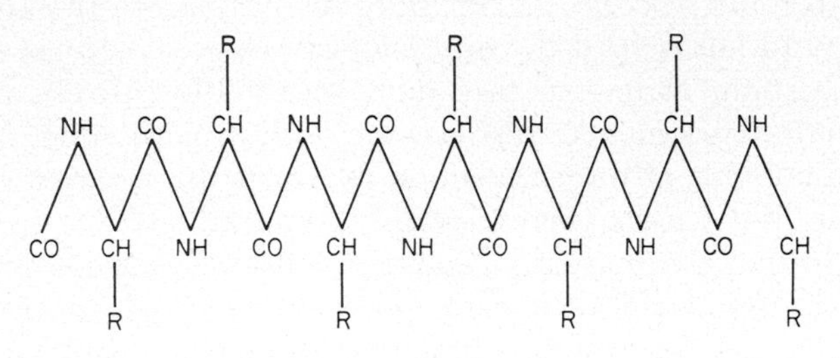

β-form straight chain

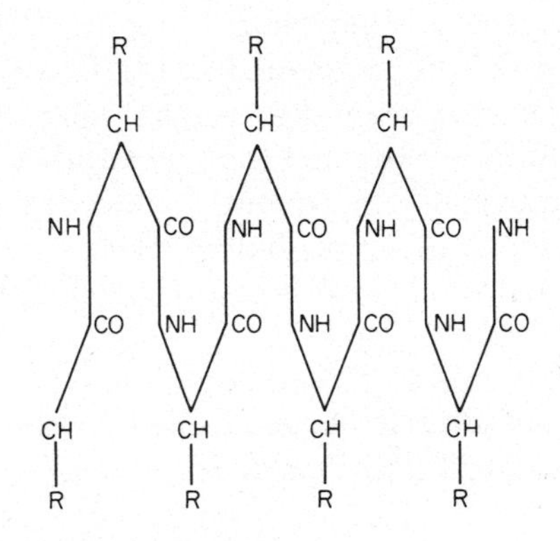

α-form folded chain (postulated)

Fig. 1.2. Two forms in which fibrous proteins can exist—the β-form in which the molecule is stretched out and the α-form when the molecule is folded and much shorter. This folding of protein molecules was used to explain muscle contraction until the use of X-ray crystallography and the electron microscope on muscle sections suggested the sliding filament model. The folding illustrated here was that postulated in the nineteen-thirties by Astbury; later work, as we shall see, has shown that the α-chain is helical rather than linear.

longitudinally, and when it contracts these fibres shorten and become fatter. An attractive theory of the structural basis of contraction was based on the assumption that the muscle fibres consisted essentially of bundles of very thin long molecules of a fibrous type of protein called myosin, and these long molecules were thought to exist in two forms, a fully stretched, or β-state and a more folded or α-state, as illustrated schematically in Fig. 1.2. Such stretched and partially folded states had been recognized to occur in other protein molecules, such as the protein of the wool fibre, keratin, the structures being recognized by the use of X-ray diffraction, the only method of deducing structure on a very small scale before the advent of the electron microscope. The protein, myosin, a prominent constituent of muscle, was shown to be able to undergo this transformation under certain experimental conditions. However, examination of the muscle by X-rays gave no evidence of this sort of change during contraction, and the true nature of the contractile process was revealed by a combination of morphological studies, involving the light and electron microscopes, and the use of X-ray diffraction.

Sliding Filaments. The studies indicated that the fibrils did not fold up but retained their original lengths; instead of folding, one type of fibre slid alongside its neighbours of another type, and as a result

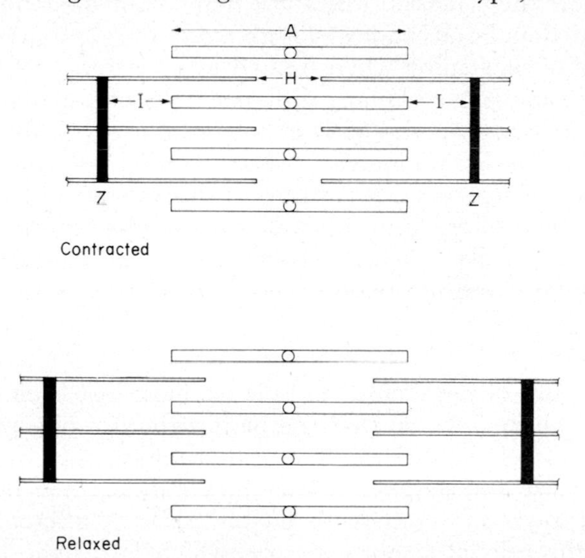

Fig. 1.3. The sliding filament model of muscle contraction. The thin filaments are muscle protein actin and are attached to the Z line. The thicker filaments are myosin, and shortening follows from a sliding action leading to increasing overlap.

the bundle as a whole shortened (Fig. 1.3). Thus the problem left for the physiologist, after this deduction from structure, was: "What makes the filaments slide?" Here, then, is an example of how morphological study has suggested a physiological mechanism, although of course, in general, the two approaches are made together so that it is never easy to say whether the structure has suggested the mechanism or the physiological phenomenon has suggested or demanded a structure.

THE FUNCTIONAL UNITS OF STRUCTURE AND BEHAVIOUR

The Cell

The structural analysis of muscle has shown us different orders of magnitude, based primarily on the limits of resolving power of the unaided eye, of the microscope and the electron microscope and X-ray diffraction. With the naked eye we may discern the fibrillar nature of many muscles, and in the light microscope we can see that the basis for this is the grouping together of long thin muscle fibres to form bundles. Now these muscle fibres are more than their name implies; they are functional units endowed with many more properties than the obvious one of shortening when treated in a certain way. The muscle fibre is a *cell*, and is one of many different types of functional unit that, working together, form the basis of the structure and function of the organism. It was Schwann who proposed what was then called the "cell theory" of structure; according to this the various tissues of the body could be resolved into units, or *cells*, of characteristic types, the behaviour of the tissue being determined by the characters of these units of structure and behaviour.

Muscle Fibre

Thus the unit of the skeletal muscle we have just been discussing is the muscle cell or fibre, so that the basic structure of a given skeletal muscle consists of bundles of these cells running side by side. These cells vary in size and shape from one muscle to another, but they have sufficient features in common to enable a clear differentiation from these and the cells of smooth muscle, which forms the basis for the contraction of the gut, the blood vessels and some other tissues.

Neurone

Another type of cell is the neurone, or nerve cell; like the skeletal muscle cell it is fibrous in type and arranged in groups, so that the bundles of fibrous extensions of these nerve cells make up what is visible to the naked eye, namely the nerve of the gross anatomist.

Cell Sheets

The surface layer of the skin and many other parts of the body, such as the gastro-intestinal tract, contain cells of another type, which are arranged in flat sheets to form what is called an *epithelium;* here they are usually columnar in shape and their tightly packed arrangement to form a sheet permits them to control the passage of material from one side of the sheet to the other; thus the epithelial sheet lining the gut controls the absorption of material from the lumen of the gut into the underlying tissue and its blood vessels. The fine vascular tubes that we have described earlier, namely the capillaries, consist of little else than a sheet of very thin cells described, this time, because of their different embryonic origin, as *endothelium.*

Blood Cells

Other cells may exist as units not arranged in a rigid or semi-rigid structure; thus the blood is essentially a suspension of individual cells of several types in a fluid medium, the plasma.

Units of Function

Some of the types of cell and their arrangement are illustrated in Fig. 1.4. In later chapters we shall have occasion to describe many of

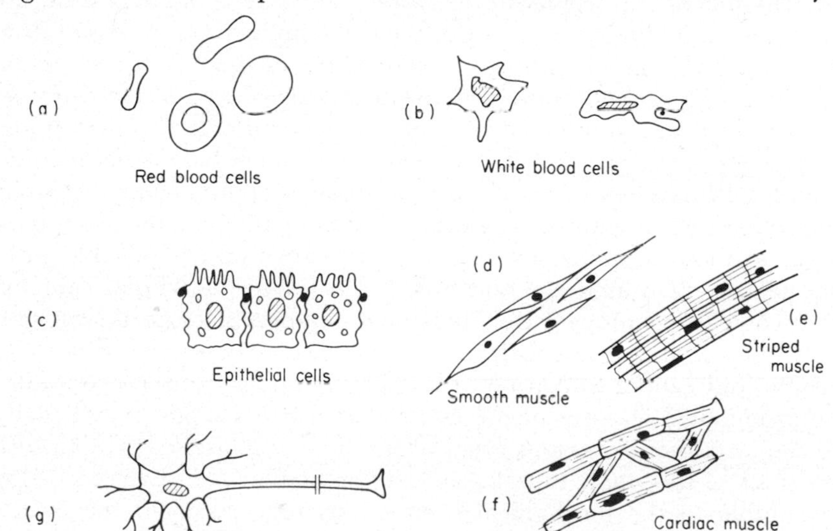

Fig. 1.4. A few examples of the variety of cell forms seen in animals: (a) red blood cells; (b) white blood cells; (c) epithelial cells joined in a sheet; (d) smooth muscle fibres; (e) striped or voluntary muscle fibres; (f) cardiac muscle; (g) a lower motor neurone.

these in more detail; it is sufficient here to appreciate that these are the physiological units of function, and that they represent degrees and forms of specialization of the primary mass of cells formed by division of the fertilized ovum. As units we may say that they are endowed with life, by contrast with subunits derived from breaking up the cell into smaller components, such as mitochondria, myofibrils, and so on.

Isolated Cells

The unitary character of the blood cells is obvious, and they can be shown to fulfil their functions independently of each other and of other structures. Thus the red blood cell, or erythrocyte, can carry oxygen and carbon dioxide when separated from its neighbouring cells, or even when taken away from its normal environment (the plasma) and placed in an appropriate salt solution. Our concept of the unitary character of cells organized in complex structures is not so easily verified, and with certain tissues, such as the epithelial sheets described earlier, must be limited in its application.

Tissue Culture

By the aid of the technique of tissue-culture it is often possible to grow cells individually. This is done by placing a piece of tissue in a medium capable of providing the cells with nutriment; if successful, the tissue will produce new cells from its edges, and these may be separated and shown to be capable of independent existence, retaining many of the features of the tissue from which they were derived. For example cultured heart muscle fibres exhibit the spontaneous rhythmic contractions, accompanied by electrical waves called action potentials, similar to those of the intact muscle. Alternatively, it is possible with some tissues, such as nerve and muscle, to isolate individual cells by microscopical techniques; thus it is now commonplace to dissect out single fibres from a frog's skeletal muscle; these may be kept in a fluid medium and caused to contract by applying an appropriate stimulus. The mechanical features of the contraction of this single muscle cell, and the responses to various types of stimulus, are essentially identical with those of the whole muscle, such differences as are observed being attributable to the mechanical interactions that occur in the intact tissue, e.g. the attachment to tendon and adjacent fibres. In this way we can justify the concept of the muscle cell as being the fundamental unit, the behaviour of the whole muscle being essentially predictable from the summated behaviours of the individual units.

This justification, however, is not complete since we may break the fibre up into smaller units, namely myofilaments, and bring about contraction of these; but these fibrillar parts of the muscle fibre do not possess the character of an intact muscle cell, e.g. they are unable to respond to an electrical stimulus, they are unable to utilize sugar to provide the energy of contraction, and so on. Thus, to demonstrate fully the unitary character of the cell we must use many criteria, two of the most fundamental being its *organization* (i.e., the existence within it of subunits of structure that work together to achieve a form of activity) and its separation from adjacent cells or fluid by a membrane, the *plasma membrane.* To appreciate the full significance of these features we must examine the structure and behaviour of the individual cell in some detail. The student of physiology must appreciate at the outset that this is no mere academic exercise unnecessary for learning the essential principles of physiology. Just the opposite; he may, indeed, grasp the mechanical basis of the circulation of the blood without knowing that a cell exists, and the same is true of many other of the grosser features of animal function, but as soon as the more fundamental mechanisms are investigated, immediately recourse is had to the analysis of behaviour in terms of the constituent cells. Let us then examine the structure of the cell in some detail and consider also some of the things that a cell, acting alone, may do.

The "Typical Cell"

From what has just been said, it must be clear that it is not easy to describe a "typical cell", since by its very nature a cell is differentiated to be able to exercise certain functions, so that to describe only those features that are common to all cells would leave us with a very bare skeleton indeed. The scheme that follows is, therefore, only a scheme, and cannot be said to characterize any actual cell at all completely.

Cytoplasm

The envelope that separates the cell from other cells and from its fluid environment is called the plasma membrane (Fig. 1.5); this is extremely thin, namely some 100 Å across, and thus well below the resolution of the light microscope; in the electron microscope it appears as a dense line if the tissue has been fixed with, say, osmium tetroxide or permanganate. With good resolution this dense line can be resolved into two dense lines sandwiching a less dense line (Fig. 1.6) suggesting that the fixative is taken up at the two surfaces. Within this envelope there is the fluid, or semi-fluid, material called *cytoplasm;* in many cells the cytoplasm is differentiated into an outer, gelatinous, *cortex* or *ectoplasm*, and an inner, fluid, *endoplasm* as revealed by the behaviour of particles in the two regions. Floating or embedded in the cytoplasm are the various intracellular organelles or functional subunits of the cell. The most obvious is the nucleus, separated from the cytoplasm by the nuclear membrane.

Nucleus

The nucleus contains the *chromosomes*, the structural elements that are responsible for transmitting the hereditary features of the organism from one cell to its daughter cells, namely the *genes*. Since every cell of the body is derived from one fertilized ovum—the *zygote*—and since

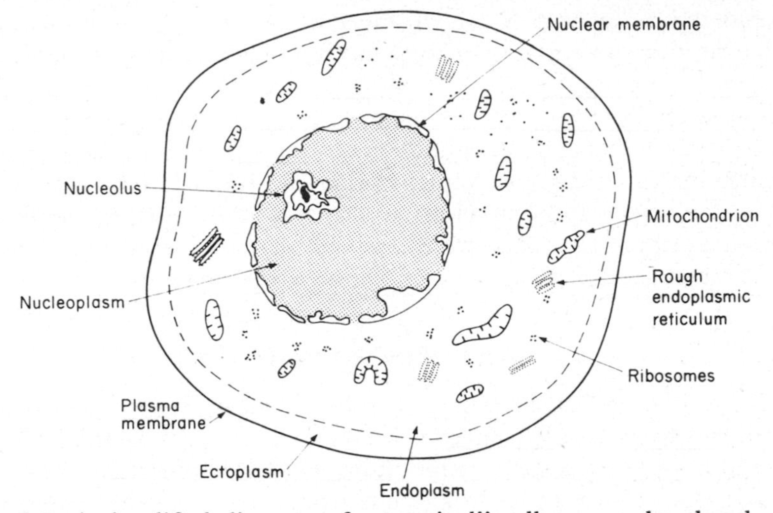

Fig. 1.5. A simplified diagram of a "typical" cell seen under the electron microscope.

the chromosomes divide to form pairs of new chromosomes, one of which is transmitted to each daughter cell, every cell in the body has a full complement of genes. The genes control the activity of the cell as a whole by virtue of the substances they contain and transmit to the cytoplasm, for example, certain nucleic acids that control the type of enzymes that the cell can synthesize. The activities of cells vary widely and we may assume that this is because, in some way, the activities of certain genes in a given nucleus are suppressed, so that only a few are actually active, and these determine the way in which the particular cell will grow, i.e. the shape it will assume, and the type of activity that it will adopt.

Fig. 1.6. A high-powered electron micrograph of the plasma membrane. The membrane can be resolved into three components: two dark regions which have taken up the electron stain and a pale region which has a much lower affinity for the stain. Each band is some 25 Å thick. (Robertson in Hendler, *Physiol. Rev.* 1966, **51,** 971.)

Nuclear Envelope. The nuclear membrane, or envelope, contains large pores which may be as much as 1000 Å in diameter; this is in marked contrast to the limiting membrane of the cell (the plasma membrane) which contains no discontinuities resolvable in the electron microscope. This envelope is double, consisting of two layers separated by an easily resolvable space; the pores pass through both layers, the rims to the pores being formed by the fusion of the two membranes, as

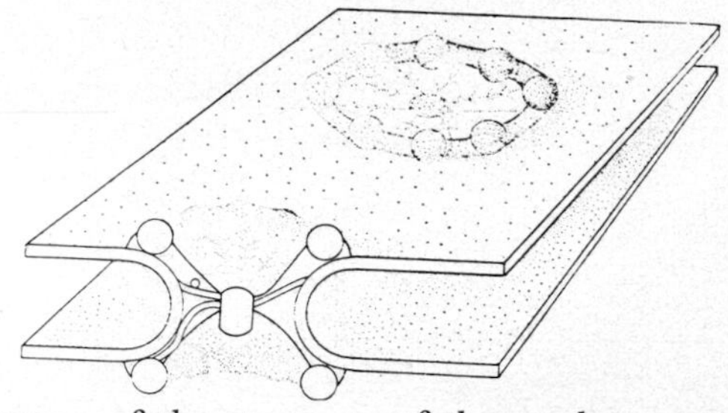

Fig. 1.7. A diagram of the structure of the nuclear pore as derived from electron microscope studies using a diversity of animal and plant cells. Embedded in diffuse material, eight regularly spaced annular granules lie upon either pore margin. A central granule or rod is located in the innermost part of the pore. These pores form pathways through the nuclear membrane. (Franke, *Z. Zellforsch.* 1970, **105,** 405.)

illustrated by Fig. 1.7. These pores undoubtedly represent the pathway for passage of large molecular weight material, such as nucleoproteins, from nucleus to cytoplasm; in electron-microscopical preparations they are by no means empty, containing material that has been described as the pore-complex. According to Fig. 1.7, the complex consists of annular granules arranged round the rim, with a central granule connected to the rest by fibrillar structures, whilst the spaces between are filled by a diffuse embedding material.

Mitochondria

The mitochondria are much smaller organelles, just resolvable in the light microscope; like the cell they are enclosed in an envelope which seems very similar in structure to the plasma membrane, but, as illustrated in Fig. 1.8, the membrane is double, the inner one being

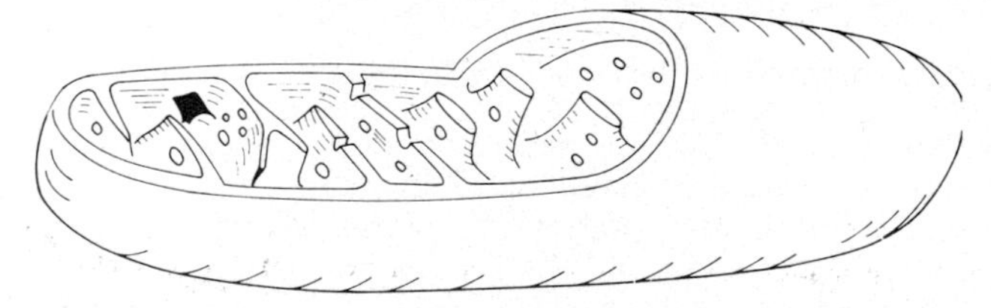

Fig. 1.8. A diagrammatic section through a mitochondrion showing the internal membranes or cristae.

folded to form the so-called *cristae mitochondriales* that doubtless serve to increase the total area of membrane. On the cristae and other regions of membrane the enzymes responsible for the conversion of foodstuffs to energy in a utilizable form are concentrated; thus the mitochondrion has been described as the "power-house" of the cell, being responsible for the oxidative metabolism that permits the cell to carry out its functions. As we shall see, these metabolic reactions are numerous and take place in an ordered sequence, each being catalysed by a specific catalyst or enzyme. The membranes of the mitochondrion represent the solid basis for the organization of many of these enzymes, so that the products of one reaction are close to the enzyme required for the next reaction in which they participate. When the mitochondrion is broken up into smaller fragments, certain "unitary particles" may be separated capable of carrying out many of the functions of the complete mitochondrion. The red blood cells of mammals have no mitochondria and thus are unable to carry out oxidative metabolic reactions; as we shall see, however, the organism can utilize sugars as sources of energy without using oxygen; this *anaerobic glycolysis* is carried out within the cytoplasm and not in the mitochondria, so that the red blood cell is not devoid of energy supplies.

Mobile Power House. In general, the mitochondrion is to be regarded as a mobile power house within the cell that takes up the basic materials required for energy production and transforms them into substances of lower energy content, the energy set free being utilized for a variety of processes, e.g. synthesis of proteins, contraction of fibrillar systems, and so on. It is not a permanent structure, being eventually broken down whilst new mitochondria are formed by reproduction from the old. The basis for the mitochondrion's motility is not always clear; there seems little doubt that in a neurone, where it may have to travel several feet to pass from its region of formation to where it is required to function, it is carried by a slow current of cytoplasm; whether or not the shape-changes that it is certainly able to undergo contribute to some form of swimming motion is not certain.

Endoplasmic Reticulum

In some cells the matrix, or cytoplasm, in which the organelles are suspended seems homogeneous even in the electron microscope; in most cells, however, the cytoplasm contains a large mass of additional material with recognizable structure; this is the *endoplasmic reticulum*, which appears as a series of tubes or flat cisterns all apparently made of the same material as that constituting the plasma membrane. On one surface of some of these membranes there are attached small particles, about 100 Å in diameter named, after their discoverer, *Palade particles*. These are, in fact, ribosomes, the nucleoprotein structural elements on which the cell's proteins are synthesized. Such regions of endoplasmic reticulum are called "rough-surfaced" to differentiate them from the "smooth-surfaced" intracellular membranes lacking these particles.

Microsomes. One way of studying the structure and function of the cell is to break it up and separate the pieces and the liberated organelles by high speed centrifugation. In this way the larger particles, namely the nuclei and mitochondria, may be separated by applying relatively small centrifugal forces (3000 G). After separating these, application of much higher forces—of the order of 20,000–100,000 G—causes much smaller particles, originally called the *microsomes*, to settle. These

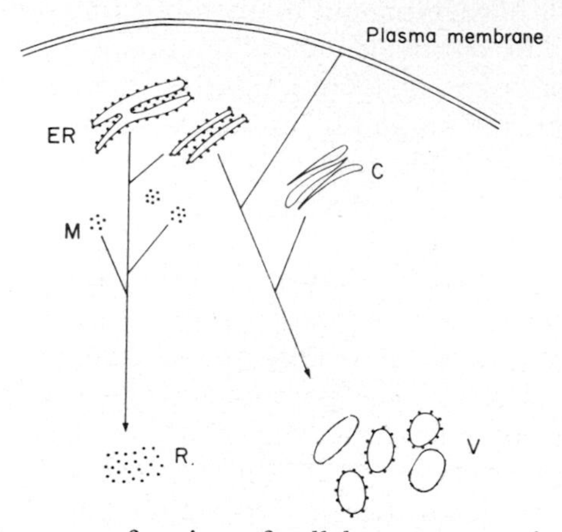

Fig. 1.9. The microsome fraction of cell homogenates is derived from the endoplasmic reticulum. This fraction can be further subdivided, by centrifugation at higher speeds, into one composed mostly of smooth vesicles formed from the membranous material of the reticulum and one composed of the much smaller Palade particles or ribosomes. ER: endoplasmic reticulum; M: microsome fraction; R: ribosomes; V: vesicles; C: Smooth surfaced cisterns.

particles are not of uniform size and are quite clearly the products of breaking up the endoplasmic reticulum, so that one of the fractions consists of the Palade particles or ribosomes. Other fractions consist of vesicles, which have been formed from the membranous material of the reticulum as illustrated in Fig. 1.9.

Golgi Apparatus

Of special interest in the membranous system of the cytoplasm is the *Golgi apparatus*, a system of vesicles or flat membranous cisterns usually occupying a characteristic region of the cell. It is especially prominent in cells that produce a secretion, such as the pancreas or mammary gland, and the function of the apparatus is revealed as one of storing the materials that are primarily synthesized on the rough-surfaced endoplasmic reticulum.

Intracellular Synthesis. Thus the material secreted by certain glandular cells, such as the parotid salivary gland or the exocrine cell of the pancreas, is an enzyme or catalyst required for digestive processes. When required, the material is expelled from the cells into small channels, or acini, whence they are carried to their site of action, e.g. the mouth for saliva, or the intestine for pancreatic secretion. Within

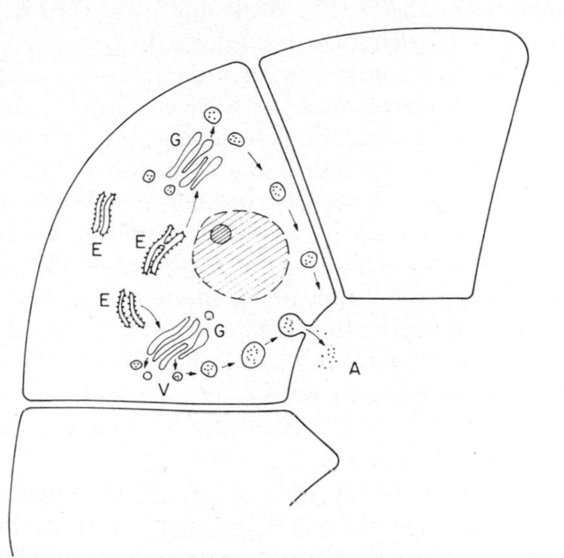

Fig. 1.10. Enzymes are synthesized on the rough-surfaced endoplasmic reticulum (E) in the form of granules. These granules are transferred to the Golgi apparatus (G) where they become condensation vesicles (V) and these finally migrate to the cell wall facing the acinus (A). The vesicle fuses with the cell wall, opens into the acinus and the contents are released.

the cell the enzymes are recognizable as well-defined granules (Fig. 1.10) and the process of synthesis of these granules can be followed in the electron microscope by the accumulation of radioactive amino acids. The presence of the radioactive material can be recognized in electron microscopical sections by the technique of autoradiography, the radioactive emissions being captured by a thin photographic film to give particles of silver which may be seen on the electron micrograph. Palade showed that, when the excised gland was exposed to labelled amino acid (^{3}H-leucine), the radioactivity first appeared above the rough-surfaced endoplasmic reticulum, indicating that synthesis occurs here; later the radioactivity moved away from the endoplasmic reticulum to the Golgi apparatus the silver grains now lying predominantly over the stacked cisternae and in expanded vesicles, these last representing the growing newly produced granules. Finally the silver grains were found predominantly in the mature secretory granules.

The Plasma Membrane

The cell membrane, seen as a dark line some 100 Å thick on the surface of the cell, is the structure that allows the cell to act as a separate unit; in many cells this is covered by an outer and often thicker envelope, but whenever experimental methods have permitted, the removal of this "extraneous coat" leaves the cell otherwise intact and capable of performing its normal functions. By contrast, if the thin plasma membrane is injured the cell rapidly disintegrates. We shall see that the internal composition of the cell is greatly different from that of the medium in which it is surrounded; this applies especially to the nature of the salts, the internal solution being made up usually of K^+ as the cation, whilst the external cation is mainly Na^+. Injury to the plasma membrane, either mechanical or chemical, results in the escape of K^+ from the cell and penetration of Na^+ into it from its environment. Thus the plasma membrane is responsible for maintaining the special composition of the cell contents.

Again, we shall see that the passage of various molecules into the cell is a highly selective matter, a sugar like D-glucose being allowed to penetrate fairly rapidly, for example, whereas one of almost identical structure, such as L-glucose, might penetrate very slowly. This selectivity depends on the intactness of the plasma membrane.

Injury. In general, then, a serious injury to the membrane leads to death of the cell, brought about by the change of its internal environment and the loss of its internal organization. Less serious injury may not kill the cell, but it may impair those functions that depend primarily on the chemical compositions of the cytoplasm and the environment. For example, we shall see that a nerve fibre conducts an electrical charge—the action potential—throughout its length, acting thereby in the control mechanisms of the organism. The conduction of this charge depends on the presence of a high concentration of K^+ in the cytoplasm, and this, in turn, depends on the selectivity of the plasma membrane. Injury to the membrane that reduces or abolishes this selectivity may thus put out of action the conducting power of the nerve fibre, but it may not be so severe as to destroy the cell; as we shall see, blockage of conduction is an important feature of the control of nervous activity; and this may be achieved physiologically by highly localized abolition of the selectivity of the membrane.

Cell as a Functional Unit

To summarize these general aspects of the cell, then, we may say that it is regarded as a unit of structure and function primarily because it is the smallest element into which we may break the tissue and still preserve life or rather some form of organized activity to which we apply this term. A tissue contains many cells, often of several quite different types, but the activity of the tissue may usually be described as the sum of the activities of the individual cells; for example, the shortening of the muscle is the effect of the shortening of many or all of the constituent muscle fibres. Morphologically the cell is recognized as a unit by its demarcation from its neighbours or its fluid environment by the plasma membrane; functionally, moreover, this membrane is all-important for the maintenance of the cell as a unit. Within the cell are structures that are responsible for many of its vital activities; the nucleus is concerned with the reproduction of the cell, when this occurs, whilst in non-reproducing phases it controls and determines the pattern of the activity of the cell as a whole. The mitochondria, by virtue of a system of enzymes attached to their membranes, are responsible for the oxidative metabolism that provides the energy that all cells require to remain alive (vegetative) and to do various types of work, be it mechanical as in a muscle fibre, or synthetic in a secretory cell. Synthesis occurs in relation to the endoplasmic reticulum, whilst the Golgi apparatus seems to be a storage house for the products of synthesis.

ORGANIZATION OF CELLS

The study of the cell as a unit is the science of cytology; the study of the organization of cells in tissues is the science of histology. It is not proposed to describe here the various cellular organizations that occur in the animal's body; as we concern ourselves with different functions we shall have recourse to the appropriate histological descriptions; at present it is sufficient to emphasize that it is relatively rarely that a cell can be studied in isolation, and that very often the behaviour of a tissue is not just a simple additive function of the behaviour of the cells.

Intestinal Epithelium

Thus the internal covering of the intestine is a sheet of cells packed tightly side by side; these cells are able to pick up the products of digestion in the intestinal fluid and transport them into the spaces adjacent to the blood vessels, thereby permitting the blood to carry them away to other parts of the body. The behaviour of a single intestinal cell has not been adequately studied, so it is by no means certain that this "polarized" type of transport, in which material is taken up at the apex of the cell (facing the lumen of the intestine) and is extruded at the base of the cell, is possible when the cell is completely

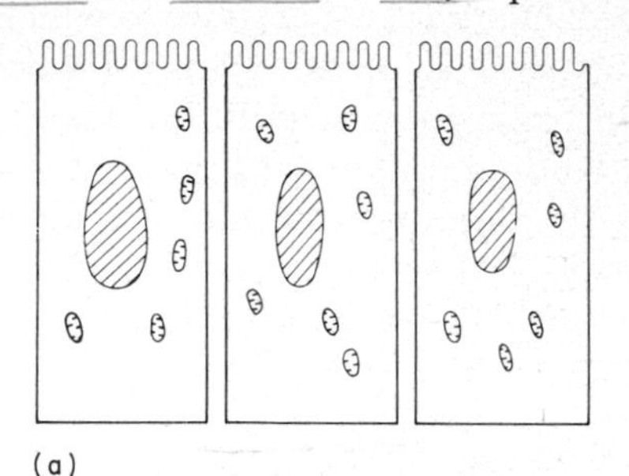

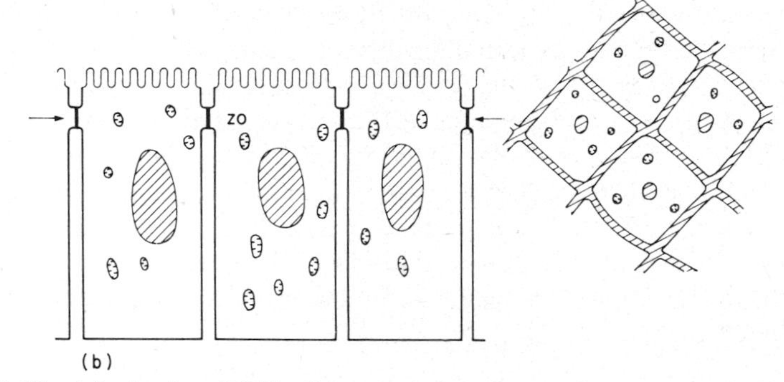

Fig. 1.11. (a) A simplified diagram of a sheet of epithelial cells with a separation of some 200 Å between cells. (b) A sheet of epithelial cells with zonula occludens (zo) sealing the space between cells close to the apical end of the cell. The inset shows a horizontal section through the region indicated by the arrows, and demonstrates the belt of occlusion across the epithelium, thus restricting free diffusion.

isolated, and it may well be that the close apposition to other cells is necessary. Even if we supposed that the individual cells were capable of this, the packing of the cells side-by-side is still of vital importance since, for the selective transport to be effective, it should not be possible for the materials selected to "squeeze" between the epithelial cells. Thus the selective process is one that the *cell* must perform, since it usually involves what is called "uphill transport" requiring metabolic energy for its success; if the transported substance is not to leak back to the place of origin, the spaces between the cells must be sufficiently small to act as a barrier (Fig. 1.11a). With a molecule like glucose, with a diameter of about 7 Å, this would require a very tight packing together of the cells indeed, and such tight packing is not easy to demonstrate morphologically, however necessary it is to postulate it physiologically.

Specialized Cell Contacts

Thus, whenever such epithelium-like arrangements of cells are examined in the electron microscope, the space between them is usually of the order of 200 Å, a space that would allow very large molecules to pass through with ease. However, careful examination of the intercellular clefts has revealed localized regions where the two cell membranes do, indeed, come very much closer together, to make what have been called *specialized contacts*. As we shall see, these contacts are of several kinds, but there is one—the *zonula occludens* or *belt of occlusion*—that seems to seal off the space between adjacent cells completely, preventing the passage of many types of molecule through the clefts from one side of the epithelial layer to the other (Fig. 1.11b). Before considering these contacts further, let us examine the structure and chemical basis of the plasma membrane in further detail.

The Plasma Membrane

It will have been noted that the whole concept of the cell as a physiological unit has depended critically on the power of its membrane to separate it from its environment and from neighbouring cells. What is this membrane like in terms of its ultimate structure?

This is a vital question, since so much of the activity of the cell depends on the proper functioning of this layer.

Electron Microscope

In the electron microscope the dark bands that separate two cells from each other can be resolved into three components—two dark regions that take up the electron stain sandwiching a light region that has a much smaller affinity for the stain; these three bands are about 25 Å thick each, so that the total membrane thickness is of the order of 75 Å (Fig. 1.12). To examine a structure in the electron microscope we must dry it very thoroughly; moreover, it may become compressed during cutting of the section, so that in the living state the membrane may be a little thicker, but not much.

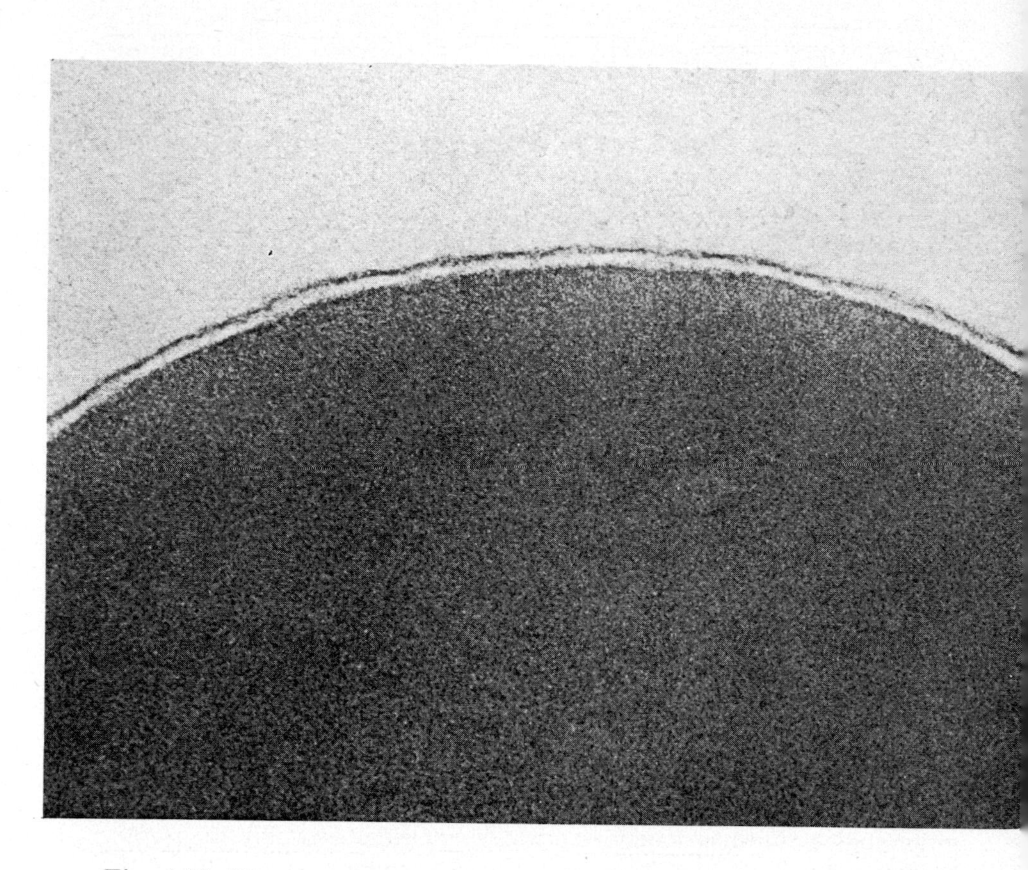

Fig. 1.12. The appearance of red cell membranes fixed with polygluteraldehyde, embedded in uranyl acetate and stained with lead. (× 270,000) Courtesy of Dr. J. D. Robertson.

Lipid Bilayer

What sort of material could be spread over the surface of the cell to act as a membrane to produce such a thin layer, a layer moreover with remarkable powers of selectivity so far as the nature of the molecules that are allowed to penetrate the cell is concerned? It is generally agreed that the fundamental basis is provided by lipoid material, since this may be extracted from cells, thereby destroying their selectivity; moreover, the types of lipid extracted (phospholipids, cholesterol, cephalins and so on, p. 22) may indeed be spread out on the surface of water to form films that are only a single molecule thick. Thus a single layer of lipid might contribute some 30–35 Å, so that it would be possible to have two layers, the molecules being arranged as in Fig. 1.13; the lipid molecules are suited to this arrangement because of their "head-and-tail" character, i.e. the possession of a hydrophilic group at one end and a long tail of hydrophobic material that orientates itself away from the water. Thus the hydrophilic ends, indicated by the

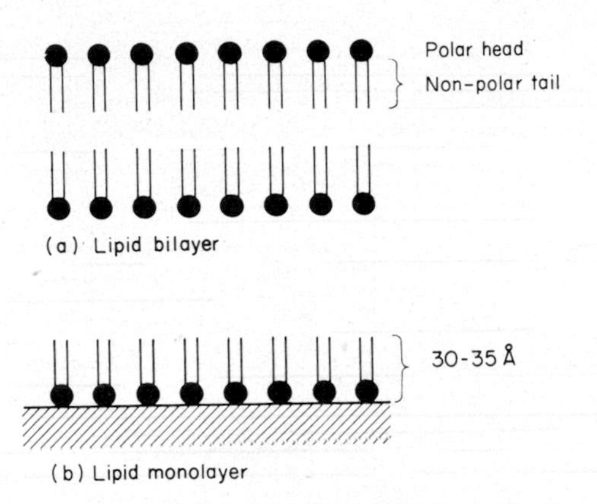

Fig. 1.13. (a) A bilayer of lipid molecules with the polar head groups facing outwards and the non-polar hydrophobic chains orientated towards the inside of the membrane away from the water. (b) Shows a lipid monolayer on the surface of water.

dark blobs, bury themselves in the watery phase facing outside and inside the cell, forming a bimolecular leaflet. The lipids extracted from preparations of cell membranes are of several types, lecithin, cephalin, cholesterol, and so on, but they have the same "head-and-tail" feature that promotes the formation of a film separating two watery phases; experimentally, moreover, it is possible to prepare artificial membranes just one or two molecules thick that have at least some of the properties of the natural cell membrane.

Protein

It is unlikely that the membrane is made up of lipoid material alone, in fact we are certain that proteins, another important class of matter in living organisms, contribute too, possibly spreading themselves out as thin 15 Å layers on the surfaces of the lipid film, as in

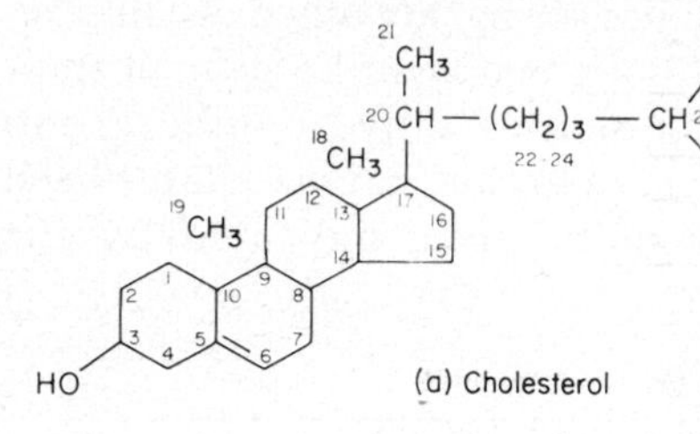
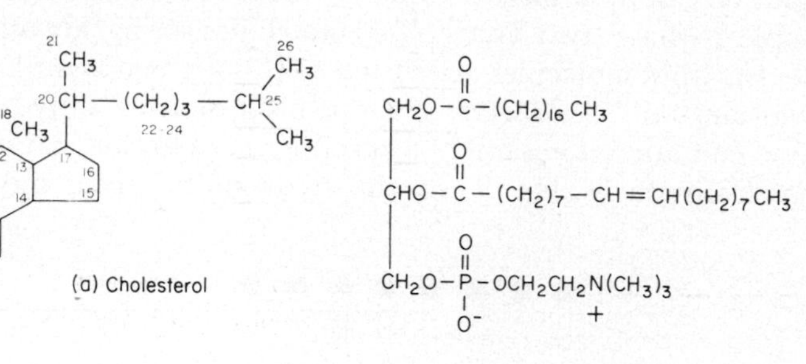

(a) Cholesterol

(b) Lecithin

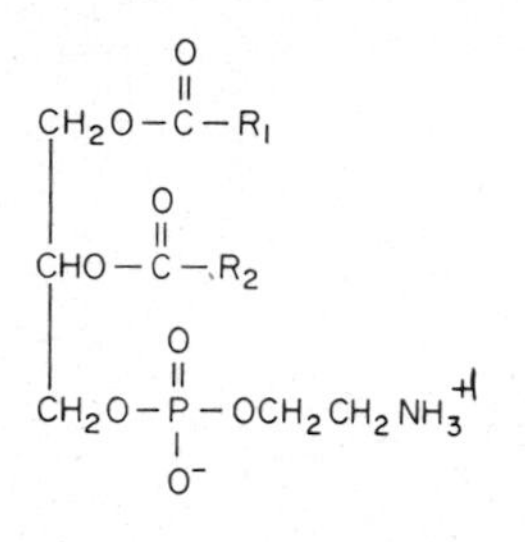

(c) Cephalin

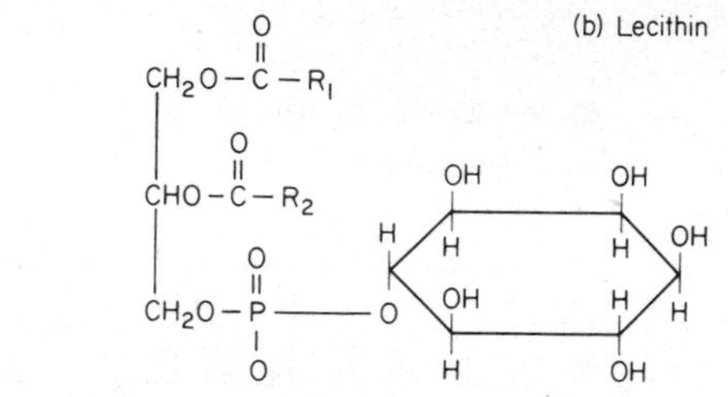

(d) Phosphoinositide

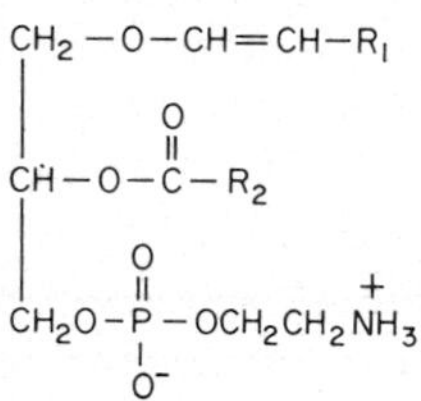
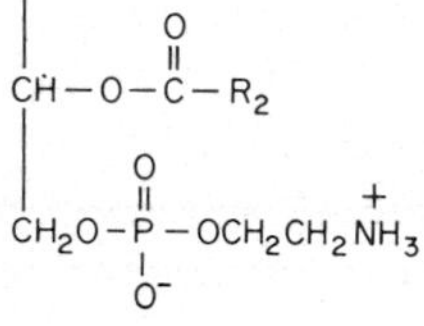

(e) Plasmalogen

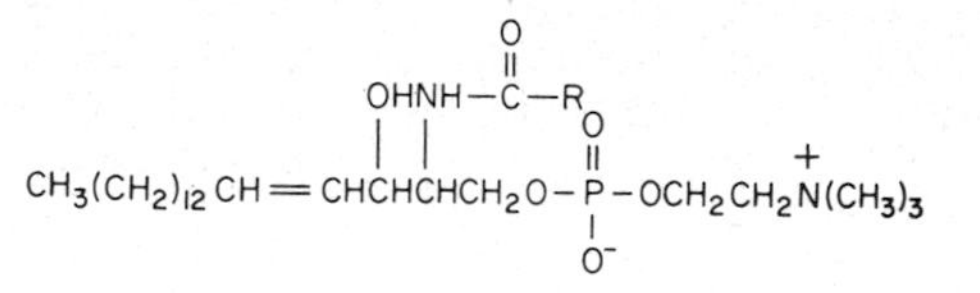

(f) Sphingomyelin

NH-C-R *Fatty acid* Ceramide

CH3-(CH2)12-CH=CH-CHOH-CH-CH2-O

Sphingosine

CH2OH

Galactose

(g) Cerebroside

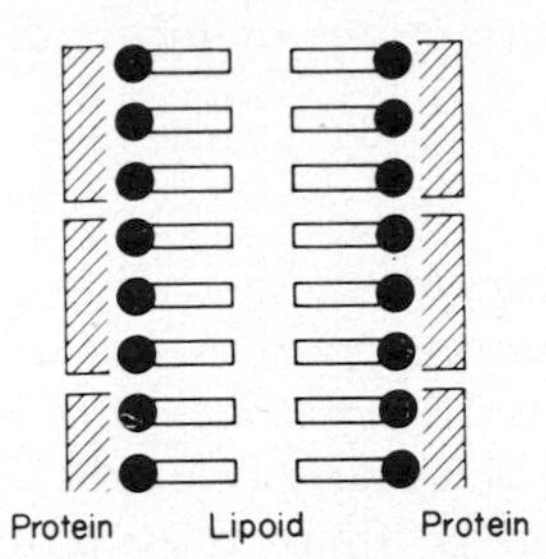

Fig. 1.14. The lipid-protein membrane proposed by Danielli and Davson. The polar head groups of the lipid molecules are covered by a thin layer of protein which may act to stabilize the membrane.

Fig. 1.14. Thus the total thickness of the membrane would amount to about 85 Å.

Triple Layered Sandwich

The picture of the cell membrane developed thus far is based on a variety of studies and was formulated long before the electron microscope revealed that there was, indeed, a thin layer of material on the surface of all cells that took up osmium and other electron stains. Unfortunately the electron microscope has revealed very little more; we have seen that it has resolved the structure into a sandwich and it may be accepted that, because the electron-stains (such as osmium or permanganate) certainly react with polar groupings, the outer dense layers represent the polar regions of the membrane molecules. Thus the unstained layer of the sandwich may well represent the saturated long hydrocarbon chains of the lipid molecules orientated as postulated in the Danielli-Davson model, but this interpretation has been questioned.

We shall defer, until we have discussed the transport of materials through the membrane, the further consideration of its ultimate structure; suffice it to say, now, that whilst many of the features of the membrane's selectivity can, indeed, be accounted for on the basis of a thin bimolecular leaflet of lipid, stabilized with protein as illustrated in Fig. 1.14, many other features demand that the membrane should have discrete regions of inhomogeneity that give it powers of selectivity that could not be achieved by a homogeneous structure.

Cell Adhesion

When cells are brought together, in epithelia for example, they appear, in the electron microscope, to maintain a separation of about

200 Å, i.e. the darkly staining membranes are separated by this distance.

Cement Substance

We may ask why, except in special regions, this separation is maintained? Opinions on this point differ; it has been argued that the separation is achieved by the presence of an intercellular cement, and it is certainly true that when many free cells are examined, such as the sea urchin's egg or protozoa, there is an outermost coat covering the functionally important plasma membrane. In bacteria and plant cells this coat is tough and is an important mechanical support, but when the cells in mammalian tissues are examined this coating is usually only seen at the high resolution of the electron microscope, being some 100–500 Å thick and appearing as a "fluffy layer". When cells are arranged in a tightly packed layer, as in an epithelium, their free or apical borders usually have this layer over them. It apparently extends into the intercellular spaces and over the bases of the cells, so that the material for some sort of cement is present. The fact that, experimentally, it is possible to separate the cells from their packed arrangement by treatment with a protein-splitting enzyme, such as trypsin, suggests that this material might be acting as a cement material holding the cells together. The fact that removal of Ca^{2+} from the medium causes cells to separate from their regular arrangement in a tissue has been cited as evidence in favour of the cementing action of the material between cells, the divalent ion acting as a bridge between the adherent fluffy layers on each cell combining with a carboxy-group on each. The chemical nature of the material is difficult to establish because there is so little of it, and there is danger of confusing it with the much thicker *basement membrane* on which an epithelial or endothelial sheet usually rests, which is probably different in composition. The basis is almost certainly a combination of sugar with protein—a *mucopolysaccharide* (Vol. 2), recognizable by its staining with ruthenium, colloidal iron, etc.

Intermolecular Forces

Alternatively, it has been argued that the separation results from a balance between the normal tendency of cells to separate through Brownian movement and a net electrical interaction due to the various forces of attraction and repulsion between surfaces. Proponents of this electrical theory point out that the so-called *dispersion*, or Van der Waals, forces between molecules might well lead to a net attraction between the cell surfaces at distances in the region of 150–200 Å, an attraction that would permit stabilization of the cells *in situ.* A much closer association would be forbidden by the large *electrostatic forces* that come into play through the presence of fixed ionic charges on the cell membrane.

Electrokinetic Potential. All cells that have been studied contain on their surfaces an electrical double layer, due to the ionization of acidic groups on their surface, in exactly the same way that colloidal

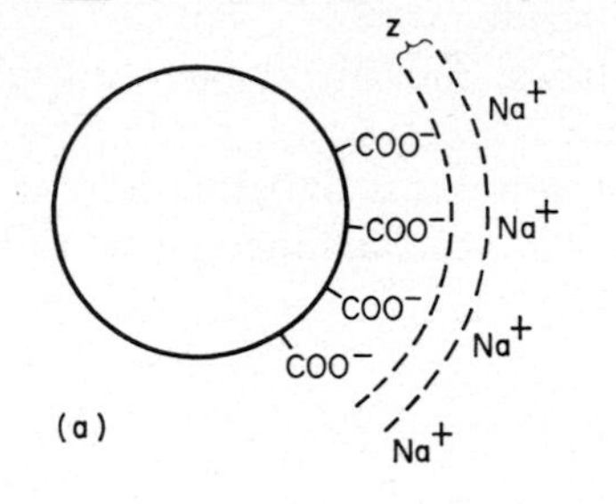

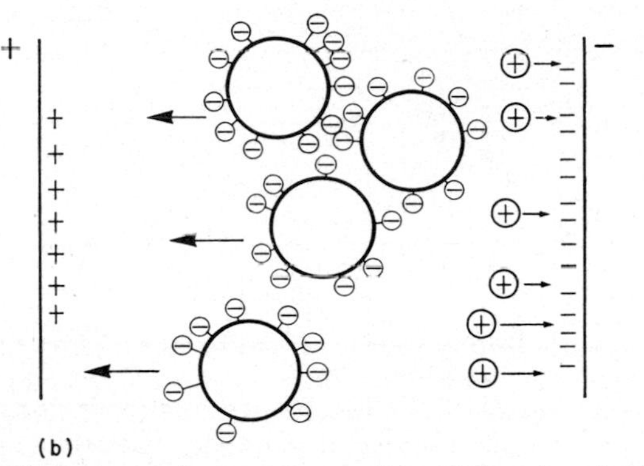

Fig. 1.15. (a) A diagrammatic representation of the electrical double layer associated with cell surfaces. This is composed of fixed negative and positive charges produced by ionization of acidic or basic groups and the selective adsorption of ions. (b) The movement of individual cells towards the positive electrode under the influence of an electric field.

particles almost invariably are charged due either to dissociation of fixed acid or basic groups, or the selective adsorption of ions of positive or negative charge. This electrical double layer, or "atmosphere", is recognized by the movement of the cell or colloidal particle through the medium when electrodes are placed in it and a difference of potential is established. Thus cells, such as the *Arbacia* egg or the erythrocyte, move towards the positive electrode by virtue of their fixed negative charges (Fig. 1.15). Measurement of the velocity of migration, called the *electrokinetic mobility*, enables the computation of the *electrokinetic potential* (zeta potential) by the formula:—

$$\zeta = (4\pi\eta/D)V$$

V being the mobility of the cell or particle in μ sec^{-1} $volt^{-1}$ cm^{-1}, η the viscosity, and D the dielectric constant.

This is the potential between the surface of the cell and the adjacent solution through which it moves.

The actual potentials, measured on red and white blood cells, are of the order of 14 mV.

The electrokinetic potential depends on the presence of fixed ionic charges attached to the cell surface; when the mobility of the cells is

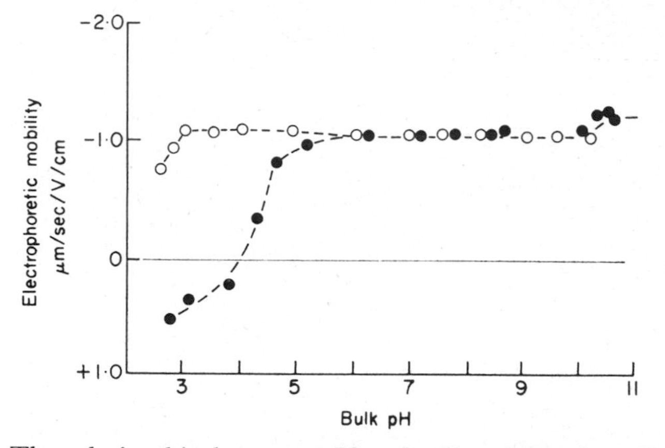

Fig. 1.16. The relationship between pH and cell mobility for saline-washed human erythrocytes (○) and Erlich ascites tumour cells (●). As the H^+ ion concentration increases with the fall in pH, the fixed negative charges on the cell membrane are protonated and made "neutral". Thus, the effect of the electric field on cell movement decreases until the isoelectric point is reached and no movement occurs as there is an equal number of positive and negative fixed charges in the cell membrane. Increasing the H^+ ion concentration further causes ionization of positively charged groups and now the cells move in the opposite direction towards the negative electrode. (Mehrishi, *Progr. Biophys.* 1972, **25,** 1.)

plotted against pH, as in Fig. 1.16, an isoelectric point is reached, on the acid side of neutrality, indicating the absence of net charge. Since the charges arise from ionization of molecules attached to the surface, the nature of these molecules can often be deduced from the isolectric point, but of course the situation becomes complicated if more than one type of molecule is present. Often a separation of effects can be achieved by treating the cell surface with a reagent that will selectively remove given components; thus the presence of ionizable phosphate was shown by measuring the change in mobility of platelets (Vol. 2) with the phosphate-splitting enzyme, alkaline phosphatase. Probably the most important surface component, so far as determining its surface charge is concerned, is *sialic acid* (*N*-actyl neuraminic acid, NANA), which is incorporated into the cell membrane as part of a polysaccharide complex; it is the COO^--groups of this sugar that are responsible for a large part of the charge of many cells, so that treatment with the enzyme neuraminidase which causes the splitting off of

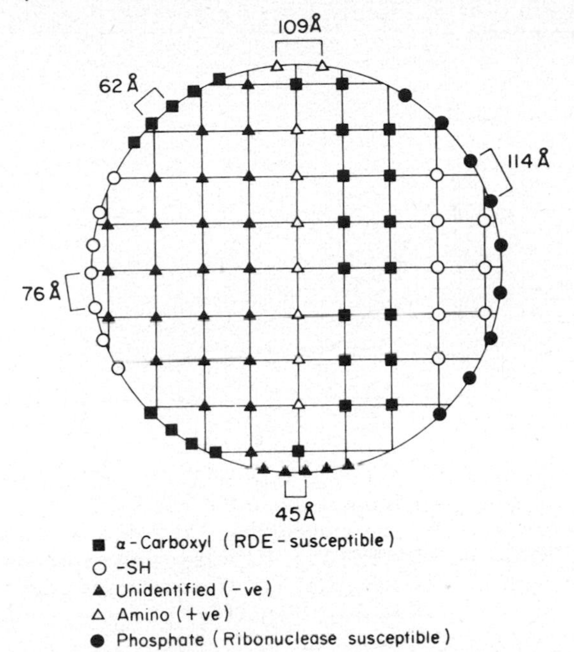

Fig. 1.17. A diagram of the distribution of various groups on the cell surface of a human lymphocyte. The cell is assumed to have a spherical shape with a diameter of 6 μm and a surface area of 113 μm^2. One symbol represents a cluster of 10^5 groups. (Mehrishi, *Progr. Biophys.* 1972, **25**, 1.)

NANA reduces the mobility. A picture of the distribution of cell ionizing groups deduced from this type of study is illustrated in Fig. 1.17.

Intercellular Forces. When the electrical forces operating between two hypothetical cells of known electrokinetic potential and intermolecular dispersion forces are computed, the results may be plotted as in Fig. 1.18 which indicates the energy of interaction at various distances, repulsion being indicated as positive values of the ordinate and attraction as negative values. It is seen that, as the cells get close, they repel each other very strongly. This is because of the preponderant effect of the electrostatic forces of repulsion between surfaces of similar charge; if this repulsion can be overcome, e.g. by reducing the electrostatic charge, then the cells will come into very close contact due to the overwhelming action of the London dispersion forces, and the result might well be fusion of the cell membranes to form the zonula occludens (p. 19). Under normal conditions, however, the cells will ber epelled and move away from each other, but because the London dispersion forces do not fall off so rapidly with distance as does the electrostatic

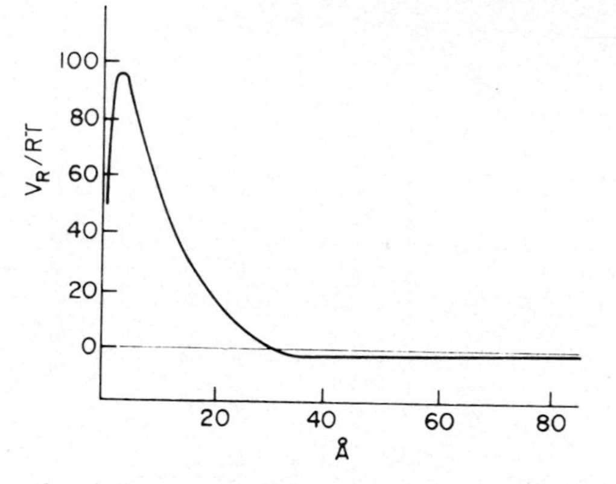

Fig. 1.18. A graph of the energy of interaction at various distances of two identical 1 μ spheres, repulsion being indicated by positive values of the ordinate and attraction by negative values. (At 25°C, assuming that attraction is due to dispersion forces [A = 10^{-4} erg] and repulsion due to ionic double layers with $\mu = 17$ mV [NaCl] = 0·145 M.) (Bangham, *Ann. N.Y. Acad. Sci.* 1964, **116,** 945).

repulsion, a point is reached—the *secondary minimum*—where the attraction is greater than the repulsion; if the net attraction is sufficiently great to resist the normal tendency of cells to disperse through Brownian movement, we may expect the cells to adhere to each other. According to Curtis' calculations it is quite likely that the 150–200 Å gap between cells, so often found, represents the position of the secondary minimum. Nevertheless it seems highly unlikely that this electrical attraction through the preponderance of London dispersion forces is the sole means of holding cells in the normal regular arrangement, and the frequent occurrence of specialized contacts, such as desmosomes (p. 31), suggests that more powerful adhesions, involving recognizable fibrillar structures, are concerned in cell adhesion.

Functions of Surface Substance

The functions of the surface molecules in the cell's economy are by no means completely understood, but they are certainly various. With cells that are normally free, such as blood cells, it is important that they remain separate from each other and also do not fuse with other types of cell when they approach closely to them. Here the surface charge is doubtless important, but by no means exclusively so.

Agglutination. For example, red blood cells are free in the blood plasma but they can be made to aggregate into tight clumps—*agglutination*—by addition of certain substances to the medium in which they are suspended. Thus Katchalsky found that long-chain basic polymers of the amino acid lysine caused the cells to clump so that, when the regions of apposition were examined in the electron microscope, the adjacent membranes appeared to be linked together by the long polymer chains. It might have been argued that the positively charged polylysine chains neutralized the charges on the red cell surfaces and thus allowed the cells to come together. Certainly the electrophoretic mobility was reduced, but the main factor promoting agglutination was undoubtedly the linkages formed between cells by the long polylysine chains which acted as bridges across the intercellular gap. Again, in the reaction to a damaged blood vessel, the loss of blood is reduced by the formation of a plug of platelets (Vol. 2); these cells, normally free, become clumped together to block the hole in the blood vessel.

This change in the surface character of the platelet is not correlated with a reduction in surface charge, so that we must attribute the increased "stickiness" to some other feature of the surface.

Contact Inhibition. Another important aspect of cell contacts is given by the phenomenon of "contact inhibition". When cells are multiplying and organizing themselves into a sheet, as in a developing epithelium, they move in relation to their environment, but when they come close to each other, normal cells will cease movement and so maintain their positions in an ordered arrangement. This arrest of movement is called by Abercrombie *contact inhibition*, and is accompanied by an almost complete inhibition of the cell's synthetic activities; pathological disturbances in the cell may abolish this inhibition and may well be a feature of tumour growth, multiplying cells being now able to push themselves between others giving rise to a disordered growth.

Implantation of Ovum. Under certain special conditions invasion of a "foreign" cell into a tissue is physiologically important; this is the case with the trophoblast of the embryo, which must "implant" itself in the maternal uterus (Vol. 2). At first, when the system is very simple consisting of just a few cells, the adhesion must be intimate, but later with greater development a separation of the embryonic and maternal tissues becomes important; this is achieved, apparently, by the secretion by the trophoblast cells of surface material—probably sialic acid—which causes a separation between embryonic and maternal cells.

Immune Reactions. We shall see that the introduction of genetically foreign cells into an organism provokes immune reactions. These reactions involve the presence of "strange" molecules in the surface of the invading cell called antigens; many of these are ionized and so contribute to the surface charge. The immobilization and ultimate destruction of these foreign cells require that the host's cells first recognize the foreigner, and they must do this through some form of surface contact; after recognition, the attack on the foreigner likewise involves approximation of cells to each other, or of antibodies produced by the host and acting at a distance on the foreign cell.

Microvilli. It will be clear, therefore, that the cell surface is an agglomerate of many specific molecules, which may be neutral or negatively or positively charged, the electrokinetic potential being the algebraic sum of the opposite charges. It is not surprising that very few correlations between cell charge and physiological behaviour have been revealed. However, one point, concerned with the close approach of cells, is worth emphasizing. It can be calculated that the charges on cells are adequate to prevent very close approximation of their surfaces, the energy provided by Brownian movement being inadequate to overcome the electrical repulsion. However, if the surface of one of the cells is drawn out into a fine spike or microvillus, then the charge interaction is reduced. Pethica calculated that fusion between the spike on the one cell and the relatively flat surface of the other becomes possible. When a phagocytic cell engulfs another, close approximation is necessary, and it is possible that this may be achieved, or at least favoured, by such spike formation.

Cell Junctions

The close approach of cells through a preliminary neutralization of the electrostatic surface charges should, theoretically at any rate, lead to a very close approximation indeed, in fact to a fusion of the two plasma membranes. Such a fusion certainly occurs and is of very great physiological importance. When some regions of close contact between epithelial cells, for example, are examined at high resolution, it is found that the two outer dark-staining layers of the apposed plasma membranes have become a single dark line, so that the usual 7-layered appearance presented by two cells in contact is replaced by a 5-layered structure, and the total thickness of the junctional region is less than the sum of two membrane-thicknesses (Fig. 1.19).

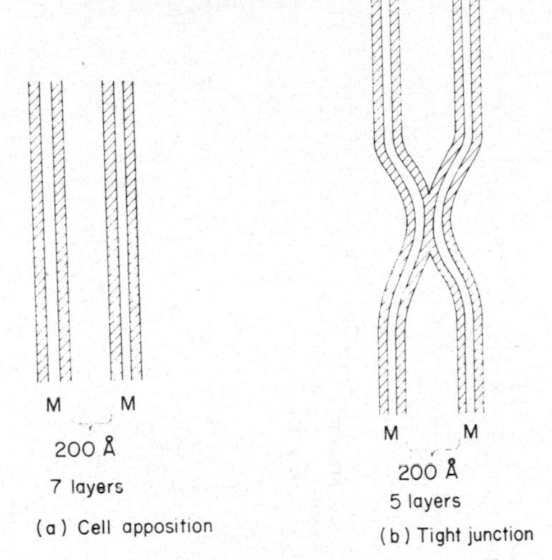

Fig. 1.19. A comparison of the appearance of two cell membranes in close apposition, showing the normal 7 layered structure, with that of the tight junction contact where the outer two layers have become fused into a single line so that the junction is now composed of only 5 layers (pentalaminar).

Junctional Complex

As described by Farquhar and Palade, the tight junction, or zonula occludens, is one of a group of junctions occurring at successive depths in the cleft, so that they spoke of a *junctional complex;* deeper to the surface of the epithelium the tight junction was usually followed by an *intermediate junction*, or *zonula adhaerens*, and deeper still was the *macula adhaerens* or *desmosome*. These last two almost certainly represent specializations of the intercellular space and the immediately subjacent cytoplasm that favour intercellular adhesion, i.e. their function is mechanical, as opposed to that of the tight junction which acts as a seal, preventing or slowing diffusion of solutes between the cells of the epithelium. Thus in the desmosome and intermediate junction there is no occlusion of the intercellular cleft, the separation of 200 Å being well maintained. The characteristic feature is the presence of densely staining material in the cytoplasm of the cells and, with the desmosome, within the intercellular cleft. When a cell has to adhere, not to an adjacent cell, but to a basement membrane, the junction is described as a hemidesmosome.*

Gap Junction. Until recently this classification of junctional types was considered to describe completely the cellular relations in a variety of tissues, but a more thorough examination of the tight junction, or zonula occludens, portion of the complex, and the use of lanthanum or

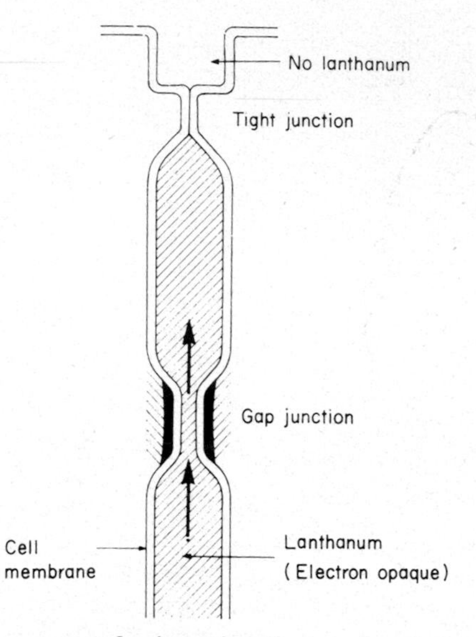

Fig. 1.20. The appearance of a junctional complex between two cells composed of a tight junction, near the apex, and a gap junction lower down. The preparation has been treated with colloidal lanthanum from the blood side, as indicated by the cross-hatching. It can be seen that the lanthanum has passed through the gap junction, but is restrained by the tight junction. (Drawing from Friend and Gilula, *J. cell Biol.* 1972, **53,** 151.)

ruthenium as electron-dense markers to show whether the junction was or was not completely occluded, have revealed another type called by Revel and Karnovsky the *gap junction;* it had been frequently described in a variety of tissues where it was called a *nexus* (for example in smooth muscle) or a *longitudinal connexion* (cardiac muscle). When it occurs in relation to a tight junction or zonula occludens, it is found deeper, i.e. farther from the apex of the cell; and the important feature is the failure of the membranes to achieve complete fusion, so that there remains a channel some 20–40 Å in diameter between the

* The macula adhaerens serves as an insertion site for the fibrous system of the epithelium described as the *terminal web.*

cells. This is demonstrated by treating the preparation with colloidal lanthanum; this shows up as an electron-dense line passing through the intercellular cleft, but being held up at a true zonula occludens (Fig. 1.20). The functional significance of this type of junction will be appreciated later, but for the moment we may emphasize two features. First, the fact that the gap junctions are some 20–40 Å in diameter means that diffusion of dissolved material between cells with only this type of junction, or else with the more common 200 Å gap between them, will be relatively unrestricted by contrast with the occludens type of junction that seals the cleft completely.

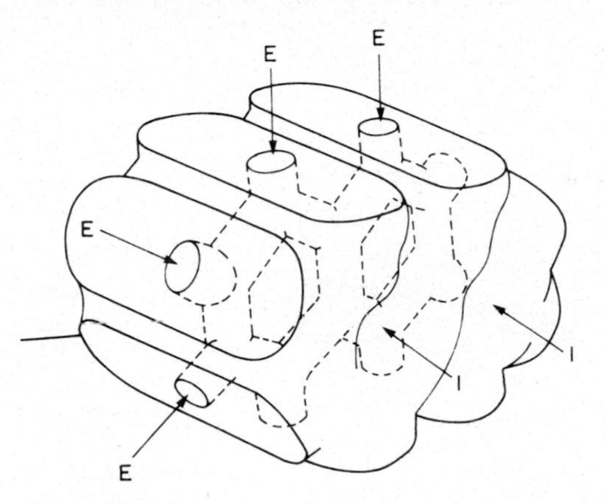

Fig. 1.21. The probable structure of a gap junction showing channels connecting the cytoplasms of the two cells, labelled I, which are responsible for electrotonic coupling between cells. At right-angles to these intercytoplasmic channels are extracellular channels (E) allowing vertical movement across the junction between the extracellular space on either side. The I channels are smaller than the E channels. The thickness of the junction is about 150 Å. (Pappas *et al.*, *J. cell Biol.* 1971, **49,** 173.)

Intercellular Connexions. Second, a number of experimental studies have shown that, when two cells are separated by the gap junction, they are in fact connected by it, in the sense that there is an apparent ready diffusion of dissolved material from the cytoplasm of one cell to the cytoplasm of the other through this region. Thus, the gap junction should probably be represented by a series of tubes running at right-angles with each other (see Fig. 1.21), one set of tubes connecting the cytoplasms of the cells and the other connecting the intercellular spaces. It is through the gap junction, or nexus, that cells are made to lose their functional identity to some extent; thus the plasma membrane is the structure that separates the cell functionally from its environment and is the barrier that enables the cell to maintain an entirely different chemical composition from that of its surrounding fluid.

Syncytia. In tissues like smooth muscle and heart muscle, the cells seem to lose their identity, the tissue behaving, not as a group of individual cells, but as a *syncytium.* It is considered that it is because of the gap-type junctions that this becomes possible. The cells are separate from their environment—the extracellular fluid which occupies the spaces between cells—and this is because the intercellular junctions communicate between cells but not between the cell and its extracellular fluid.

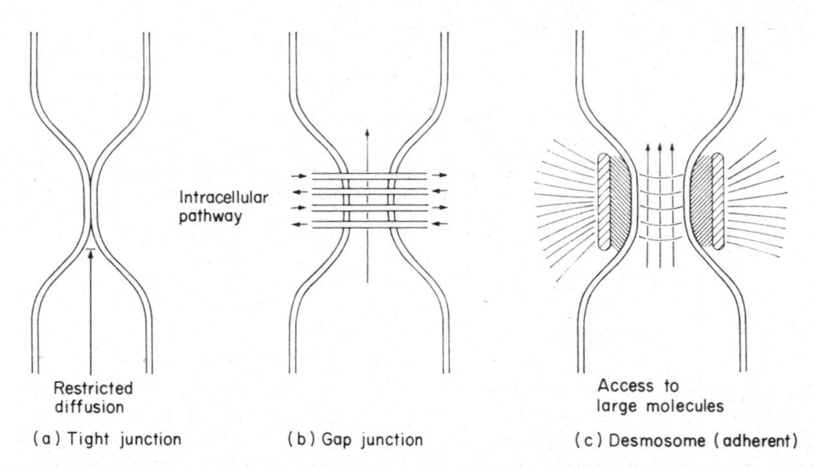

Fig. 1.22. The three classes of junction found between cells. (a) The occluding type of tight junction which restricts movement between cells and enables an epithelium to act as a barrier. (b) The communication or gap junction which permits electrotonic coupling between cells and allows diffusion across the cellular layer. (c) The adherent or desmosome junction which allows free diffusion across the cell layer and has a structural role.

To summarize, then, the junctions fall into three classes (Fig. 1.22):

(1) The occluding type, or tight junction, that bars movement between cells, and enables an epithelium to act as a barrier to diffusion.

(2) The "communicating" type or gap junction or nexus, that permits communication between cells and across the cellular layer.

(3) The "adherent" type, exerting a structural role, neither interfering with passage between cells nor permitting transport from one cell to the other.*

Whether this desmosome constitutes the exclusive adhesive force binding cells together in a tissue, or whether electrical forces and cement material contribute, is not clear.

SOME FUNCTIONAL CHANGES IN THE MORPHOLOGY OF THE CELL

When a cell has become highly differentiated its structure usually remains fairly rigidly fixed throughout its life, and it loses many of the capacities of its more primitive state. Thus it usually no longer reproduces itself and it shows little or no movement in relation to other cells in its tissue. Hence, to appreciate the potentialities of change in the structural features of a cell it is common to examine primitive cells, e.g. the protozoa, or cells in tissue-culture. In this latter preparation the highly differentiated cells of a tissue, such as skin or the heart, on being allowed to grow in a liquid medium as an *explant*, revert to more primitive forms and in this condition exhibit the capacity to change shape and move over the solid surface of the vessel in which the explant is contained. These changes are of some general interest in the study of the complex animal since they reveal certain potentialities of the protoplasm within the cell which may appear in a less extreme form in the more differentiated cells of the animal's tissues. Thus the streaming of protoplasm from one part of the cell to another is a prominent feature of many primitive cells, and of cells in tissue-culture, but is not usually observed in highly differentiated cells; nevertheless, modern studies of the transport of materials from one part of a neurone to another, where the distances travelled may be very large, leave us in little doubt that the transport could not be the result of random diffusion but must be the result of a directed flow of the fluid contents of the neurone from one region to another, analogous with the protoplasmic streaming of primitive cells.

* The problem of the junction will be discussed further in Chapter 2; it is worth pointing out here that it has been argued that communication between cells is achieved by the tight junction, so that the individual cells of an epithelium may be considered to act as one through these intercellular junctions at the region of membrane fusion. This view has been questioned however. A second problem concerns the pathway of dissolved material when it crosses an epithelial layer; does it go through the cells or through the junctions or both? The evidence suggests that it is easier for many molecules to pass through the sealed region rather than pass into the cell and subsequently out of it; i.e. the passage across the region of membrane fusion is simpler than a double trip through cell membrane.

Amoeboid Movement

An amoeba, as is well known, is a unicellular organism; its structure varies from moment to moment by virtue of its tendency to send out extensions from its body called *pseudopodia*. The basic structure of *Amoeba proteus* is illustrated in Fig. 1.23; its outermost surface consists of an elastic pellicle; this gives it mechanical strength and protects the underlying plasma membrane from damage. Beneath the plasma membrane and pellicle is a layer of solid cytoplasm, the *cortex*, or cortical gel (plasma gel), whilst the bulk of the contents are fluid

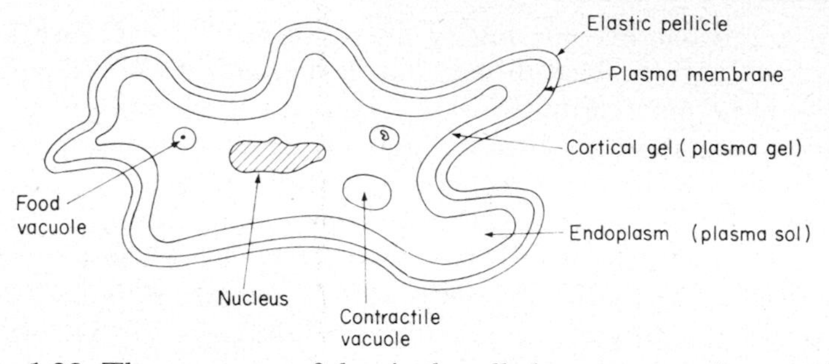

Fig. 1.23. The structure of the single celled protozoan, the amoeba.

constituting the *endoplasm* or plasma sol. It is because the solid cortex and liquid endoplasm are to a large extent interconvertible that changes in cell form, and amoeboid movement, are possible.

Pseudopod Formation

According to the classical studies of Mast, which have been largely confirmed by more recent work, the formation of the pseudopod

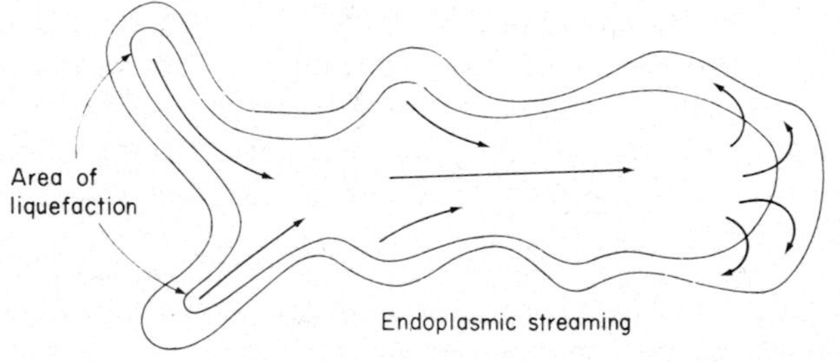

Fig. 1.24. The formation of pseudopodia by the amoeba. The pseudopod is formed by a local liquefaction of the cortical gel, and liquid endoplasm is forced into this region causing a bulge. The contractility of the cortical gel causes streaming of endoplasm to the new pseudopod from the end where pseudopodia are being retracted by liquefaction.

results first from a localized liquefaction of the cortical gel at the point on the body where the pseudopod is about to form; liquid endoplasm is then forced into this region causing a bulge, this endoplasm being propelled by an active contraction of the body of the cell, due to an inherent contractility of the cortical gel. When the streaming endoplasm, which can be recognized in the microscope by virtue of the granules carried with the stream, reaches the weakened region, it solidifies above and below as it comes into contact with the solid cortex, so that in this way a solid tube of cortex, enclosing liquid endoplasm, is built up. The liquid endoplasm flowing into the tube is derived from the body, or "tail", of the *Amoeba*, and results from liquefaction of part of the previously solid cortex in this region. Fig. 1.24 illustrates the general scheme.

Locomotion

It is not difficult to envisage an actual bodily movement of the cell by virtue of this stretching out of a pseudopodium; this could occur as a walking movement, the pseudopodium, after forming, becoming

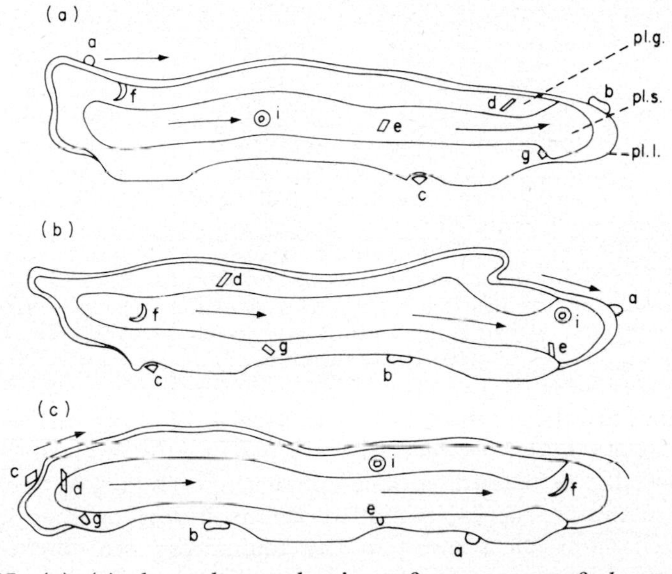

Fig. 1.25. (a)–(c) show the mechanism of movement of the amoeba at different stages The movement is thgouht to be brought about by attachment of the pseudopod to the substrate and then contraction of the body to this point. Further extension of a new pseudopod allows movement to proceed. The letters a–g indicate particles which permitted identification of the streaming of the plasma gel (Pl.g.), the plasma sol (Pl.s.) and the plasma lemma (Pl.l.). (De Bruyn, *Q. Rev. Biol.* 1947, **22,** 1.)

attached at its tip to the substrate whilst a new pseudopodium forms above it and, owing to the contraction at the posterior end of the body of the *Amoeba*, moves over the first point of attachment (Fig. 1.25).

Fibrocyte Movement. In other types of cell it may well be that the mechanism of movement is fundamentally different; thus the fibrocyte that grows out from an explant of embryonic heart muscle develops a characteristic fan-like pseudopodium on its leading edge; this spreads along the glass substrate of the explant, while ruffles or undulations of its membrane are formed continuously near the leading edge of the cell. According to the microscopic studies of Ambrose, these

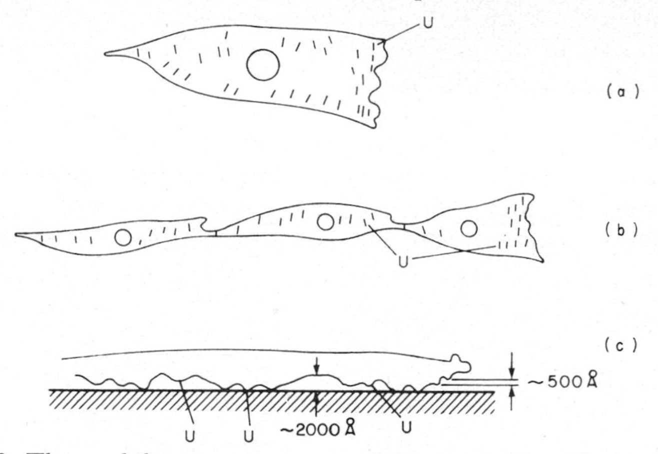

Fig. 1.26. The undulatory movement of fibrocytes. Fan-like pseudopodia develop along the leading edge and contact with the substrate. In addition to this, localized regions of the pellicle expand and undulate to carry the fibroblast forward. (a) Shows the sites of undulations (U) and the pseudopodia forming on the leading edge; (b) a streaming group of fibroblasts; (c) is a side view showing the undulating contacts. (Ambrose, *Expl. Cell Res.* 1961, suppl. **8,** 62.)

undulations are the result of a vertical expansion of the pellicle. When the regions of contact between the cell and its substrate are examined by a specially developed surface-contact microscope, it seems that contacts with the surface occur at localized regions that are continually changing as a result of the undulatory movements of the membrane. These undulatory, or wave-like, changes in the surface may well be the consequence of a wave-like propagated compression, running in the opposite direction to that of the movement of the organism, and the result is to bring about a worm-like mode of progression (Fig. 1.26). This compression, presumably taking place in the surface layers of the cell, could be due to the shortening of contractile filaments that have been identified immediately beneath the surface of many cells including these motile fibroblasts.

Cell Division

When a cell divides into two daughter cells, usually two processes are necessary, namely a division of the nucleus into two (*karyokinesis*)

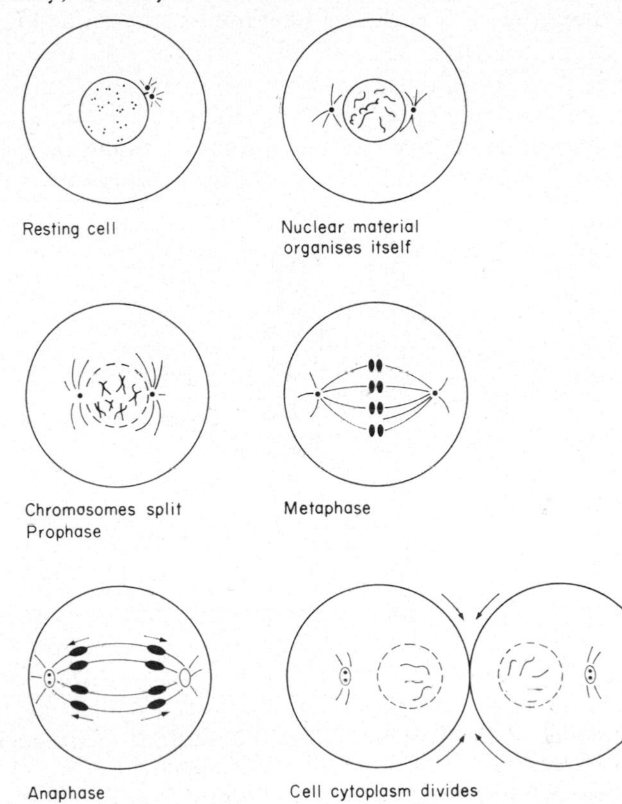

Fig. 1.27. The stages of mitotic cell division. Nuclear re-organization occurs and the chromosomes become visible. Prophase—the spindle mechanism develops and the chromosomes undergo division. Metaphase—the divided chromosomes become orientated along the equator of the spindle mechanism. Anaphase—the daughter chromosomes migrate towards the poles of the spindle system and then the cytoplasm divides to complete the process of cell division.

and a division of the cytoplasm into two in such a way as to include the new nuclei in the daughter cells, or blastomeres (*cytokinesis*). Division of the nucleus in a somatic cell is described as *mitosis*, and is illustrated in Fig. 1.27; essentially it represents the division of the individual chromosomes, on which the genes are arranged in linear order, by longitudinal splitting, and the migration of the individual daughter chromosomes along a *spindle*, which forms at the time of this splitting, to the separate poles of the cell. Associated with this separation is the process of cytokinesis, as a result of which a cleft in the cytoplasm of the mother cell divides the spindle and cytoplasm in such a way as to produce two daughter cells with one new nucleus in each (Fig. 1.27).

Microtubules

The main problems presented to the cytologist are the mechanism by which the chromosomes are caused to migrate along the spindle to opposite poles of the cell, and the mechanism whereby a cleavage

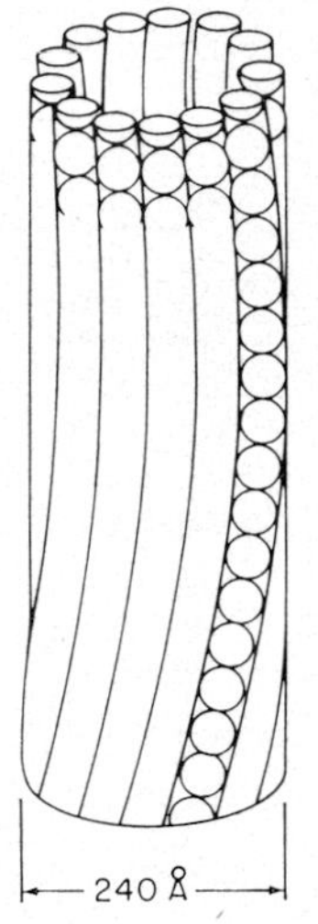

Fig. 1.28. A model of the structure of a microtubule built up from 12 laterally apposed thin filaments. Each filament appears in turn to be built up from spherical subunits of the protein, tubulin. (Schmitt, *Symp. Int. Soc. Cell Biol.* 1969, **8,** 95.)

furrow forms in the cytoplasm so as to divide the cell into two. The processes are to some extent independent, so that it is possible to inhibit cell cleavage without preventing nuclear division; in this case cells with several nuclei can be formed. The structural basis for these changes of shape and of internal movement within the cell are only just beginning to be understood thanks to refinements in electron microscopical technique that have permitted the demonstration of very fine fibrillar elements within the cytoplasm of cells that had previously escaped detection. Of these, the *microtubules* are probably the most ubiquitous, but they become especially prominent when changes in shape or internal structure of the cell are taking place. Thus many cells form small villous projections from their surfaces, called *microspikes* or *microvilli;* these may form and disappear, but they exhibit some rigidity, and this is provided by the formation of microtubules which constitute a skeleton for the newly formed structure. The microtubules may be as long as $1\cdot5\,\mu$ and on cross-section they appear hollow (see Fig. 1.28), the external diameter being some 250 Å. The tubules can be shown to be built up of some 12 laterally apposed thin filaments of diameter 25 Å evenly arranged to enclose the hollow centre; these filaments themselves seemed to be built up by an aggregation of spherical subunits of a protein called *tubulin.* Thus the temporary structural elements appearing during alterations in shape of the cell seem to be formed *ad hoc* by assembly of units of tubulin existing in solution in the cytoplasm.

Microtubules and Chromosomal Movement

So far as movements of the chromosomes during mitosis are concerned, there is no doubt that the mechanical basis of the spindle, and

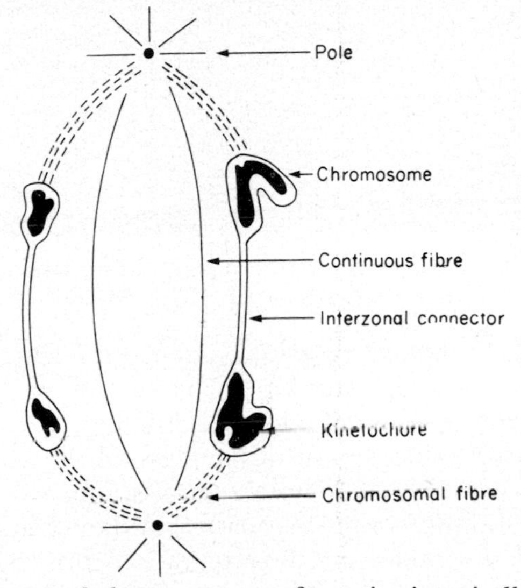

Fig. 1.29. Diagram of the structure of a mitotic spindle. Two types of spindle are seen: long continuous fibres which run from pole to pole and chromosomal or kinetochore fibrils, to which the chromosomes are attached and act as guides directing the daughter chromosomes towards the poles of the spindles. (Inoué, *In* Primitive Motile Systems in Cell Biology, (Allen and Kamiya, eds.) 1964. Academic Press, New York and London.)

the fibres connecting the chromosomes with it, is one of microtubules. Thus in some cells the tubules can be seen growing in the cytoplasm towards the nuclear envelope during prophase and passing into the nuclear substance, either in association with the dissolution of the nuclear membrane, or before this event. The fibrils of the mitotic apparatus may be divided into two classes as illustrated in Fig. 1.29. The long *continuous fibrils* run from pole to pole, so that the elongation of the spindle during mitosis is associated with a lengthening of these;

Fig. 1.30. An electron micrograph of a microtubule. (Zickler, *Chromosoma*, 1970, **30,** 287.)

the *chromosomal* or *kinetochore fibrils* connect the chromosome to the continuous fibres through the kinetochore, and we may regard the continuous fibres as guides directing the ordered motion of the chromosomes. The microtubules constituting these fibrils appear as a result of a process of assembly of tubulin subunits, and we must regard the state of a microtubule at any moment as a dynamic condition where the process of assembly or distintegration predominates, leading to lengthening or shortening. When the mitotic apparatus is associated with centrioles, then it seems likely that the assembly of the continuous fibrils is organized by these bodies (Fig. 1.30); it is considered that the kinetochore of the chromosome exerts a similar role in regard to the kinetochore or chromosomal fibrils.

As to the mechanism through which the microtubules are able to direct the movement of the chromosomes in their orderly progression to the metaphase plate and subsequently to the poles of the cell, opinions differ. On the one hand it may be argued that the mechanical process is one of pushing through elongation of fibrils by assembly from subunits, or alternatively the fibrils might be formed first and subsequently caused to contract.

Colchicine. Many of the biological phenomena in which the assembly of microtubules seems to be a basis may be inhibited by the drug colchicine; thus the movement of the chromosomes during mitosis may be inhibited by colchicine, and this is accompanied by dissolution of the microtubules. Cooling and high pressures likewise inhibit microtubule assembly and also the mitotic movement of chromosomes.

Ciliary Motion

The formation of a microspike on a cell is a temporary phenomenon; more permanent structures are the flagella of motile cells such as the spermatozoon (Vol. 2) and the cilia of fixed cells whose flail-like oscillations cause movement of fluid across the cell's surface.

The Cilium

The basis of the cilium is a central rod, or *axoneme*, built up of a characteristic 9 + 2 arrangement of microtubules, as illustrated in cross-section by Fig. 1.31. The nine peripheral subfibres are in effect

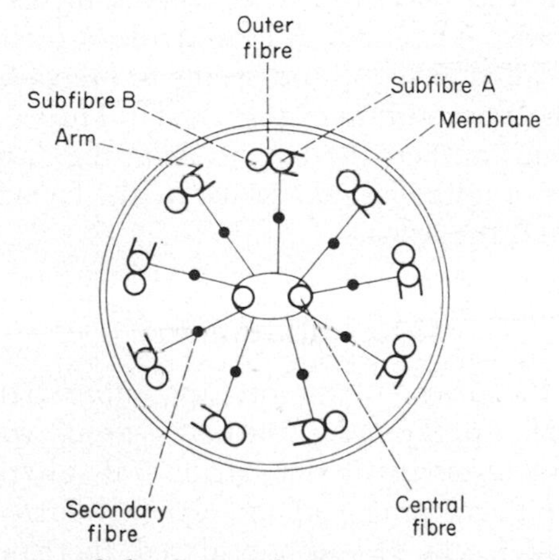

Fig. 1.31. A diagram of the internal structure of a cilium showing the 9 peripheral subfibres plus two forming the axis. Each peripheral subfibre is composed of two microtubules joined by protofibrils. (Gibbons, *Arch. Biol.* 1965, **76,** 317.)

double microtubules sharing some protofibrils over the region of junction. The fibres of the cilium or flagellum are embedded in a matrix and the whole is surrounded by plasma membrane. At its base the cilium merges with the basal body, a DNA-containing structure related to the centriole around which the mitotic spindle becomes organized. As with karyokinesis, ciliary motion is inhibited by colchicine presumably by reducing the stability of the microtubular basis of the cilium.

ATP and Movement

The mechanism by which a cilium or flagellum carries out its movements is by no means clear; the flagellar movement, e.g. of the tail of a spermatozoon, consists in the passage of a circular type of wave along the whole flagellum, and seems to be the result of a sliding of the individual tubules along each other, analogous with the sliding of myosin and actin filaments in skeletal muscle, which will be described later (Vol. 2). As with this latter process, the contractile machinery can operate in the dead cell, i.e. with its plasma membrane removed. Thus addition of ATP (p. 167) to a muscle fibre with its contractile machinery intact but otherwise dead will cause it to contract, the structural protein of the fibre acting as an enzyme that splits ATP to ADP and inorganic phosphate, i.e. it is an ATPase. In a similar way dead spermatozoa, with their membranes removed by treatment with a detergent, can be made to swim almost normally by adding ATP to the medium, and it is found that the protein basis of the microtubules also acts as an ATPase; the protein exhibiting this activity has been called dynein, and seems to be confined to the arms projecting from the microtubules (Fig. 1.31). Thus the sliding of one pair of tubules along another, thought to lie at the basis of the bending of cilia and flagella, might well take place by the formation of chemical attachments through the arms.

Microfilaments

Application of the drug colchicine, and other treatments such as cooling, that break up the microtubular system will inhibit many other cellular processes involving structural alterations, such as cytokinesis, i.e. the development of the cleavage furrow that divides a cell into two daughter cells. It was considered, therefore, that the basic feature of the formation of this furrow was the development of a system of microtubules that would ultimately contract in the form of a ring (Fig. 1.32). However, more recent work has shown that, in addition to

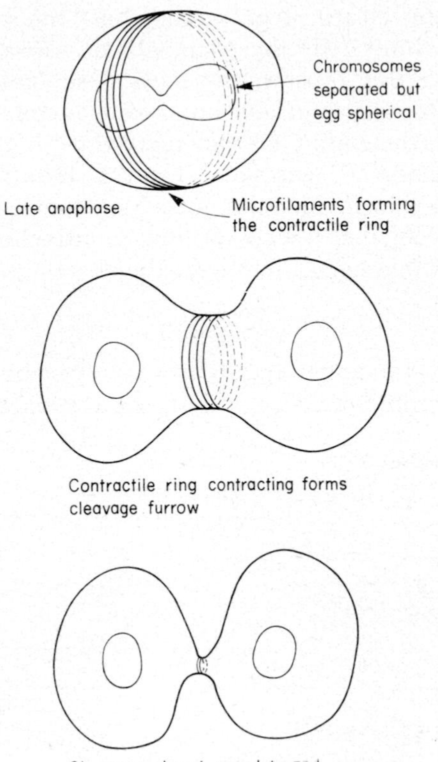

Fig. 1.32. The development of the cleavage furrow after karyokinesis (mitotic nuclear division). Towards the end of anaphase microfilaments develop in an equatorial position as shown in the upper diagram. These filaments form a contractile ring which gradually divides the cytoplasm in half to form the two daughter cells. On completion of this process the microfilaments disperse.

microtubules, cells contain either as temporary or permanent structures a system of *microfilaments*, much finer than the microtubules with a diameter of about 40–60 Å; and it is the appearance of these in the cleavage furrow that is the decisive event.

Cleavage Furrow

Electron microscopical studies of the cell membrane in the region of furrowing show that, before this begins, the cytoplasm immediately below the plasma membrane contains no microfibrils; as soon as the furrow shows signs of appearing, the cytoplasm immediately below the plasma membrane clears of granules whilst microfilaments appear; if the process is inhibited, the filaments disappear. Thus it is considered that with this process, the basis is the microfilament rather than the microtubule, and an important means of differentiating between processes depending on the one or the other is provided by the substance *cytochalasin B*, extracted from a mould; this specifically inhibits processes that involve the microfilaments leaving microtubules intact, whilst colchicine, which specifically attaches itself to tubulin, inhibits activities involving the microtubules.

Axon Growth

Many structural changes apparently involve co-operation between the two, the microfilaments providing the contractile power and the

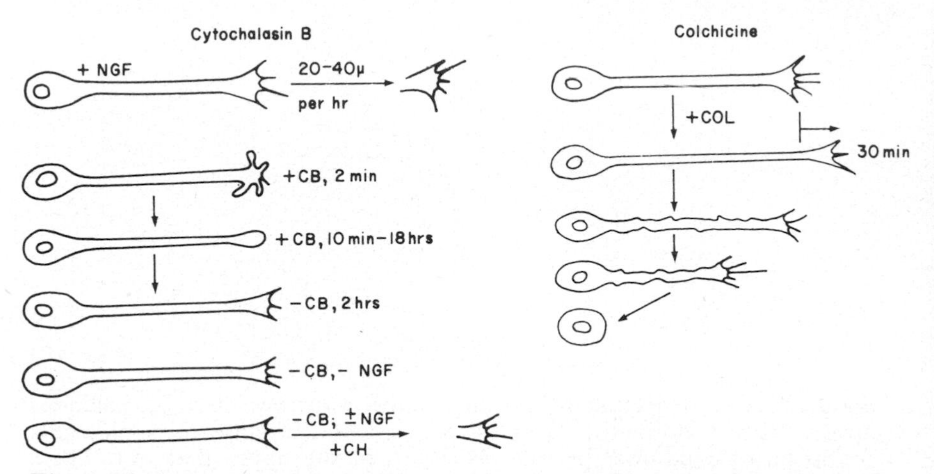

Fig. 1.33. A summary of the effects of cytochalasin B, which inhibits microfilaments, and colchicine, which breaks up the microtubules. Axon growth would appear to depend on the microfilaments for elongation and on microtubules for rigidity. Cytochalasin B halts growth cone function and elongation but is reversible even in the absence of nerve growth factor (NGF) or in the presence of cycloheximide (CH). Colchicine only inhibits elongation indirectly after the axonal tip has developed and the dispersion of the microtubules causes retraction of the axon into the cell body. (Wessels, *Science*, *N.Y.*, 1971, **171,** 135. © 1971 Am. Assoc. for the Advancement of Science.)

microtubules the structural rigidity, or *cytoskeleton*, so that either drug may inhibit the process. This is true of cytokinesis and also of the growth of an embryonic nerve axon; the elongating tip of such an axon is tipped by an expanded region called a *growth-cone*, from which slender microspikes protrude and wave about; the organelles actually within the growth cone are microfilaments, and it is the primary activity of these that seems to cause elongation. Microtubules and larger filaments are involved probably in providing rigidity for the newly developed part so that both cytochalasin and colchicine inhibit growth although the structural consequences are different. Cytochalasin causes rounding up of the axon tip and disorganization of the microfilaments; removal of the cytochalasin allows the cone to form with new microspikes appearing. With colchicine, on the other hand, growth-cone and microspike activities continue for some time, but ultimately the axon shortens and finally sinks back into the body of the cell (Fig. 1.33), suggesting that the dissolution of the microtubules has destroyed the structural basis for growth. Thus in this case, and presumably in many others, the two systems work in co-operation.

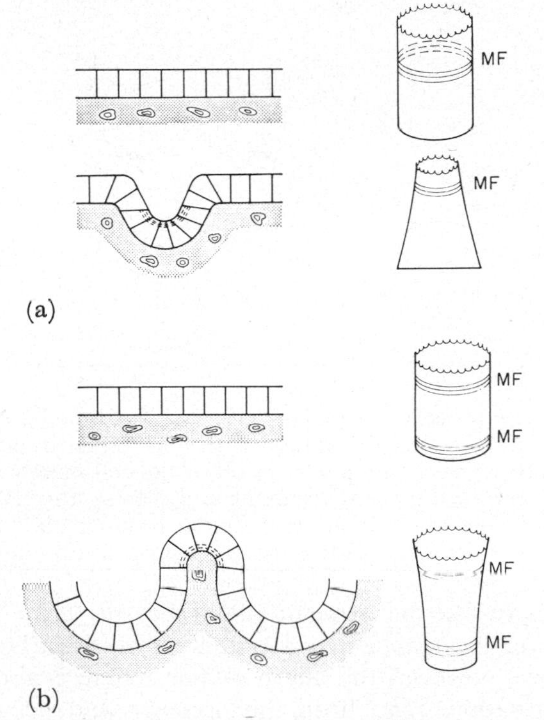

Fig. 1.34. Contraction of microfilaments (MF) during embryological development can cause sheets of cells to invaginate as shown in (a) or bulge outwards as shown in (b). The insets show how contraction of a ring of microfilaments at the apex of the cells causes invagination, whereas the microfilaments must contract at the base of the cells to cause a bulge outwards. (Wessels, *Science*, *N.Y.* 1971, **171,** 135.)

Changes in Cell Shape

The changes in cell shape that occur during development of a tissue are probably brought about by contraction of microfilaments lying in a dense array immediately under the plasma membrane; as illustrated by Fig. 1.34, the contraction of a ring of microfilaments at the apical end of cells arranged in an epithelium could result in the development of a pit, or invagination, the morphological basis for the development of a gland. It is likely that the changes of shape within

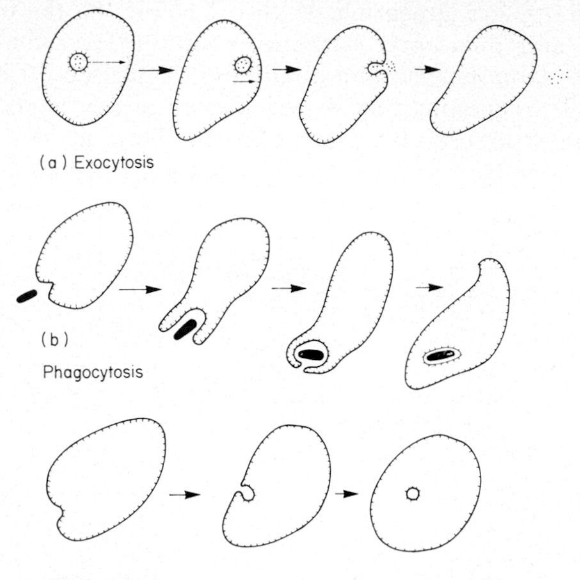

Fig. 1.35. (a) The process of exocytosis. A vesicle formed within the cytoplasm migrates to the cell wall and fuses with it to release its contents outside the cell. (b) Endocytosis is the pinching off of the cell membrane to enclose a region of the extracellular environment and passing this as a vesicle into the cytoplasm. Phagocytosis is the engulfment of particulate material such as bacterium; cell "drinking" is termed pinocytosis if fluid only is engulfed.

a given cell may also be brought about by this layer of filaments, including the endocytosis and exocytosis to be described below. As indicated above, however, the microtubular system is also involved in many transformations, and until the presence and possible function of the microfilaments were established, the morphological changes in cells during development were attributed to the assembly of microtubules, through given organizing centres, and their subsequent disassembly at later stages.

Endocytosis and Exocytosis. Two very important morphological changes in the cell are those involved in the engulfment of material from the outside—*endocytosis*—and the extrusion of material out of the cell—*exocytosis*. The two processes are illustrated in Fig. 1.35, endocytosis involving the pinching off of the cell's plasma membrane to form a vesicle whilst exocytosis involves the fusion of a membrane-covered vesicle with the cell membrane leading to the pouring out of the vesicle's contents. If the cell merely engulfs fluid from the outside medium the process is called *pinocytosis*—cell drinking—whereas if particulate material, such as a bacterium or food particles is absorbed, the process is called *phagocytosis* (Fig. 1.35). The phenomenon of exocytosis is characteristically seen when a secretory cell ejects its specific secretion. Thus, we have seen that the enzymes of the pancreas or parotid acinar cells are in the form of granules; these are membrane-bound vesicles tightly packed with the enzyme, the contents of which are ejected by a fusion of the granule with the cell's plasma membrane, and subsequent emptying of the vesicle (Fig. 1.35).

Phagosomes and Lysosomes

Some aspects of phagocytosis will be discussed when we consider the organism's defence mechanisms against invasion by bacteria, so here we may describe a general feature of cellular activity, namely its internal digestive apparatus. Thus a cell may have to destroy material that may be outside it—such as cellular debris—or within it, when, for example, it is changing shape during morphogenesis. In either case the material to be destroyed is engulfed within a vesicle, which is called a *phagosome* or *autosome* according to whether the material engulfed was outside or inside the cell. In order that the material engulfed by membrane should be broken down, certain digestive enzymes must be brought into the phagosome or autosome; and this is done by fusion with another body called a *primary lysosome*, typically seen microscopically as granules in such phagocytic cells as the polymorphonuclear leucocytes of the blood.

The primary lysosome contains digestive enzymes so that the *phagolysosome*, or *autolysosome*, so formed, is equipped for the breakdown of the absorbed material. The final fate of the phagolysosome or autolysosome probably varies with the nature of the cell and of the absorbed material; as indicated in Fig. 1.36, it may become a dense body (*tele-lysosome*) and finally a *residual body*, which may be excreted from the cell by exocytosis. These processes of endocytosis and exocytosis are of especial interest from the point of view of cell function; thus the enzymes required for breakdown of cellular material are insulated from the normal cytoplasm by enclosure within a membrane in the

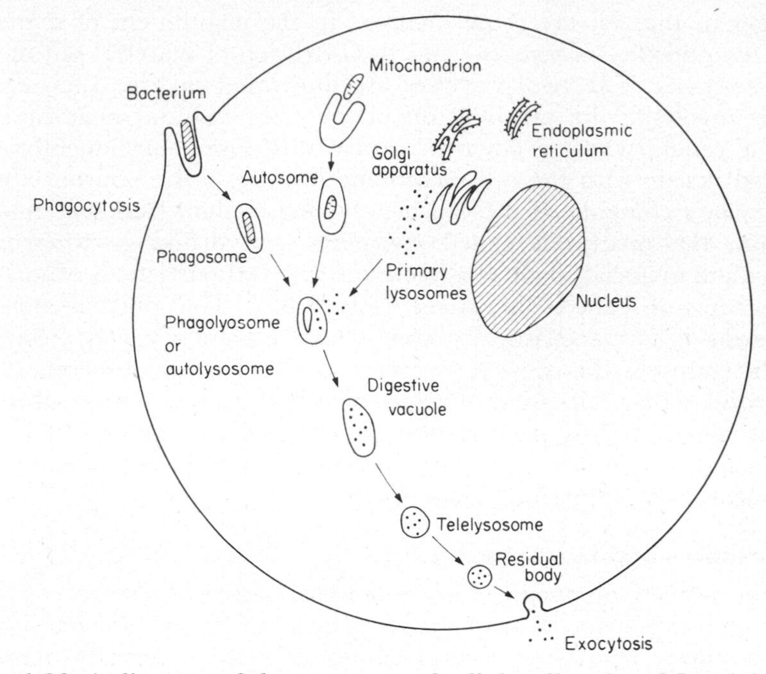

Fig. 1.36. A diagram of the sequences of cellular digestion. Materials from outside the cell are ingested by phagocytosis to form a vesicle termed a phagosome (or autosome if the material is from inside the cell). Enzymes are transferred to these vesicles by primary lysosomes, granules formed by the endoplasmic reticulum and the Golgi apparatus. Digestion proceeds until only debris is left which forms a residual body or telelysosome. This body usually passes out of the cell by exocytosis. (Modified from Gordon *et al.*, *J. Cell Biol.* 1965, **25,** 41.)

form of a lysosome, and thereby prevented from damaging the cell's own structures, such as mitochondria.

Cell Contact

The process of ingestion of foreign material raises interesting problems of cell contact; thus we have seen that strong forces of electrical repulsion enter when two cells approach each other to within, say, 50 Å, so that when a typical phagocytic cell, such as a polymorphonuclear leucocyte (Vol. 2) absorbs a bacterium, we must envisage some mechanism for reducing this repulsion. It has been argued that, because the force is reduced very considerably when the surface of the absorbing cell is drawn out into a villus or spiky process, this is the initial event establishing the intimate close contact necessary for engulfment; certainly the surface of a phagocytic cell changes shape continuously.

Microtubular Forces

Finally, there is once again the force brought into play to cause the changed surface during endocytosis or exocytosis, and here the microtubule or microfibillar systems have been invoked. More probably it is the microtubule system, since many processes requiring exocytosis, for example the secretion of enzymes by the emptying of vesicles into the exterior of the cell (p. 49), are inhibited by colchicine, the drug that specifically binds with microtubule protein. The secretion of the thyroid gland into the blood-stream relies on a preliminary endocytosis of iodine-containing material, and this is also inhibited by colchicine.

Nucleo-cytoplasmic Relations and Protein Synthesis

The interphase nucleus, i.e. the nucleus that is not undergoing mitotic (or meiotic, Vol. 2) division, appears as a more or less homogeneous body when observed in the light microscope, but the appearance of the fibrillar chromosomes in association with the spindle during mitosis reveals its true character as the site of storage of the chromosomal material, the basis of the genes. We may assume that the basic structure of the chromosomes is present throughout the life of the cell and that their prominence during mitosis is the result of a change of state, e.g. a more intense coiling of the fundamentally fibrillar structure, together with the synthesis of new material, or the condensation of previously synthesized material on the pre-existing fibrillar scaffolding. Only on this basis can the chromosomes play their fundamental role in transmission of hereditary characteristics from the mother to daughter cells; it is not conceivable that these characteristics could be retained by material that completely dissolved during interphase.

Chromatin as DNA

The classical histologists demonstrated the specific nuclear material with the Feulgen staining technique which is specific for desoxyribonucleic acid, DNA; they called it *chromatin* and the characteristic feature of mitosis was the appearance of these chromatin-bearing bodies, the chromosomes. During interphase, only small concentrations of chromatin—*heterochromatin*—were present, suggesting that in some way the Feulgen-staining material had disappeared. The use, however, of more refined methods of assay of DNA has shown that DNA does not disappear from the nucleus, but actually doubles in amount prior to the appearance of the chromosomes and spindle. In the electron microscope the resolution of the chromosomes in the resting nucleus has not been easy, because of the densely electron-staining character of the material.

However, when special techniques are employed for breaking up the structure of the nucleus, it may be shown that the chromatin material consists of a complicated web of fibres some 250 Å thick, the fibres doubtless representing a complex of DNA fibrils coated with protein. Thus treatment with chelating agents that remove Ca^{2+} and Mg^{2+} tends to break up the complex, yielding fibres some 100 Å thick, and treatment of these with enzymes that split off protein leads eventually to the naked DNA fibrils of some 25 Å diameter. As indicated above, however, the chemical basis of the chromosome responsible for transmission of hereditary characteristics is DNA; this material not only transmits information to the rest of the cell but also—and this is essential for its function—it reproduces itself in its own image prior to cell division.

Functions of the Nucleic Acids

By reproducing itself the nuclear chromatin or DNA acquires the ability to transmit the genetic character to the daughter cells following division, but this of course is insufficient; if the genetic material is to exert its influence on the manner in which the cell develops and subsequently behaves, the nuclear DNA must transmit information to the cytoplasm, which is the seat of the great bulk of the metabolic activity, containing as it does the mitochondria, endoplasmic reticulum and Golgi apparatus.

RNA and Protein Synthesis

This influence is achieved by the synthesis of another type of nucleic acid, *ribonucleic acid* or RNA; this material passes out of the nucleus through the nuclear membrane into the cytoplasm, where it controls the synthesis of specific proteins that give the cell its peculiar character, e.g. haemoglobin for the erythrocyte, myosin and actin for skeletal muscle, and so on. Important amongst these synthesized proteins are the enzymes, which act as catalysts for the chemical reactions taking place, including those actually involved in the synthesis of RNA and the proteins. Thus the basic functions of the nucleus may be summarized schematically as in Fig. 1.37. DNA is the indestructible component that retains its integrity throughout the life of the cell; prior to division, it replicates; during the life of the cell it organizes the manufacture of ribonucleic acids (RNA); these nucleic acids are

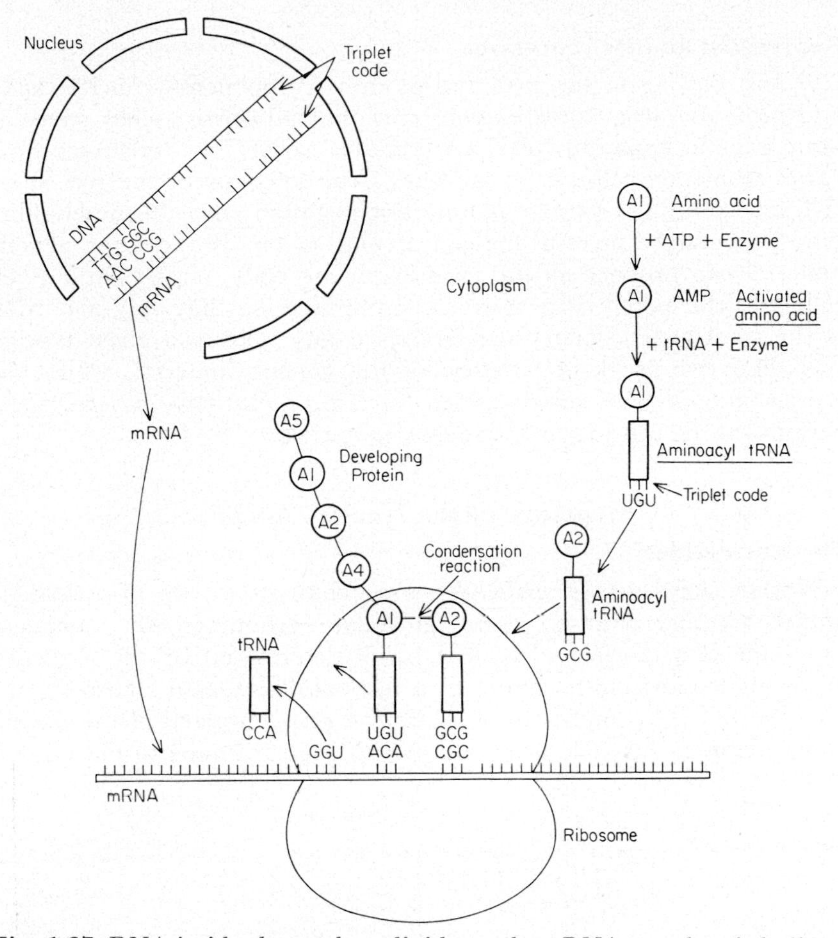

Fig. 1.37. DNA inside the nucleus divides and an RNA template is built up onto the DNA. This RNA template, messenger-RNA (m-RNA), now diffuses out of the nucleus and links onto a ribosome. The m-RNA has all the necessary information coded on its surface to control the amino acid sequence of a specific protein. Amino acids are linked to transfer-RNA molecules (t-RNA) each specific to its amino acid. The t-RNA amino acid complexes pass to the ribosome m-RNA complex, and the amino acids are linked on to each other, the t-RNA reading the m-RNA code so that the correct sequence of amino acids is added to the new polypeptide chain. When the chain is complete, it is released from the ribosome.

of different molecular structure and, according to the type, a given protein will be synthesized by its aid. The sum total of proteins synthesized will govern the behaviour of the cell through its life, largely by virtue of the enzymes that are produced.

Repression and De-repression

Before passing to the structure of these components—nucleic acids and proteins—it is worth emphasizing that all somatic cells have the same genetic apparatus of DNA, yet the variety of protein synthesis varies from one cell-type to another. The ancestor of the red blood cell, for example, produces mainly haemoglobin; skeletal muscle fibre produces mainly myosin and actin, and so on. If the same genetic material can produce all the proteins in the body, clearly, in a given cell, this pluripotentiality must be suppressed, so that only a portion of the total genetic material is active at any time in a given type of cell. Thus we speak of *repression* of the genetic material, whilst the repression may be lifted under certain conditions—*de-repression*—permitting the cell to synthesize new materials.

Structure of the Nucleic Acids

The Nucleotides

Nucleic acids are *polynucleotides*, long chains built up of *nucleotides*, linked together through their phosphate groupings, the chemical character of a given nucleic acid being determined by the sequence of the nucleosides on the chain. By a nucleotide is meant a base–sugar–phosphoric acid complex: thus *adenylic acid* is composed of the purine base, adenine, linked to a pentose which itself is linked to phosphoric acid.

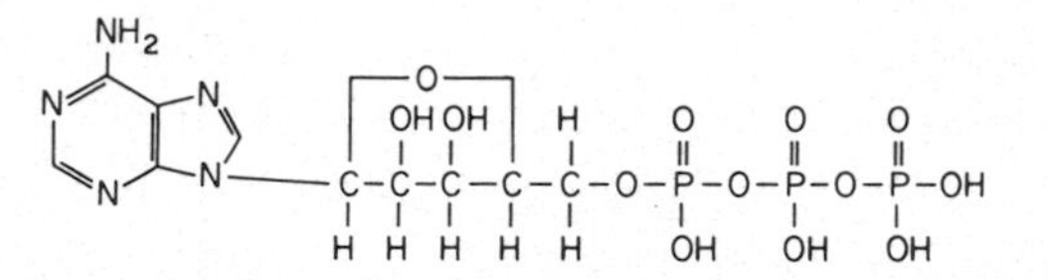

Adenosine 5' triphosphate (ATP)

The base may be either a *purine*, such as adenine or guanine, or a *pyrimidine*, such as cytosine or uracil:

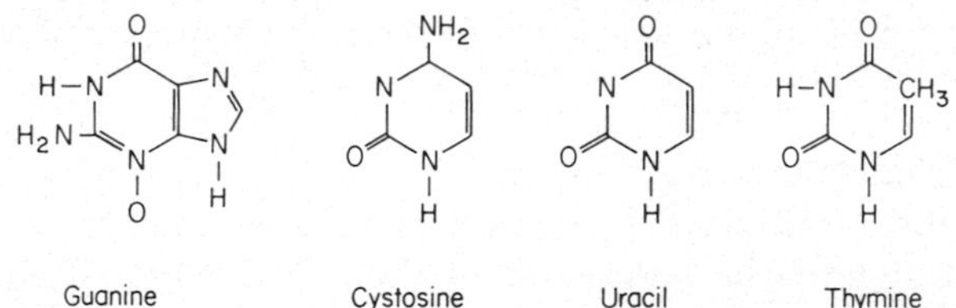

Guanine Cystosine Uracil Thymine (methyl uracil)

and the sugar may be either D-ribose or a deoxy-derivative, D-2-ribodesose:

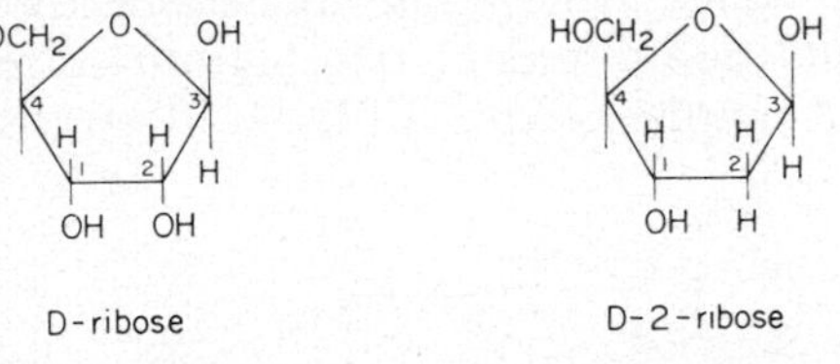

Bases and Sugar. The nucleic acids containing the deoxyribose are called deoxyribonucleic acids (DNA) and those containing ribose are called ribonucleic acids (RNA). In most ribonucleic acids (RNA) the four bases are two purines—adenine and guanine—and two pyrimidines—cytosine and uracil; in the DNA series the bases are usually guanine, adenine, cytosine and methyl uracil, or thymine. Because of this difference in base-composition it is easy to label DNA specifically with radioactive thymine, or its precursor, leaving RNA unlabelled.

Phosphate Links. In the adenylic acid molecule illustrated overleaf, the phosphate is linked to the 3′-carbon atom. This is characteristic of the nucleotides derived from ribonucleic acids (RNA). Nucleotides derived from DNA, containing of course deoxyribose, can have the phosphate linked at either the 3′- or the 5′-position, the adenylic acid with the phosphate in this last position being illustrated below:—

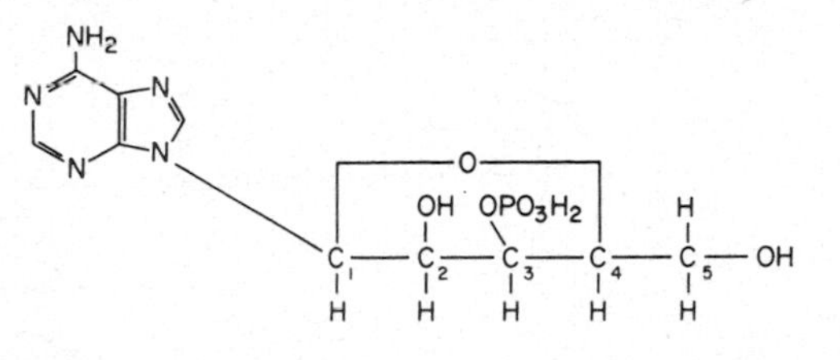

Adenylic acid (Adenosine 3′ phosphate)

The nucleotides link with each other through the phosphate group which joins the two sugar residues of adjacent molecules; this is illustrated by the tetranucleotide of Fig. 1.38 derived from DNA.

Terminology

This may be confusing; the *nucleotide* is the base–sugar–phosphoric acid complex, adenylic acid, and so on; when the phosphoric acid residue is split off, e.g. as a result of treatment with a nucleotidase, the result is a base–sugar complex, called a *nucleoside;* thus adenylic acid gives the nucleoside *adenosine*, cytidylic acid gives *cytosine*, and so on. Consequently adenylic acid may also be called adenosine monophosphate, abbreviated AMP. The nucleoside di- and tri-phosphates are formed by addition of one or two phosphate residues respectively; thus adenosine-5′-tri-phosphate (ATP), is illustrated below:

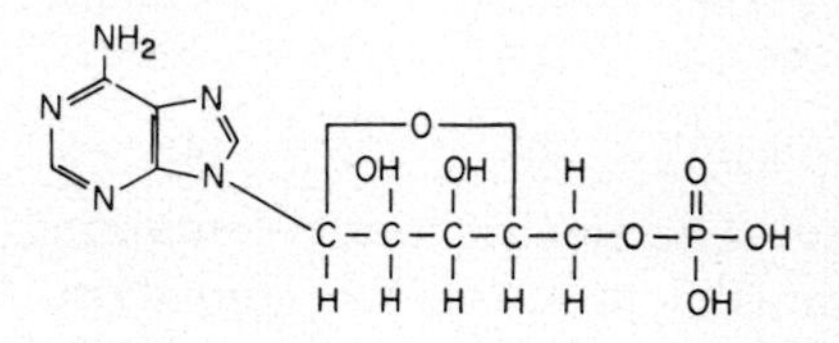

Adenylic acid (Adenine 5′ phospodeoxyribonucleotide)

Nucleoside Triphosphates

It is clear from Fig. 1.38 that polynucleotides of indefinite length

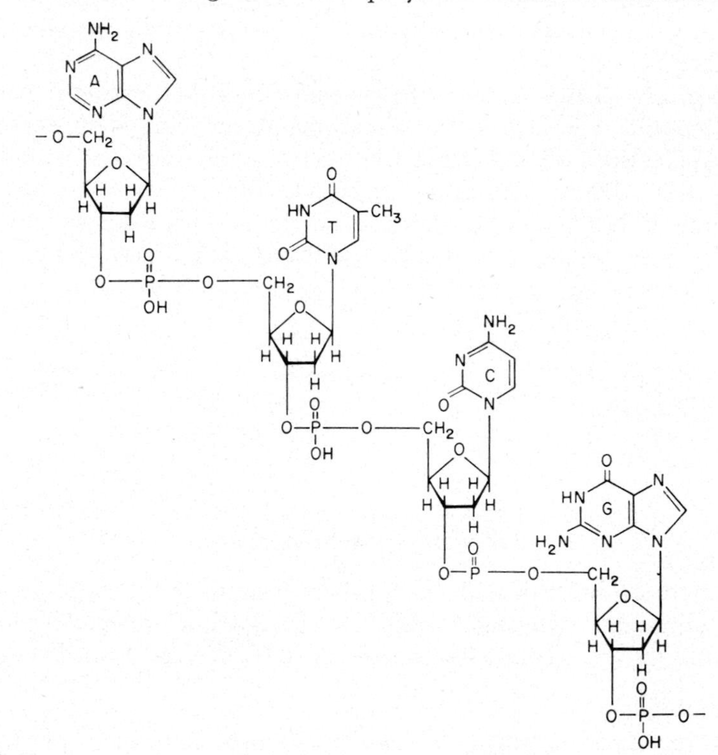

Fig. 1.38. Part of a single strand of the DNA molecule composed of adenine (A), thymine (T), cytosine (C) and guanine (G) deoxyribonucleotides. The nucleotides link with each other through the phosphate group which joins the two sugar residues of adjacent molecules. (Harper, *Rev. Physiol. Chem.* 1973, 14th Ed. (Lenge) Blackwell.)

can be built by addition of successive nucleotides. The actual building materials are the triphosphates, ATP, GTP, etc., so that, for example, in the synthesis of a DNA, adenosine triphosphate, containing a deoxyribose and with its phosphate residues linked at the terminal, 5′, position, forms a link with the OH-group at the 3′-position of the terminal nucleotide, casting off two of its phosphate groups in the process. In general, the DNA-type polynucleotides are built with this 3′-5′ linkage, but in the RNA's both 3′-5′ and 3′-3′ linkages are possible.

Nucleoproteins

In their natural state the nucleic acids are usually associated with protein, linked to this chemically to give *nucleoproteins*. In the nucleus

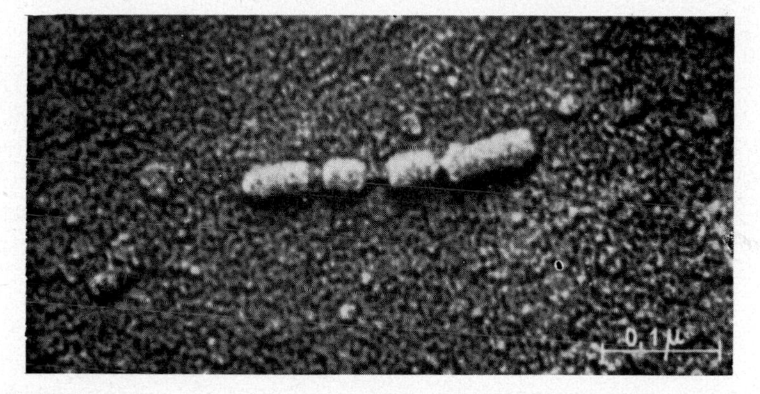

Fig. 1.39. An electron micrograph of the tobacco mosaic virus. ×150,000. (Schramm, Schumacher and Zillag, *Z. Naturf.* 1955, **10b,** 493.)

of the cell, the DNA occurs in combination with proteins of low molecular weight called histones or protamines (usually histones) from which they may be separated by fairly simple treatment.

Molecular Weights

The molecular weights of the nucleic acids are usually very high; this is especially true of the DNA's of nuclear material; thus it is considered that the whole of the nuclear material of the bacterium *Escherichia coli* consists of a circular chain some 1·3 mm long with a molecular weight of some 2.10^9. In eukaryotic cells, i.e. in cells of higher organisms with a well defined nucleus, the basic unit of DNA structure may well be the chromatid, rather than the gene, in which case the chain of DNA would be about 1 metre long.

The tobacco mosaic virus responsible for causing disease in tobacco leaves has a molecular weight of $40.7 . 10^6$; the virus consists of a fibre of nucleic acid within a protein shell (Fig. 1.39) the protein representing some 94 per cent of the total molecular weight; this gives a molecular weight for the nucleic acid of about 2 million. So far as the genes of the eukaryote nucleus are concerned, it may well be that each one is a single molecule of DNA linked to histone; and it is considered that the histone acts as an inhibitor to the DNA's powers of organizing synthesis, i.e. it is responsible for the repression discussed above.

The very high molecular weights of nucleic acids mean that, if the sequence of bases can be altered at will, the total number of molecules that may be made up on this basis is enormous; in fact, some limitations are imposed by the requirements of stability of the chain, but still the potentialities for variation in structure are enormous.

The DNA Helix

Largely on the basis of the X-ray diffraction pattern of DNA, Watson and Crick suggested the helical structure for DNA that has been accepted since first proposed; the basic structure is illustrated schematically in Fig. 1.40, where the two ribbons are chains of phosphate–sugar, whilst the rods represent the bases, which interact with each other through hydrogen-bonds, and thus act as links stabi-

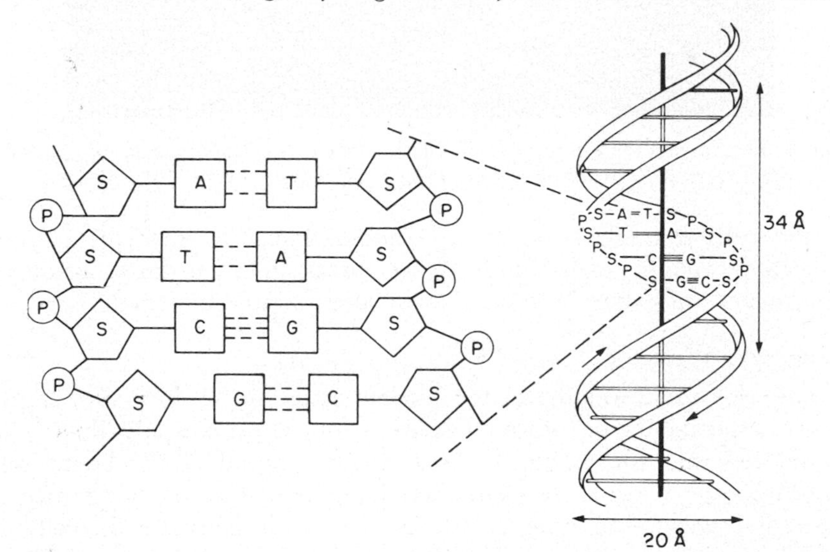

Fig. 1.40. The Watson and Crick model of the double helical structure of DNA. A: adenine; C: cytosine; G: guanine; T: thymine; P: phosphate; S: sugar (deoxyribose). (Harper, *Rev. Physiol. Chem.* 1973, 14th Ed. (Lenge) Blackwell.)

lizing the two helices. If the bases are to be able to react with each other, through hydrogen-bonds, then certain limitations are imposed on the system; namely that if a pyrimidine is on one chain then a purine must be on the other; thus if adenine (purine) is on one chain its partner must be thymine (pyrimidine); if guanine is on one chain its partner must be cytosine, and this accounts for Chargaff's finding that, in all DNA's, there are equal molal amounts of adenine and thymine and of guanine and cytosine. There is no restraint on the sequence of bases, this permitting a large variety of molecules built up of relatively few building blocks; but once one chain has been fixed, then the sequence of its partner, or *complementary chain*, is also fixed. This at once suggests the manner in which the chains can reduplicate, provided we can conceive of a mechanism for their separation during the process, each chain producing its complementary chain, and the pairs subsequently uniting.

Ribonucleic Acids

In general, the regularity in base-sequence necessary for a two-stranded helix is not present in the RNA's, so that they are usually single-stranded, although some looping of chains with cross-linking between the partners of the loop is by no means excluded. The RNA of the tobacco mosaic virus is a single strand, and it is able to remain in this uncurled condition by being held within a protein shell; in other situations, e.g. the ribosomes, the ribonucleic acids are linked to a variety of proteins which presumably act to stabilize the particular structure required.

Nucleic Acids and Protein Synthesis

The fundamental characteristic of the nucleic acids is their ability to organize the assembly of units to form new chains, using themselves as templates on which the synthesis occurs. Thus experimentally a preparation of DNA can be provided with the various sugar–base–phosphate complexes, in the form of ATP, GTP, and so on, and provided an appropriate enzyme, called a DNA-polymerase, is present, new DNA or new RNA can be formed according to the type of sugar in the nucleosides. Thus DNA can be made to replicate itself in the test-tube, and it can also be made to organize the synthesis of RNA's, the RNA so formed having a so-called base-complementarity to the DNA which acted as the template. The sequence of bases in the synthesized RNA is similar to that of the DNA which acted as a template, with the difference that thymine (methyl uracil) was replaced by uracil. It is well established that many of the RNA's of the cell, whether they are found in the cytoplasm or the nucleus, are derived from the DNA of the nucleus, exhibiting base-complementarity with that part of the DNA in the nucleus that presumably acted as the template. What, then, may we ask, is the function of the RNA? Certain RNA's can behave like DNA and replicate themselves, e.g. the RNA-containing viruses, but in the cell the main, or exclusive, function of the RNA is to organize the synthesis of protein, acting in one of three basic manners and being called, according to its mode of operation, *messenger RNA*, *ribosomal RNA* and *transport RNA*. To appreciate these roles we must first briefly consider the structure of proteins.

Protein Structure

Proteins are formed by the linking of amino acids to form chains, the linkage being called *peptide*, and the chains being called *polypeptides*. A protein may consist of a single chain, as with myoglobin, or of more than one linked together at certain points; thus insulin, containing some 150 amino acid residues (M.W. 6000), contains two chains, linked together through the SH-groups of cysteine residues, as

NH_2 NH_2

Phe. Val. Asp. Glu. His. Leu. Cy. Gly. Ser. His. Leu. Val. Glu. Ala. Leu. Tyr. Leu. Val. Cy. Gly. Glu. Arg. Gly. Phe. Phe. Tyr. Thr. Pro. Lys. Ala

S S

NH_2 S NH_2 NH_2 S NH_2

Gly. Ileu. Val. Glu. Glu. Cy. Cy. Ala. Ser. Val. Cy. Ser. Leu. Tyr. Glu. Leu. Glu. Asp. Tyr. Cy. Asp

S —— S

Fig. 1.41. The structure of the insulin molecule. (Sanger, *Br. Med. Bull.* 1960, **16,** 183.)

illustrated by Fig. 1.41. Insulin is a small protein; much larger molecules may be made by increasing the lengths and numbers of the individual polypeptide chains, so that molecular weights of over a million may be achieved, e.g. with haemocyanin, the invertebrate respiratory protein.

Amino Acids

The formulae of several naturally occurring amino acids are shown below, and it will be seen that they may be written basically:

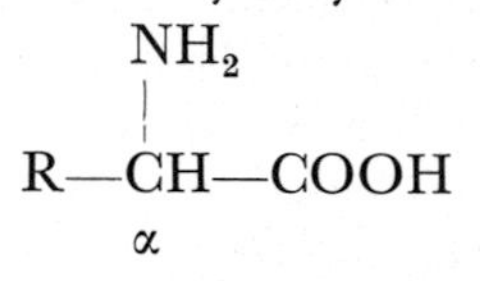

$$\begin{array}{c} NH_2 \\ | \\ R\text{—}\underset{\alpha}{CH}\text{—}COOH \end{array}$$

an amino group occurring on the α-C-atom. R is called the side-chain, and thus may be the H-atom in glycine, CH_2SH in cysteine, $NH_2(CH_2)_4$ in lysine, and so on. It will be noted that glutamic and aspartic acids have two carboxyl groups and are called *acidic* amino acids, by contrast with the *basic* lysine and arginine, with an additional NH_2-group, and the *neutral* glycine and alanine.

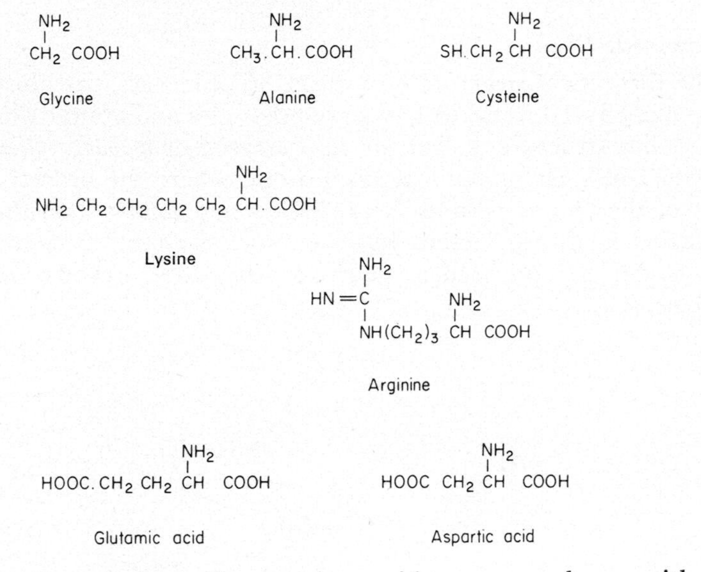

Peptide Formation. Two amino acids may condense with the elimination of water to form the dipeptide:

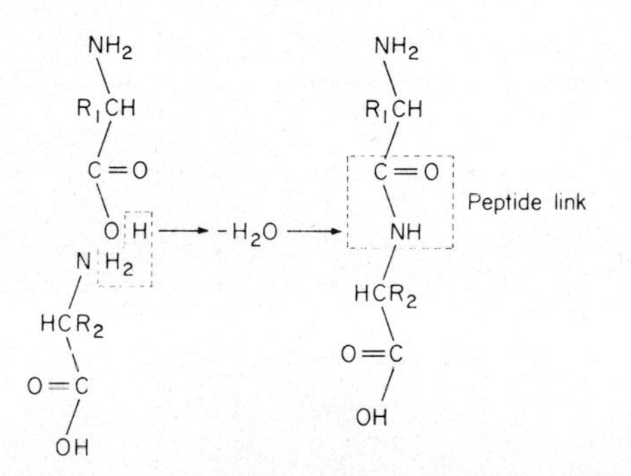

containing at one end a *terminal amino group* and the other a *terminal carboxyl group*. It is clear that this process may be repeated indefinitely, the ensuing polypeptides retaining this character of terminal amino and carboxyl groups. The number of naturally occurring amino acids in proteins is some 26, so that the possibilities of variation in composition are enormous. Thus a genetically determined alteration in the nature of only a single amino acid in haemoglobin, the vertebrate respiratory protein, can produce a diseased condition, that of sickle-cell anaemia. Here glutamic acid has been replaced by valine to give *Haemoglobin S;* in another mutation, the glutamic acid is replaced by lysine (*Haemoglobin C*).

Polypeptide Chains

The theoretical structure of a chain of amino acids is illustrated by Fig. 1.42 deduced from the known bond-angles and atomic dimensions; and such a structure, called the *fully extended* or *β-configuration*, can be identified in such proteins as silk fibroin, where the orderly arrangement of the chains permits X-ray diffraction analysis. Without some stabilizing factor, or factors, however, such a chain on its own would tend to curl up on itself to give what has been called a *random-coil arrangement*.

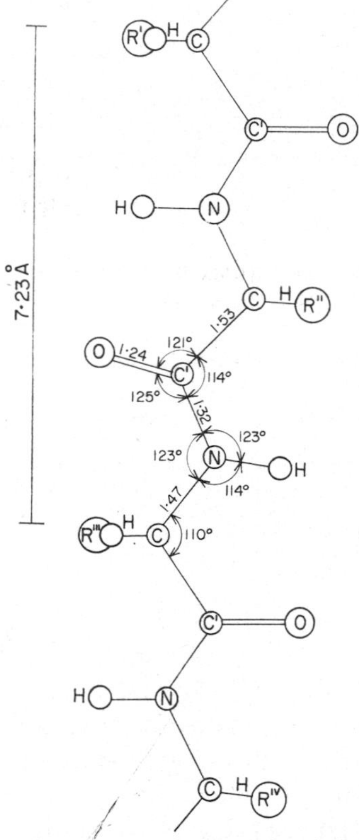

Fig. 1.42. A diagrammatic representation of a fully extended polypeptide chain with the bond lengths and bond angles derived from crystal structures and other experimental evidence. (Corey and Pauling, *Proc. Roy. Soc.* B 1953, **141,** 10.)

Pleated Sheet

The essence of analysis of the structure of proteins, once the amino acid composition and the sequence of the amino acids have been determined, is to assess the manner in which the chains have twisted and the factors that have stabilized them in a given arrangement, which may often be as regular as the arrangements of molecules in a crystal. In silk fibroin, the chains retain this extended configuration, coiling being prevented by the parallel arraying of the chains in such

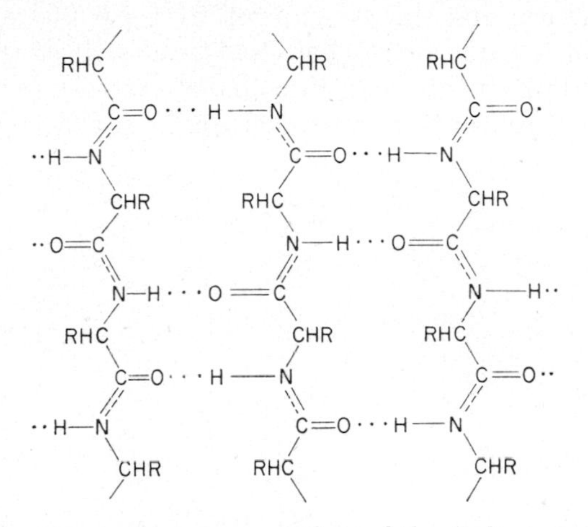

Fig. 1.43. A diagrammatic representation of the "pleated sheet" structure of protein. Polypeptide chains held together by hydrogen bonding. Alternate chains oppositely orientated. (Pauling and Corey, *Proc. Nat. Acad. Sci. Wash.* 1951, **37,** 251.)

a way that hydrogen-bond links between adjacent C=O and NH-groups of the chain are stereochemically possible. This "pleated sheet" structure is illustrated in Fig. 1.43.

α-Helix

Another arrangement involves a certain degree of curling of the chain to form the *α-helix*, so that three amino acids only extend some 5·1 Å along the length of the fibre-axis as against 7·23 Å for the fully extended β-configuration (Fig. 1.42). This helical arrangement is complicated by the tendency of the helix itself to coil with a much larger pitch to give a "coiled coil", whilst the system is stabilized by the helices winding round each other. In wool keratin there are some six, winding round a seventh central one; with such an arrangement the side-chains tend to project out at right-angles to the direction of the long axis, and where these have chemically active groups they tend to react with each other, e.g. the SH-groups of cysteine tend to form disulphide bridges; OH- and NH_2-groups form hydrogen bonds, etc., and in this way the whole arrangement is stabilized.

Triple Helix

Collagen, the basis of vertebrate connective tissue, is a protein; as with fibroin and keratin its chain tends to remain extended to give a fibrous structure, and this is achieved by a parallel arrangement of three chains in a triple helix (Fig. 1.44), the helices stabilizing each other by cross-linkages through side-chains. As with keratin, the helix tends to coil on itself to form a "coiled coil".

Fig. 1.44. The triple helix of collagen. Each chain is held to the others by cross linkages through side chains.

Globular Proteins

The proteins we have so far considered are called *fibrous* because the long polypeptide chains are held in a parallel array to constitute the submicroscopical basis of the visible fibres that are made up of bundles of these units. The maintenance of this extended type of chain, whether it be in silk fibroin, hair keratin or connective tissue collagen, is a rare condition, so that the more common one is for the chains to roll on themselves, but usually in a highly ordered and crystalline manner, to give *globular proteins*. The rolling up is carried out in such a way that polar residues, e.g. OH-groups, NH_2-groups, etc., tend to face outwards, thereby imparting to the molecule its water-solubility; if the protein is an enzyme, i.e. a catalyst for a specific chemical reaction, then the special sequence of amino acids required for this enzymatic activity may be expected to be on the surface of the globular molecule, and therefore accessible to the chemical reactants.

Primary Structure

The sequence of amino acids in the polypeptide chain, or chains, of a protein is described as its *primary structure*. It is now considered that the orientations of the amino acids in the chains and the manner of coiling are governed entirely by this sequence of amino acids, so that if the amino acids of, say, insulin, were strung together artificially, the protein so formed would adopt the globular configuration actually found; similarly, if the three separate chains of the collagen helix were synthesized, they would tend to aggregate into a triple helix, which itself would tend to form a coiled coil. To a large extent this is understandable; the stabilization of a given structure relies on interaction between side-chains of amino acids, and only if these occur with appropriate regularity, and in appropriate places along the chains, will such interaction be possible.

Denaturation. The importance of this stabilization of structure is shown in the phenomenon of *denaturation*, when a protein loses some of its characteristic features due to the breakdown of links that had previously held the chains in a definite pattern. Thus the enzyme ribonuclease relies for the maintenance of its globular form on the presence of disulphide linkages between chains; when these are ruptured, the enzyme is denatured and loses its power to catalyse its specific chemical reaction. When the linkages are allowed to re-form, by appropriate chemical treatment, the protein readopts its native structure and its enzymatic activity returns.

Secondary Structure

By the secondary structure is meant the orientation of the individual amino acids in relation to each other; thus silk fibroin and keratin differ in their secondary structures by virtue of the β- and α-arrangements. With many of the proteins a given molecule may be made up of different amounts of α-helix, the remainder of the chains having adopted a more randomly coiled arrangement. Thus the muscle protein tropomyosin contains some 90 per cent of α-helix, whilst the globular serum albumin only 45 per cent.

Tertiary Structure

By the tertiary structure of a protein we mean the manner in which its polypeptide chains have been arranged, e.g. the coiled coil structures of collagen and keratin, the globular structure of serum albumin, and so on.

Quaternary Structure

Many proteins can be broken down by relatively mild treatment into subunits of lower molecular weight; the manner in which these units are built up defines the quaternary structure.

Thus the length of chain that the cell can synthesize is limited, so that when we are dealing with proteins of molecular weight of the order of hundreds of thousands, these must of necessity be built up of subunits, these last being the chains synthesized separately. The collagen molecule is large, with a weight of 340,000, but it may be resolved into subunits of molecular weight about 10,000, hence the *quaternary structure* consists of the arrangement of these subunits to give the highly asymmetrical tropocollagen molecule. Again, the protein of the tobacco mosaic virus has a molecular weight of some 40 million, and can be shown to be built up by the assembly of protein subunits of molecular weight 35,000 arranged in a helical manner around a fibre of nucleic acid. The growth of large protein molecules has been likened to the growth of crystals, but in this latter case there is no limit to the size of the final product, whereas there is some factor that limits the number of subunits in the protein molecules. When the protein molecules are organized in a fibre, however, the end-to-end aggregation of molecules occurs by some form of physical or chemical interreaction, so that in a sense we may think of "giant molecules" with lengths corresponding to that of the microscopically visible fibre.

Linkages

The linkages between polypeptide chains necessary to sustain the tertiary structure are brought about by interaction between side-chains when these come close enough together, and when the chemical or physical basis for reactivity is present. Thus if a glutamic acid side-chain comes opposite an arginine or lysine side-chain, a *salt-linkage* may be possible (Fig. 1.45a). As indicated earlier, a linkage between two SH-groups on cysteine residues gives rise to a disulphide linkage (Fig. 1.45b); this, being covalent, is exceptionally powerful and requires chemical agents, such as mercury or its derivatives, to break the link. A very important type of link is the *hydrogen-bond;* this is something mid-way between the covalent link and the salt bridge, the hydrogen atom of, say, an OH-group tending to be shared with the N-atom of an amino-group, or with the O-atom of another OH-group (Fig. 1.45c). This type of bond can be weakened by making the aqueous solution concentrated in respect to salt (high ionic strength); high concentrations of urea and other hydrogen-bond breaking agents are also effective.

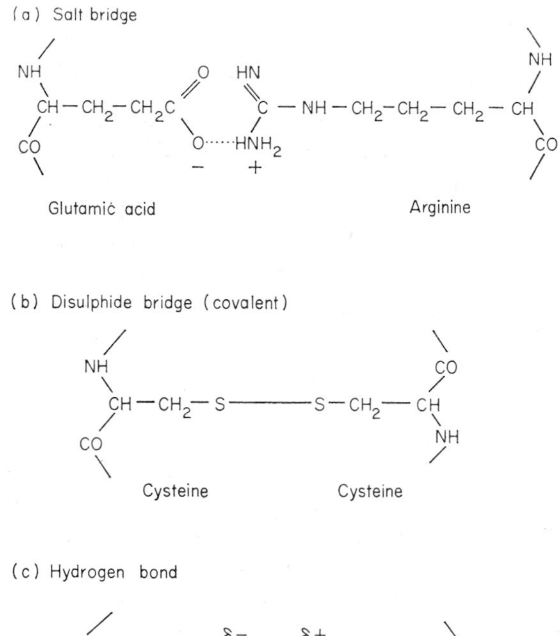

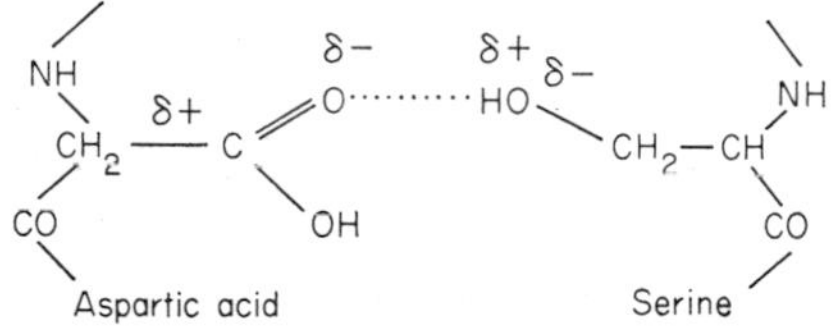

Fig. 1.45. The three-dimensional or tertiary structure of proteins stabilized by the interaction of side chain groups. (a) Salt linkage—electrostatic weak; (b) disulphide linkage—covalent strong; (c) hydrogen bonding—moderately strong depending on the conditions.

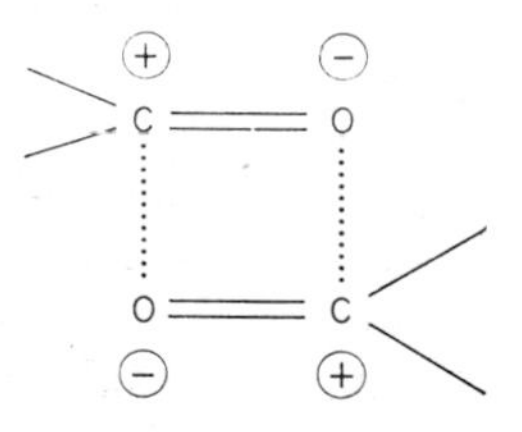

Fig. 1.46. Dipoles are produced when covalent links are formed and the electron shell of one partner is attracted to the more electropositive atom. These bridges give adhesive forces between molecules which can be moderately powerful.

Two other possibilities of interaction are through permanent dipoles and *dispersion*, or *Van der Waals*, forces. Because of a tendency for one atom of a pair of covalently linked atoms to attract the electron shell of the other strongly, the bond becomes polarized to form a dipole; thus the C=O group in the carboxyl group tends to become a dipole with the C-atom positive and the O=atom negative (Fig. 1.46). Dipoles tend to interact producing forces of adhesion. Again non-polar substances develop quite significant forces of adhesion; thus the myoglobin molecule consists of a single chain which is wound into a spherical shape with the non-polar side-chains facing into the centre. This structure is maintained entirely by virtue of the Van der Waals dispersion forces operating between the non-polar side chains in the centre.

The Specificity of Proteins

The proteins encountered in the animal or plant tissue may be separated by a variety of techniques that depend on their molecular weights and other physical and chemical characteristics, e.g. their electrical charges during electrophoresis, their tendency to adsorb during chromatographic separations, and so on. When these separations are carried out, a variety of classes may be established, such as the albumins and globulins of the blood plasma, collagen of connective tissue, actin and myosin of muscle, etc. Within a class, such as the albumins, more subtle physical and chemical differences can be detected, and through these further separations may be achieved. All these differences are the results of differences in amino acid composition, and can usually be detected by the relative proportions of the amino acids, as estimated by hydrolysis of the proteins and subsequent chemical separation.

Amino acid Sequence. Once a pure specimen of the individual protein has been separated it is possible, by modern techniques, to establish the sequence of amino acids in its chains. Thus Fig. 1.41 (p. 60) shows the sequence of amino acids in insulin, the first protein for which this was established (by Sanger). Insulin is a hormone, i.e. a compound exerting a highly specific action on the metabolism of the organism. It exerts this action by virtue of a certain sequence of amino acids within its molecule, so that it is possible to find a variety of insulin molecules with the same special chemical activity but with slightly different compositions because the changes in composition have not affected the important sequence.

Variations in Primary Structure. Thus the insulin depicted in Fig. 1.41 was prepared from cattle pancreas; insulin prepared from other species exhibited variations in the S–S linked ring, which in cattle insulin contains Ala.Ser.Val. in that order; in the sheep it is Ala.Gly.Val.; in the pig, Thr.Ser.Ileu; and in the horse, Thr.Gly.Ileu. These variations, occurring by genetic mutations, have not affected the activity of the hormone because this depends on other parts of the amino acid sequence. As indicated earlier, however, variations in the amino acid composition of haemoglobin, involving only a single amino acid, can have serious consequences; and this is because the variations occur in a critical region.

Antigenic Activity. The most sensitive ways of detecting subtle variations in amino acid sequence are those involving enzymatic or hormonal activity, where the change occurs in the sequence or sequences responsible for this; and also in the *immune response*. By the latter we mean the ability of a protein derived from a different species, to stimulate the formation of antibodies to it; once again this "antigenic" capacity depends on only a small portion of the molecule. It is in this way that the insulins from different species can be readily differentiated, because the sequence of amino acids governing its hormonal activity is different from the sequence governing its antigenic activity Thus we may inject pig insulin into a rabbit, say, and it will evoke the production of anti-pig insulin antibodies. These are proteins of the globulin class and may be separated, and they will react chemically with pig insulin but not with horse insulin, the small difference in amino acid sequence being sufficient to govern the production of an antibody that is specific for this insulin but not for others.

Ordered Assembly. If it is recalled that the secondary and tertiary structures of the protein are governed by the amino acid sequence, and if we consider these further aspects just mentioned, it becomes clear that the synthesis of protein molecules must require a very careful ordering of the assembly of the amino acids, a single error in the formation of a chain perhaps having a disastrous effect on the physiological activity of the protein molecule as finally assembled. We may now pass to a consideration of this assembly process.

Synthesis of Proteins

Template RNA

When twenty or more different amino acids are in solution, the orderly linking into a polypeptide of unique sequence requires some sort of template on which the assembly is to take place; and this is provided by RNA, the sequence of bases on this long-chain molecule serving to direct the sequence of amino acids through the operation of

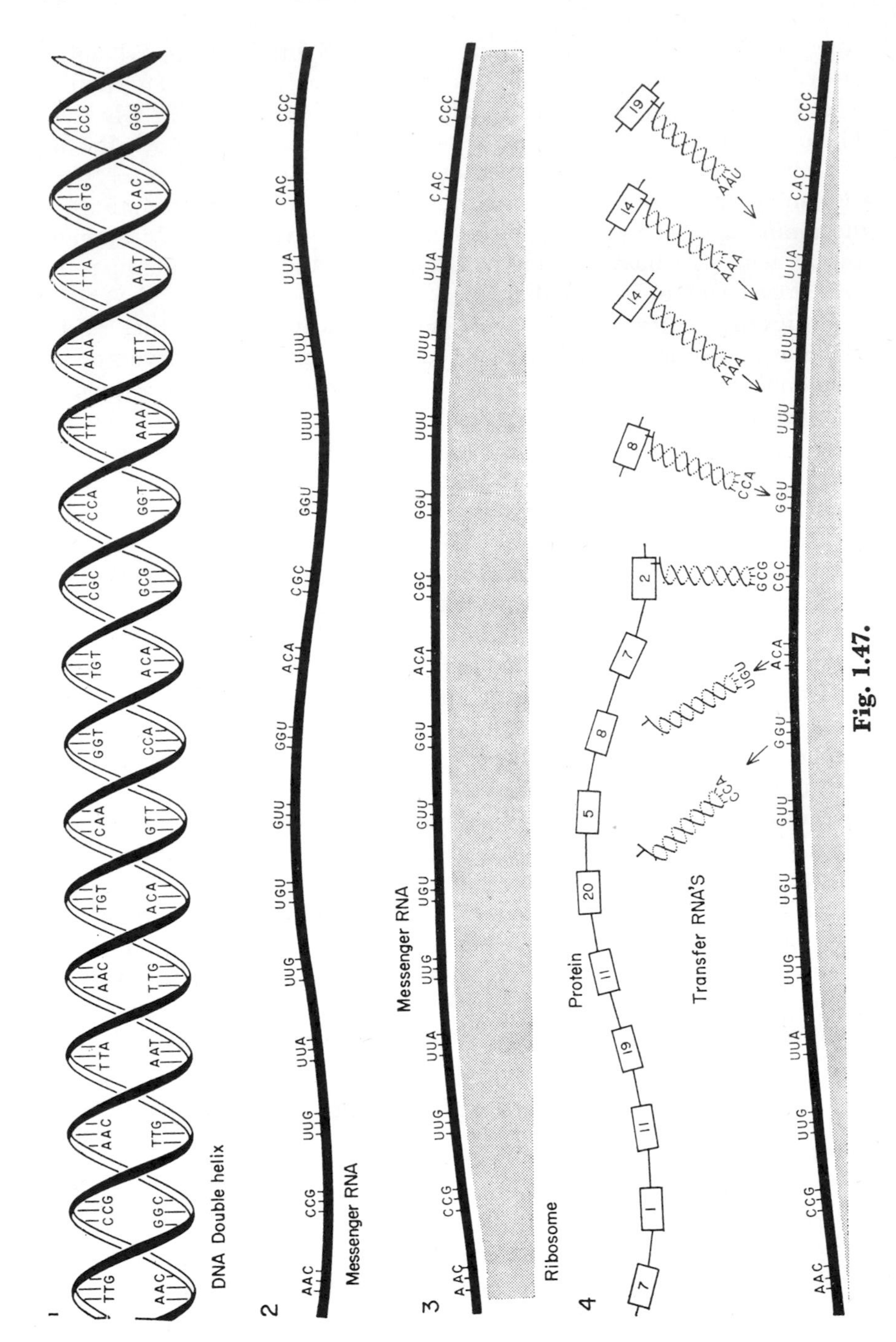
1
DNA Double helix
TTG CCG AAC TTA AAC TGT CAA GGT TGT CGC CCA TTT AAA TTA GTG CCC
AAC GGC TTG AAT TTG ACA GTT CCA ACA GCG GGT AAA TTT AAT CAC GGG
2
Messenger RNA
AAC CCG UUG UUA UUG UGU GUU GGU ACA CGC GGU UUU UUU UUA CAC CCC
3
Ribosome
Messenger RNA
AAC CCG UUG UUA UUG UGU GUU GGU ACA CGC GGU UUU UUU UUA CAC CCC
4
Protein
7 1 11 19 11 20 5 8 7 2
Transfer RNA'S
8 14 14 19
AAC CCG UUG UUA UUG UGU GUU GGU ACA CGC GGU UUU UUU UUA CAC CCC

Fig. 1.47.

the so-called RNA-code. The RNA that acts as a template is called *messenger-RNA*, because it has been synthesized on the nuclear DNA, and it passes as a messenger from the nucleus to the ribosome attached to the endoplasmic reticulum, i.e. the Palade particle. It attaches itself to this ribosome and, by virtue of the sequence of nucleotides along its length, it governs the order in which the amino acids in solution join up with each other.

Transfer RNA

The individual amino acids first react with specific molecules of RNA, called *transfer-RNA* because their function is to carry the amino acid to its appropriate part of the messenger, or template, RNA. Each transfer RNA carries a specific amino acid, and it is able to "recognize" the appropriate spot on the messenger RNA through a triplet of nucleotides, a triplet that is complementary in some way to a triplet carried by itself. Thus in Fig. 1.47 the letters UUU, UUA, CAC, etc., on the messenger RNA are triplets corresponding to three nucleotides, e.g. UUU = three uridylic acids in sequence, CAC = cytidylic acid, adenylic acid, cytidylic acid; and so on.

Ribosome

The 70s ribosome is a nucleoprotein which may be broken down into two subunits, a large one of molecular weight $1{\cdot}8 . 10^6$ and a smaller one of $0{\cdot}85 . 10^6$—the 50s and 30s subunits respectively.* The ribosomes serve as non-specific structures to which the protein assembly system attaches; when prepared with care they are found to be aggregated into polyribosomes, which is probably their natural condition since, on purely geometrical conditions, a long polypeptide chain containing, say, 150 amino acid residues would require a length of

---

Fig. 1.47. Synthesis of protein begins with the genetic code embodied in DNA (1). The code is transcribed into messenger-RNA (2). In the diagram it is assumed that the message has been derived from the DNA strand bearing dark letters. The messenger-RNA finds its way to a ribosome (3), the site of protein synthesis. Amino acids, indicated by numbered rectangles, are carried to proper sites on the messenger-RNA by molecules of transfer RNA. Bases are actually equidistant, not grouped in triplets, and mechanism of recognition between transfer-RNA and messenger-RNA is hypothetical. Linkage of amino acid subunits creates a protein molecule. (After Nirenberg, *Scientific American*.)

* The Svedberg Unit, indicated by s, defines the rate of sedimentation of the particle under a centrifugal force, and is closely related to the molecular weight of the particle. The animal ribosome is larger than the bacterial, 74s, and dissociates into a 62s and a 45s subunit; yeast ribosomes are 80s, and dissociate into 40s and 60s.

some 1500 Å of template RNA if, for every amino acid, three nucleotide residues are required for encoding the sequence. Hence the assembly apparatus consists of a string of RNA connecting several ribosomes, the number of the latter probably being determined by the size of protein chain that has to be synthesized. Thus the relatively small haemoglobin chains seem to be synthesized on pentamers whereas the large protein of the polio virus requires polyribosomes containing some 50 individual ribosomes.

Amino acid Assembly

The full description of the sequence of events in linking of amino acids together, and the subsequent release of the completed chain, cannot be entered into here; a brief summary of modern views is as follows:—

The amino acid to be incorporated must first react with t-RNA; this is brought about through an activating enzyme and ATP, to form

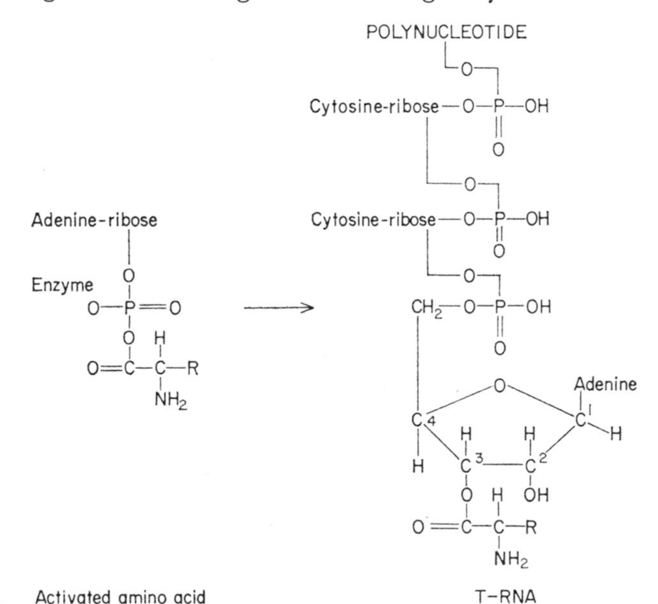

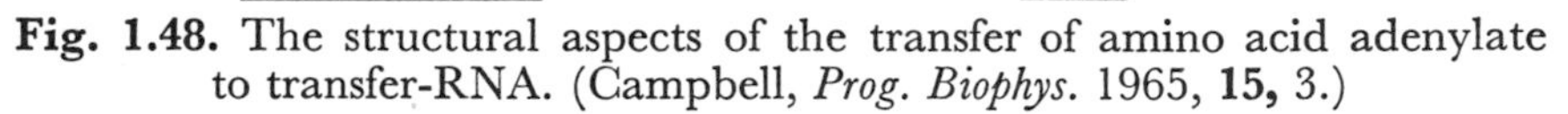
Fig. 1.48. The structural aspects of the transfer of amino acid adenylate to transfer-RNA. (Campbell, *Prog. Biophys.* 1965, **15,** 3.)

an aminoacyl-t-RNA (see Fig. 1.48). The aminoacyl-t-RNA, carrying this new amino acid, approaches the ribosome to which the growing chain is attached at a *receptor site* on the 50s subunit, the attachment being through the -RNA belonging to the last amino acid to be

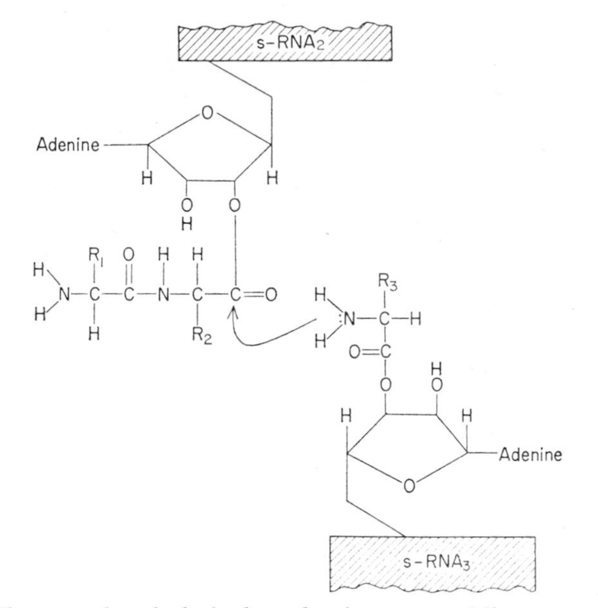

Fig. 1.49. The postulated chain lengthening step adding on a new amino acid. (Zamecnik, *Biochem. J.* 1962, **85,** 257.)

incorporated. The incoming aminoacyl-t-RNA "recognizes" its triplet codon on the messenger RNA attached to the 30s subunit. It is now ready to react with the terminal amino acid of the polypeptide chain, and it does this through its amino group which condenses with the COOH-group of the attached amino acid (Fig. 1.49), having attached itself first to the *donor site* in close proximity to the acceptor site (Fig. 1.50). As a result of this reaction, the t-RNA attached to the

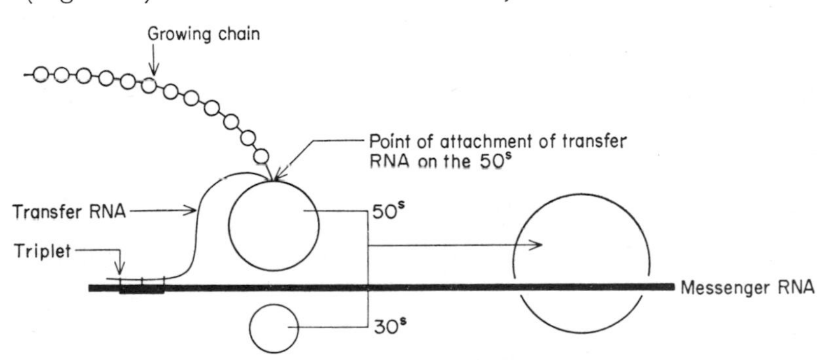

Fig. 1.50. Diagrammatic representation of the relationship between ribosome, messenger-RNA, transfer-RNA and polypeptide chain. (Campbell, *Prog. Biophys.* 1965, **15,** 3.)

polypeptide chain is released, so that the lengthened chain is now attached by the t-RNA of the latest arrival. Finally, the terminal t-RNA, with its attached polypeptide chain, moves across to the acceptor site, and the messenger ribosome is ready to receive a new aminoacyl-t-RNA, and the cycle begins again.

Tape Mechanism. The system of a polyribosome made up of a tape of RNA linking individual ribosomes suggests a static arrangement

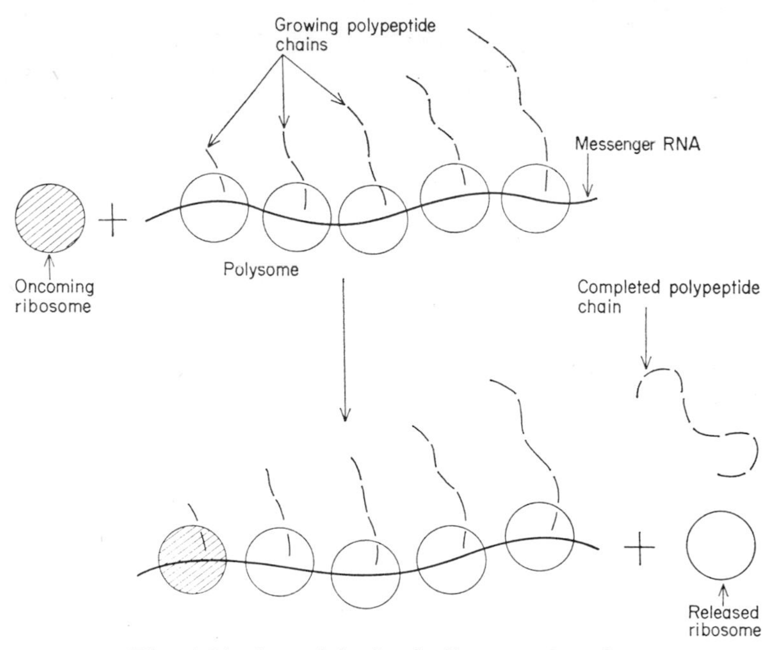

Fig. 1.51. A model of polyribosome function. (Goodman and Rich, *Nature, Lond.* 1963, **199,** 318.)

of the ribosomes; in fact, however, it is found that ribosomes are cast off the tape of messenger-RNA as the chain lengthens; thus a given ribosome passes along the length of the messenger-RNA tape, reading off its codons as it attaches each new amino acid; when it has reached the end of the tape its polypeptide chain has achieved the requisite length and it casts off, releasing itself from its ribosome as illustrated by Fig. 1.51.

Inhibitors of Protein Synthesis

Synthesis of protein is vulnerable at two main points, namely at the early stage of synthesis of messenger-RNA on the DNA template, called the *transcription process;* and also at the stage of assembly of amino acids on the messenger-RNA and ribosome. The former process may be inhibited by *actinomycin D*, which specifically inhibits the DNA-polymerase, i.e. the enzyme that catalyses the synthesis of messenger-RNA on a DNA-template, whilst *puromycin* inhibits the translational process; it does this by virtue of its chemical similarity to an aminoacyl-t-RNA that enables it to be carried into the assembling polypeptide chain, but it is sufficiently dissimilar to interfere with further growth of the chain.

Replication of Cytoplasmic Organelles

It is now well established that mitochondria are DNA-containing organelles, and through this DNA they are able to replicate themselves. The mitochondria contain ribosomes which doubtless form the basis for protein synthesis, a synthesis that may be inhibited by actinomycin D, which presumably blocks the transcription of messenger RNA from the DNA. Another organelle is the centriole; two of these organize the mitotic apparatus during nuclear division; these are DNA-containing bodies that replicate themselves, so that after cell division the new cells have the necessary centriolar apparatus. The centriole also exerts a function in relation to cilia or flagella where it is called a *basal body* or *kinetosome.* Reproduction of the centriole is through the formation of a procentriole which then develops into a new centriole and operates to organize the mitotic apparatus during cell division, or it may become a kinetosome, or basal body, of a cilium. In some organisms the centriole may perform both functions, one end organizing the astral rays of the mitotic figure, and the other end producing the new procentriole from which are ultimately derived the basal bodies for the flagella of the daughter cells.

Bacteria

These are a primitive type of organism and are plant cells, having a tough outer coat in which is contained the plasma-membrane lined cell, or *protoplast.* There is no well-defined nucleus as in higher plant cells, however, the genetic material being concentrated in more vaguely defined clumps of nucleoprotein. Protein synthesis takes place on ribonucleoprotein particles—ribosomes—and in fact most of the pioneering studies on protein synthesis have, in fact, been carried out on bacterial preparations. One of the complex organism's main defences against bacteria is to engulf them by endocytosis (phagocytosis), but its ability to do this depends on the nature of the surface coat of the bacterium, so that an important aspect of *invasiveness* is the surface coat secreted on to the rigid cell wall. If this is "unwettable" the phagocyte is unable to make the necessary intimate contact and so fails to engulf it. Thus the "encapsulated" strain of a bacterium is more likely to be virulent than a non-encapsulated one, as with the pneumococci. The nature of the surface also governs the bacterium's antigenicity, i.e. the provocation of antibody formation. Replication of bacteria is by division, the process being similar to that of a plant cell where a new tough outer coat must be synthesized at the site of division. It may also be "sexual" in the sense that a bacterial cell may transmit genetic material to another cell.

The Virus

The virus is a nucleoprotein molecule that has the power of reproducing itself, but only if it is in an environment that provides the necessary metabolic machinery for this; within the animal or plant cell it diverts the cell's metabolic machinery into multiplication of the virus's protein and nucleic acid, and in this way it exerts its harmful effects. The nucleic acid may be either RNA, as with the tobacco mosaic virus, or DNA, as with the vaccinia group; it is enclosed in a protein shell, or *capsid*, which is usually polyhedral, being built up on a semicrystalline arrangement of subunits. As we have seen, the tobacco mosaic virus is tubular, the nucleic acid being a fibre contained in the tubular shell. The virus is absorbed into the cell by endocytosis, and it exerts its synthetic effect after removal of its protein core, i.e. the initial process consists in replication of its nucleic acid; at the same time, protein is synthesized, using the viral nucleic acid as a template for the messenger-RNA, and new virus particles are assembled. Thus the infective agent is the nucleic acid and, in fact, transmission of disease, such as tobacco mosaic, can be effected experimentally by transfer of the nucleic acid alone; this is difficult usually, because the protein acts as a stabilizing shell. Since the virus carries no enzymes with it, the first step it must take before it can be reproduced is to induce the synthesis of the DNA or RNA polymerase necessary to assemble the nucleotides on the DNA or RNA template; this polymerase is called the *replicase*. Thus the cell's DNA is not employed in the production of the replicase, and in fact the virus inhibits synthesis on the cell's DNA, diverting the nucleotides that would be required to its own template.

Phage

This is the name given to viral particles that infect bacteria; they are DNA-containing nucleoproteins and, like the viruses, they organize the cell's (bacterial) enzymes into the synthesis of new phage nucleic acid and protein, leading finally to the death of the bacterium. The phage is a tadpole-like particle which attaches itself to the outside of the bacterium and injects its nucleic acid into this, the protein shell, or capsid, remaining outside. It is the tail that attaches itself and presumably, after some enzymatic degradation of the bacterial surface, this is punctured, enabling the infective nucleic acid to penetrate.

Interferon

Cells are able to develop a defence against viral infection independently of immune reactions; and this is through the induction, by the viral RNA or DNA, of the synthesis of a protein, *interferon.* Interferon is thought to be produced on a messenger-RNA synthesized with the host's DNA, possibly by the derepression of a portion of the genome caused by the invading viral nucleic acid. The action of interferon is to inhibit viral synthesis, possibly by inducing the synthesis of a translation-inhibiting protein that blocks the translation of the viral messenger RNA, i.e. it blocks the synthesis of the protein moiety of the virus by attachment to the host's ribosomes. This attachment does not block translation of the host's RNA.

Replication of DNA

We may conclude this chapter by returning to the most fundamental problem of all, namely the manner in which the DNA-double helix can reproduce itself, as in the duplication of the chromosomes of the cell.

Semi-conservative Duplication

The basic fact of division, first demonstrated by Taylor *et al.*, is that it is *semi-conservative*, in the sense that, when the double helix reproduces itself, the two new strands do not join together to form a new double helix of their own, but instead each new strand combines with one strand of the original helix. This was shown by allowing cells to grow in a medium containing tritiated thymidine, a label for their DNA, and preparing autoradiographs of the chromosomes in their first division; the chromatid-pairs, each consisting of a double helix, were both labelled, as we should expect, by the new strands which would be the labelled ones, paired with the old strands. Subsequent division when the cells were placed in an unlabelled medium should then be

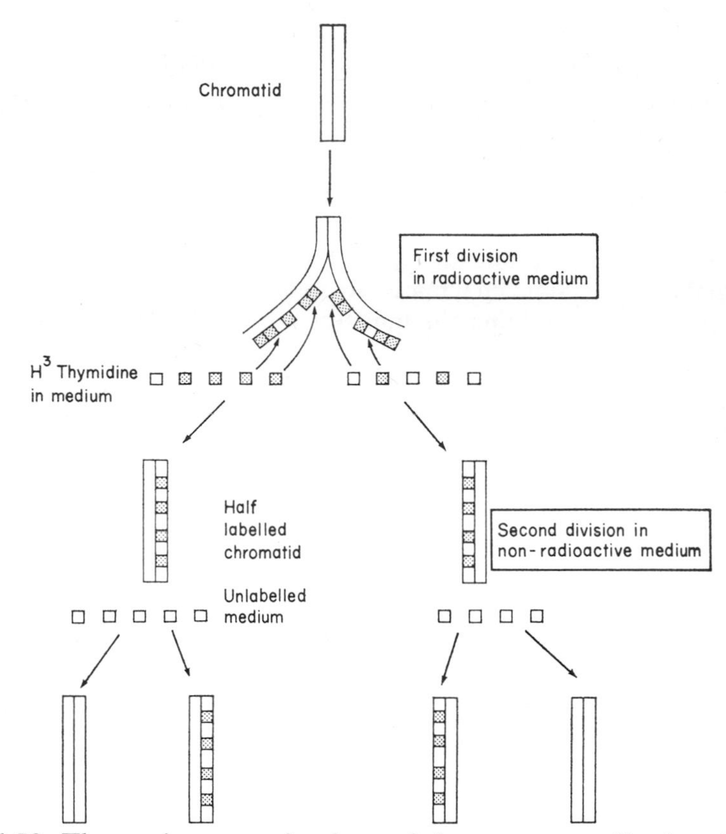

Fig. 1.52. The semi-conservative form of chromosome replication. Half of each parent in each pair of daughter chromosomes.

followed by the appearance of chromatid pairs with one chromatid labelled and one unlabelled, and this is what was observed (Fig. 1.52).

Circular Double Helix

Modern studies have shown that the DNA molecule in the genome of a bacterium, or that constituting the chromatid, is a single circular double helix which may be as long as a metre. Within the cell it is wound on itself. Replication must consist in somehow breaking the circle of double helix, unwinding so as to allow the individual strands to act as templates to organize the formation of a complementary strand, and subsequently rewinding in the semi-conservative manner just described, the new strand pairing with its parent. Reproduction is rapid, perhaps as much as 20 μm of bacterial DNA being reproduced per minute, and this rapidity of reproduction is achieved by the

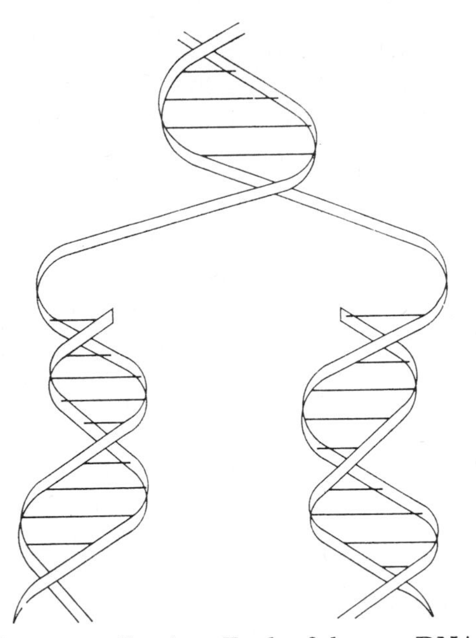

Fig. 1.53. Discontinuous replication. Each of the two DNA chains replicates but does so in opposite directions not like the method shown in this diagram. (Cove, Genetics, C.U.P. 1972.)

institution of numerous initiation points for reproduction along the length of the molecule, which may amount to 1000 on a given chromatid; thus at these points, which may be separated by a few μ from each other, there is a local unwinding with replication (Fig. 1.53).

Discontinuous Replication

An interesting discovery was that, at each point, the two chains replicate in opposite directions and not in the same direction as envisaged in Fig. 1.53, so that after a given period there will be ribbons of new DNA strands waiting to be united, a process that is organized by a sealing enzyme, called *ligase*. The reproduction in opposite directions seems an unnecessary complication but is actually necessary if the same enzymatic process is to be employed in forming the new chains. Thus there is only one form of nucleoside triphosphate employed, namely with the phosphate on the 5′-position of the deoxyribose (p. 54); this is bound to link with the 3′-position of the sugar of the nucleotide with which it is joining, so that synthesis will take place in the progression: 3′-5′-3′-5′ . . . . Now the two chains of DNA are such that the order of linkage in one runs in the opposite direction to that of the other (Fig. 1.54) so that replication of the two chains must take place in opposite directions if the same arrangement is to persist in the new pairs of stands. Such an arrangement is possible with the discontinuous synthesis that takes place at large numbers of initiation points.

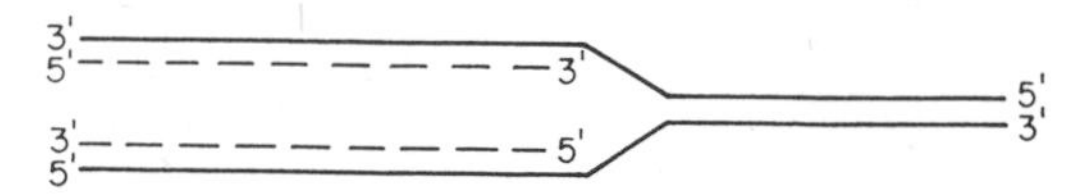

Fig. 1.54. The diagram shows the opposite polarity of the two chains of the DNA double helix and as a result the opposite polarity of the two new chains at the growing point. (Cairns, *Br. Med. Bull.* 1973, **29,** 188.)

Primer RNA

In artificial systems synthesis of DNA can be brought about provided that, in addition to the nucleoside triphosphates ATP, GTP, etc., three other factors are present, namely the *template DNA* which organizes the nucleotide sequence, the *primer*, i.e. a portion of DNA on to which the successive nucleotides are added, and the DNA-polymerase that catalyses the reaction, more accurately called a *nucleotidyltransferase*. We may assume that the existing DNA strands of the double helix in a chromosome act as templates for their replication, and the main problem has been to define the nature of the DNA-polymerase responsible, and to discover how the synthesis of each new piece of DNA begins, i.e. whether a piece of priming material is necessary or not. To consider this last question, attempts at finding a polymerase that would act without a primer have so far been fruitless, and it now seems that pieces of RNA rather than DNA act as templates at the growing points on a replicating chromosome, the RNA being newly synthesized at the growing point.

DNA-Polymerases

So far as the catalysis of the assembly of nucleotides is concerned, it appears that the DNA-polymerase, active in the classical studies on bacterial DNA synthesis, is only one of several; it is called DNA-polymerase I and it functions in the repair of DNA chains, a necessary process since many cells are normally exposed to damaging influences, such as ultraviolet light, which cause localized chemical modification in the DNA chain and which must be repaired by first excising the damaged region and, later, synthesizing a new piece. It is understandable that this enzyme would be active in artificial systems derived from cell homogenates, since the DNA extracted would almost certainly be damaged. Subsequent work has shown that other polymerases are required for normal reproduction, notably polymerase III.

CHAPTER 2

The Cell in Relation to its Environment

The Plasma Membrane

Before we plunge into the study of the whole animal we must continue our examination of the fundamental unit in behaviour, namely the individual cell. We have seen that it is defined morphologically by its limiting envelope—the plasma membrane—which separates it from its neighbours in a tissue or from its fluid environment. This envelope may well be surrounded by a tougher structure, or by what the electron microscopists call *basement membrane material*, but all the physiological evidence points to the relative functional unimportance of these outer layers which, under appropriate experimental conditions, may be removed from the cell without seriously altering its function or its power to remain alive.

Energy Requirement of Cell

In order that it may survive, the cell must absorb "energy-giving materials", i.e. substances such as glucose and oxygen, and it must give up products of metabolism such as carbon dioxide. As indicated earlier, the control over these and other exchanges between the cell and its environment is exercised by the plasma membrane. It was mentioned, too, that the special internal composition of the cell, particularly its high concentration of potassium, is dependent on the proper functioning of the plasma membrane, so that damage to this causes loss of potassium to the environment, often with a corresponding gain of sodium. Associated with these changes in internal composition there is often a swelling of the cell, due to imbibition of water, which, if not checked, can lead to rupture of the cell membrane, a process that is normally irreversible, so that the cell dies. Again, we have indicated that certain specific functions, such as the conduction of electrical impulses by a nerve or muscle fibre, depend on the difference in the ionic concentrations between the inside and the outside. Damage to the membrane, leading to alterations in the internal composition of the cell, will therefore impair these specific functions. Finally, we may note that when a cell is deprived of its supply of energy, often quite drastic changes occur in it, which include loss of internal ions such as K^+ and swelling due to imbibition of water, in fact changes similar to those described following damage to the plasma membrane. Thus deprivation of energy, be it lack of oxygen or of other substrates, does more than cause the cell to stop working, and it looks at first sight as though the integrity of the cell's membrane depended on a continuous supply of metabolic energy.

Ion Pumps. However, we shall see that the supply of energy in this particular case is not required for maintaining the structural integrity of the membrane, since this is a reasonably permanent part of the cell's structure; the energy is required for certain dynamic processes, called "ion-pumps" that are involved in maintaining the composition of the cell, including its water content, and thus its size, constant. Clearly, then, before we pass to the study of the whole animal and its tissues we must pursue the study of the fundamental unit in isolation; only when we know how it is that a cell exchanges material with its environment will we be in a position to understand the fundamental principles involved in its nutrition and its special activities, be these conduction of impulses in a neurone, contraction of a muscle fibre, secretion of milk by a mammary gland cell, and so on.

Internal Composition of the Cell

The contents of the cell are largely water and salts, of which the major cation is usually potassium; the surrounding medium is likewise made up mainly of salts and water but the major cation is now sodium. In Fig. 2.1 the internal composition of a "typical" mammalian cell is indicated together with that of the blood plasma, and by this latter we mean the fluid part of the blood obtained by centrifuging the cells down. It is the medium through which all the cells of the body are ultimately supplied with nutrition and although, as we shall see, the real medium bathing the cells is not blood plasma but the extracellular fluid that forms a thin film over their surfaces, we may in fact treat the plasma as the essential bathing medium since any change in the composition of the blood plasma is usually rapidly reflected in a corresponding change in the extracellular fluid.

The important feature of the internal composition of the cell is, as indicated above, the high concentration of K^+ and low concentration of Na^+; the concentrations of the common anions, Cl^- and HCO_3^-, are quite low in most mammalian cells although, as we shall see in Chapter 7, the red blood cells constitute an important exception. In order that the two solutions be electrically neutral, the sums of the concentrations of cations and anions must be equal; the "deficiency" of Cl^- and HCO_3^- in the cell is made good by a variety of organic anions, largely phosphates. In passing, we may note that the total

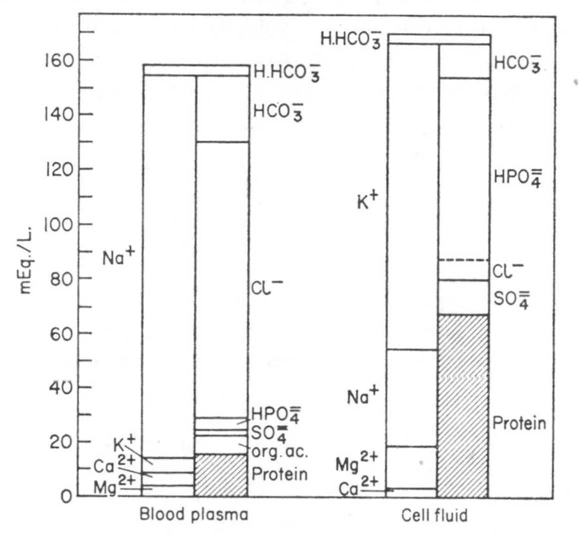

Fig. 2.1. The internal ionic composition of a typical cell shown on the right and that of mammalian blood plasma on the left. The principal cation inside the cell is K^+ (ca 110 mEq/L) and of the plasma is Na^+ (ca 140 mEq/L). (Harper, "Review of Physiological Chemistry" 14th ed. Lange 1973.)

concentrations of solutes in the cell and plasma are about equal, an important point with respect to osmotic exchanges between the two, as we shall see below.

OSMOSIS

When a cell is placed in its normal medium, its volume remains constant; when the medium is diluted with water it swells, due to the uptake of water—this swelling is called *osmosis*. It is a reflexion of the tendency of water to pass from a solution of lower to one of higher concentration of solutes, or, to express the matter more technically, we may say that the water is passing from a condition where its *chemical potential* is higher to one where it is lower.

Erythrocyte

The situation is illustrated schematically in Fig. 2.2a, in which an erythrocyte is imagined to be floating in blood plasma, which for simplicity has been indicated as a solution equivalent to 160 milliequivalents per litre of NaCl whilst the internal composition is indicated as being equivalent to 160 milliequivalents of K-salts. Initially the internal concentrations of solutes within and without the cell are equal, and if we assume, for the sake of argument, that the plasma membrane is impermeable to the ions of the system, but permeable to water, we may say that the system is in osmotic equilibrium, the tendency for water to pass in one direction across the membrane being the same as

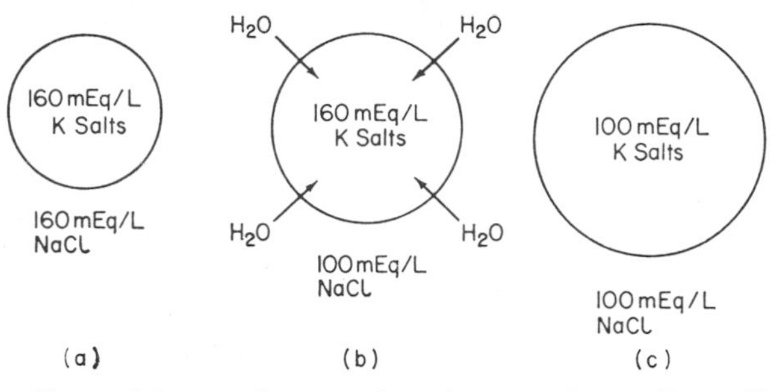

Fig. 2.2. Swelling of the erythrocyte in a hypotonic medium. For simplicity the internal composition is represented as 160 mEq/L of K^+ salts and the external medium 160 mEq/L of NaCl. At (b) the outside medium, which is very large compared with the cell, has been diluted to 100 mEq/L. A new equilibrium is established (c) with the internal solution of the cell diluted to the same osmolality as that of the outside, the cell having increased in volume.

the tendency for it to pass in the opposite direction. In Fig. 2.2b, the cell is now surrounded by diluted plasma, which has been put equivalent to 100 millimoles/litre of NaCl. Now the system is no longer in osmotic equilibrium, the tendency for water to pass into the cell being greater than to pass out, or the *chemical potential or escaping tendency* of water is higher in the outside medium than inside the cell. Water therefore passes into the cell until the inside has been sufficiently diluted to make its concentration equal to the outside concentration. At equilibrium, the concentrations will be equal, i.e. the chemical potential of the water will be the same inside and outside the cell. In Fig. 2.2 the volume of the medium surrounding the cell has been imagined to be very large by comparison with that of the cell, so that the outside concentration has not changed measurably, and equilibrium is reached with the cell's internal solution diluted to 100 millimolar of K-salts (Fig. 2.2c). It is easy to see that if the outside medium is made more concentrated than 150 milliequivalents per litre, the chemical potential of the water in the cell will be higher than that outside, and the cell will lose water, i.e. it will shrink, until the total concentrations of salts inside and outside are equal.

Osmotic Pressure

The process of water movement in response to altered concentration of solute is called *osmosis*, and the force determining it is called the *difference of osmotic pressure* between the inside and outside of the cell. It is given approximately by a simple formula:

$$\pi = RT(C_{In} - C_{Out}) \qquad (1)$$

the *Boyle-Van't Hoff Law.*

Lysis

It will be noted that the cell's adaptation to this difference of osmotic pressure has been a change in its shape that has permitted it to take up, or lose, water so as to reach the same concentration of solutes inside and outside. If the outside medium is diluted more, there comes a point at which the cell can no longer increase in volume without bursting its limiting membrane; with an erythrocyte this occurs at a concentration of about 90 millimoles NaCl per litre, but in any given suspension of erythrocytes there is a continuous spectrum of concentrations at which this bursting, or *osmotic lysis*, occurs, so that the most resistant cells require an external concentration as low as 60 millimoles/litre, whilst the least resistant or most *fragile* will burst after taking up very much less water, i.e. in a solution containing about 90 or even 100 millimoles NaCl per litre.

Concentrations

In this discussion we shall have occasion to use the term "concentration" frequently, so let us define our units for measuring this. The concentration, as employed in thermodynamic formulations, is expressed as a mole fraction:

$$\text{Mole fraction} = \frac{\text{Moles of solute}}{\text{Moles of solute} + \text{Moles of solvent}}$$

the mole being the molecular weight of the substance in grammes often called the Dalton.

Since in dilute aqueous solutions the number of moles of solute is small compared with those of solvent, we may use an approximate mole fraction:

$$\text{Moles of solute/Moles of solvent.}$$

The more practical unit of concentration is the *molality*, where the unit for defining the amount of solvent is the kilogramme of water:

$$\text{Molality (M)} = \text{Moles of solute/wt. of } H_2O.$$

Thus, a solution containing 58·5 g of NaCl in 1 kg H_2O is described as 1 M NaCl.

Molality and Molarity. The *molarity* of the solution is a more commonly used quantity and is often very little different in magnitude; it is defined as the number of moles of solute in 1 litre of *solution;* thus if we add 58·5 g of NaCl to 1 kg of H_2O we obtain a 1 molal solution, but the number of moles in 1 litre of the solution will be rather less, because the litre of water will have increased in volume by the addition of the 58·5 g NaCl. In quantitative analysis the molarity of solutions is invariably used because it is easy to make, say, a molar solution by dissolving a gramme-mole in a litre volumetric flask and making the volume up to 1 litre exactly. When discussing biological fluids, however, it is necessary to employ molalities because the differences between molarity and molality can be quite large owing to the presence of high molecular weight materials that occupy a considerable volume of the solution. Thus, blood plasma contains some 80 g of solids, mainly protein, in a litre so that the amount of water in a litre is only about 920 g. The *molarity* of sodium in plasma is some 150 milliequivalents/litre, whilst the *molality* is 150/920 = 163 milliequivalents/kg H_2O. It is the latter figure that is important for comparison with other fluids, e.g. lymph, cerebrospinal fluid and so on, so that it is desirable to use molalities routinely in the discussion of solute relations between fluids.

The Change in Shape

The erythrocyte has been illustrated in Fig. 2.2 as a simple spherical cell and its swelling or shrinking indicated as a simple increase or decrease in diameter, which would involve a stretching or shrinking of the cell membrane. This picture would be true of a great many cells, such as the eggs of many marine animals, and of some white cells of the blood; the erythrocyte, however, has normally a very characteristic shape, like a dumb-bell in section but correctly described as a biconcave disk; it is the shape of a squashed tennis-ball with a leak in it (Fig. 2.3a). When the cell takes up water it becomes more and more like a sphere, and the point at which the membrane ruptures seems to be just beyond the point at which the perfect spherical shape has been attained (Fig. 2.3b, c). It appears that the membrane has no "stretch" to it, so that holes appear as soon as a strain is placed on it. With the spherical sea-urchin egg, on the other hand, swelling may occur to about double its volume, so that the surface area must increase by about 59 per cent, so here the membrane has considerable stretch. Certain other cells have tough coats surrounding their membranes; this is true of bacteria and plant cells; if these tough membranes are

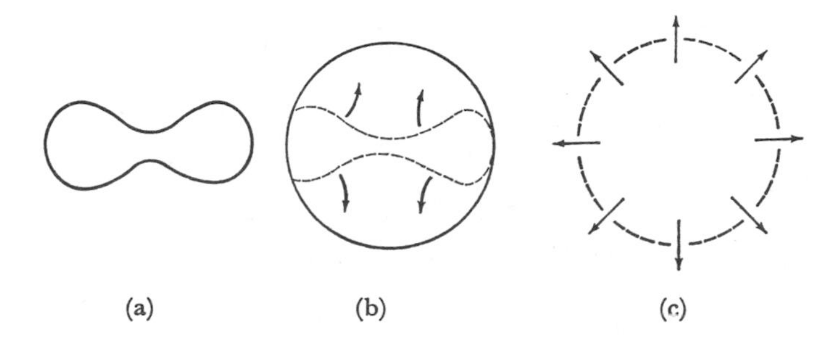

Fig. 2.3. (a) The normal shape of an erythrocyte is a biconcave disc. (b) Osmotic penetration of water causes the cell to increase in volume but its area remains the same. The membrane cannot stretch appreciably, however, so that when the spherical shape has been reached further osmosis causes the membrane to rupture (c).

removed, then the "protoplasts" so isolated are spherical, and respond to osmotic changes by swelling or shrinking. In their natural state, however, the coat is so strong that swelling does not occur, and it is of interest to examine the physical state of affairs.

Pressure in Cell

We imagine a cell (Fig. 2.4) with a tough inextensible coat, placed in a medium of lower total concentration than that within it. Water tends to enter the cell by virtue of the difference of chemical potential, and this tendency is reflected in an increase in the pressure of the cell; if a needle were placed in the cell and connected to a pressure-measuring device (e.g. a pressure-transducer), the measured pressure would be given ideally by:

$$\pi = RT(C_{In} - C_{Out})$$

i.e. the pressure developed is what we have called the difference of osmotic pressure between inside and outside In the case of the cell with the inextensible surface, the difference of pressure actually develops, and if it were high enough it would rupture the tough outer coat. With the protoplast, or many animal cells, there is no tough outer coat and the cell offers little or no resistance to the pressure-difference that could develop; instead the cell swells, thereby reducing the difference of osmotic pressure. If the extensibility of its outer

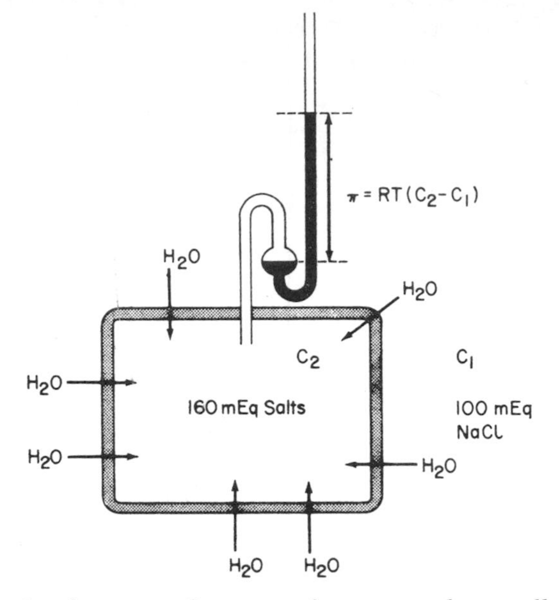

Fig. 2.4. The development of an osmotic pressure by a cell with a tough inextensible membrane. The tendency for water to penetrate the cell is expressed by a pressure within it, measurable by a manometer.

membrane is adequate, it takes up the amount requisite to give equal concentrations, and therefore equal osmotic pressures inside and outside, and the cell remains intact. If its extensibility is exceeded, the envelope breaks and we have *osmotic lysis*.

Difference of Osmotic Pressure

We have spoken of *osmotic pressures* and *differences* of osmotic pressure. What do we mean by these terms? We may define the osmotic pressure of a solution as the pressure it will develop when separated by a semi-permeable membrane from pure water; if the concentration is C_1 its osmotic pressure is given by:

$$\pi_1 = RT[C_1]$$

Thus, in the case of the cell with the tough envelope, illustrated by Fig. 2.4, the pressure developed, if it were placed in pure water, would be RT (Internal Concentration). The osmotic pressure of another solution of concentration C_2 is likewise given by:

$$\pi_2 = RTC_2$$

The pressure developed in a cell with internal concentration C_2 and outside concentration C_1 is given by:

$$\pi = RT(C_2 - C_1)$$

in other words, it is the *difference of osmotic pressure* of the two solutions of concentrations C_2 and C_1.

Osmotic Coefficient

The osmotic pressure depends on the concentration of solute, i.e. the number of moles per kg H_2O, and is independent of the nature of the solute, so that ideally there would be no difference of osmotic pressure between solutions of 1 molal urea and 1 molal sucrose, provided that the membrane used to separate the solutions was impermeable to both urea and sucrose and permeable to water. Where ions are concerned, the unit is now the ion and not the molecule, so that 0·5 molal NaCl would ideally have the same osmotic pressure as that of 1 molal urea, or sucrose. In actual fact the measured osmotic pressures, although they are linearly related to the concentration of solute over the range encountered in most biological work, are usually less than those computed on the basis of the simple equation: $\pi = RTC$, so that an empirically determined *osmotic coefficient*, *g*, is employed, such that the osmotic pressure can be predicted from the equation:

$$\pi \text{ actual} = RTgC$$

Thus, if we call the osmotic pressure computed from the "ideal equation", π ideal, *g*, is the ratio of π_{ideal} over the actual osmotic pressure, π actual, i.e.

$$\text{Osmotic coefficient} = g = \pi_{actual}/\pi_{ideal}.$$

Activity and Activity Coefficients

The osmotic coefficient is a particular example of the employment, in thermodynamic formulations, of the concept of *activity* in place of concentration. Because of mutual interaction, the molecules of solute fail to behave as individual units, and the net result is that the effective concentration is usually reduced, although in some instances it may be enhanced. This deviation from "ideal behaviour" is allowed for by the use of activity in place of concentration in all such formulations as those describing osmotic pressure, depression of vapour pressure, and so on. Thus we have:

$$\pi \text{ actual} = RTa$$

a being the *activity* of the solute molecules as it is related to the concentration, C, by the equation:

$$a = \gamma C$$

γ being the *activity coefficient.*

Thus $$\pi_{actual} = RT\gamma C$$

Hence the activity coefficient, here, is identical with the osmotic coefficient.

An important thermodynamic relation is that between the depression of freezing point of a solution and its concentration; thus, adding a solute to pure water lowers its freezing point, and the depression is given by the relation:

$$\Delta T_f^{\circ} = K_f a$$

where K_f is the *molal depression constant,* and *a* is the activity. Empirically, for water, and with not too concentrated solutions, the value of K_f is 1·858°C/mole/litre.

Hence $$\Delta T_f^{\circ} = 1{\cdot}858\ a$$

Whence the activity, *a*, is given by $\dfrac{\Delta T_f^{\circ}}{1{\cdot}858} = a = \gamma C$.

Thus by measuring the depression of freezing point of a solution and the concentration of solutes in it, we may deduce the activity coefficient, γ. Finally we may note that the actual osmotic pressure of a solution may be expressed in terms of ΔT_f°:

$$\pi_{actual} = RT\ a = \frac{RT\Delta T_f^{\circ}}{1{\cdot}858}.$$

Osmolality

It is now necessary to introduce the concept of osmolality; it is like the activity, an effective concentration that indicates the actual osmotic pressure that the solution will develop when separated from water by a semipermeable membrane, i.e.

$$\pi_{actual} = RT\ Osm.$$

It is an especially useful concept when biological solutions containing many solutes are considered. It is, in effect, the total concentration of solute particles, be they molecules or ions, each multiplied by their osmotic coefficient. Thus if we have a single solute, say μ moles of urea dissolved in W kg of water, its concentration, C_{urea}, is μ/W, and its osmolality is

$$\frac{\mu \times g_{urea}}{W} \text{ or } g_{urea} \times C_{urea} \text{ Osmoles/kg.} H_2O.$$

If to the same solution we add s moles of sucrose, the total osmolality now becomes:

$$\frac{\mu \times g_{urea} + s \times g_{sucrose}}{W} \text{ Osmoles/kg.} H_2O$$

and so on for any number of solutes.

Depression of Freezing Point. Experimentally it is not necessary to analyse a mixed solution, such as blood plasma, to determine its osmolality, since we may measure, instead, the depression of freezing point, ΔT°_{f}, this being related to the osmolality.

Thus the osmolality is defined by the equation:

$$\pi_{actual} = RT \text{ Osm}$$

Since, also,

$$\pi_{actual} = \frac{RT\Delta T^{\circ}_{f}}{1{\cdot}858}$$

we have

$$\text{Osm} = \frac{\Delta T^{\circ}_{f}}{1{\cdot}858}$$

the number 1·858 being the *molal depression constant* for dilute aqueous solutions. Thus human blood plasma with a depression of freezing point, ΔT°_{f} of, say, 0·54°C, would have an osmolality of 0·290 osmoles/kg H_2O or 290 milliosmoles/kg H_2O. In this range of concentration the osmotic coefficients of the main constituents of plasma are about 0·9, so that the total concentration of solute would be about 290/0·9 = 322 millimoles/kg H_2O, and this would be approximately equivalent to 160 millimolal NaCl.

Osmolarity. Just as with the distinction between molality and molarity, so the *osmolarity* of a solution is expressed in terms of osmoles per *litre of solution* instead of osmoles per kg water.

Artificial Extracellular Fluids

The osmolality of amphibian body-fluid is considerably less than that of mammals, being equivalent to some 0·11 Molal NaCl; fishes fall into two main classes, the bony fishes or teleosts having osmolalities rather similar to those of the mammal, whilst the elasmobranchs have very much higher osmolalities comparable with that of sea-water which is about 900 milliosmolal, i.e. equivalent to about 0·5 Molal NaCl. Thus in experimental studies in which it is necessary to maintain a tissue in an artificial medium it is important that the osmolality be correct, otherwise there will be exchanges of water that will cause swelling or shrinking of the cells. The saline solution required to maintain a frog muscle in osmotic equilibrium would have to have a total osmolality of about 200 milliosmoles, whilst for mammalian muscle it would have to be about 290 milliosmoles. For some purposes a simple solution of NaCl may be adequate, for example in studies of erythrocytes, but as Ringer showed many years ago, sustained function cannot be maintained in such a simple solution, and one containing K^+, Ca^{2+} and Mg^{2+} in the proportions found in the animal's blood is necessary. Such a solution is called a Ringer's solution; since Ringer's time, many additions have been made to meet the requirements of the

TABLE I

Physiological Salt Solutions (composition in millimoles/litre). (Perry, "Pharmacological Experiments on Isolated Preparations", Livingstone, 1970)

	Human serum	Ringer Locke (1900)	Krebs (1950)	Frog plasma	Frog Ringer
Na^+	142·0	155·8	141·0	103·8	113·7
K^+	5·0	5·6	5·9	2·2	1·9
Ca^{2+}	2·6	4·3	2·6	2·0	2·2
Mg^{2+}	1·5	–	1·2	1·2	–
Cl^-	103·0	163·9	104·8	74·3	115·3
$H_2PO_4^-$	2·0	–	2·2	3·1	0·1
HCO_3^-	27·0	1·8	24·9	25·3	2·4
SO_4^{2-}	0·5	–	1·2	1·9	–
Glucose	variable	5·0	10·0	3·9	10·0
Amino Acids		–	4·9	6·9	–
Other Acids		–	15·7	>3·3	–
Total Organic Anions	22·0	–	20·8	–	–
Aerating Gas		O_2	O_2 + 5% CO_2		Air

tissue less inadequately. Thus mammalian Tyrode solution has glucose added together with $NaHCO_3$ and NaH_2PO_4, the two latter salts constituting a buffer that maintains the pH in the normal range of 7·2–7·4. Its actual composition is indicated by Table I. In all these solutions the total osmolality is fixed so as to agree approximately with that of the animal's body fluids.

Cell Penetration

The osmotic phenomena so far described have depended on the existence of a membrane supposed to be impermeable to the solutes inside and outside it, but permeable to water; the classical *semi-permeable membrane*. Since truly semipermeable membranes are very rare in Nature, it is important to understand what happens when this impermeability to solute is only partial, in the sense that one or more of the solutes may actually diffuse across the membrane.

Figure 2.5 shows a cell with internal osmolality of, say, 300 milliosmoles/kgH_2O made up of molecules to which the membrane is completely impermeable. The medium contains 300 milliosmoles/kgH_2O of a solute like urea which can penetrate the cell fairly easily. Initially the chemical potential of water is the same on both sides; however, the urea will diffuse into the cell because its chemical potential is higher in the medium, so that we imagine a stage (b) where the concentration of urea has become equal to that outside. It is clear from Fig. 2.5 that the osmolality within the cell is now much greater than that of the medium, so that water should penetrate; this leads to stage (c) with the cell swollen, but now the concentration of urea is less inside than outside, so more urea must diffuse in; this leads to a further increase in internal osmolality with the consequent osmosis of water, and eventually the cell bursts.

Rates of Penetration

This phenomenon is easily demonstrated with the aid of test-tubes and a few drops of blood; if we place 10 ml of 160 milliequiv./litre NaCl in one tube and 10 ml of 300 mM urea in the other. A drop of blood is added to each; the suspension in NaCl will remain opaque due to scattering of light by the intact cells, but in a short time the suspension in the urea solution will become clear due to the bursting of the cells. If glycerol is used in place of urea the same phenomenon will occur, because glycerol can penetrate the cells, but the process of bursting the cells will be slower because glycerol penetrates them more slowly than urea, and it will be evident from Fig. 2.5 that the

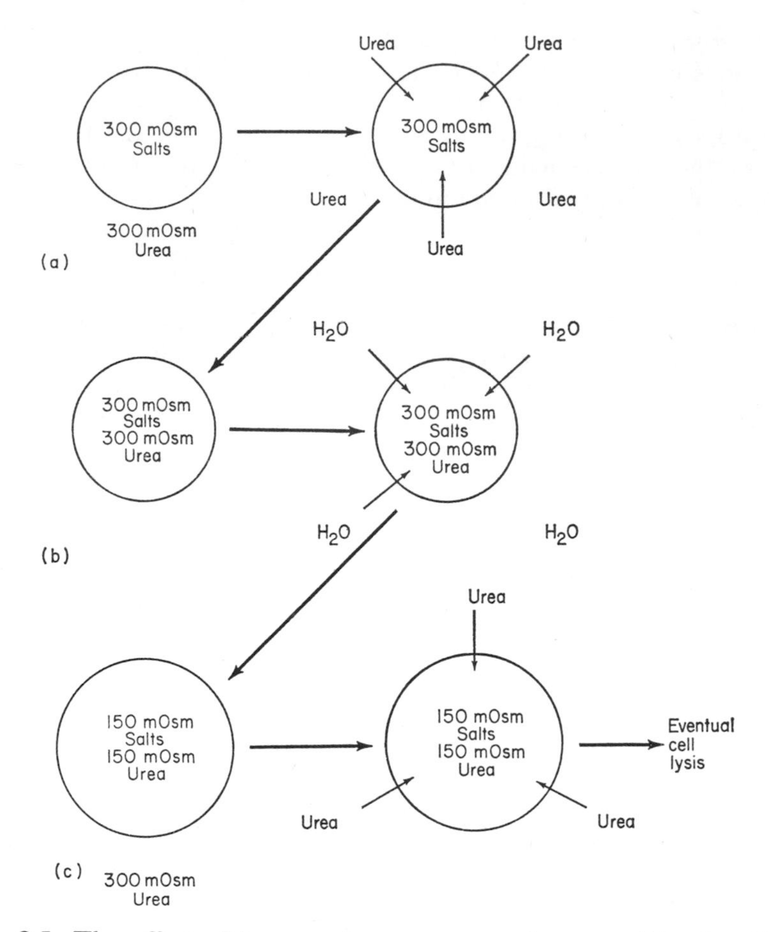

Fig. 2.5. The effect of iso-osmotic urea on erythrocytes. If red cells are added to a 300 mOsm solution of urea, the red cells swell since urea, unlike sodium, can cross the cell wall easily. In (a) the urea enters the cell until it has equilibrated with the external concentration. However, as can be seen in (b), this effectively increases the osmolality within the cell so that water now enters and the cell becomes swollen. (c) As the volume of the cell increases, the concentration of urea inside the cell falls and more enters from the outside, followed by more water until the cell bursts.

process of swelling and bursting waits on the primary penetration of the solute. If a variety of different solutes are taken, some measure of their relative rates of penetration of the cells is given from the reciprocals of the times taken to cause bursting of the cells and clarification of the suspension.

Reflexion Coefficients

We see from Fig. 2.5 that in stage (b), when the cell contains urea as well as its 300 milliosmolal salt, the presence of the urea causes an osmotic flow of water. We may presume that if we prevented the swelling an osmotic pressure would develop, although from the point of view of urea the membrane is not semipermeable. In fact, a pressure does develop but not the full pressure expected were the membrane impermeable to urea. Thus the ideal osmotic pressure developed in the case of complete semipermeability would be:

$$\pi = \frac{RT}{1000}(600 - 300)$$

The actual pressure is less, because the urea molecules may penetrate the membrane, and the degree to which the osmotic pressure is reduced is given by Staverman's *Reflexion Coefficient;* if the solute is completely impermeable, to give a perfectly semipermeable membrane, the coefficient is unity and the osmotic pressure is given by multiplying the osmolality by RT; if the reflexion coefficient is zero, penetration of the cell is so rapid that no osmotic pressure develops; with intermediate permeabilities, the reflexion coefficient is between 1 and 0.

Salt Substitution

This consideration of cell penetration, or reflexion coefficient, has an important bearing on the use of artificial solutions for maintaining animal tissues; thus in order that the solute, such as NaCl, be able to exert its osmotic effects, balancing the osmotic pressure of the inside of the cell, it must be effectively impermeant, i.e. have a reflexion coefficient of unity. Urea penetrates most animal cells fairly rapidly and so an isotonic solution made up of urea instead of NaCl would be useless. Sucrose behaves as an essentially impermeant molecule, so that this is frequently employed when a substitute for the salts of a Ringer's solution is required, e.g. to discover the effects of reducing the Na^+-concentration in the medium. Because sodium constitutes such a large fraction of the total, merely leaving it out from a Ringer's solution would make the solution very hypotonic; instead the NaCl is *replaced* by the equivalent number of osmoles of sucrose.

Osmotic Transient. An interesting situation is given when the cell is surrounded by its normal environment, and a solute is added to this to raise its total osmolality above that in the cell, as in Fig. 2.6. The response is a shrinking of the cell due to osmosis outward (Stage b). If the solute penetrates the cell, however, this shrinking will be only

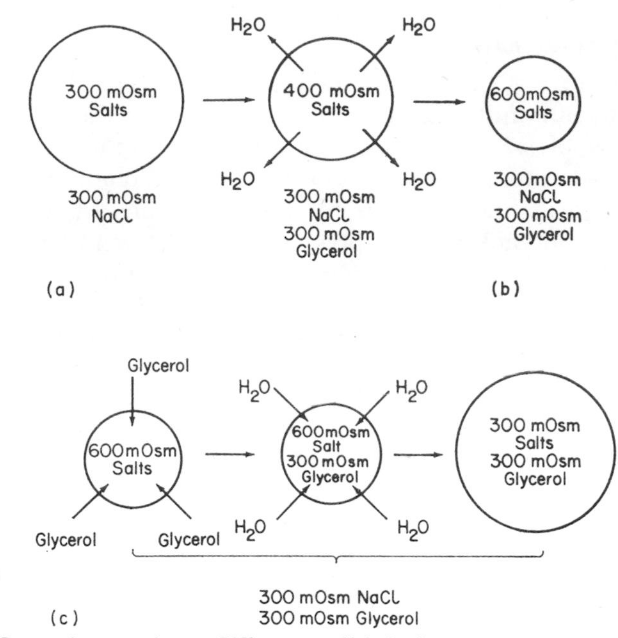

Fig. 2.6. Osmotic transients. When a cell is in its normal environment and a solute is added to increase the osmolality of the extracellular fluid, water moves out of the cell, as shown in (a), and the cell shrinks until there is equal concentration across the cell wall (b). However, if the solute itself can enter the cell situation (c) occurs and the entry of solute will now cause the cell to become hypertonic to its environment. This increase in osmolality will in turn cause water to re-enter the cell which expands to its original size.

transient; thus at Stage c the solute has reached equal concentration inside and out; the cell is now hyperosmolar and water must penetrate from the medium. The final situation is given with the cell back at its normal volume. The processes of escape of water, penetration of solute and return of water will occur continuously, and the magnitude of the shrinkage will depend on the degree of impermeability of the membrane to the added solute—if the solute penetrates very slowly then the shrinkage will be large and the return to normal volume will be slow; if the solute penetrates very rapidly indeed, the shrinkage may not be detectable. This, then, gives us another method of measuring permeability of the cell membrane, at least qualitatively.

Loss of Semipermeability

Suppose the membrane separating the cell from its environment loses its semipermeability in respect to salts; this semipermeability not only permits the osmotic exchange described above, but it also ensures that the specific composition of the cell's interior be maintained; further, and this is not so easy to predict, it ensures that the cell can maintain equal osmolality inside and outside. Thus we imagine the cell to contain the equivalent of 300 milliosmolal KCl and outside the equivalent of 300 milliosmolal of NaCl (Fig. 2.7). Loss of semi-

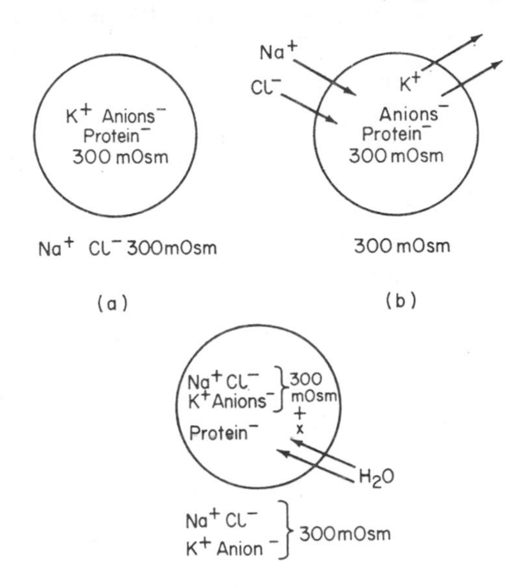

Fig. 2.7. When a cell (a) loses its effective impermeability to salts, the ions re-distribute themselves, K^+ escaping and Na^+ entering, as in (b). This equilibration of solute concentrations should cause no change in volume. However, the large protein molecules within the cell remain, and there is thus a net difference of osmolality, x, favouring the penetration of water. (c) Water enters the cell, but this now lowers the internal concentration of salts so that more now penetrate from outside; however, this penetration of salt raises the internal osmolality above that outside, causing more water to enter. The two processes (water and salt penetration) continue until the cell bursts—*colloid osmotic lysis*.

permeability means that the KCl diffuses out to the medium and is replaced by NaCl from it, so that ultimately the inside and outside media become the same. This should not result in any permanent change in volume of the cell, although if the KCl leaked out faster than the NaCl penetrated, it would shrink at first but recover its volume as the NaCl continued to penetrate the cell. In actual fact it is observed with all cells that loss of semipermeability leads to an eventual swelling that finally causes lysis.

Colloid Osmotic Swelling. This is because the cells contain colloidal material that fails to escape, so that even when the ions and other non-colloidal solutes have come into equilibrium on each side of the cell membrane, the osmotic pressure of the medium inside the cell is greater than that outside by virtue of this additional colloidal material. This causes penetration of water, but the penetration of water reduces the concentration of salt below that outside, and so some salt penetrates, and this now leaves the cell hyperosmolar and more water goes in, more salts then penetrate, and so on. This *colloid osmotic swelling* is a very significant factor in the cell's relations to its environment since, as we shall see, the semipermeability of cell membranes, vis-à-vis ions, is only effectively maintained by metabolic processes that pump the ions out of the cell or into it; if these processes fail, then the effective semipermeability is lost and the consequence is, indeed, this colloid osmotic swelling.

PERMEABILITY OF THE CELL MEMBRANE

Osmosis depends on the relative ease with which water and a given solute can pass across the cell membrane, the maximum osmotic pressure being developed when the cell has no permeability to the solute; it is essentially a measure of the relative restraints to passage of solute and water, given by the reflexion coefficient. It is important that we understand the concept of permeability, its measurement in quantitative terms, and the results of measurements on different solutes as well as water.

Permeability Coefficient

For a given non-electrolyte, the rate of penetration of a cell is measured in terms of the amount entering in unit time, and this is governed by the simple "permeability equation":

$$\text{Amount entering in Unit Time } (dS/dt) = kA(C_{Out} - C_{In}) \quad . . . (2)$$

thus the amount entering in unit time depends on the area of the cell, A, the concentration difference, and the *permeability coefficient*, *k*. Experimentally we must measure the area of the cell and the amount entering in unit time when a given concentration difference between inside and outside has been established.

Lipid-Solubility

Coefficients measured in this way vary greatly from one solute to another, but when different cells are studied, the relative rates are often very similar suggesting that the barrier to penetration has certain features common to all cells. Thus the most important generalization is that of Overton, to the effect that the more lipid-soluble a substance is, the more rapidly it will penetrate cells; this is illustrated by the classical study of Collander and Bärlund in Fig. 2.8. The lipid-solubility (B) is measured by the partition coefficient of the substance between a typical fat, such as olive oil, and water:

$$B = \text{Concn. in Fat/Concn. in water.}$$

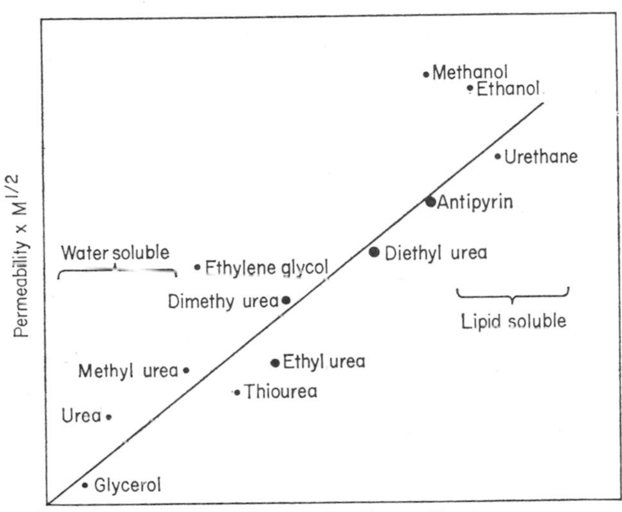

Fig. 2.8. The relation between lipid-solubility, as represented by the oil/water partition co-efficient and the permeability of a variety of solutes through the cell membrane of the giant cells of the alga *Chara*; the greater the solubility of a compound in oil, the larger the partition coefficient, the more easy is it for the compound to pass through the cell membrane.

Thus ethyl alcohol, with a partition coefficient of about 0·03, would be described as a relatively lipid-soluble substance, and it penetrates all cell membranes easily; glycerol, with a partition co-efficient of 0·0001 is relatively lipid-insoluble, and its rate of penetration is usually slow.

Lipid Membrane. The factor of lipid-solubility is probably easily explained on the basis of the lipid nature of the cell plasma membrane, as revealed by chemical analysis. Thus the diffusion into the cell may be represented (Fig. 2.9) as the establishment of a difference in concentration within the cell membrane ($C'_{Out} - C'_{In}$), accompanied by

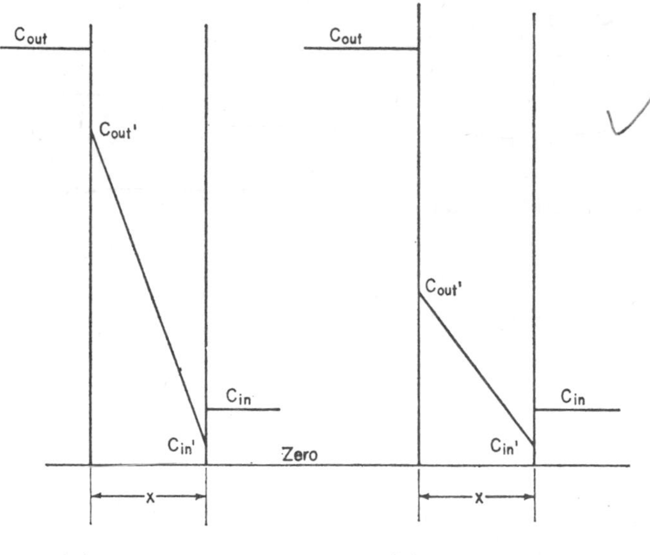

Fig. 2.9. The effect of lipid solubility on the concentration gradient of compound across cell membranes. (a) High lipid solubility (high oil/water partition coefficient). (b) Low lipid solubility (low oil/water partition coefficient). The ease with which a compound travels across the membrane is increased with increase in lipid solubility since the effective concentration gradient (which is the driving force to diffusion) is much steeper. C_{Out} or C_{In} are the concentrations of the compound outside and inside the cell. x is the membrane thickness, C' values are the effective concentrations within the membrane at the water interfaces.

the movement of solute down this gradient. We may assume that it is essentially the delay in diffusion down this gradient that gives rise to the slow penetration, diffusion in the aqueous medium being rapid by comparison with diffusion in the membrane. Clearly, at any moment, the concentration in the membrane nearest the outside, C'_{Out}, will be given by the partition coefficient, or lipid solubility, so that $C'_{Out} = BC_{Out}$ and the inside membrane concentration will be given by $C'_{In} = BC_{In}$. Now the passage through the membrane will be given by Fick's Law which states that the amount diffusing through an area, A, will be given by the product of the *diffusion coefficient*, D, and the *concentration gradient*, i.e. the difference of concentration divided by the thickness of the layer being considered. Since the concentrations in the membrane are given by the partition coefficient times the concentrations in the aqueous medium, we have:

$$dS/dt = DAB\frac{(C_{Out} - C_{In})}{x} \quad (3)$$

where x is the thickness of the membrane, and the situation is represented by Fig. 2.9a for a substance with a relatively high partition coefficient and by Fig. 2.9b for a substance with a relatively low coefficient. Quite clearly, the concentration gradient is higher with the high partition coefficient, so that it is evident that, other things being equal, high lipid-solubility will favour penetration through the membrane.

Diffusion Through Membrane. It will be seen by comparing the permeability Equation (2) with Equation (3) that the permeability coefficient is equivalent to DAB/x, so that if the thickness of the membrane and the partition coefficient in the membrane are known, we may estimate the diffusion coefficient within the membrane and compare this with the diffusion coefficient in aqueous solution. For example the permeability of the erythrocyte to glycerol is $1{\cdot}7.10^{-8}$ cm/sec and if we assume a membrane thickness of, say, 200 Å, i.e. 2.10^{-6} cm, and a partition coefficient of 1.10^{-4} we have a diffusion coefficient of $1{\cdot}7.10^{-8} \times 2.10^{-6} \times 1.10^{4} = 3{\cdot}4.10^{-10}$ cm²/sec. This compares with a value of about 1.10^{-5} cm²/sec for diffusion in water, and gives some idea of the restraint on passage occasioned by the presence of the membrane.*

Molecular Size

It will be seen from Fig. 2.8 that the size of a molecule is important as well as its partition coefficient, so that smaller molecules tend to penetrate more rapidly than larger ones of the same lipid-solubility (this point is well illustrated by a consideration of the urea series in Fig. 2.8). This is understandable, since small molecules tend to move more rapidly by thermal collisions than large molecules at the same temperature, since the kinetic energy of the molecules is given by:

$\frac{1}{2}m\bar{v}^2$ when m is the mass and $\bar{v}$ the mean velocity; thus with two substances of molecular weights M_1 and M_2 we have:

$$\tfrac{1}{2}M_1\bar{v}_1^2 = \tfrac{1}{2}M_2\bar{v}_2^2$$

whence we have:

$$\frac{\bar{v}_1}{\bar{v}_2} = \frac{M_2^{\frac{1}{2}}}{M_1^{\frac{1}{2}}}$$

* It will be clear from this simple treatment that we may expect the permeability coefficient to be linearly proportional to the partition coefficient. This is only true with rapidly penetrating substances, as Danielli showed, the relationship being more complex where relatively slowly penetrating substances are concerned, because under these conditions Fick's diffusion equation is not strictly applicable.

Thus the average velocities will be inversely proportional to the square roots of the molecular weights of the molecules, so that if we wish to compare the permeabilities of different substances, and make allowance for the advantage of the small molecules over the large, we should really plot $PM^{\frac{1}{2}}$ against partition coefficient.

Porous Membrane

In fact, this has been done in Fig. 2.8, so that the fact that there is still a strong bias in favour of small molecules means that an additional factor comes into play, and it has been suggested that the membrane contains small pores, of about 4 Å in diameter, filled with water; these pores would be available for the smaller molecules, and so reduce the barrier they have to overcome by comparison with large molecules that must pass entirely through the lipid phase. However, when dealing with dimensions of this order, which correspond perhaps to the spaces between packed lipid molecules, and are only about twice those of a water molecule, the concept of a "water-filled pore" must necessarily be vague.

Experimental Measurement

In the earlier days of studying cell permeability, the experimenter would rely on chemical analysis either of the cells or outside medium. For a single cell this was usually impracticable except with large multinucleate plant cells; as illustrated in Fig. 2.10 this cell contains a large vacuole from which as much as 1 ml of sap may be obtained. Thus the cell is placed in a medium to which the solute to be studied is added in a suitable concentration, preferably sufficiently low to produce a minimal osmotic disturbance. After a time the sap is removed and the amount of solute entering is determined. Here, we are measuring passage through the protoplast which is the cytoplasm of the cell; the solute passes through two layers of plasma membrane—that adjacent to the cellulose wall and that adjacent to the vacuole.

Plasmolysis

A less direct method, applicable to small cells, would be to place them in an isotonic medium and add a relatively high concentration of the solute to the medium, making it hypertonic. As a result, as indicated earlier in discussing osmotic transients, water will pass out causing shrinkage of the cell, or plasmolysis; if the solute penetrates the cell this plasmolysis will gradually reverse, the rate being governed by the permeability of the solute. In this way qualitative estimates of

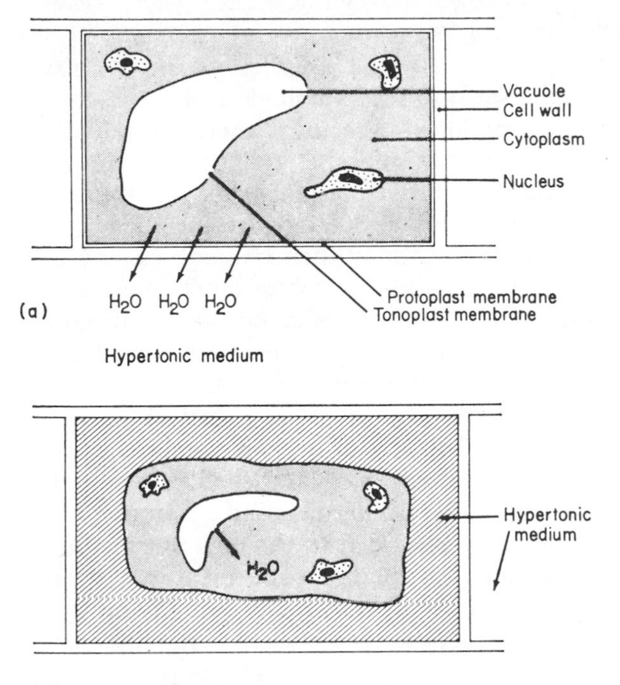

Fig. 2.10. The effects of placing a plant cell in a hyperosmolal medium. In (a) the cell is in its normal state; the cytoplast or protoplast (dotted) is bounded by the protoplast and tonoplast membranes. Both of these are effectively semi-permeable, allowing net movements of water but not of salts. The cell wall is freely permeable to salts and water. In the hypertonic medium the first step is the osmosis of water from the cytoplasm into the medium due to the semipermeability of the protoplast membrane. (b) The osmotic shrinking of the cytoplast has taken place; loss of water has caused the cytoplasm to become hypersomolal in respect to the vacuole which loses water and shrinks. The space made available by shrinkage of protoplast and vacuole is replaced by the hypersomolal medium, which passes without restraint across the cellulose wall.

relative permeability can be obtained. This technique formed the basis of the classical studies of Overton, leading to the concept of lipid-solubility.

Streaming Potential

More recently it has been modified to permit very rapid measurements, advantage being taken of the fact that, when water flows across the cell membrane, it creates a streaming potential which is proportional to the rate of flow.

Cell Suspensions

Chemical techniques can be applied to small cells, such as the erythrocyte, but instead of using a single cell a suspension containing many millions is employed. We may, therefore add the penetrating solute to blood directly; this consists mainly of erythrocytes suspended in about an equal volume of plasma. After an appropriate time the cells are centrifuged at about 3000 g, causing them to sediment, and the cells or plasma or both may then be sampled and analysed. Thus we may measure the amount of solute entering a known volume of cells in a given time; the area of a known volume of cells can be measured approximately and the measurements necessary for calculating the permeability coefficient obtained.

Radioactive Labels. The use of radioactively labelled substances, be they ions such as ^{24}Na, ^{42}K or organic compounds such as tritium-labelled glucose (^{3}H-glucose) or ^{14}C-labelled glucose (^{14}C-glucose) has revolutionized the study of permeability. Minute amounts of the labelled substance may be added to the medium surrounding the cells and the amount entering in unit time measured by measuring the radioactivity taken up. As with so many simplifying techniques, however, there are often pitfalls; thus glucose is rapidly converted by cells to other products such as CO_2 or lactate, and these compounds will contain the radioactive label and give rise to misleading conclusions.

Facilitated Transfer

When the permeability characteristics of a given type of cell are examined, it is usually found that certain substances stand out as exceptional in having a much higher permeability coefficient than expected on the basis of their lipid-solubility and size; thus the human erythrocyte allows some sugars, like glucose and mannose, to penetrate rapidly yet their lipid-solubilities are very low and on this basis would hardly be expected to penetrate the cell at all. The striking feature of this type of permeability is its specificity, so that a sugar of very closely related structure may penetrate very slowly; e.g. L-glucose penetrates slowly by comparison with D-glucose. Associated with this specificity is mutual interaction between sugars, so that the presence of, say, D-galactose reduces the permeability coefficient for D-glucose, and vice versa. Furthermore, a given sugar will "compete with itself" in the sense that, as the concentration at which the measurement is made is raised, the computed permeability coefficient decreases, whereas, as Equation (2) indicates, by definition the permeability coefficient should be independent of the concentrations. Where ions are concerned this type of permeability may show an optimum temperature and pH; in general, raising the temperature should, on kinetic grounds, increase the permeability coefficient—a rise of 10°C being usually sufficient to multiply the permeability by 2 to 3. However, in the cat erythrocyte the permeability to Na^+ is found to decrease if the temperature is increased from 37·5°C to 40°C; similarly the permeability shows a marked maximum at pH 7·4, the pH of the blood.

Carrier Hypothesis

All these features are reminiscent of enzyme-catalysed chemical reactions, in which the reaction depends on binding to a specific site on the enzyme molecule. In terms of the permeability process, it seems that the penetrating molecule has to attach itself to a specific grouping on a "carrier" situated in the cell membrane. Compounds unable to attach to this site would be unable to cross the membrane easily, if at all, whilst those molecules that were able to attach to the site would compete with others that could also do so; hence we may speak of *relative affinities* of the molecules for the carrier site. A given cell may behave as though it had several types of carrier, one, say, for transporting sugars, one for amino acids, one for fatty acids, and so on. In this event there would be no signs of competition between the types of molecules since a sugar does not interfere with the transport of an amino acid. Amongst the amino acids, moreover, there seem to be different groups that compete with each other but not with those in other groups, e.g. the neutral amino acids like glycine and alanine compete with each other but not with basic amino acids like lysine or with acidic amino acids like glutamic acid.

Isolated "Carriers" As to the nature of the carrier in molecular terms, it is not easy to be specific. The cell membrane is so thin that it is not possible to envisage a carrier molecule moving bodily across the membrane. Instead, we must envisage certain groupings of atoms on and within the membrane, and it is by virtue of their presence that a given type of molecule gains ready access to the interior of the membrane and thence is able to cross from one side to the other. Certainly in modern work, in which radioactively labelled sugars have been studied, it has been possible to isolate from the membrane material to which the radioactive sugar is attached. This may well be a complex between the sugar and the membrane representing the intermediate state of the sugar during its passage.

Physiological Significance

The significance of this type of transport is immediately obvious;

the membrane has developed primarily as a barrier permitting the separation of the internal medium of the cell from the outside. A thin lipid-protein membrane serves this function, but if it were completely undifferentiated it would exclude important metabolites (such as the sugars and amino acids) from penetrating. Thus the development of specific sites or carriers is a necessary corollary to the development of a lipid plasma membrane.

Anion Exchanges in Erythrocyte. Of special interest is the exchange of anions across the erythrocyte membrane. Thus we have up

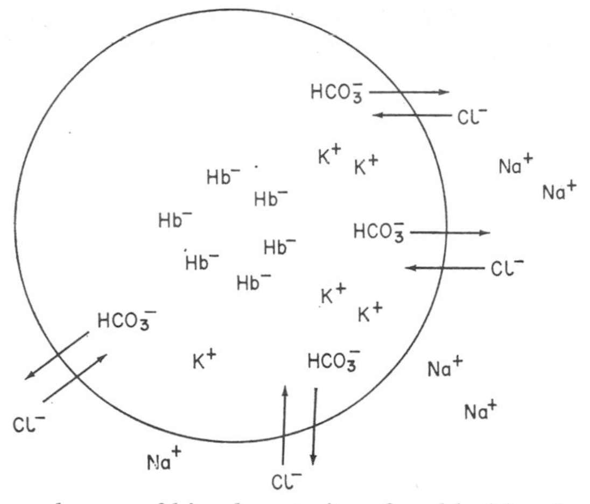

Fig. 2.11. The exchange of bicarbonate ions for chloride when erythrocytes are placed in isotonic saline. The red cell membrane is permeable to anions, and these move down their concentration gradients. Cations are less permeable and the active exchange of sodium for potassium maintains the distribution difference. No water movements occur during this process since the cells are in osmotic equilibrium.

to this point treated the erythrocyte membrane as semipermeable, so far as KCl and NaCl are concerned. However, because of the requirements for electrical neutrality, a membrane impermeable to the cations Na^+ and K^+ but highly permeable to the anions Cl^-, HCO_3^-, etc. would behave as a semipermeable one. In fact when the erythrocyte, containing as its internal salts largely KCl and $KHCO_3$, is placed in a solution of NaCl of the same osmolality, there is no osmotic movement of water, but the HCO_3^- ion inside the cell exchanges for the Cl^- outside, as in Fig. 2.11.

This exchange is extremely rapid and requires special methods for measuring its speed; as with the permeability to sugars the process

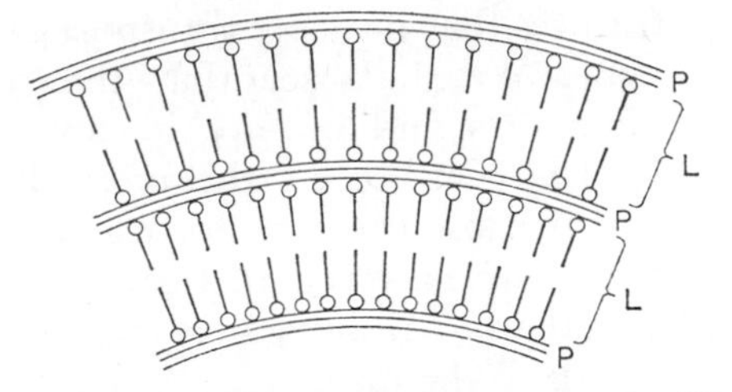

Fig. 2.12. The arrangement of lipid (L) and protein (P) in myelin sheaths as deduced from X-ray diffraction studies. The lipid is arranged in a bilayer with the hydrocarbon hydrophobic chains inside the membrane and the polar groups on the outside. A layer of protein covers the inside and outside polar groups of the lipid molecules.

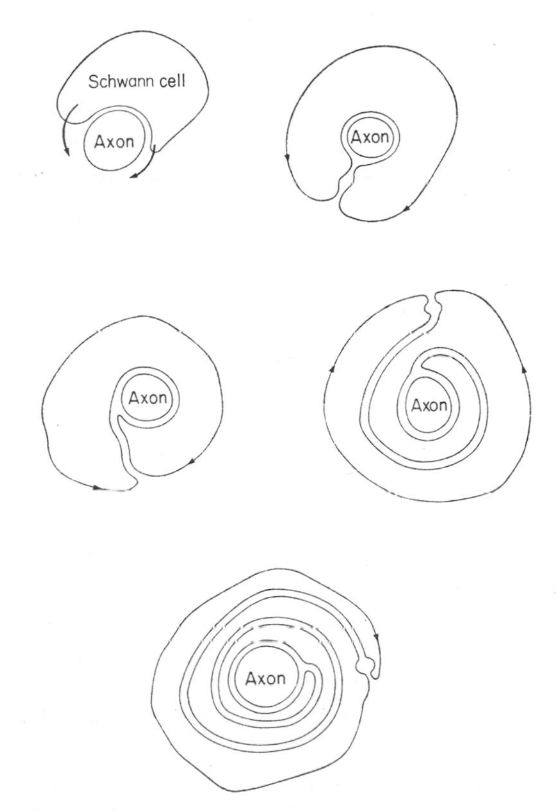

Fig. 2.13. The development of the myelinated nerve sheath. The Schwann cell gradually envelops the axon and then proceeds to grow in a rotational fashion around the nerve to form finally the several layers of the Schwann sheath.

reveals a number of features that indicate transport of the facilitated or "carrier-mediated" type. We shall see that the transport of CO_2 from the tissues to the lungs and out to the air is associated with movements of HCO_3^- and Cl^- across the red cell membrane; this transport must be rapid because the time for exchange of gases in the lungs is very short, so that we may regard this very high permeability to anions, probably peculiar to the erythrocyte, as an adaptation to meet the respiratory needs of the animal.

Structure of the Plasma Membrane

Bimolecular Leaflet. In Chapter I the plasma membrane was represented as a bilayer of lipid with layers of protein on the external and internal surfaces (Fig. 1.14, p. 23). This basic structure was first formulated in 1935 by Danielli and Davson as being compatible with the known extreme thinness of the membrane, the results of permeability measurements, and the known tendency for lipids to form layers made up of well-orientated molecules with their polar groups facing the aqueous phase and their hydrophobic hydrocarbon tails orientated away from the water, and therefore adopting a tail-to-tail arrangement when forming a bilayer.

Myelin Sheath. This concept was greatly strengthened by Robertson's electron microscopical studies of the surface layers of a variety of cells, and by X-ray diffraction studies on the thick lipid sheath of myelinated nerve. These latter X-ray studies, pioneered by Schmitt, confirmed that the lipid molecules in myelin were indeed arranged in bilayers, with layers of protein sandwiched between, as indicated in Fig. 2.12. The myelin sheath represented, therefore, a number of theoretical Danielli–Davson plasma membranes piled on top of each other. The electron microscopical studies showed, moreover, that this is exactly how myelin sheaths are formed, the axon of a neurone being covered by membrane of the enveloping Schwann cell. Figure 2.13 illustrates the embryonic development of a myelinated nerve, the bare axon being enveloped by the Schwann cell, which ultimately, through rotation and formation of the Schwann cell membrane, is wrapped by successive layers of membrane.

Electron-microscopical examination of the individual membranes of myelin, or of the limiting layers of many individual cells, showed that each one could be resolved into a triple layered structure, a pale unstained layer being sandwiched between two densely staining layers, (Fig. 1.12, p. 20); and there now seems little doubt that the densely staining layers, representing regions of uptake of osmium or other electron-stain, do indicate the location of the orientated polar groups of the membrane molecules, facing towards the outside and inside of the cell, whilst the unstained layer represents the hydrophobic lipid tails which would tend not to react with osmium.

Subunit Hypothesis

However, the basic concept of the bilayer membrane has been attacked from several quarters, and alternative structures have been put forward. The most serious claim is that the membrane is built

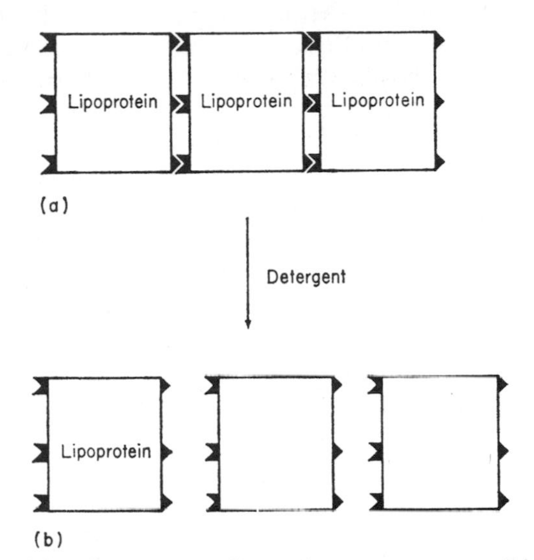

Fig. 2.14. The subunit concept of membrane structure. Lipoprotein units are linked together as shown in (a) to form a stable membrane. Detergents disrupt mitochondrial membranes into subunits (b) which can be reassembled again into a membrane structure; and it is this type of evidence that has been adduced in support of this theory.

up of blocks, or subunits, of lipoprotein, which may be dissociated experimentally with a detergent, and the membrane subsequently reassembled by removing the detergent. The essential difference between the Danielli–Davson concept and these subunit theories is that in the former the lipid bilayer provides its own stability by virtue of the tendency of its molecules to orientate to give a close-packed assembly; the layers of protein on each side contribute to the stability but are not chemically bound to the lipid to form lipoprotein molecules. The subunit hypothesis, on the other hand, imputes the stability of the system to the tendency of the lipoprotein blocks to link together, as in Fig. 2.14a, b. The main argument in favour of this subunit hypothesis was that membranes could be broken up and reassembled, and this was especially true of the mitochondrial membrane which appeared to have a basic structural protein in its membranes which could be dissociated from the lipids and reassembled. However it has now been established (e.g. by Engelman using the membranes of the cells of *Mycoplasma laidlawii*) that the "subunits" are not lipoproteins but the individual lipids and protein of the membranes. Therefore, the separation and reassembly are by no means incompatible with the Danielli–Davson hypothesis, the subsequent reassembly of the separated lipid and protein being, in effect, a mimicry of the natural event, since the lipid and protein must have been synthesized separately.

Bilayer Hypothesis Sustained

Considerable evidence has been adduced supporting the concept of an ordered arrangement of lipid molecules. It is known, for example, that such an arrangement would have a characteristic "melting point" when heated or cooled, depending on the types of fatty acids in the lipid layer: below the melting point the lipid molecules would be well orientated side by side; and above, the arrangement would be less regular (Fig. 2.15) but the changes would be quite reversible. These changes can be detected in various ways; for example Steim *et al.* used a differential scanning calorimeter, which records the power output required to heat the sample to a given temperature and compares this simultaneously with the amount required to heat an inert sample not showing the transition. With such a device they showed that the isolated membranes of the organism *M. laidlawii*, or the membranes on the intact cell, exhibited this phase-transition indicating that the lipids were, indeed, arranged in a bilayer since such a transition would

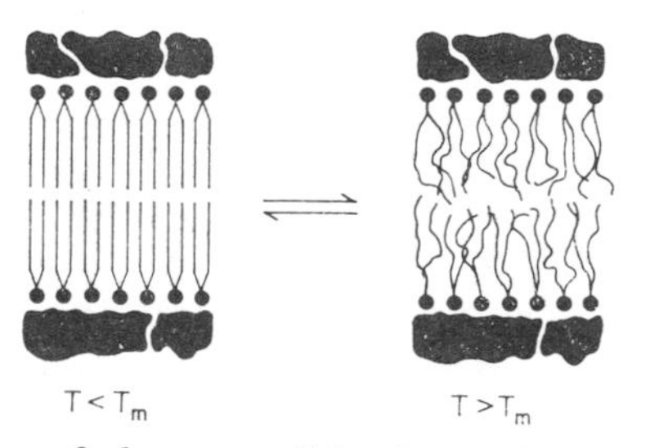

Fig. 2.15. A diagram of the reversible thermal transition postulated for *Mycoplasma laidlawii* membranes, showing the orderly crystalline state of hydrocarbon chains below the transition temperature and the liquid-like state above the transition. Protein is shown covering the outside surfaces of the membrane. The polar ends of the lipid are shown as black dots. (Steim *et al., Proc. Nat. Acad. Sci.*, 1969, **63,** 104.)

not occur with lipoprotein subunits. In a similar way the X-ray diffraction pattern could be shown to change reversibly indicating, once again, this type of phase transition.

Discontinuities in Membrane. Thus the basic concept of the lipid-bilayer appears to be sustained, but of course the facts of permeability and active transport demand the existence of discontinuities that would allow localized regions of high permeability to selected substances. Such discontinuities might be provided by polypeptide chains traversing the whole thickness of the membrane as in the Lenard–Singer model of Fig. 2.16. The main requirement of any such postulate, however, must be that the system has inherent stability and retains the basic arrangement of lipid molecules indicated by X-ray and other studies.

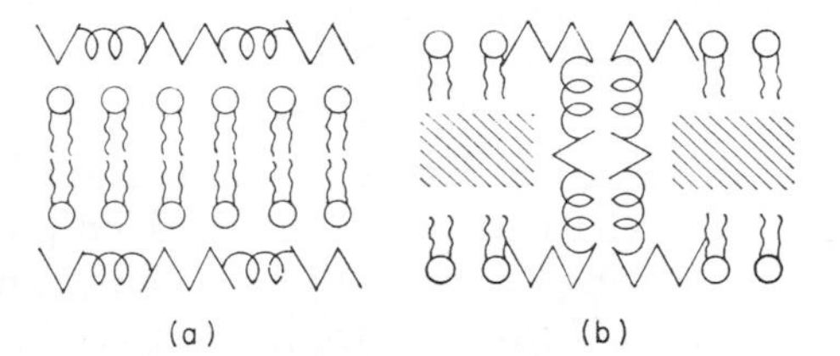

Fig. 2.16. Schematic basis of membrane structure. (a) The modified Danielli–Davson–Robertson unit membrane; the proteins on the outer surface consist of helical and random-coil parts. The polar lipids are orientated in a bimolecular leaflet with their polar heads (O) facing out. (b) The structure suggested by Lenard and Singer; the cross-hatched areas are assumed to be occupied by the relatively non-polar constituents (hydrophobic amino acid residues and lipids). Single polypeptide chains are drawn to traverse the entire membrane. (Lenard and Singer *Proc. Nat. Acad. Sci.*, 1966, **56,** 1828).

Freeze-Etched Membranes. It will be noted that in this model there is a region in the middle of the membrane of undefined composition separating the two layers of orientated lipid. The freeze-etch technique has enabled the membrane to be cloven in a plane at right-angles to the orientated lipid molecules, i.e. tangentially to the surface of the cell, and in this way the lipid bilayer seems to have been split. When the new surfaces, so revealed, are examined, globular units embedded in the matrix of the membrane have been identified. In highly specialized membranes, such as those of the photoreceptor, the embedded units are probably the photochemically sensitive pigment, and in the photosynthetic cell of the green plant they may represent the complex of chlorophyll and protein that has been called a "quantasome". The units are present in the membranes of non-specialized cells, however, and may well be related in some way to transport processes. But it is by no means certain that, in the natural state of the membrane, they were indeed located between the lipid monolayers. Such a location would detract very strongly from the stability of the bilayer, and it may well be that the appearance of these globular bodies between the lipid monolayers in freeze-etched specimens is an artefact, the material having migrated from other regions during preparation of the specimen.

Metabolism and Transport

Stored Blood

When blood is stored in the cold the erythrocytes tend to lose their K^+ and take up Na^+ from the plasma, i.e. they seem to lose their apparent semipermeability so far as Na^+ and K^+ are concerned. If the process goes on for long enough the cells do, in fact, swell up and burst, as predicted earlier (p. 96). If, after 24 h say, the blood is re-warmed, the Na^+ taken up is extruded and the lost K^+ is recovered. Thus in the cold the cell tends to come into equilibrium with its environment, the K^+ moving "downhill" from cell to environment and the Na^+ running "downhill" from environment to cell. Such behaviour is understandable thermodynamically, the ions going down their gradient of chemical potential, and we would only have to postulate the development of a permeability to the ions caused by the lowered temperature. The reverse phenomenon on re-warming is harder to understand since the ions are moving uphill against gradients of chemical potential.

Active Transport

This movement is given the name of *active transport* and is a process requiring a supply of metabolic energy on purely thermodynamic grounds. Thus the cell in its normal metabolizing condition appears to be impermeable to Na^+ and K^+ because the concentrations of these ions inside and outside remain constant in the face of gradients of concentration or chemical potential. In fact it *is* permeable to these ions, but no net movements occur because, as soon as a K^+-ion is lost and a Na^+-ion is gained, the active transport reverses the process, extruding one Na^+-ion and accumulating one K^+-ion in exchange.

Energy Requirement. This requires the continuous expenditure of energy, and is the predominant cause of the metabolic consumption of energy by the erythrocyte, which carries out very little activity apart from this. When the cell is cooled, the metabolic processes are slowed and the active transport process is deprived of its energy, with the result that the K^+ and Na^+ ions are now allowed to run downhill. Robbing the cells of their glucose supply has an essentially similar effect, and certain metabolic poisons, such as fluoride, will likewise deprive the process of its energy.

ATPase. One very interesting class of inhibitors of active transport are the cardiac glycosides, typified by ouabain. These inhibit the action of an enzyme intimately connected with the active transport of Na^+ and K^+, namely *ATPase*, the enzyme that converts adenosine triphosphate (ATP, see p. 56) to adenosine diphosphate and water. There is more than one ATPase in the cell, and this one is characterized by the fact that it is only active if both Na^+ and K^+ are present in the medium; it is called the *Na+K-activated ATPase.* The mechanism by which ATP becomes involved in the transport of ions against gradients of chemical potential is by no means elucidated. It is interesting that when an erythrocyte has ceased to carry out active transport by virtue of loss of metabolites, the artificial incorporation of ATP into the cell allows the active transport to proceed. As we shall see, ATP is a substrate concerned in a variety of metabolic processes in addition to that of active transport; in fact we have already encountered its involvement in activation of amino acids in protein synthesis (p. 72).

Cell Swelling. We have seen that the loss of the cell membrane's effective semipermeability, so far as Na^+ and K^+ are concerned, leads to colloid osmotic swelling; in general, metabolic inhibitors, in so far as they block active transport of Na^+ and K^+, do cause the cell to swell, and this is the basis of the "*intracellular oedema*" accompanying lack of oxygen or metabolic poisoning in the animal.

Before pursuing the subject of active transport further it is desirable that we consider the mechanism by which gradients of electrical potential across cell membranes are established and maintained.

BIOELECTRIC POTENTIALS

Injury Potential

It has been realized for many years that the inside of most, if not all, cells is at a different electrical potential from that of the medium surrounding it. This was manifest as the so-called *injury potential*, measured when one electrode was placed on the intact part of a tissue, such as a muscle or nerve, and the other was placed on a damaged part (see Fig. 2.17). The electrode on the injured part was found to be negative in relation to the other, and it was correctly assumed that the electrode on the injured part was making electrical contact with the insides of the cells of the tissue, and that the measured potential

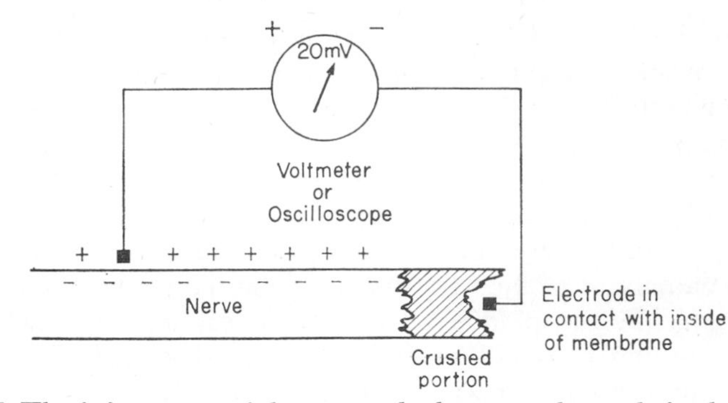

Fig. 2.17. The injury potential measured when one electrode is placed on the intact portion of a nerve or muscle and the other electrode is placed on a damaged part. The electrode on the injured portion is found to be negative in relation to the other, and it was assumed that this electrode was making electrical contact with the inside of the cells of the tissue.

was essentially the potential between insides and outsides of the injured cells. Such a method of recording gave values of the order of 20 mV for nerve and muscle, but the possibility of internal short-circuiting suggested that the true potential between the inside and outside of an individual cell would be larger.

Micro-electrode Measurements

In fact, modern micro-electrode techniques have permitted the puncture of many types of single cell, and the recording of the potential between inside and outside; the first nerve fibre to be so punctured was the giant axon of the squid, a large nerve running to the mantle of the squid that was shown by J. Z. Young to be, in fact, a single axon instead of a bundle of axons as would be found in a mammalian nerve. Subsequent improvements in technique have allowed the puncture of much smaller cells, e.g. single muscle fibres, the motor neurones of the mammalian spinal cord, and so on. In general, the membrane potentials of nerve and muscle fibres are in the region of 70 to 100 mV, the inside being negative. The understanding of the origin of the potential across a given cell is fundamental to the understanding of all nervous phenomena, and it is important, therefore, that at this early stage we devote adequate space to the physico-chemical principles governing the maintenance of this type of potential.

Diffusion Potential

If we place recording electrodes in a solution of salt we normally record no difference of potential because, on average, any accidental accumulation of, say, negative ions at one electrode is very quickly neutralized by an accumulation of positive ions; thus the ions of the solution are free to move, and do so under thermal motion, but as soon as a few positive ions separate from a few negative ions the local potential so created causes positive ions to flow towards the accumulation of negative ions and vice versa. Statistically, then, the solution is *isopotential*, and to record the actual local fluctuations of potential would require electrodes of atomic dimensions and a recording apparatus with very rapid resolution in time. The situation becomes different when two solutions of different concentrations of a salt such as NaCl are placed in contact as in Fig. 2.18; now there will be net

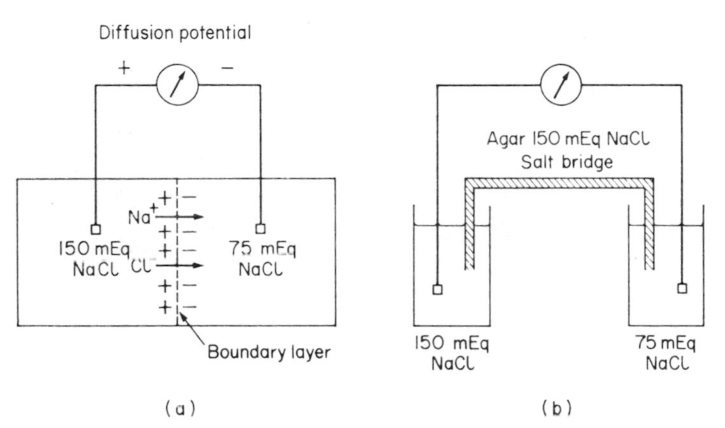

Fig. 2.18. Diffusion potentials are produced when two solutions of sodium chloride are placed in contact with each other. The chloride ion has a greater mobility than sodium, so that the Cl^- ions tend to move ahead of the Na^+ ions and a small charge separation is achieved. (a) Illustrates the theoretical concept and (b) how this may be measured experimentally.

diffusion of Na^+- and Cl^--ions from the more concentrated side on the left to the less concentrated side on the right. Under these conditions an electrode in the less concentrated (right-hand) side does become negative in relation to the electrode in the more concentrated (left-hand) side. This is due to the fact that the Na^+-ion has a smaller mobility than that of the Cl^--ion, and so the Cl^--ions tend to get ahead leaving the positive Na^+-ions behind.

Boundary Potential

It will be clear that if this separation of charges continued to any great extent the right-hand solution would become strongly negatively charged and the left-hand solution strongly positively charged. In fact, this does not occur because of the strong electrical interaction that would result, leading to the pulling of oppositely charged ions together again. The bulks of the two solutions have, on average, equal numbers of negative and positive ions, but at the junction between the two there is, indeed, a separation of charge due to a higher concentration of Cl^--ions at the right of the junction and a higher concentration of Na^+-ions at the left, as illustrated in Fig. 2.18. Because of this localized *boundary potential*, the slower-moving Na^+-ions are accelerated across the boundary and the faster-moving Cl^--ions are decelerated, so that in unit time equal numbers of both cross the boundary.

Potential as a Compromise

Thus the potential can be seen to be a compromise between two tendencies: the one is for the ions to move at their own speeds (mobilities) and the other is to retain electrical neutrality in the bulks of the two solutions. The tendency of the ions to move at their own inherent speeds gives rise to the separation of charge, whilst the tendency for the solutions to remain electrically neutral confines the separation of charge to the boundary. Clearly, if a salt whose ions have equal mobilities is studied, e.g. KCl, then no diffusion potential occurs.

Transience of the Potential

Again, the potential depends on the net movement of salt from one compartment to the other; when the concentrations have equalized the potential must disappear. Thus the *boundary*, or *diffusion potential* is a transient one, depending on net movement of salt. If, by some special mechanism, the salt diffusing to the more diluted solution could be recaptured and put back in the more concentrated compartment, then the potential would be permanent as long as this special mechanism operated. As we shall see, it is largely through this special mechanism that potentials across cell membranes are, in fact, held constant. Essentially they are diffusion potentials which would run down in the absence of special mechanisms, or active transport. This does not mean, however, that more permanent potentials always require special mechanisms for their maintenance.

Membrane Potential

The boundary potential discussed above depends on the different mobilities of the ions of a salt, and on a difference of concentration between two compartments. The potential can be exaggerated by the use of an ionically selective membrane that enhances the difference of mobility of positive and negative ions. If the difference of mobility is enhanced to the point that one type of ion cannot move through the membrane at all, then we have the basis for a permanent membrane potential. Let us consider such a membrane separating a solution of KNO_3 from one containing a mixture of NaCl and KCl as in Fig. 2.19 and assume that the membrane is impermeable to all ions except K^+. The concentrations have been chosen so that there is no difference of osmolality. Net movements of salt cannot take place because the only ion that can cross the membrane is K^+ and in the interests of electrical

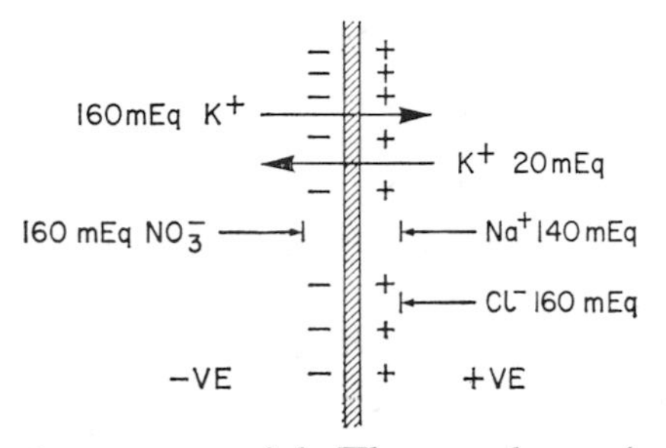

Fig. 2.19. The membrane potential. The membrane is permeable only to the positive K^+ ion, which tends to move from left to right down its concentration gradient. Since negative ions cannot accompany it, and positive Na^+ ions cannot pass in the opposite direction, the movement of the K^+ ion leaves the left-hand side negative, generating a potential, localized at the membrane, which tends to oppose further movements of K^+ ions from left to right. Equilibrium is achieved when the magnitude of the potential just balances the diffusional forces favouring outward net diffusion of K^+ ions. In other words, although only one K^+ ion strikes the membrane from the right-hand side, for every eight ions striking from the opposite side, the acceleration from right to left, caused by the potential, ensures that equal numbers of ions actually *cross* the membrane in each direction in unit time.

neutrality of the bulk of the solutions any loss of K^+ must be accompanied by either a NO_3^--ion or by the exchange with a Na^+-ion. In fact there are no measurable changes in concentrations of ions in the two compartments, but a potential is established, such that the KNO_3 compartment is negative in respect to the other; this is due to the tendency for K^+-ions to diffuse down their concentration gradient. At the membrane boundary an excess of K^--ions does, indeed, accumulate on the right-hand side and this leaves a corresponding excess of negative ions on the left-hand side. The resulting potential across the membrane is of such a magnitude that it retards the passage of K^+-ions down their concentration gradient and accelerates them up it.

Therefore, in unit time, although some eight K^+-ions will hit the membrane from left to right against one hitting it from right to left, the single ion passing from right to left will have eight times the probability of crossing the membrane because the potential favours it to that extent.

Nernst Equation

The magnitude of the potential is simply calculated on thermodynamic grounds, and is given by the Nernst Equation:

$$E = RT/nF \ln \frac{[K^+]_1}{[K^+]_2} \tag{4}$$

R being the gas constant, *T* the absolute temperature, *n* the valency *F* Faraday's constant, and the square brackets indicating concentrations, or, more accurately, activities.

With a univalent ion and at a temperature of 20°C, the equation becomes:

$$E = 58 \log_{10} \frac{C_1}{C_2}$$

where concentrations (or rather activities) are in moles per litre and the potential is in millivolts. Thus, for a tenfold difference in concentration, there is a potential of 58 mV.

The potential of Fig. 2.19 is stable because the system is prevented from running down by the diffusion of ions down their concentration gradients; this is in contrast to the boundary potential discussed earlier. If the membrane were to become permeable to all ions in the system, the potential would become unstable and would finally become zero as the concentrations of ions equalized.

Muscle and Nerve

The situation in most, if not all, cells is similar to that discussed above, in the sense that a membrane enhances the differences of mobilities of the ions, but it is also similar to the simple diffusion potential in that there is, in fact, a tendency to run down by the net movement of salt.

Resting Potential

Let us examine the situation in muscle and nerve as illustrated by Fig. 2.20. Within the cell there are high concentrations of K^+ and low concentrations of Na^+, Cl^- and HCO_3^-; the ions represented by A^- are large organic anions to which the cell membrane is impermeable.

Outside, the concentrations of Na^+, Cl^- and HCO_3^- are large, and that of K^+ small. The membrane behaves as though it were impermeable to Na^+ as well as to A^-, and the result is that there is a tendency for K^+ to diffuse down its concentration gradient out of the cell, and for Cl^- to diffuse into the cell. Net movements of salt are prevented because K^+ is unable to exchange with Na^+ by virtue of the impermeability of the membrane to Na^+, whilst K^+ cannot move in company with A^- because of the impermeability to A^-. Net movement of KCl out of the cell would require that Cl^- move up a steep gradient

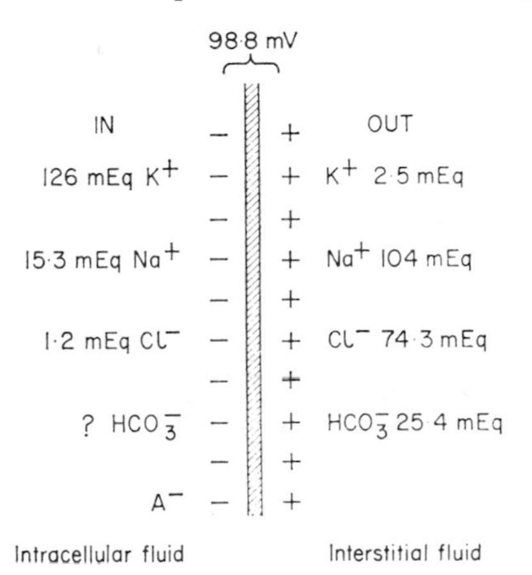

Fig. 2.20. The distribution of ions and potential across nerve or muscle cell membranes. The membrane is permeable to K^+, Cl^- and HCO_3^- but effectively impermeable to Na^+ and the organic anions, A^-. The relative concentrations of the permeable ions are theoretically governed by the Nernst equation relating these to the membrane potential.

of concentration, and when the system is in equilibrium the concentrations of the ions will be such that the tendency for K^+ to move out and Cl^- to move in are just balanced by a potential, the inside being negative (see Fig. 2.20). Thus the potential, called the *membrane* or *resting potential*, is given by the Nernst Equation, and since it may be viewed as being created either by the K^+- or the Cl^--ion, it will be given by:

$$-RT/nF \ln \frac{[K^+]_{In}}{[K^+]_{Out}} \text{ or } -RT/nF \ln [Cl^-]_{out}/[Cl^-]_{In}$$

Gibbs–Donnan Distribution

It follows at once that the ratio of K^+ concentrations

$$[K^+]_{In}/[K^+]_{out} \text{ is equal to } [Cl^-]_{out}/[Cl^-]_{In},$$

and this is an expression of the Gibbs–Donnan Equilibrium, which will be discussed in more detail in another Chapter. Essentially, it tells us that when a membrane, impermeable to one or more species of ion, separates two solutions, the products of the concentrations of the permeable positive and negative ions will be equal, i.e.

$$[K^+]_{In} \times [Cl^-]_{In} = [K^+]_{Out} \times [Cl^-]_{Out};$$

hence $$[K^+]_{In}/[K^+]_{Out} = [Cl^-]_{Out}/[Cl^-]_{In} = [HCO_3^-]_{Out}/[HCO_3^-]_{In} = r.$$

Extrusion of Na^+

The potential across muscle and nerve, as envisaged above, depends on an impermeability of the cell's membrane to Na^+ and the organic anions, A^-. Now, by the use of the isotope K^+, it is quite easy to confirm that the membrane is permeable to K^+, so that on adding ^{42}K to the outside medium, after a period of some hours the proportions of ^{42}K to inactive K^+-ions become the same on each side of the membrane. The same, however, happens with ^{24}Na, although rather more slowly. Clearly, the membrane is, in fact, permeable to Na^+, in which case we would expect the K^+-ions to diffuse out of the cell in exchange for Na^+-ions entering, i.e. we should expect the obliteration of the concentration gradients of Na^+ and K^+. That this does not occur is because of the active transport process that "pumps" Na^+ out of the cell. Thus there is, indeed, a tendency for K^+ to diffuse out and Na^+ to diffuse in, thereby leading to a fall in potential, but this is countered by the active extrusion of the Na^+ associated with an exchange for an external K^+-ion. Hence, as indicated earlier, the bioelectric potential shares the characters of the transient diffusion potential illustrated by Fig. 2.18 and the permanent membrane potential illustrated by Fig. 2.19.

Electrochemical Potential

When two solutions of, say, sucrose are placed in contact, there will be a net movement of sucrose from the more concentrated to the more dilute solution; the system is said to *approach equilibrium.* This is achieved by passage of solute from a condition of higher *chemical potential* to one of lower, the chemical potential, or *escaping tendency*, being in this situation defined by its concentration, or activity. When movements of ions are concerned we cannot say unequivocally that a given ion will move from a region of higher concentration, or activity, to one of lower; thus in the situation described by Fig. 2.19, the system is in equilibrium and yet the concentration of K^+ is higher on one side of the membrane than on the other. The reason for this is the potential that restrains the movements of K^+ in one direction and accelerates them in the opposite, so that to define equilibrium, i.e. the situation in which there will be no net movement of an ion, we must use a measure of chemical potential, or *escaping tendency*, that takes into account both concentration and potential. This measure is the *electrochemical potential* of an ion and is defined as the sum of two potential terms, i.e. $RT \ln C$, which takes into account its concentration, and πzF, where π is the electrical potential of the solution in relation to some standard, z is the valency of the ion, and F is the Faraday constant.

Non-electrolyte Solutions

Thus with two non-electrolyte solutions of C_1 and C_2 separated by a membrane permeable to the non-electrolyte (Fig. 2.21), the two solutions will be in equilibrium when the chemical potentials of both solutions are equal, i.e. when

$$RT \ln C_1 = RT \ln C_2$$

i.e. when

$$C_1 = C_2$$

or

$$RT \ln C_1/C_2 = 0.$$

Ions

With two ions of concentration C_1 and C_2, and potentials in the solutions π_1 and π_2, they are in equilibrium if we have:

$$RT \ln C_1 + \pi_1 zF = RT \ln C_2 + \pi_2 zF \tag{5}$$

i.e. if

$$RT \ln C_1/C_2 = (\pi_2 - \pi_1) zF \tag{6}$$

Where $\pi_2 - \pi_1$ is the difference of potential across the membrane separating the two solutions.

If we divide both sides of Equation (6) by zF we have:

$$\frac{RT}{zF} \ln C_1/C_2 = \pi_2 - \pi_1 = \text{P.D.} \tag{7}$$

In other words, the ion is in equilibrium if its ratio of concentrations on each side of the membrane is governed by the Nernst equation. Thus in Fig. 2.21 K^+ and Cl^- are in equilibrium, but Na^+ is not, so that if the membrane is permeable to Na^+ we must invoke some special mechanism for maintaining it at a much higher electrochemical potential on the right-hand side of the membrane.

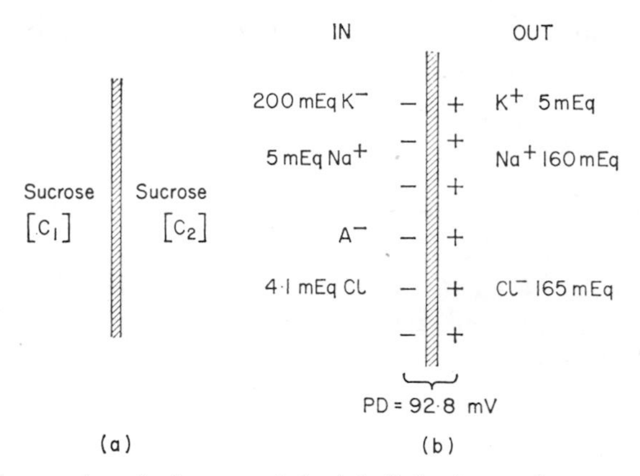

Fig. 2.21. Electrochemical potential. (a) Solutions of sucrose at different concentrations separated by a permeable membrane will approach equilibrium through diffusion from the higher to the lower concentration; at equilibrium C_1 will equal C_2. (b) Here we imagine the membrane to be permeable to all ions of the system. Although the concentration of K^+ is much higher in the left-hand solution and that of Cl^- much lower, these two ions are in electrochemical equilibrium because the potential across the membrane is just sufficient to maintain these differences of concentration, accelerating K^+ ions from right to left and negative ions from left to right. We can say that K^+ and Cl^- are at equal electrochemical potential. The Na^+ ion is clearly not at electrochemical equilibrium since both concentration and potential favour movement from right to left.

Capacity and Resistance

The cell membrane, with its low permeability to ions and its separation of two solutions of high electrical conductivity, may be likened to a condenser carrying charges on opposite surfaces represented by the potential. As such, we may expect to be able to measure its electrical capacity, for example by placing electrodes inside and outside the cell. Because there is a permeability to ions, the membrane has a finite resistance, and may be represented as a leaky condenser with a resistance, R, in parallel with the capacity, as in Fig. 2.22. By suitable electrical measurements the membrane's *impedance*, which is a function of both resistance and capacity, can be resolved into its component elements. In general, the capacity is remarkably constant from one cell to another, being of the order of one microfarad per square centimetre. The transverse resistance is more variable, as we might expect, since it depends on ionic permeability; values ranging from 170 ohm.cm^2 for frog's egg to 8000 ohm.cm^2 for the axon of crab nerve have been found.

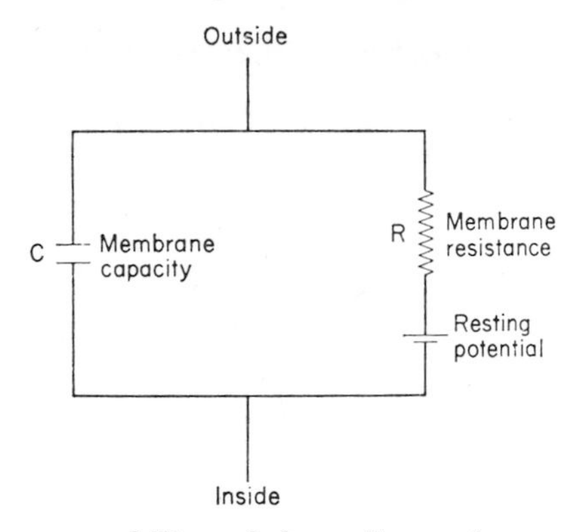

Fig. 2.22. The low permeability of the cell membrane may be represented by a charged condenser *C* with a potential across the plates. However, since there is a finite permeability to ions there is a constant loss of charge, and this can be represented by the resistance *R* in parallel with the condenser.

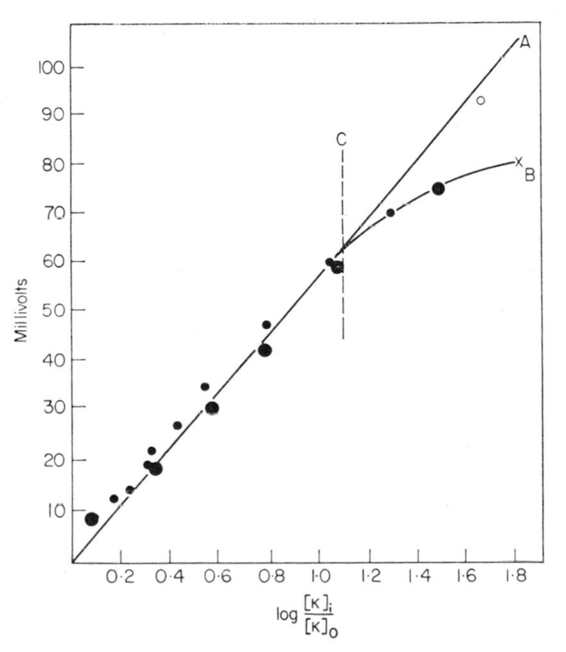

Fig. 2.23. Variation of resting potential of isolated muscle with the ratio $[K^+]_{In}/[K^+]_{Out}$, which was altered by changing the concentration of K^+ in the external medium. The straight line through A is the theoretical line from the Nernst equation. The vertical line C at 1.11 on the abscissa represents the level of 10 mM K^+ in the external fluid. (Conway, *Physiol. Rev.* 1957, **37,** 84).

Membrane Transients

The membrane potential of a cell can vary considerably; in fact, it is the rapid changes in this that constitute the messages carried by nerve fibres. We may therefore consider, in general terms, how the potential can be influenced. Clearly a failure of the active transport mechanisms, extruding Na^+ from the cell, will lead to a decline in the potential; this will be a slow process depending on the gradual build-up of a high internal Na^+ concentration and a low internal K^+; such a slow decline is found in nerve during oxygen-lack. Acute changes in the concentration of ions outside the membrane can have large effects. Thus if the medium is replaced by pure isotonic KCl, the gradient of K^+-concentration is abolished, the potential is abolished and the membrane is said to be *depolarized*. If the concentration of KCl outside is increased in steps, the potential falls progressively, and the changes may be predicted with some accuracy by the Nernst equation, as illustrated in Fig. 2.23.

Constant Field Equation

Finally, we may ask how changes in ionic permeability are likely to affect the potential. So far as Na^+ is concerned, the Na^+-pump ensures that the system behaves as though the membrane were impermeable to this ion; the membrane, in fact, is permeable to Na^+ and the passive influx associated with the active outflux can modify the potential, causing deviation from the magnitude predicted on the basis of absolute Na^+-impermeability, so that a more general expression, involving permeabilities of the major ions in the system should really be employed to describe the potential. The equation is:

$$E = -\frac{RT}{F} \ln \frac{P_K[K]_i + P_{Na}[Na]_i + P_{Cl}[Cl]_o}{P_K[K]_o + P_{Na}[Na]_o + P_{Cl}[Cl]_i}. \qquad (8)$$

the so-called constant-field equation of Goldman. If Cl^- is distributed passively this simplifies to:

$$E = -\frac{RT}{F} \ln \frac{P_K[K]_i + P_{Na}[Na]_i}{P_K[K]_o + P_{Na}[Na]_o}. \qquad (9)$$

Increasing the permeability to Na^+ will clearly reduce the value of the potential; as we shall see, during the action potential the Na^+-permeability becomes so high that the potential is governed by the Na^+-concentrations, and thus adopts an opposite polarity.

Potential and Permeability

If we return to Fig. 2.20 we see that the resting potential of muscle or nerve may be described in terms of the relative concentrations of K^+ or Cl^-, the membrane being permeable to both ions, and the equilibrium conditions demanding that their ratios of concentrations be reciprocals of each other. If the permeability of the membrane is very much greater to, say, K^+ than to Cl^-, as occurs with nerve, then it is customary to refer to the potential as predominantly a *K^+-potential.* Of course, provided we are at equilibrium, it can just as correctly be described as a Cl^--potential. When a deviation from equilibrium is brought about, e.g. by increasing the concentration of K^+ outside but leaving that of Cl^- unchanged, then a change in potential will take place, but the magnitude of this will depend very much on the relative permeabilities of the membrane to K^+ and Cl^-. If the permeability to K^+ is much the higher, then we shall see a depolarization of the membrane due to the altered K^+-ratio; the new potential will clearly not fit the Cl^--contentration ratio, and the system will readjust itself by a movement of Cl^- into the cell, tending to lower the ratio and to bring it equal to the K^+-ratio. The final result will depend on the movements of all ions and need not be entered into here. The important point is that the immediate change in potential is governed by the changed K^+-concentration because of the assumed high permeability to K^+. With skeletal muscle the situation is different, the Cl^--permeability being greater, so that the potential is more sensitive to changes in this ion's concentration than to those of K^+.

Hyperpolarization

In many excitable systems, because of active transport processes, the actual resting potential is not equal to the theoretical Nernst potential for any ion, e.g. the giant axon of the squid, when isolated, usually has a lower potential than that given by the Nernst formula for the K^+-concentrations. This potential can be increased (*hyperpolarized*), however, by an increase in permeability to K^+; in other words we are taking the potential nearer to a pure K^+-potential by increasing the mobility of the K^+-ion through the membrane by comparison with those of other ions. This becomes evident when we look at the Goldman equation; clearly, increasing P_K will tend to increase the product $P_K \times [K]$ and so increase the importance of these terms in the fraction. When P_K is very high compared with the other permeabilities, these last may be equated to zero and the potential becomes a pure K^+-potential.

Active Transport Potentials

The potentials we have discussed so far have been the predictable consequence of the establishment of certain gradients of concentration between the inside and outside of the cell. The active transport process, involving the extrusion of Na^+, has only been of significance in so far as it prevented the system from running down; apart from this it had no influence on the membrane potential. There are some situations in which the active transport process contributes strongly to the potential. An example is the frog-skin, the first system to be adequately investigated (by Ussing).

Frog-skin Potential

The frog has the capacity to absorb Na^+ from its fresh-water surroundings across its skin. The concentration in the fresh-water is very low whilst that in the animal's extracellular fluid is high, so that the Na^+ is moved against a high gradient of concentration. When an

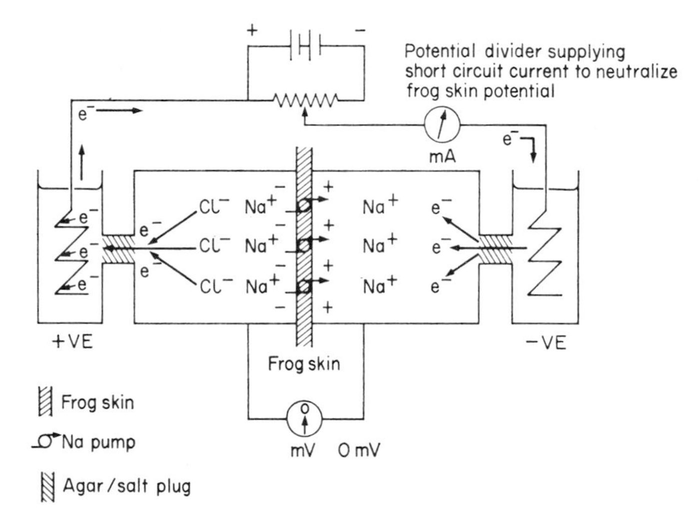

Fig. 2.24. Apparatus suitable for measuring the short-circuit current across the frog skin. The active transport of Na^+ from left to right creates a potential across the membrane due to the slower movement of Cl^- ions. If a potential is applied from a battery (top) opposing this potential, a current must flow in the external circuit such that for each ion of Na^+ transported one electron passes to neutralize it. Thus, electrons in the external circuit replace the Cl^- ions that would have moved from left to right accompanying the actively transported Na^+ ions. If, under these short-circuit conditions, the only ion being transported is Na^+, the short-circuit current is a measure of the transport of Na^+.

electrode is placed under the skin and another in the water, it is found that there is a potential across the skin, the inside being positive. Thus the transport of Na^+ is not only against a concentration gradient but also against a potential gradient, so that active transport must be invoked. The potential is conveniently studied by stretching a piece of isolated skin across a hole separating two chambers containing isotonic Ringer's solution as in Fig. 2.24. The studies of Ussing have left little doubt that the cause of the potential is the active transport of Na^+. The mechanism for this is by no means clear, but if we imagine that the Na^+-ion forms a complex within the skin and, in this un-ionized form, is rapidly transferred to the inside of the skin and set free as an ion, it is clear that, unless a negative ion is transported equally rapidly, or there is an exchange with a positive ion such as K^+ on the other side, the inside of the skin will become positive. Thus the rapid transport of Na^+-ions causes the inside to become positive up to the point where the potential accelerates Cl^--ions sufficiently to provide an equal number of Na^+- and Cl^--ions passing across the skin in unit time. As the potential builds up to its steady-state value, negative ions will be accelerated more and more while the movement of Na^+-ions will be decelerated. On this basis, then, we may say that the potential has been caused by the active transport of Na^+ whilst the movement of Cl^- and other anions might well be the passive consequence of the movement of Na^+.

Short-circuit Current. Ussing devised a very ingenious method for studying this transport; he argued that, if he applied a counter potential across the skin so as to make the potential zero, there would be no force driving Cl^- from one side of the membrane to the other if he started with equal concentrations of NaCl on each side of the membrane. The potential would presumably not affect the active transport of Na^+, if it was being transported as an un-ionized complex, so that the neutralization of the positive charges brought on to the inside of the skin by the transported Na^+ would have to be done by electrons in the external circuit (see Fig. 2.24). Thus the current flowing in the external circuit—the *short-circuit current*—would be equivalent to the ions carried by the active transport mechanism if this transport of Na^+ were the sole process in generating the potential. In fact, by measuring the next flux of Na^+ across the skin by the use of isotopic Na^+, Ussing and Zerahn showed that this corresponded nearly exactly with the current flowing in the external circuit. In general, this short-circuit current technique has been applied to the study of many actively transporting systems, and useful information has been obtained regarding the actual ion, or ions, that are being actively transported. To return to the frog skin, therefore, it can be argued that, because the short-circuit current is equivalent to the net flux of Na^+, it is unlikely that the Cl^- ion is actively transported by a similar mechanism, since if this happened some of the Na^+-ions carried across would be neutralized by Cl^--ions carried at the same time.

Independence Relationship. Another test applied by Ussing is the so-called *independence relationship;* if an ion is not being actively transported, its fluxes, defined as the number of equivalents crossing unit

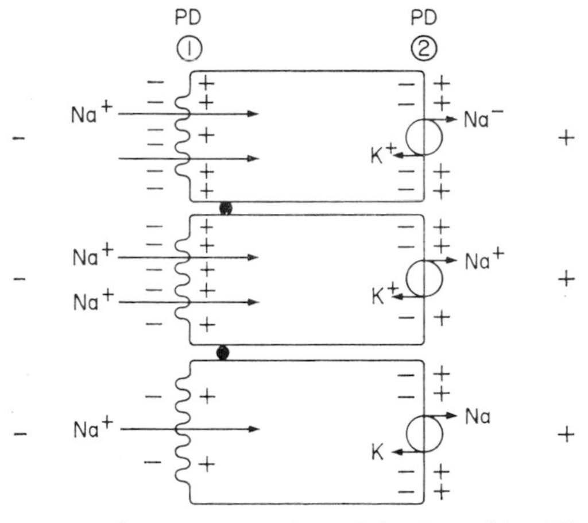

Fig. 2.25. A diagrammatic representation of the frog skin. The skin, although composed of many layers, behaves as if it consisted of a single layer of epithelial cells joined by a continuous belt of tight junctions close to the mucosal side of the skin. The outside-facing membranes of the cells behave as if they are highly permeable to Na^+ with an Na^+/K^+ exchange pump on the inside-facing membranes.

area of membrane in unit time, will be simply determined by the potential across it, and the concentrations on each side:

$$M_1/M_2 = C_1/C_2 \exp - FE/RT$$

M_1 being the flux from solution 1 to solution 2, and M_2 the flux from solution 2 to solution 1.

The fluxes of the ion can be measured by the use of isotopes, and the potential measured at the same time. Ussing found that, in the frog skin, the fluxes of chloride followed the independence relationship fairly closely, whereas those of Na^+ not at all.

Epithelial Asymmetry. The potential across the frog skin can be described theoretically in terms of a single layer of epithelial cells with their membranes fused laterally to give a seal preventing significant diffusion of ions between the cells and therefore restricting exchanges to a pathway across the cells (Fig. 2.25). The cells as a whole transport Na^+ actively through their cytoplasm, taking it in at the outside membrane and extruding it at the inside-facing membrane. Ussing's studies indicated, moreover, that the outside-facing membrane behaved as though it were highly Na^+-permeable, so that changing the concentration of Na^+ in the outside medium had a large influence on the potential across the skin; the inside-facing membrane behaved as though it were highly K^+-permeable. The active transport of Na^+ would then begin as a passive leakage into the cell of Na^+ from outside, by virtue of the high Na^+-permeability. At the inner membrane there would be a tendency for K^+ to leak out passively down its gradient of concentration but this would be counteracted by a Na^+—K^+-pump, i.e. an active process operating across the inside-facing membrane causing Na^+ to be extruded from the cell in exchange for K^+ which was being pumped uphill because of the high resting concentration of K^+ in the cell. Thus, as Fig. 2.25 shows, the potential across the skin is the sum of two steps, so that if we equate the outside potential to zero, then there is a first positive step on crossing the outward facing membrane due to the passive influx of Na^+ which

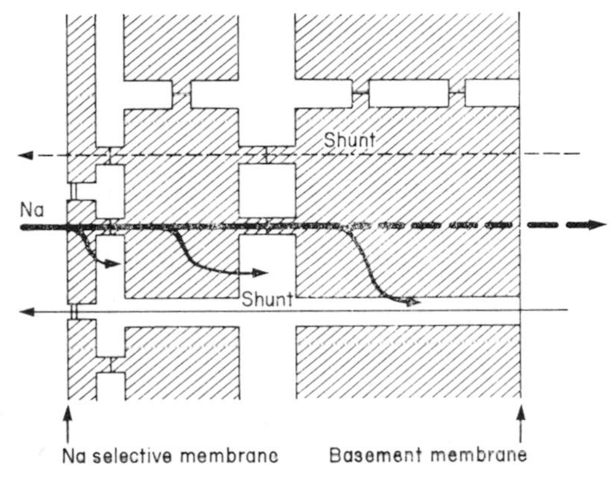

Fig. 2.26. The model proposed by Ussing to explain the observed properties of frog skin. The sodium-selective membrane is on the outermost layer of cells joined by tight junctions. Cells are interconnected electrically by gap junctions and behave as a syncytium. Sodium can enter the outer cells and pass out into the intercellular spaces or via the gap junctions into cells deeper in the skin and then into the intercellular spaces. The Na/K pump is located on the inward facing membranes of all cells and on their lateral facing membranes facing the intercellular clefts. (Ussing,' 'Electrophysiology of Epithelial Cells", Schaffauer Verlag Stuttgart, 1971).

occurs more rapidly than that of the anions, mainly Cl^-. At the other membrane there is a jump to greater positivity because of the high internal and low external concentrations of K^+. Looked at in this way the potential is the algebraic sum of two diffusion potentials, one dominated by Na^+ and the other by K^+, the whole system being prevented from running down by a compulsory link between Na^+-extrusion and K^+-entry.

Anatomical Basis. A great deal of experimental evidence supports this general concept, including the finding that, when a microelectrode is pushed into the skin, the potential-profile is one of two main jumps in positivity. However, anatomically the skin consists of several layers of cells so that we must envisage transport of Na^+ as taking place in a more complex manner, much of it along the intercellular clefts. Thus Fig. 2.26 illustrates schematically a more realistic arrangement, the Na^+-selective membrane being the outside-facing membrane of the outermost layer of cells; these are connected with the inner layers probably by gap junctions (p. 33) that convert the whole epithelium virtually into a syncytium, so that Na^+, transported into the outermost layer, can pass from one cell to the next, or else it may pass into the intercellular spaces which make ultimate connection with the inside fluid. The spaces between the cells of the outermost layer are sealed by true tight junctions that therefore channel the Na^+ first of all into the outermost cells. The site of the Na^+-K^+ pump, formerly located at the inward-facing membrane of a single layer of cells, now becomes the inward-facing membranes of all cells, together with the lateral membranes facing the intercellular spaces. In fact, recent work, in which the enzyme ATPase, activated by Na^+ and K^+ and associated with this form of pump activity, has been localized histochemically, shows that the enzyme is confined to the lateral surfaces of the epithelial cells, where there is considerable interdigitation of the cell membranes.*

Isoelectric Pump

The existence of a potential across a structure such as the frog skin is the sign of active transport processes taking place; similar potentials are observed across the toad and turtle urinary bladders, and these also may be attributable to active transport processes involving the

* As indicated in the text, when a microelectrode is inserted into the skin the potential develops in two main steps; more critical examination shows that this is too simple a view. There is an immediate large step, following penetration of the outward-facing membrane of the first cell layer, and subsequently there are smaller steps (due to penetration of deeper cells) until the full potential develops, when the electrode has passed through the innermost layer of cells. This potential profile is in accord with the scheme of Fig. 2.26.

movement of ions and water. It must be appreciated that the absence of a potential across a membrane, such as an epithelium, does not necessarily mean absence of active transport, nor yet does the presence of a potential mean active transport. The actual potential developed depends on many conditions; thus the potential across the frog-skin has been attributed to two diffusion potentials—a Na^+-potential governed by the high permeability of the outward-facing membrane to Na^+ compared with that to anions such as Cl^-, and a K^+-potential due to a high permeability of the inward facing membrane to K^+ by comparison with anions. The system is stabilized by the active transport process, but since this can be brought about by the linking of movement of one Na^+-ion in one direction with the movement of one K^+-ion in the opposite direction, the pump itself is not contributing to the potential; it can be called "neutral".

Independence of P.D. and Na^+-transport

If Ussing's delineation of the frog skin potential is correct, then we need not expect variations in the active transport of Na^+ to affect the magnitude of the potential, since the function of the active transport is merely to maintain the level of Na^+ at a low value within the cell. So long as this is maintained, the potential depends on the *permeability characteristics* of the cell, a high Na^+-permeability of the outward facing membrane and a high K^+-permeability of the inward facing membrane. In fact, the transport of Na^+ across the skin is largely independent of the potential, so that when different skins are compared, large variations in potential may be found but not of sodium transport. Again, we may accentuate the potential by lowering the cell's permeability to anions, e.g. by treating the skin with copper salts. Such increases in potential are not accompanied by parallel changes in transport of Na^+.

Shunting

With a number of epithelia, such as that of the gall bladder, there is vigorous active transport of salt but a very low potential. This was attributed by Diamond to a linked transport of Na^+ and anions, the movement of Na^+ being compulsorily linked with that of Cl^-, e.g., as if both had to be present together on a "carrier" for the carrier to operate. More recently Diamond has concluded, however, that the reason for the low potentials is the "shunting" through low-resistance pathways. Thus Na^+ is assumed to be transported, say, into the intercellular clefts, but the permeability of the junctional regions to

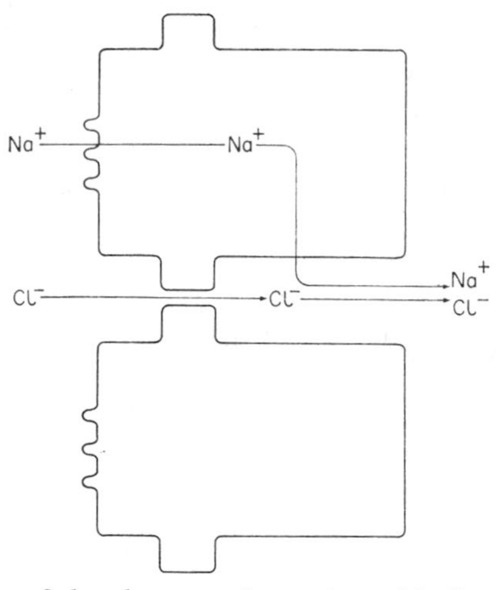

Fig. 2.27. A diagram of the shunt pathway in epithelia that do not develop a potential. Sodium is transported into the intercellular clefts, as in other tissues, but the junctional complex is very permeable to chloride ions, which rapidly pass through and, by preventing charge separation, shunt out any potential that might be developed.

Cl^- is considered to be sufficiently high to neutralize any tendency for charge-separation (Fig. 2.27).

Non-isoelectric Pumping

In describing the maintenance of the resting potential of muscle or nerve we have invoked the active Na^+-pump as a mechanism for maintaining the concentration gradients of K^+ and Cl^- that determine the potential. Thus no part of the potential has been attributed to the active transport of Na^+, as such, although, by analogy with frog skin, we might expect the active extrusion of Na^+ to contribute to the internal negativity of the cell. In fact there are some situations (i.e. with muscle and nerve, when the cell has been allowed to accumulate a considerable excess of Na^+ and the pump is working at full capacity) in which the active pumping of Na^+ does, indeed, modify the resting potential, i.e. in which it is not *isoelectric*.

Under these conditions, we may presume, the Na^+-ion is driven out of the cell more rapidly than its accompanying anion, giving rise to a diffusion potential with the inside of the cell becoming more negative. If Na^+-extrusion were tightly linked, in a one-for-one fashion, with entry of K^+-ions, then, of course the pump would be isoelectric or neutral, so we must assume that this coupling, which probably occurs normally, breaks down when the Na^+-pump is working hard.

FURTHER CONSIDERATIONS ON ACTIVE TRANSPORT

The Work of Active Transport

When two solutions of different concentration of the same solute are brought into contact, the equalization of concentration occurs spontaneously, and we may say that the system passes from one of greater to lesser free energy or lesser to greater entropy. To separate the solute from the solvent, so that the original difference of concentration is re-established, requires the expenditure of energy. Active transport processes are, by definition, processes of separation whereby ions and molecules are forced to move against their spontaneous tendency to pass from high or low chemical potential, and therefore active transport requires metabolic energy to sustain it. The work can be calculated by simple thermodynamic procedures which need not be entered into here; suffice it to say that the greater the difference of chemical potential established by the process, the greater will be the work required. Another factor will be the leakiness of the membrane across which transport occurs. If we suppose that the erythrocyte maintains its high internal concentration of K^+ and low Na^+ by a process of actively extruding Na^+ and accumulating K^+, then the amount of osmotic work it will have to do to maintain a steady state will depend on the permeability of the cell membrane to these ions. If they are highly permeable the work will be very high, since in a given time many K^+ ions will leak out of the cell, and many Na^+ ions leak in under their gradients of electrochemical potential. It will be like trying to carry water in a very leaky sieve.

Oxygen Consumption

The work of active transport can often be measured experimentally by the oxygen consumed by a tissue while it is actively transporting, over and above that consumed while active transport is inhibited. Thus Zerahn computed the extra energy consumed, per equivalent of Na^+ transported by the frog skin, at 1340–2680 calories, and this represented an efficiency of 27–54 per cent, in the sense that the theoretical amount of work necessary to do this osmotic work, calculated on thermodynamic principles, was some 27–54 per cent of the energy made available by the extra oxygen consumption.

ATP

The level of ATP in a cell, such as the erythrocyte, seems to be a measure of the energy available for active transport, and it is interesting that Whittam has shown that failure of active transport is usually associated in this cell with diminution of its ATP stores. The strong involvement of this substrate in active transport is emphasized by the presence of a specific enzyme, ATPase, that hydrolyses ATP, in the membranes of actively transporting cells. This ATPase is peculiar in that Na^+ and K^+ must both be present simultaneously for it to work efficiently, and this marks it off from the ATPase concerned in the contraction of muscle. Potent inhibitors of active transport of ions are ouabain and other cardiac glycosides (such as digitalis) and their action is due to the inhibition of the specific ATPase concerned in active transport of Na^+ and K^+.

Non-specific Inhibition

Other inhibitors of active transport are usually of a less specific type, in so far as they merely attack the metabolic processes that provide the energy for active transport, e.g. dinitrophenol (DNP) uncouples the oxidative phosphorylation processes in aerobic metabolism thereby preventing the synthesis of ATP; cyanide and azide have a direct attack on the electron transport chain of the mitochondrial oxidative metabolism, whilst fluoride and iodoacetate inhibit anaerobic glycolysis. Since the mammalian erythrocyte can only utilize anaerobic metabolic mechanisms, it is these last named inhibitors that have a striking inhibition of its active transport.

Active Transport and Water Flow

The system represented by the frog-skin, separating two compartments of, say, isotonic Ringer's solution, provides a mechanism, not only for the active transport of NaCl from the outside to the inside, but also for the transport of water, since the transport of NaCl will create a difference of osmolality that will lead to movement of water to the inside of the skin. In fact we must regard the system as both a salt- and water-conserving device, permitting the animal to take both up when required. It is not surprising, moreover, that such a process is controlled, the hormone *pitressin* (Vol. 2), or *antidiuretic hormone*, causing a remarkable increase in water permeability that would clearly favour uptake of water.

Permeability Asymmetry

In general, then, the basic requirement of a water transporting system is a layer of cells with a permeability asymmetry such that Na^+ is allowed to flow into the cells at one face and is actively extruded at the other face, so that the Na^+-pump will be on the side of the cell facing the direction of water-flow. The Na^+-pump is linked with a K^+-accumulating system, and the membrane at this side shows a relatively high permeability to K^+ as demonstrated by Ussing. The presence of this pump is apparently revealed by the high concentration of ATPase in the membrane. Thus, in frog-skin this is accumulated on the basal and lateral aspects of the secreting cells, the presumed site of the Na^+–K^+-pump, whilst the apical region is the site of high Na^+-permeability.

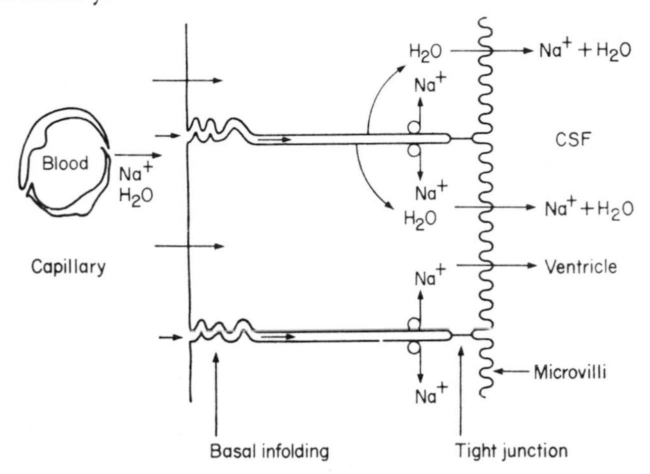

Fig. 2.28. The choroid plexus. The epithelial cells of the choroid plexuses secrete cerebrospinal fluid (CSF) into the ventricle of the brain. Movement of salts and water is in the opposite direction to that in the gall bladder in the sense that it is from base to apex of the epithelial cell. According to Diamond, the Na^+ ions are actively "pumped" out of the intercellualr clefts into the cells whence they pass into the ventricle.

Choroid Plexus

In some other systems the asymmetry is present but in the reverse direction; for example, the choroid plexuses of the brain ventricles are responsible for a flow of fluid from the blood-side of its epithelial lining to the ventricle side, and, as illustrated schematically in Fig. 2.28, this means that if the same model is applied, the region of high Na^+-permeability is the basal aspect of the cell, and the location of the Na^+–K^+-linked pump is on the apical aspect.

Conditions for Flow

If the flow of water is governed by the primary transport of salt we may expect the osmolality of the fluid, as it comes away from the transporting membrane, to be higher than that on the opposite side; for example we might expect the osmolality of the cerebrospinal fluid in the ventricle, illustrated in Fig. 2.28, to be higher than that of the interstitial fluid within the choroid plexuses and thus of the blood. In fact, it is difficult to detect any difference of osmolality, the fluid being described as an *iso-osmolal secretion*; and the same is true for quite a number of fluids transported across cellular membranes through what appears to be a primary active transport of salt. Now the question arises as to how water can be made to follow the active transport of salt and yet leave no evidence of the difference of osmolality that presumably caused its net movement. It could be argued that the permeability of the membrane to water was infinitely high by comparison with the transport of salt, so that the primary difference of osmolality that drove the water across the membrane was obliterated immediately. Unfortunately for this explanation, however, the measured permeability to water, though high, is not high enough to

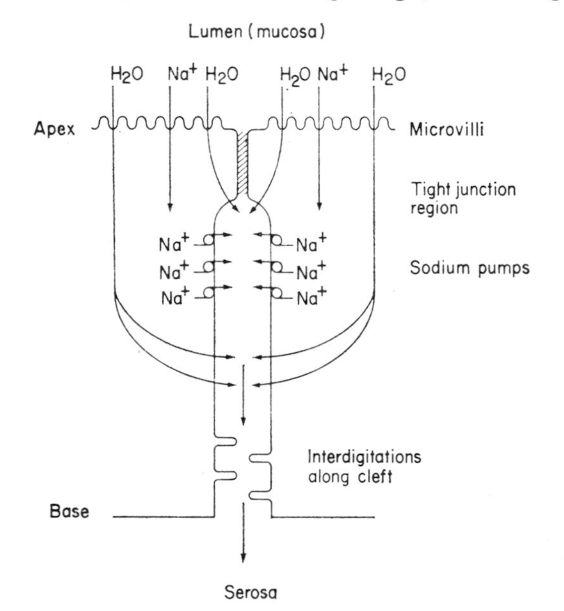

Fig. 2.29. The Diamond and Bossert concept of solute-coupled water movement in the gall bladder. Sodium is pumped into the clefts between epithelial cells close to the tight junctions. This movement of sodium causes a local increase in osmotic pressure and water is drawn from the lumen into this region. The entry of water in turn causes a rise in hydrostatic pressure and movement of salt and water down the cleft. Since the clefts are long, osmotic equilibration occurs by the time the fluid emerges on the serosal surface.

account for this iso-osmotic flow, and several models have been put forward to avoid the dilemma.

Intercellular Clefts. The currently accepted view is that of Diamond and Bossert based on the observation that the filling of the intercellular clefts of transporting epithelia, such as that of the gall-bladder (Chapter 9), varied with the flow of fluid across. Thus inhibition of active transport of Na^+, and hence of flow, caused the clefts to become thinner, suggesting that the primary process of active transport of Na^+ was into the intercellular cleft, as illustrated by Fig. 2.29. On the basis of Fig. 2.29., salt passes passively into the cell and is transported actively into the cleft where it creates a hypertonic fluid. Osmosis of water takes place simultaneously, to create a gradient of osmolality along the cleft, and it is possible to calculate that the fluid finally emerging from the cleft could have an osmolality very little different from that of the fluid on the inside of the epithelium, i.e. that transport could be iso-osmotic. It could be shown that this iso-osmolality of the transported fluid was just one of several possibilities, so that a hypo-osmolal or hyperosmolal fluid could be transported on the assumption of different parameters, such as rate of Na^+-transport and permeability to water. Some epithelia produce strongly hyperosmolal fluids, such as that produced by the salt gland of birds, whilst others may produce a hypo-osmolal fluid, such as the sweat and some salivary glands.

The Mechanism of Active Transport

We have so far spoken of active transport as an uphill pumping of an ion, such as Na^+, against a gradient of electrochemical potential, but we have not suggested any possible mechanism for the process, and with reason since there is very little that is definite which can be said. It must be emphasized that active transport is not by any means confined to ions; we shall see that the intestinal absorption of sugars and amino acids takes place against large gradients of concentration, and the same is true for the absorption and elimination of these metabolites in the kidney. The high degree of specificity of active transport, through which the transporting system can discriminate between such similar ions as Na^+ and K^+, and Ca^{2+} and Mg^{2+} and (where non-electrolytes are concerned) between dextro- and laevo-isomers of sugars and amino acids, demands the existence of specific carrier sites in the membrane to which the transported solute has special access. If the number of sites is limited the phenomenon will show self-inhibition and mutual competition between solutes (just as with facilitated transfer) and these phenomena are, indeed, observed.

Energy Link

This access to a carrier may favour transport across the membrane but of course it cannot of itself cause accumulation against a gradient of electrochemical potential. Such a process demands the intervention of a metabolic supply of energy, so the problem resolves itself into how the metabolic process is linked to transport. We shall see that a similar problem arises in relation to contraction of muscle. Once again the substrate ATP is concerned, and there is a specific enzyme, ATPase, in the contractile machinery which catalyses the chemical process that involves ATP in the contractile process, although the intimate nature of the link is still a matter of speculation.

Active Transport Model

In general terms it is easy to build a model of active transport of an ion or uncharged molecule, as in Fig. 2.30; here the first step is the linkage of the transported solute to the membrane "carrier" C. On this carrier complex the solute moves across the membrane and at the opposite side it meets with a chemically reacting system that reduces the affinity of the carrier for the solute. As a result the solute is released and escapes from the membrane. Thus the energy-requiring link is the chemical change in the carrier-solute complex that alters its stability and favours release. The enzyme ATPase would be involved in cata-

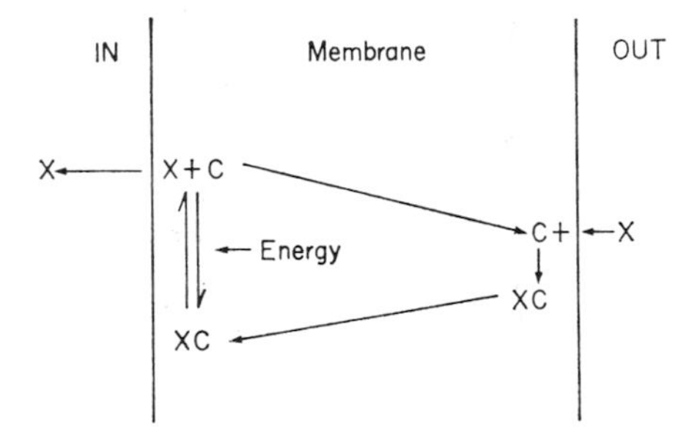

Fig. 2.30. Basic model of possible mechanism by which active transport works. The first step is the linkage of the actively transported solute to the membrane carrier, C. This diffuses down its concentration gradient to the opposite side of the membrane. Here it becomes involved in a chemical reaction that results in the release of the solute, which now diffuses out of the membrane. The free carrier diffuses to the original side of the membrane and acquires another solute molecule to repeat the cycle. The chemical reaction may involve a change in shape of the carrier, leading to a reduction in its affinity for the solute; moreover the carrier is probably fixed in the membrane, so that "diffusion" of the complex in the membrane really consists in shuttling the solute from one part of the carrier to another.

lysing the reaction where this was sensitive to cardiac glycosides, and of course ATP would be a substrate. The fact that ATPase is probably a fixed component of cell membranes at the site of active transport suggests that this protein molecule provides a suitable "nest" within its globular structure, in which the reaction can occur, just as with the oxygenation of haemoglobin (Chapter 7). It may be that the microenvironment provided by this "nest" accentuates the chemical difference between similar ions, e.g. between Na^+ and K^+, and thus provides the basis for the selectivity of active transport. The ATPase may be isolated from the membrane fraction of homogenized cells, and it may be solubilized by treatment with detergent. In this soluble condition, however, it loses its activity, and it is interesting that Whittam has shown that this may be returned by addition of a lipid, phosphatidyl serine, to the preparation. Thus the activity of the enzyme depends on its association with a lipid constituent of the cell membrane. As we have indicated earlier, the concept of the carrier as a moving molecule is only a useful abstraction; the thinness of the membrane excludes any large translatory movements, and it seems likely that the carrier is fixed, and that the transported molecule moves along it to be finally released when the change in affinity is brought about at the appropriate end.

CHAPTER 3

The Energy Requirements of Life

THE CONSUMPTION OF ENERGY

The living organism consumes energy so long as it is alive; death is the sign that energy has ceased to be utilized. The energy is required for many different processes, some of which are obvious such as the locomotory movements, the contraction of the heart, and so on, whilst the nature of others requires more elaborate physical and chemical analysis for its demonstration, e.g. the conduction of a nerve impulse, the elaboration of special secretions, the operation of the "sodium pumps", and so on.

Sources of Energy

Higher organisms make use of the energy derived from chemical reactions involving the oxidation of three classes of compound, namely carbohydrate, protein and fat. Thus the carbohydrate glucose, when oxidized to CO_2 and H_2O, is capable of liberating a certain amount of energy, which we may measure in ergs or, more usually, in calories; the amount of energy that is theoretically made available for useful work is called the *free energy* of the reaction, and is fairly close to, but less than, the heat of combustion of the sugar. Thus, by burning the glucose in a calorimeter we can measure the total heat of the reaction, and by suitable thermodynamic calculations we may estimate the free energy available to the organism when it utilizes, say, one mole.

Efficiency

As to whether the organism can utilize all this available energy in the performance of useful work will depend on the efficiency of its chemical processes. Thus when a muscle contracts a certain amount of potential energy is used up from its chemical stores, but also some heat is evolved which, from the point of view of the useful work done by the muscle, is useless. It is found, in this case, that even when the muscle is working at maximum efficiency, only about 20 per cent of the energy liberated by the chemical reactions is utilized as useful work, the rest being dissipated as heat. Thus the efficiency of the skeletal muscle is about 20 per cent.

First Law of Thermodynamics

In general, the *First Law of Thermodynamics* tells us that energy may exist in many forms, including heat, but that in physical processes there is no loss of energy on converting from one form to another. Hence, with heat being regarded as a form of energy, we have the simple equation:

$$E = W + H.$$

E is the total energy made available by the chemical reaction we are considering, W is the work done, and H is the heat liberated. If the organism does no work, then all the energy made available appears as heat.

Oxygen Consumption

The oxidation of glucose is the most important source of energy in higher organisms, so that if we knew how much oxygen was used up by the animal in a given period we should have a fair approximation to the total utilization of energy. Thus according to the reaction:

$$C_6H_{12}O_6 + 6O_2 \rightarrow 6CO_2 + 12H_2O + 673 \text{ k calories}$$

it is easy to calculate that 1 litre of oxygen used up should provide 5·05 k calories of useful energy plus heat; thus a knowledge of the rate of consumption of oxygen, which is easily measurable, enables us to estimate the rate at which energy is made available to the animal. Actually, fat and protein are made use of too, and since the calorific value of a litre of oxygen used in the oxidation of these compounds is different from that of oxygen used in the oxidation of carbohydrate, the estimate of available energy would be wrong without some correction.

Respiratory Quotient. However, the amount of protein utilized during the period of study can be determined from an analysis of the amount of urea in the urine, whilst the amount of fat consumed can be estimated from the *respiratory quotient*, i.e. the ratio of the volume of CO_2 produced divided by the volume of O_2 consumed. Thus the respiratory quotient (RQ) of an animal oxidizing carbohydrate alone would be given, from the stoichiometric equation above, by the ratio:

$$\frac{6 \text{ volumes of } CO_2}{6 \text{ volumes of } O_2} = 1$$

With fat, the ratio is about 0·71, so that if only carbohydrate and fat were being utilized the proportions could be estimated from the measured respiratory quotient. By making use of the estimated amount of protein consumed, a *non-protein RQ* may be computed, and from this the calorific value of a litre of O_2 may be calculated. Some values are shown in Table I.

TABLE I

RQ (Non-protein)	Calories per litre of O_2
0·71	4·69
0·80	4·80
0·85	4·86
0·90	4·92
0·95	4·98
1·00	5·05

Basal Metabolic Rate

When the oxygen-consumption of an animal is measured at rest, i.e. when there is no obvious mechanical work being performed beyond that involved in beating of the heart, breathing, and so on, we have what is called the basal metabolic rate. As one might expect, the magnitude of the basal metabolic rate depends on the size of the animal and, in man at any rate, it seems to be linearly related to the area of the body. It is as though the amount of energy used up were related to the amount of heat that was lost from the surface; the bigger the surface the bigger the loss of heat and therefore the greater the amount of energy consumed.

Heat Production

This is too simple a view of the matter, especially when animals of different species are compared, but it does emphasize one fact, namely the *production of heat* by the animal. How is this related to the utilization of energy? Under these so-called basal conditions and, indeed, under many other conditions, the heat liberated by the animal is equal to the energy consumed; if the O_2 consumed is a precise measure of the energy made available by chemical reactions, i.e. if no anaerobic processes contribute to the energy consumption, then the heat liberated from the animal is equal to the calorific value of the O_2 utilized. Why is this? Simply because the animal is not changing what we may call its *energy status* during the period of study. During this time very many chemical reactions are taking place, each providing useful energy which is employed in a variety of functions, e.g. the contraction of the heart, the transmitting of nerve impulses, the synthesis of secretions, the active transport of ions, and so on.

Energy Dissipation. At first thought we might say that the heat liberated in the animal would be less than the total heat of the chemical reactions, because work was being done. However, over any lengthy period all these manifestations of useful employment of energy lead to heat; and thus the energy originally manifest, say, as pressure in the heart is dissipated into heat because the pressure is converted into an increased flow of blood, which subsides again because of the frictional resistance—the kinetic energy being converted into frictional heat. In a similar way other processes, initially causing the potential energy of the animal or its parts to increase, are followed by the dissipation of this potential energy to heat; thus the active transport of ions leads to the maintenance of concentration gradients, but these are continuously being dissipated by downhill leakage, so that the energy continuously being expended appears finally as heat.

Work and Heat. A living organism, then, is utilizing energy continuously in order to maintain its *status quo*, and it is this rate of energy consumption that is its basal metabolic rate. This means using oxygen, which is involved in certain chemical reactions; during the chemical processes some of the energy appears immediately as heat, but a part appears as useful work which, however, is dissipated finally as heat. Hence all the energy made available to the organism by its chemical reactions, during a period when its energy status on average remains unchanged, may be expected to appear as heat; and the classical experiments of Rubner, who measured the heat liberated by an animal in a calorimeter and compared this with the calorific value of the O_2 consumed, confirmed this within the limits of experimental error. If the energy-status of the animal does not remain on average the same during the period of study, the equality between energy consumed and heat produced will obviously no longer hold. Thus, if an animal diverts some of its energy intake to the laying down of stores of fat, as with the goose, then some of the energy involved in synthesizing the fats remains in the fat, and the heat liberated is less than that predicted from the oxygen consumption. If mechanical work is done, such as lifting weights, then the heat produced, although greater than the basal value, will not be equivalent to the chemical energy utilized. Even in this case, however, the heat production may equal the energy liberated if the weights are lifted and then brought back to their former level; the muscles relax, and the heat liberated during this process is equal to the potential energy of the weight. If the weight is lifted on to a shelf and left there, then the potential energy remains in the weight and the heat production is less.

Body Weight and Cell Density

The basal metabolism represents the sum total of the energy liberated by the metabolic reactions of the animal when it is in a steady state; clearly the two main factors determining the magnitude of this parameter will first be the weight of the animal—the larger it is the more cells will it contain and hence the greater the heat production—and second, the rates of chemical reactions taking place in the different tissues; these will vary considerably according to the density of cells in the tissue, being very low in such supporting structures as bone and cartilage and very high in muscle and nervous tissue; thus the lung, with a relatively large amount of connective tissue, has a Q_{O_2} of 8·6; the corresponding Q_{O_2} for brain cortex, with a dense accumulation of neurones, is 26·3. Here Q_{O_2} is defined as the number of microlitres (10^{-3} ml) consumed per mg of dry-weight of tissue per hour. Thus, when we compare one class of animals with another, we may expect to find differences due to the average weight of the animal and due to the different preponderance of tissue types; within a class, e.g. the mammals, we may expect the main variable to be the body-weight, since the proportions of the different tissues will not vary greatly.

Homeotherms and Poikilotherms

In so far as the differences between animal classes are concerned, the animal kingdom can be divided into two groups, the warm-blooded animals, or *homeotherms*, and the cold-blooded animals, or *poikilotherms*. The former have a generally higher rate of metabolism per unit weight than the latter; and this is manifest in the generally much higher body-temperature of this group. However, the important distinction between the two is not so much the difference of temperature and metabolic rate, but the fact that the warm-blooded animal, or homeotherm, tends to maintain a constant body-temperature in the face of wide fluctuations in its environmental temperature. By contrast, the cold-blooded animal, or poikilotherm, has a body-temperature that is approximately equal to that of the environment, varying with changes in this passively. Thus the "cold-blooded" fish, *Leuciscus thermalis*, living in the hot springs of Ceylon at a temperature of 50°C, has a body-temperature of 50°C and therefore one considerably greater than that of the warm-blooded man or dog.

Temperature and Metabolic Rate. In general, the rate of a chemical reaction increases with temperature; and this is certainly

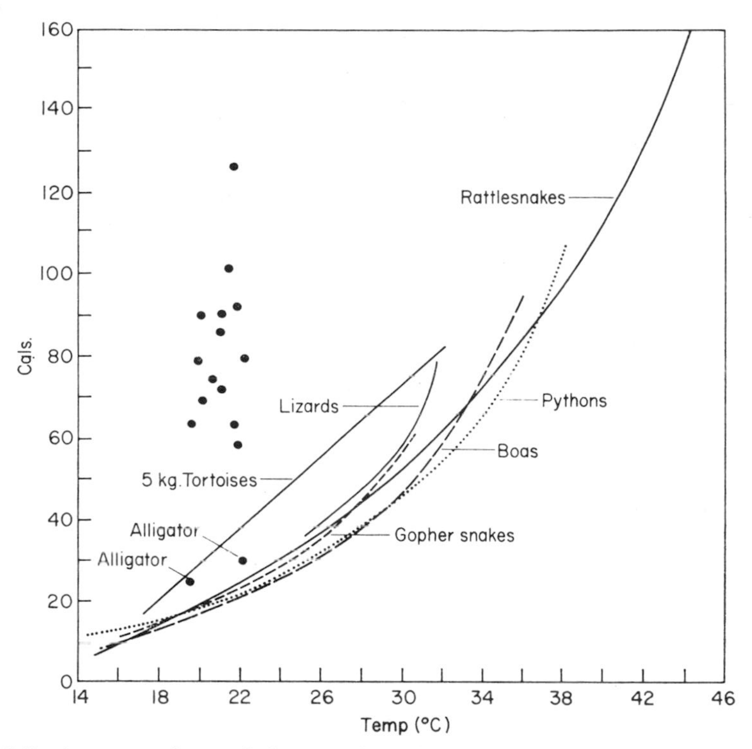

Fig. 3.1. A comparison of the standard heat production per square metre of body surface per 24 h, with reference to environmental temperature for a variety of poikilotherms. (Plotted points are for giant tortoises.) (Benedict, *Carnegie Inst. Wash. Rept. No. 425.*)

true of the metabolic reactions taking place in the living cell, so that to some extent the lower metabolic rates of poikilotherms are a result of their lower body-temperatures; in fact, however, there is a real difference in the metabolic rates when allowance is made for this; thus we may plot a curve of the metabolic rate for poikilotherms against environmental temperature and extrapolate this to 37°C; when the average for many species is taken, the curve extrapolates to

100 kcal/m^2/24 h (Fig. 3.1), whilst the corresponding value for homeotherms calculated on the same basis is 800 k calories. Again, it is possible to measure the metabolic rates of homeotherms, such as the dog or rabbit, at low body temperatures, such as 25–30°C when they are in a state of hypothermia; and these rates may be compared with those of poikilotherms of comparable size. Always the hypothermic homeotherm had a metabolic rate some 10 times greater. Thus, in the transition from poikilothermy to homeothermy there has been a fundamental change in the metabolic rates of the individual cells of the animal, quite apart from changes in the anatomy and physiology that serve to conserve or dissipate heat from the surface of the body (*physical thermoregulation*).

HOMEOTHERMS, OR WARM-BLOODED ANIMALS

Body-temperature

The body-temperature of the homeotherm is commonly stated to be about 37°C, and by this is meant what the physiologist calls the *core-temperature*, the temperature of the blood in the deep regions of the body, by contrast with the skin-temperature, which quite obviously varies over a wide range in accordance with the environmental temperature and other factors. Thus, with a bird standing on ice the skin-temperature may well be below 0°C, and that of the blood flowing through the foot will also be close to this; again, the skin-temperature of a man in hot sunshine may well be above his core-temperature, at any rate for some time. When different species are compared, it is found that the core-temperature varies between about 36° and 38°C; birds are said to have temperatures in the region of 41° to 42·5°C, whilst certain primitive mammals, like the egg-laying platypus and the spiny ant-eater (echidna), have temperatures of 30° and 28·5°C respectively; and it is interesting that their metabolic rates are smaller, that of echidna being only 300 kcal/m^2/24 h compared with 800 k calories for mammals of higher temperature and comparable size. Perhaps more interesting is the fact that these primitive mammals do not maintain their temperatures constant in the face of thermal stress, be it cold or heat, so that it would seem that the development of homeothermy runs parallel with the development of a high body-temperature.

Value of High Body-temperature

We may now ask why it is that a body-temperature normally higher than that of the environment is a physiological prerequisite for efficient regulation, i.e. for homeothermy? One factor is that we then have a system that allows accurate regulation; with a temperature higher than the environmental, a steady production of heat is necessary, and the temperature may be maintained constant (in the face of varying external temperatures and humidities) either by varying the rate of loss of heat from the surface, or by modification of the rate of production of heat.

It will be clear that with the temperature of the body fixed at a value close to the average environmental temperature, say 25°C, the possibilities for control are much more restricted; a rise in environmental temperature would have to be combated by a decrease in metabolism, unless the loss of heat from the surface could be increased, but this would be difficult without a large rise in body-temperature. It could well be that the reduction in metabolism, already low because of the normal body-temperature of 25°C, would reduce the activity of the animal and impair the chance of survival.

In addition to permitting more delicate thermoregulation, the higher body-temperature permits a more intense metabolic activity—and of course requires it—and this doubtless permits the animal to adapt and respond to its environment much more efficiently, e.g. the rate of conduction of a mammal's large A-type nerve fibre is some 100 m/sec; in a comparable nerve fibre of a frog it is some 40 m/sec. The rapidity with which a warm-blooded muscle can contract is also higher.

Ideal Body-temperature

It would be very difficult, without a great deal more knowledge, to say what is an ideal body-temperature, but we may conclude that at 37°–38°C a balance has been struck in which various factors have had their influence. Thus, too high a temperature might well make too great a demand on the energy supplies of the animal, and this factor is well seen in the phenomenon of *estivation* when an animal will allow its temperature to fall if its energy supplies are failing; too low a temperature, on the other hand, may restrict the powers of thermoregulation and also the physical and chemical activities of the cells, making the power of the animal to react to its environment too limited.

Variations in Basal Metabolic Rate

Surface Area

In man we have seen that a good prediction of his basal metabolic rate may be made on the basis of his surface area, which is closely related to his height and weight. This finding, and comparable studies on some lower animals, gave rise to the concept of the *Surface Area Law*, according to which it was argued that the heat production of the animal was adjusted to the rate at which it could lose heat from its surface. This concept, especially when applied to the metabolic rates of different species of animals, poses too many questions to be of value. Thus, are we to agree that the heat production of a polar bear is the same as that of a man of equal surface area? If so, under what conditions are we to make the comparison—with the man naked and both exposed to the same environmental temperature? What temperature should be used—the polar bear's normal winter one or one of 20°C? Thus, only by choosing some rather arbitrary experimental conditions will comparisons be feasible, and when made, their significance will lie in relation to the more complex problems that the species has to solve, namely that of matching its heat production to its heat loss consistent with maintaining a constant body-temperature.

Metabolic Rate and Environmental Temperature

The concept of a basal metabolic rate is essentially an ideal; it represents the heat production in unit time when the organism is not changing its potential energy status; it is essentially the minimum energy consumption while at rest. If the metabolic rate of an animal is measured at different temperatures, it is found that this is by no means constant, but tends to increase at low temperatures, a phenomenon described as *chemical regulation*, since it is a response that favours

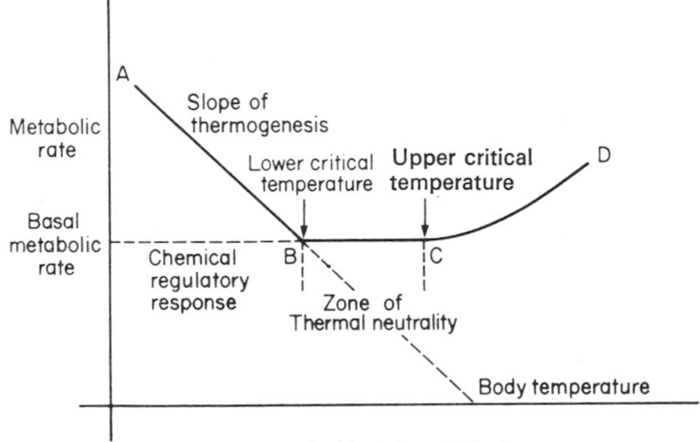

Fig. 3.2. A curve of metabolic rate against ambient temperature. Metabolic rate is constant only over a limited range, the zone of thermal neutrality. Metabolic rate is increased below the lower critical temperature (B) as chemical regulation is necessary to balance heat loss, and it is also increased above the upper critical temperature (C) when there is increased sweating, respiration, etc., as the animal strives to combat the rise in temperature.

maintenance of a constant body-temperature in the face of increasing loss of heat to the environment. When metabolic rate is plotted against ambient temperature, the graph has a typical form, and is illustrated schematically in Fig. 3.2. Over a limited range, called the *zone of thermal neutrality*, the rate remains constant; below this, it rises linearly, the *chemical regulatory response*. The point at which the metabolic rate begins to rise is called the *lower critical temperature* and varies with the species; it is essentially the temperature at which chemical regulation becomes necessary and is obviously related to the insulation of the animal's coat. Thus, as Fig. 3.3 shows, with a poorly insulated animal

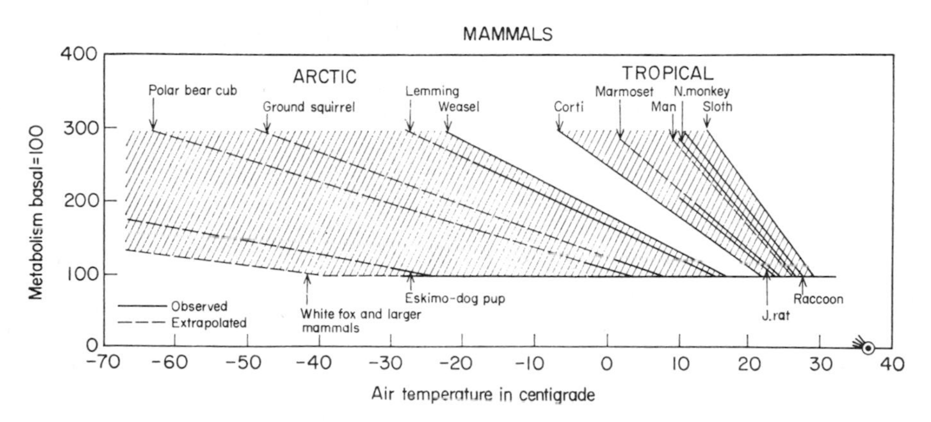

Fig. 3.3. Heat regulation and temperature sensitivity in arctic and tropical mammals. Note that the fox only needs a small increase over its basal metabolic rate in order to be able to withstand the coldest temperatures. (Scholander *et al.*, 1950, *Biol. Bull* 1950, **99,** 237.)

like the sloth, chemical regulation begins as soon as the ambient temperature is reduced below 30°C, but in the eskimo dog pup the metabolic rate remains constant until the ambient temperature is below —20°C; moreover the *slope of thermogenesis*, i.e. the steepness of the line relating metabolism to temperature, is small for the well insulated animal and steep for the poorly insulated animal. Figure 3.3 illustrates very well the nature of the metabolic response of the animal to cold. The metabolic response to heat is not a reduction in heat production, in fact, above the zone of thermoneutrality, i.e. beyond the upper critical temperature of Fig. 3.2, the metabolic rate rises because of the extra physiological activity that comes about while the animal is fighting against a rise in body temperature, i.e. sweating, increased respiration, flapping wings, and so on.

Species Variations

When comparing different species with respect to their basal metabolic rates, the experimenter quite arbitrarily chooses as temperature for measurement the zone of thermal neutrality, the region in which the animal is neither fighting against cold nor against heat. When this is done, and the metabolic rate is plotted as a function of the animal's weight, as in Fig. 3.4, it is seen that the metabolic rate per unit weight of the animal increases as the animal becomes smaller; thus

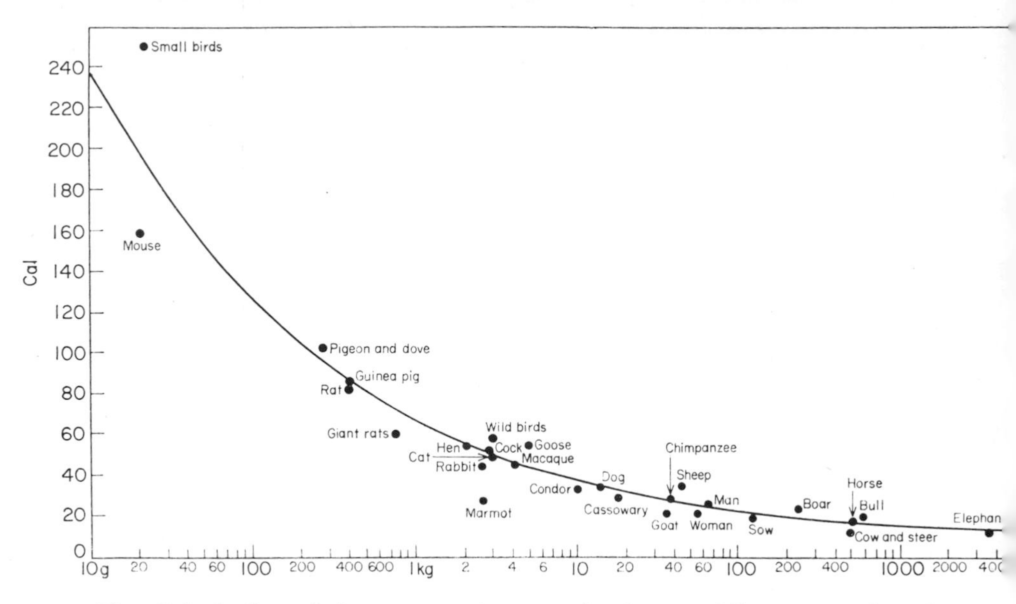

Fig. 3.4. A plot of the average heat production per kilogram against the logarithm of body weight for a variety of homeotherms. The metabolic rate per unit weight of animal increases as the animal becomes smaller. (Benedict, *Carnegie Inst. Wash. Rept. No. 503.*)

1 gramme of mouse metabolizes at about ten times the rate as 1 gramme of elephant. Now the area of body-surface exposed by 1 gramme of mouse is much greater than the area exposed by 1 gramme of elephant, so that on simple thermodynamic grounds we must expect the mouse to show the greater metabolic rate per unit of weight if the two animals are to maintain the same temperature, provided the physical conditions favouring dissipation of heat from the animals are the same.

"Surface Area Law". If the Surface Area Law applied, i.e. if heat production at thermal neutrality did, indeed, match the body area, then we should expect the metabolic rates, expressed as calories of heat liberated per sq. metre of surface, to be the same. Figure 3.5 shows that this is not true, although it is very nearly so for quite a large group of species whose basal rates are of the order of 800 kcal/m²/24 h; as the animal gets very large or very small the basal rate becomes greater or lesser than this, showing that the surface area law does not apply strictly. This is understandable, as indicated earlier, since the surface area is only one factor determining loss of heat. Moreover, the weight of the animal is not the only factor determining the total

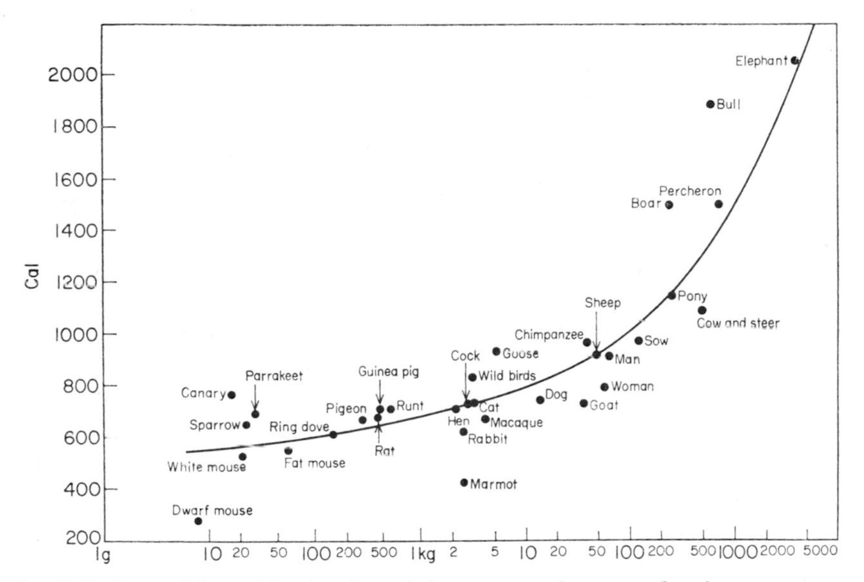

Fig. 3.5. A semi-logarithmic plot of the average heat production per square metre surface area for 24 h against body weight in a variety of homeotherms. Note that the "surface area law" is true for a large group of animals, but that for the very large and very small, basal heat production deviates from 800 kcal/m²/24 h. (Benedict, *Carnegie Inst. Wash. Rept. No. 503.*)

metabolism, since as an animal gets larger the proportions of slowly metabolizing supportive structures to actively metabolizing tissues probably change. As a general rule, all we can say is that, as an animal gets bigger, its metabolism per unit weight decreases (Fig. 3.4); on geometrical grounds its area per unit weight decreases too, so that the loss in body-temperature that the large animal might encounter due to its lower metabolic rate per gramme is countered to some extent by the reduced area of heat dissipation. However, these two factors do not match each other, with the result that metabolic rate is not a simple linear function of body area (Fig. 3.5).

Empirical Relationship. The actual relation between metabolic rate and body weight is given by the experimentally derived equation of Brody and Procter:

$$\text{Log metabolic rate (kcal/24 h)} = \log 89 + 0{\cdot}64 \log \text{wt}$$

i.e. the metabolic rate is proportional to the body weight raised to the power of 0·64. Since the most probable estimate of the surface area of an animal, irrespective of species, is proportional to the weight raised to the power of 0·66, it is easy to see how closely the surface area law may be applicable over a limited range of body weights.

Very Small Animals. When the animal is very small, such as humming birds and bats, the heat production per unit of weight is very large, and it has been estimated that in an animal with a weight as low as 2·5 g (small humming bird) survival would not be possible without some special "legerdemain" simply because the food supplies would not be available. The "legerdemain" consists in the abandonment of homeothermy at suitable periods; thus bats in activity have the normal homeotherm's body-temperature but when they retire to their caves their temperatures fall to that of the environment so that they become temporarily poikilotherms. Similarly the humming bird, when active, maintains a typical homeotherm's body-temperature, but at night it abandons temperature control and passes into a torpid state. Arousal from torpidity occurs well before daylight, and so is not triggered off by a rise in environmental temperature, but is rather the consequence of an inbuilt diurnal rhythm.

Chemical Regulation

Chemical regulation has been defined as the increase in metabolic rate in response to a reduced ambient temperature. As the generalized curves of Fig. 2.3 indicate, this is only brought into play when the thermal stress reaches a certain value, so that the animal usually relies on reducing its heat loss as the initial response, i.e. by the process of physical thermoregulation. When man is subjected to reduced ambient temperatures, his metabolic rate remains remarkably constant until a point is reached when shivering begins, and the muscular effort involved in this increases heat production. Thus, in man, chemical regulation consists in shivering, but in lower species there is no doubt that cold causes an increased metabolic rate long before any shivering is manifest; this is especially prominent in very young animals and in human infants, and it is interesting that in these cases the extra energy is provided by the oxidation of reserves of brown fat, which disappear as the animal matures.

Physical Thermoregulation

Let us consider some of the basic physical principles concerned in the regulation of heat loss by the warm-blooded animal. Heat flows from a region of higher to lower temperature by conduction, convection and radiation.

Conduction

By conduction we mean the direct transfer of heat from one molecule to another or one atom to another; for example, this occurs in a liquid, by the circumstance that in the hotter region the molecules vibrate and move with greater energy than those in the cold region, and the "hotter" molecules transfer some of their heat energy, by impact, to the "colder" molecules. In a solid it is a question of transmission of vibrational energy, whilst in a fluid, be it liquid or solid, it may be a matter of vibrational or translational energy or both.

Convection

Heat may be transmitted to a fluid not only by this conductive process, but by *convection*, which involves a definite flow of the fluid carrying the heat with it. Thus, when a current of cold air is passed over a hot body, the main loss of heat may be due to this convective process, but of course conduction from the surface to the air takes place by the transmission of vibrational energy from the solid to the air molecules. In the absence of any forced current, there will still be some convection because of the pull of gravity on the fluid which results in the rising of warm air and falling of cold air; this *natural convection*, as it is called, is a necessary component of conductive losses when fluids are concerned. Because the mechanisms of heat transfer in conduction and convection are so similar, it is often convenient to describe conductive plus convective heat loss simply as "convective".

Radiation

Finally heat may be lost by radiation from the surface of a body, the energy being transmitted in the form of infra-red electromagnetic radiations.

Evaporation

An additional and often very important means of heat loss is through evaporation; here no difference of temperature between surface and ambient air is necessary, the heat transfer—often called *insensible heat*—being associated with a change of state of water, so that, approximately, the heat loss is given by the weight of fluid evaporated times its latent heat of evaporation.

Cooling Equation

In general, the sensible heat transferred by all three processes of conduction, convection and radiation may be approximately described in terms of simple equations of the form:

$$C = kS(T_s - T_a)$$

where C is the rate of transfer of heat, e.g. in kcal/sec, T_a is the ambient air temperature, T_s is the surface temperature, S is the surface area, and k is a "cooling constant" whose meaning will depend on the mode of losing heat. When heat is being lost by all processes together it may be described, then, by Newton's empirical equation

$$\text{Rate of loss of heat} = H' = kS(T_s - T_a)$$

or, expressing heat loss in terms of k calories per unit area of body surface:

$$H_s/S = H = k(T_s - T_a).$$

Conductance and Insulation

The cooling constant defined in this way is often called the *body conductance*, indicating as it does the speed with which heat is carried away from the body to its environment, but strictly speaking the term conductance should be confined to the cooling constant when transfer of heat is by conduction alone. The reciprocal of the conductance is called the *insulation*, and represents the resistance to the flow of heat, being analogous with the resistance to the flow of electricity through a circuit.

The loss of heat from the animal's body takes place in three steps, i.e. transfer from core to skin, from skin to the layer of still air on the surface and finally from the layer of still air to the freely moving air adjacent to this. Each of these steps may be described by the simple Newton equation, with its appropriate cooling constant or conductance. Thus loss from core to skin may be written:

$$H_1 = k_t(T_c - T_s)$$

k_t being the conductance of the tissues and T_c the core-temperature.

Again, we have, $$H_2 = k_{Cl}(T_s - T_{Cl})$$

where k_{Cl} represents the conductance of the fur or clothing. Finally, the flow of heat from the clothing or fur will be given by:

$$H_3 = k_a(T_{Cl} - T_a)$$

where k_a is the conductance of the ambient air.

The three conductances, or insulations, are in series, and the total conductance, or transfer from core of the body to the ambient air, is given by $1/k_{Tot} = 1/k_t + 1/k_{Cl} + 1/k_a$ or the total insulation,

$$I_{Tot} = I_t + I_{Cl} + I_a.$$

Calorimetry

Modern techniques of calorimetry applied to man and animals have been devoted to estimating these three insulations or conductances, and variations in these. Just as important, if not more so, is the separation of the several forms of heat exchange from each other; thus heat may be lost or gained by radiation, and to separate this heat exchange from that due to conduction, convection and evaporation often requires some experimental skill. When it is desired to study losses uncomplicated by radiation we may adopt the simple expedient of enclosing the animal in a chamber with its walls kept at the same temperature as the skin temperature. We shall not enter into details regarding the mechanisms of measurement beyond pointing out that modern techniques permit the extremely accurate and rapid measurement of the sensible heat lost from the body, whilst the loss resulting from evaporation can also be assessed rapidly and easily, especially if conditions are chosen such that all sweat secreted is evaporated, in which case the loss of body-weight is a good index to water loss over a short period of time.

Changes in Conductance or Insulation

The conduction of heat from the core to the periphery involves two main factors, namely conduction in its true sense, and transfer of heat by warm blood flowing into cooler regions, or *vice versa*. It is this latter factor that may be varied physiologically and constitutes a principal means of heat dissipation during severe exercise, acting in cooperation with the evaporative process that requires a high skin temperature for its effectiveness. Changes in the blood flow to the periphery are important for heat conservation too; the blanching of the human skin in the cold is an obvious example, whilst the arrangement of many arteries and veins in a countercurrent system is an extremely interesting aspect of heat conservation.

Countercurrent Flow

Thus the blood returning from a cold limb along *venae comitantes* that run parallel with, and close to, the arteries, tends to cool the arterial blood before it reaches the cold extremity and so prevents it from giving up as much heat as it would have done had it passed to the surface some distance from the cold venous blood. The situations for the limb exposed to cold (10°C) and warm (30°C) environments are

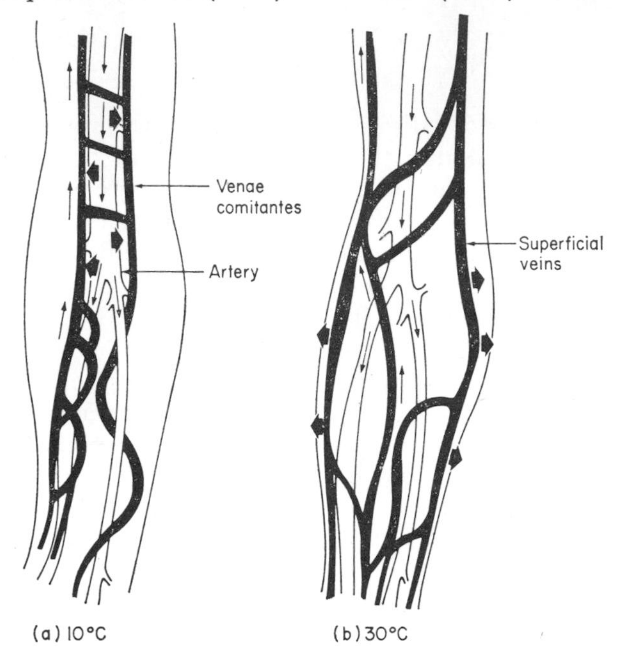

Fig. 3.6. Heat conservation in the human forearm. Blood returning from the periphery can follow two routes. If heat is to be conserved, the venous return is via the venae comitantes as in (a) so that heat is gained from the arteries and the core temperature is maintained. If heat is to be dissipated, a more superficial route can be taken as in (b) and the blood returning to the heart will be cooled. Thick arrows indicate heat exchange. (Selkurt, E. E. (Ed.), "Physiology", Little Brown & Co. Boston, 1966).

shown in Fig. 3.6. In (a) the return blood flow is mainly through the venae comitantes so that the arterial blood is cooled on its way to the extremity by giving heat up to the returning venous blood; thus the arterial blood loses less heat to the environment than it would have done if it had arrived in the extremity at its core-temperature of 37°C. In (b) the returning blood has been shunted into peripheral veins (so that the arterial blood retains its high core-temperature as it passes through the extremity) and thus tends to dissipate heat.

If it is appreciated that an Arctic bird standing on ice must have a skin temperature in its feet of around 0°C, the loss of heat to the feet would be very great if blood at 37°C were reaching the skin; instead it is "pre-cooled" by the returning cold venous blood. Clearly the system could be used in reverse to assist heat dissipation if blood, normally running in venae comitantes, were diverted to other veins.

Layer of Still Air

In animals with hairy or feathered coats, and in clothed man, the insulation, I_{Cl}, is of some interest since it, too, is variable; essentially the insulation of fur is equal to that of the layer of still air that it can contain and this depends on the length of the hairs and the number per unit area. By preventing a bird from fluffing its feathers it may be shown, experimentally, that its capacity to fight against cold is considerably reduced. Similarly, the wetting of the fur, by driving out the enclosed air, reduces insulation; with furry animals that normally live in water, such as seals, the young pups do, in fact lose their insulation in water; older animals, however, develop a layer of dense water-repellent underfur, and above this lie wettable guard hairs that form a sleek covering giving little resistance to swimming. Thus young seals are very ineffective in coping with cold.

The insulation of the ambient air depends on its velocity; hence in adjusting itself to its thermal environment the animal may increase the insulation of its ambient air by avoiding draughts or huddling with others; alternatively the insulation may be decreased by fanning with the wings, as with bats.

Evaporation

The important factor governing heat loss by evaporation is the humidity of the atmosphere. In fact the heat loss can be represented by the simple equation:

$$E = k_e(P_s - P_a)$$

P_s being the vapour pressure of the sweat and P_a that of the atmosphere. In a hot and dry climate the sweat will evaporate quickly and so permit ready loss of heat; by contrast, when humidity is high the loss of heat by this mode is restricted, so that a man can tolerate only a temperature of about 32·5°C (90°F) at high humidity compared with 45°C (113°F) at very low humidity.

Sweat Secretion. Loss of heat by evaporation occurs continuously, in the absence of sweating, as a basic *insensible perspiration;* at a certain critical point of heat-loading, this evaporative heat loss increases abruptly as a result of the secretion by the sweat glands of the skin (Fig. 3.7). When the humidity of the ambient air is very high, evaporation is seriously reduced, and the animal must rely on the losses by radiation and conduction.

Other Mechanisms. Many animals, typically the dog, have no sweating apparatus, and they rely on other means of evaporative heat dissipation, such as panting, the saliva being evaporated from the tongue whilst the tidal flow of air (p. 383) is sufficiently small as not to interfere with gas exchanges in the lungs. Bats flap their wings and lick their fur, whilst marsupials, besides panting, will lick their fur to cause evaporative heat loss.

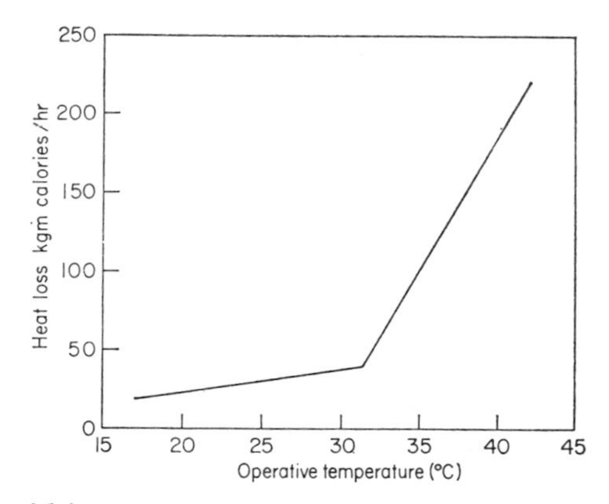

Fig. 3.7. The rapid increase in heat loss as a result of secretion of sweat by the sweat glands of the skin.

Radiation

So far as animals and man are concerned, radiation is a two-way process, the animal either gaining heat from its environment or losing heat to it by this mechanism. The commonly employed equation governing exchanges between two surfaces, e.g. the surface of the skin and the surrounding walls of a room, is that of Stefan:

$$R = k_r A_r (T_S^4 - T_W^4)$$

where k_r is a constant depending on the emittances of the surfaces involved in the radiational exchanges and A_r is the effective area of the body from the point of view of receiving or giving out radiational energy; this is by no means equal to the geometrical area. Thus an animal rolled up into a ball radiates over a much smaller area than one stretched out. The importance of radiational exchanges in determining the comfort of man is easily seen by comparing rooms with identical air temperatures but warm or cold walls. In general the energy absorbed by animals is that from the sun, in which case the energy extends over a large band of wavelengths from the ultraviolet to the far infra-red; the energy exchanged with, for example, the walls of the room is exclusively in the far infra-red.

Skin Colour. So far as this latter form of energy is concerned, the absorbance, or emittance, is independent of the colour of the skin;

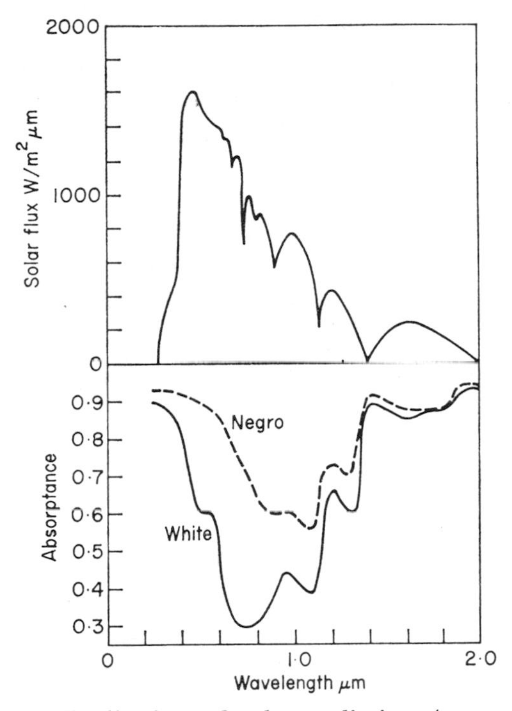

Fig. 3.8. Spectral distribution of solar radiation (upper curve) and of absorption of white and negro skin (lower curves). (Kerslake, "Stress of Hot Environments", CUP, redrawn from Gates and Jacquez *et al.*)

when sunlight is concerned, however, the black skin absorbs some 80 per cent of the sun's radiant energy compared with 60 per cent by the white skin (Fig. 3.8) so that the pigmented skin increases the heat-load but has the advantage of absorbing the ultraviolet light more effectively and thus protecting the skin from its deleterious action.

Operative Temperature

As indicated above, when a subject is in a room with its walls at a different temperature from that of the ambient air, the thermal demands may be markedly different according to the wall temperature, and for this reason an *operative temperature* has been defined that expresses, by a single value, the effective ambient temperature; thus with an air temperature of 20°C and a wall temperature of 40°C the operative temperature will be higher than with a wall temperature of 10°C. This operative temperature, T_o, is such that the sensible heat transfer is governed by a Newton-type equation:

$$H = k_o(T_s - T_o)$$

H being the heat transferred by convection and radiation, and k_o being an *operative heat exchange coefficient*, being made up of corresponding coefficients in the convection and radiation processes.

Heat Storage

When an animal is thermally in a steady state with its environment, the heat produced in unit time, M, is equal to the heat lost by all routes:

$$M = E + R + C + K$$

where E is the evaporative heat loss and R, C and K are those due to radiation, convection and conduction respectively. If M is given a positive value, then clearly heat losses must be given positive values too* to make the equation balance, whilst a gain of heat from the environment must be indicated by a negative quantity (it is a negative loss).

When the animal is not in a steady state it will be gaining or losing heat from the environment, so that we must introduce into the above equation a *storage term*, to make the two sides balance:

$$M = E + R + C + K + S.$$

It is clear that if the total heat loss is greater than that produced, the animal will be losing heat and the storage term will be negative. To choose an example, the metabolic rate is 100 kcal/m²/h and the heat exchanges are: E = 40; R = −20 and C + K = 90 kcal/m²/h. Thus the animal loses 130 kcal/m²/h to its environment and gains 20 kcal/m²/h by radiation. The total loss is thus 110 kcal/m²/h compared with a production of 100 kcal/m²/h and the storage is −10 kcal/m²/h:

$$100 = 40 - 20 + 90 + S$$
$$S = -10.$$

* In adopting this convention regarding signs we have followed Kerslake's authoritative monograph *The Stress of Hot Environments*. Unfortunately another convention treats net gain of heat by the animal as a negative heat storage, and correspondingly heat lost by the animal is made positive.

Work

When the animal is doing work, then unless the effects of this work are converted into heat which is measured in the calorimeter, we must deduct this from the metabolic term:

$$M - W = E + R + C + K + S.$$

Total Conductance

The facility with which an animal may lose heat measures its total conductance:

$$H = k'(T_c - T_a)$$

where T_c is its core-temperature and T_a is the ambient temperature. Clearly, if the animal is in a steady state and is not doing mechanical work, we may equate the heat loss, H, with the metabolic rate, M, when the animal's total conductance is given by:

$$\text{Metabolic rate}/(T_c - T_a) = k'.$$

Core-temperature

It will be recalled that, when speaking of body-temperature, we must distinguish between the *core-temperature*, which represents the temperature of the blood in the depth of the body, and the skin temperature, measured by a thermocouple on the surface of the skin. The latter varies widely according to the ambient temperature and physiological state of the animal, whilst it is the former—the core-temperature—that is held within fairly narrow limits of about plus or minus a degree centigrade. It has been common to use the rectal temperature as an

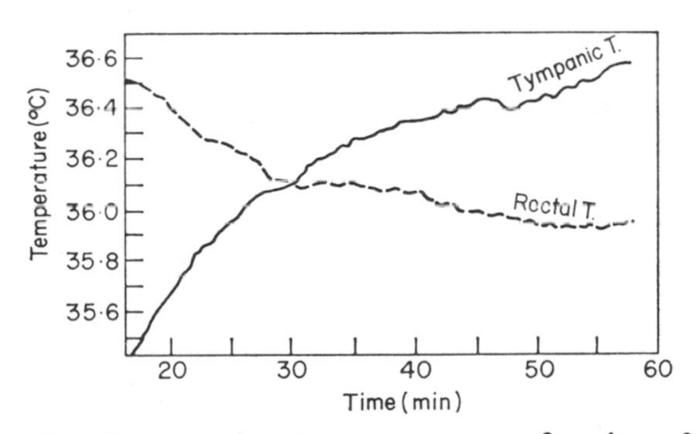

Fig. 3.9. Rectal and tympanic temperatures as a function of time after a diver, kept in a cold tank for 80 minutes, is removed into warm air at 40°C. Note how divergent they can be although both are often described as "core temperatures". (Benzinger, *Physiol Rev.* 1969, **49,** 671.)

index to core-temperature, but there is no doubt that this may be significantly different from the temperature of the blood in the large arteries, and Benzinger considers that the temperature recorded by a thermistor in the external auditory meatus, close to the tympanum—the *tympanic temperature*—is a far better index, especially because it is situated close to the internal carotid artery and should thus reflect the temperature of the blood passing to the brain. As we shall see, the main "sensor" for blood temperature is in the brain so that, in order to investigate the relation between, say, sweating rate and body-temperature, the temperature likely to give the most meaningful relation would be the intracranial temperature. The way in which rectal temperature may differ from tympanic temperature is illustrated in Fig. 3.9.

Temperature Lability

The constancy of the body's core-temperature is one aspect of homeostasis that permits the animal to function smoothly. As with other parameters, however, the constancy is not absolute, so that animals permit themselves a certain degree of hyperthermia when under heat stress, and hypothermia in cold stress; thus physiologically man's temperature may fluctuate by about a degree above and below the mean, and this is especially manifest in the hyperthermia of exercise.

Camel

In certain animals, much larger variations are permitted. The ability of the camel to withstand desert conditions so much better than man is due to its labile body-temperature; it allows its core-temperature to rise during the heat of the day to 40·7°C, and to cool during the night to 34·2°C, thereby conserving the water that would have been necessary in the evaporative cooling required to keep the temperature "normal" during the heat of the day.

African Ungulates

In general, when African ungulates were studied, two main "strategies" of defence against heat and conservation of water were found; large animals, like the eland, had very low body-temperatures in the morning—as low as 32·8°C—which increased slowly during the course of the day, so that they might never need to increase evaporation. Smaller animals, such as Grant's gazelle, allowed their temperatures to rise rapidly to as high as 42°C without either panting or sweating.

Animals of intermediate size, such as the oryx, used a combination of both methods. When the desert species were dehydrated, temperatures as high as 46·5°C were tolerated.*

Hibernation

We have already mentioned that very small animals, like bats and some mice, will abandon homeothermy, and their high body-temperature, in the interests of energy conservation; this abandonment is especially prominent in the "winter-sleep" or *hibernation* of such animals as the marmot, hamster, ground-squirrel, etc. In the laboratory, a marmot may be caused to go into hibernation by reducing the environmental temperature. During hibernation the body-temperature may fall to very little above the ambient temperature; thus with an ambient temperature of about 1°C, the core-temperature is about 2°–3°C; associated with this there is a profound fall in metabolic rate which may be as low as 20 kcal/m^2/24 h. Heart-beat and respiration are maintained, but are very slow and are adequate to meet the energy demands of the animal just because they are so low at this body-temperature. Arousal may be induced by an appropriate stimulus, indicating the effectiveness of the nerves and brain which, in a normal homeotherm, would have been blocked at the low temperature. Arousal is accompanied by an enormous burst of metabolic activity, which may reach 3000 kcal/m^2/h, and it is interesting that it is the special depots of brown fat that are metabolized at this time.

Brown Fat

This is the name given to depots of adipose tissue prominent mainly in the intercapsular region of newborn animals, but also in hibernating animals where it has been called the *hibernating gland.*† Thus, if thermocouples are placed in the brown fat of a hibernating marmot it is found that, during the burst of metabolism accompanying arousal, the temperature of this fat rises far more rapidly than that of other regions. The actual contribution of the fat to the whole metabolism is probably small, and it is its strategic location in relation to the blood vascular system (being well placed to warm the blood returning from the body to the heart) that accounts for its value. In newborn animals there

* It may be that the brain is spared from such high temperatures, the arterial blood running on its way to the brain through a *rete* or network of veins carrying blood from the cooler nasal region.

† It is also called *multilocular fat* because the fat is stored in the individual fat cells in the form of many droplets as opposed to the single large globule in white fat.

are stores of brown fat which may disappear later, e.g. in the rabbit it is replaced by white fat within seven days of birth. When newborn animals are exposed to cold, this brown fat shows the higher temperature and is responsible for a part of the increased metabolic rate that occurs, often in the absence of shivering. It is possible to acclimatize rats to the cold, in the sense that, after being maintained in a low-temperature environment, their basal metabolic rate at normal environmental temperature remains unusually high; in this condition it is found that the metabolism of brown fat is increased, so that the total energy provided by brown fat is some 5–8 times the control value, in spite of the fact that the animal is not shivering.

Physiological Mechanisms

The thermal adjustments that lead to the usual maintenance of a steady core-temperature involve definite physiological changes in the organism, which may include altered metabolic rate (with or without altered muscular activity), altered blood flow in the periphery, altered state of the hairs or feathers, altered respiratory rate and secretion of sweat. Obviously these physiological responses require a high degree of central coordination, and these will be discussed in Volume 3.

Adjustments in Man

Some idea of the responses of nude man to alterations in the thermal demands of his environment may be obtained by a careful study of Fig. 3.10 which shows the effects of altered operative temperature on several parameters, plotted as ordinates. Thus curve A indicates metabolic rate, and it is seen that, over the period studied, the subject made no attempt to alter this over the whole range in spite of very considerable alterations in the actual heat lost to the environment, i.e. the human subject did not exhibit the chemical regulation so manifest in many lower animals. Curve B shows that the heat exchanges due to radiation and convection varied almost linearly with the operative temperature. Curve C is the curve describing evaporative heat loss, which becomes serious as the environmental temperature rises, the sudden change occurring when the skin temperature was about 32°–33°C (not shown). Of most interest is the heat storage curve D which indicates whether the man was gaining heat (positive storage according to the present convention) or losing it. It will be seen that at low operative temperatures the storage term was negative, indicating "chilling" and thus showing the inadequacy of man's

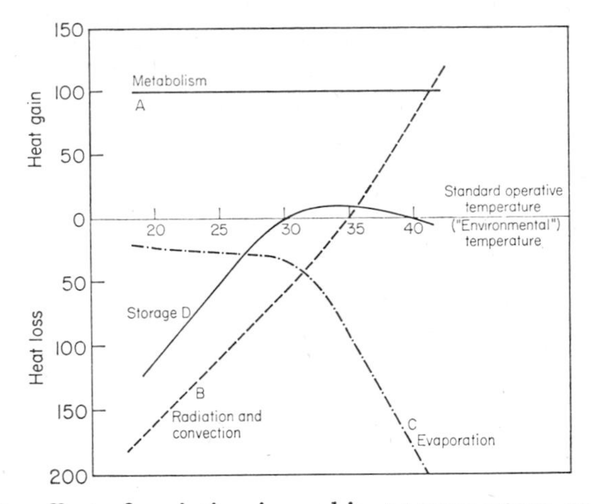

Fig. 3.10. The effect of variation in ambient temperature on an unclothed subject. Curve A is the basal metabolic rate which did not alter during experiment. Curve B showed that heat exchange due to radiation and convection varied almost linearly with temperature. Curve C is evaporative heat loss. Curve D is the storage curve and at low temperatures this is negative indicating chilling whilst it is almost zero at higher temperatures, when evaporative heat loss is well able to cope with the heat load. (Replotted from Winslow *et al.*, *Am. J. Physiol.* 1937, **120,** 1.)

physical adjustments to cold. By contrast, the storage term is nearly zero in the higher range showing the effectiveness of the evaporative mechanism. Thus it is mainly because the evaporative term has no "positive counterpart" that man is unable to adapt to a cold environment, therefore indicating his tropical origin. His ability to spread to temperate and frigid zones is due to the adoption of "artificial" means of reducing heat loss (as with clothes) or increasing heat gain (through fires, as used by the nude Tierra del Fuegans).

CHAPTER 4

The Supply of Energy

INTERCONVERSION OF ENERGY

The characteristic feature of life is the utilization of energy or, more technically put, the interconversion of forms of energy; this consists in the unlocking of the energy bound up in certain substances and its employment in the production of useful work or heat. This "unlocking" involves the interaction of compounds with each other in such a way that, as a result of the rearrangement of atoms, the energy-content of the resultant molecules is less than that of the reacting ones.

Endergonic and Exergonic Reactions

Thus if we consider the reaction: $X + Y \longrightarrow Z$, this will result in the liberation of energy if the rearrangement of atoms involved in converting X and Y to Z, e.g. $C + O_2 \longrightarrow CO_2$, is such that what the thermodynamicians call the "internal energy" of Z is less than that of $X + Y$. Some of this liberated energy is available for useful work, i.e. the *free energy*, and the rest appears as heat. In general, chemical reactions may be classed as those that proceed with the liberation of energy—*exergonic*—and those that require to absorb energy if they are to proceed, when they are called *endergonic*. The fundamental energy-providing reactions of the body are thus exergonic, whilst the great variety of reactions that permit the organism to carry out its normal functions—transport of ions, synthesis of proteins and carbohydrates, contraction of muscle, and so on—are endergonic. The biochemists have found it convenient, then, to consider the exergonic reactions as those that provide the energy that drives the endergonic reactions necessary to life, but to the thermodynamician this is a rather loose way of putting it; the energy of one reaction cannot drive another.

Synthesis and Breakdown of ATP

It is preferable to regard the fundamental exergonic reactions as being linked to the reversal of the reactions that convert adenosine triphosphate (ATP) to adenosine diphosphate (ADP). Thus the first stage in the breakdown of glucose involves its phosphorylation by ATP (Fig. 4.1)

$$\text{Glucose} + \text{ATP} \longrightarrow \text{Glucose-6-phosphate} + \text{ADP}$$

and a new reaction is required that will reconvert the ADP to ATP, and make it available for further phosphorylating reactions. Thus, by abstraction, the biochemist speaks of the *synthesis* of ATP by the fundamental exergonic reactions that convert carbohydrate to CO_2

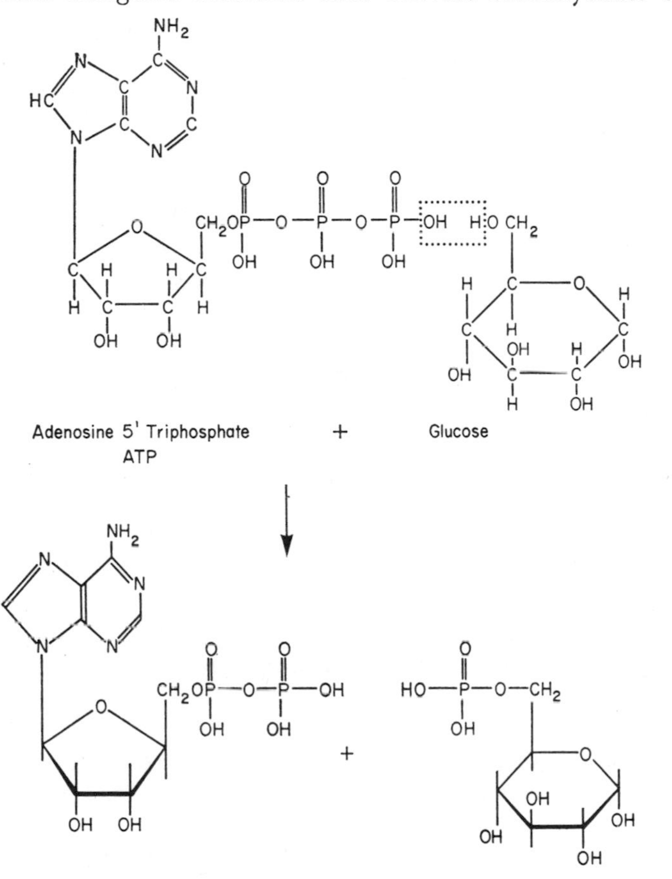

Fig. 4.1. The phosphorylation of glucose by adenosine triphosphate (ATP) to form glucose-6-phosphate and ADP.

and H O, but this is, indeed, an abstraction, since the synthesis of ATP occurs at the same time as its breakdown to ADP, so that, when the animal is in a steady state metabolically, its total supply of ATP remains stationary.

Steps in Utilization of Foodstuffs

The chemical processes that lead to the production of ATP have been divided into three phases. In phase I the foodstuffs absorbed into the body are prepared for radical change by hydrolysis—carbohydrates such as glycogen or starch are hydrolysed to glucose, proteins to amino acids and fats to glycerol and fatty acids. These reactions are exergonic, but the liberated energy appears only as heat. Phase II represents a partial combustion of the relatively simple molecules

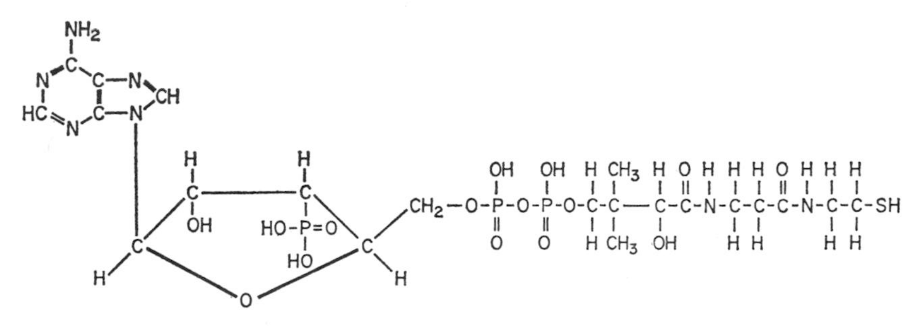

Fig. 4.2. The structure of coenzyme A.

formed by hydrolysis, which leads ultimately to the production of three substrates, i.e. the acetylated form of coenzyme A (Fig. 4.2) (AcCoA); α-ketoglutarate acid; and oxaloacetate. The chain of reactions leading to the conversion of glucose to pyruvate, called the Embden-Meyerhof pathway, is indicated in Fig. 4.3; by subsequent steps the pyruvate is converted to acetyl coenzyme A; this, together with α-ketoglutarate and oxaloacetate can enter a complicated series of reactions called the *tricarboxylic acid*, or *Krebs cycle.*

Krebs Cycle

By this cycle is meant a series of reactions linked together so that the product of one forms the substrate for the next (Fig. 4.4). To take an example, acetyl coenzyme A results from, say, the breakdown of glucose to pyruvic acid; the increased concentration of acetyl coenzyme A in the cell causes the reaction with oxaloacetate to proceed:

$$\text{Acetyl coenzyme A} + \text{Oxaloacetate} \longrightarrow \text{Citrate} + \text{Coenzyme A}$$

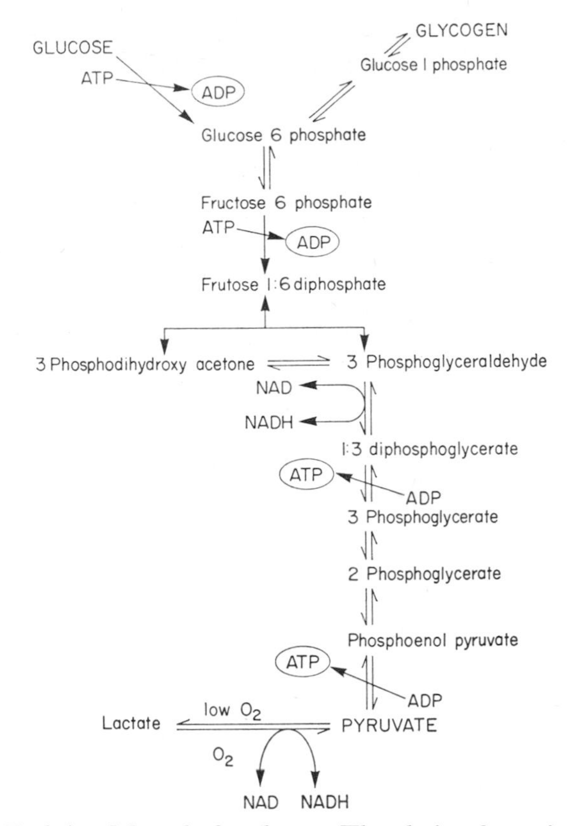

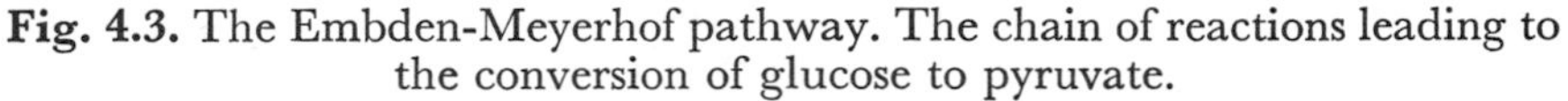
Fig. 4.3. The Embden-Meyerhof pathway. The chain of reactions leading to the conversion of glucose to pyruvate.

The higher concentration of citrate, resulting from this, causes the conversion to iso-aconitate + water to proceed, and so on.

At certain critical stages in the cycle, oxygen is required, and this involves the liberation of considerable amounts of free energy; thus the oxidation of an α-ketoglutarate to succinate and CO_2 liberates 70 k calories per mole compared with only 0·88 k calories for the non-oxidative reaction converting fumarate to malate. It is during these oxidative reactions, which are themselves represented by chains of reactions in series, that ATP is formed in considerable quantities, a process that is described as *oxidative phosphorylation.*

Electron-transport Chain

These oxidative reactions do not occur in the cytoplasm of the cell but in the mitochondria, which have therefore been described as the

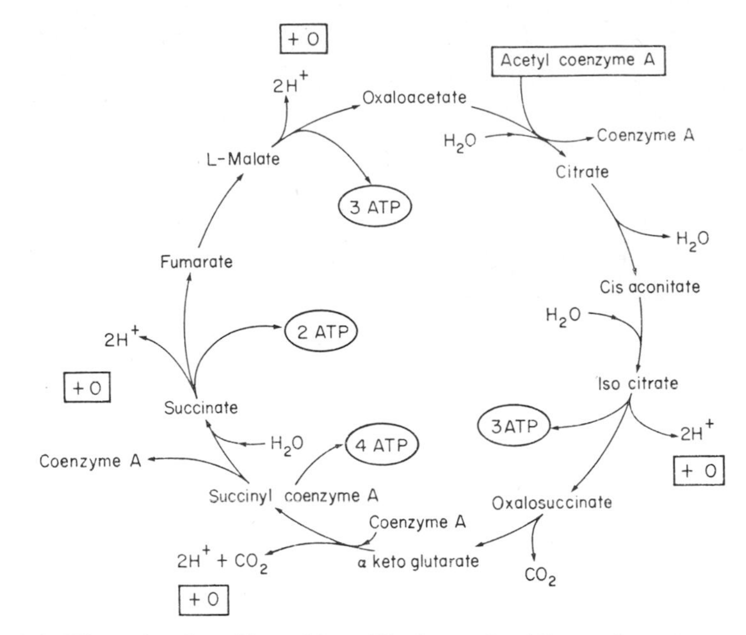

Fig. 4.4. The tricarboxylic acid or Krebs cycle. The series of reactions is linked so that the product of each step forms the substrate for the next reaction. Thus acetyl coenzyme A, derived from pyruvic acid, for example, enters the cycle to react with oxaloacetate to form citrate plus free coenzyme A; the increased concentration of citrate leads to the formation of more aconitate, and so on. During one turn of the cycle, one acetic acid equivalent is completely oxidized, whilst the four pairs of H atoms that arise ultimately react with oxygen to form water, and twelve molecules of ADP are converted to ATP.

power houses of the cell (p. 13). Thus the substrate, e.g. α-ketoglutarate, enters the mitochondrion and there it is dehydrogenated, i.e. oxidized, by first reacting with a coenzyme, *nicotin amide adenine dinucleotide*, indicated by NAD, giving up an atom of hydrogen to it:

$$MH_2 + NAD \longrightarrow NADH_2 + M.$$

From NADH the hydrogen atom is passed along a series of enzymes fixed in the wall, or cristae, of the mitochondrion, called the *electron-transport chain*, as indicated by the reactions:

$$NADH_2 + \text{Flavoprotein} \longrightarrow \text{Reduced Flavoprotein} + NAD$$

$$\text{Reduced Flavoprotein} + 2\ \text{Ferricytochrome} \longrightarrow \text{Flavoprotein} + 2\ \text{Ferrocytochrome}$$

$$2\ \text{Ferrocytochrome} + \tfrac{1}{2}O_2 \longrightarrow 2\ \text{Ferricytochrome} + H_2O.$$

The enzymes are probably attached to the walls and cristae of the mitochondrion in a definite spatial array so that the products of one reaction may pass to the enzyme responsible for the next. The reactions are described as the "electron transport chain", since they consist effectively in the removal of a hydrogen atom from the substrate molecule (oxidation) and the conversion of this to a hydrogen ion, H^+, which is a hydrogen atom minus its electron. Thus the electron from the hydrogen atom provided by the substrate is passed along the series of enzymes constituting the electron transport chain. At three points in this chain, ADP reacts (i.e. adenosine diphosphate, the hydrolytic product of ATP) to be converted into ATP as indicated schematically in Fig. 4.5, and it is supposed that hypothetical intermediate compounds M_1, M_2 and M_3 are involved in this conversion.

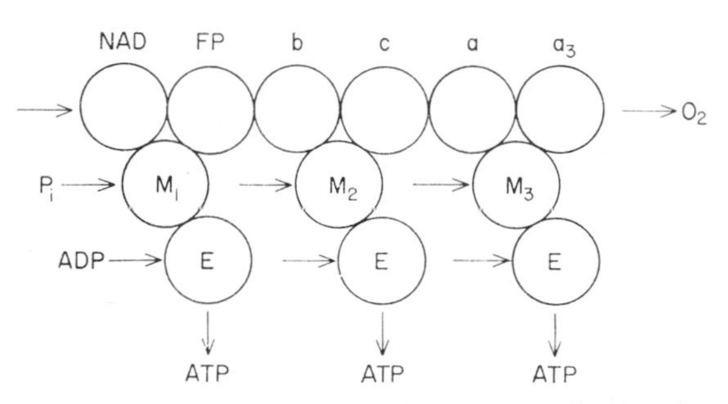

Fig. 4.5. Oxidative phosphorylation. At three stages in the electron transport chain the energy of electron transfer is trapped through the mediation of hypothetical compounds M_1, M_2, M_3. (Lehninger, *Physiol Rev.* 1962, **42,** 467.)

Since the involvement of ATP in the vital reactions of bodily processes is followed by its conversion to ADP, we may look upon this oxidative phosphorylation process as the refuelling of the metabolic system involved in vital activities. Thus if the contraction of muscle resulted in formation of ADP from ATP, the reconversion of the ADP to ATP could be looked upon as renewal of the fuel for muscular contraction, or better, as the replacement of one element in the fuel.

Electron Transport Particle. By fragmenting the mitochondrion, particles of molecular weight of about five million may be obtained capable of carrying out the enzymatic activities of the electron transport chain, and they have been called by Green *electron transport particles*, or *ETP*. It is possible that these particles are the same as the "elementary particles" described in the electron microscope which are some 70–80 Å in diameter and are attached to the cristae of the mitochondrion. Figure 4.6 illustrates the role of NAD as a link between the dehydrogenases that remove H-atoms from the various substrates of the citric acid cycle, e.g. α-ketoglutarate, and transfer these to the electron transport chain.

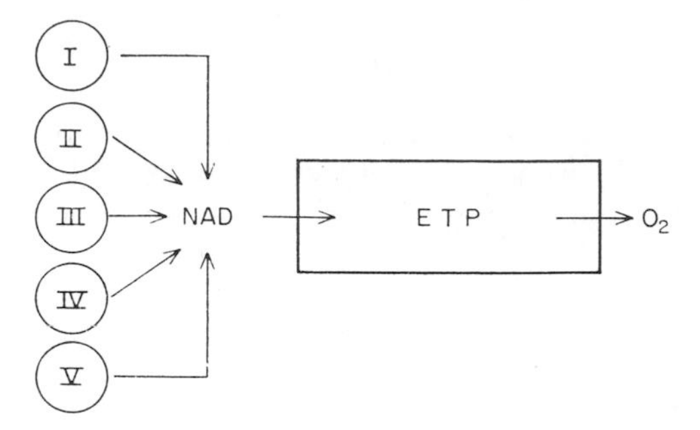

Fig. 4.6. Illustrating the relation between the electron transport particle (ETP) and the various dehydrogenating complexes. Bound NAD is the link between the two segments of the chain. I: hydroxybutyric; II: pyruvic; III: ketoglutaric; IV: malic isocitric; V: fatty acid dehydrogenating complexes. Arrow indicates direction of flow of electrons. (Green and Oda, *J. Biochem.* (*Tokyo*) 1961, **49,** 742.)

Anaerobic Glycolysis

By virtue of the reactions of oxidative phosphorylation, relatively large amounts of ATP are formed from the conversion of, say, 1 mole of glucose to CO_2 and H_2O, namely 36:

$$C_6H_{12}O_6 + 6O_2 + 36\ ADP + 36\ H_3PO_4 \longrightarrow 6CO_2 + 36\ ATP + 42\ H_2O.$$

In the absence of oxygen, metabolism of glucose may proceed to the point of production of pyruvic acid, which is then converted into lactic acid instead of to acetyl coenzyme A and is thus debarred from entering the tricarboxylic acid cycle (Fig. 4.7).

The conversion of glucose to pyruvic and lactic acids is called *glycolysis* (Fig. 4.3), and, since it occurs in the absence of O_2, it is described as *anaerobic glycolysis*. In the presence of adequate O_2, however, the conversion of pyruvate to lactate by $NADH_2$ is inhibited, the pyruvate entering the tricarboxylic acid cycle.

Energy Liberation. During anaerobic glycolysis energy is made available, since the change in free energy on converting one mole of glucose to lactate is of the order of 56 calories. During the process, some two moles of ATP are formed for each mole of glucose converted to lactate. Thus, from the point of view of production of ATP, anaerobic glycolysis is not so effective as oxidative phosphorylation. The process is also not so effective from the point of view of provision of free energy and work, only 56 k calories per mole of glucose being available compared with some 686 k calories per mole for oxidative conversion to CO_2 and H_2O. However, in certain cells such as the erythrocyte anaerobic glycolysis leading to the formation of lactate is the only metabolic energy-supplying reaction. Again, when a muscle is doing

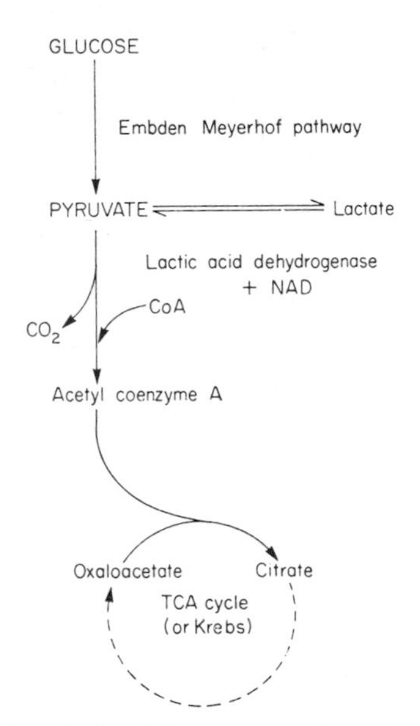

Fig. 4.7. Anaerobic glycolysis. The conversion of glucose to pyruvate is anaerobic and oxygen is only required in the tricarboxylic acid cycle. If oxygen is not available, pyruvate is converted to lactate instead of acetyl CoA and the reaction cannot proceed further.

very rapid and sustained work its oxygen supply becomes inadequate so that the energy from anaerobic glycolysis is necessary, the muscle being said to go into *oxygen-debt*.

Glycogen Synthesis

An example of the involvement of ATP in a chemical reaction that, in the absence of ATP's involvement, would be endergonic is given by the synthesis of glycogen from glucose:

$$\text{Glucose} + \text{ATP} \longrightarrow \text{Glucose-6-phosphate} + \text{ADP}$$
$$\text{Glucose-6-phosphate} \longrightarrow \text{Glucose-1-phosphate}$$
$$\text{Glucose-1-phosphate} \longrightarrow \text{Glycogen} + \text{Phosphate}$$

Thus, in essence ATP, because of the reactivity of its terminal phosphate group, is able to convert glucose into a phosphate; this in turn makes the glucose reactive, so that it may polymerize with the loss of water to form a polysaccharide, glycogen:

$$\text{n Glucose} - \text{n}H_2O \longrightarrow \text{1 Glycogen}$$

It is customary to state that the energy involved in the conversion of ATP to ADP and phosphate (hydrolysis of ATP) is used to drive the conversion of glucose to glycogen, which, viewed simply, is an endergonic reaction. Thermodynamically this is an unsound way of expressing the matter. The conversion of glucose to glycogen can, indeed, be viewed from a stoichiometric point of view as the sum of several separate reactions in which the hydrolysis of ATP to form ADP and H_3PO_4 occurs, but these reactions do not occur separately; the common factor is the entry of ATP into the system to phosphorylate glucose, and its exit as ADP.

Other Metabolic Pathways

Photosynthesis

Glycolysis and oxidative phosphorylation are the two main reactions available to the complex animal to provide energy for its vital processes. In many other organisms different fundamental energy-providing reactions take place.

Thus in green plants the fundamental process is photosynthesis, in which CO_2 and H_2O are converted to carbohydrate by a photosynthetic mechanism, the energy of light being utilized to permit the highly endergonic reaction:

$$6CO_2 + 6H_2O \longrightarrow C_6H_{12}O_6 + 6O_2$$

to go from left to right. When synthesized, the carbohydrate may be used by the cell in characteristic reactions, such as glycolysis, that it has in common with animal cells. Stoichiometrically the synthesis of carbohydrate in this reaction can be written as the reverse of the oxidation of glucose to CO_2 and H_2O, but, just as with the oxidative process, the reduction of CO_2 involves chains of interlinked chemical reactions each requiring its separate enzyme. Quite a number of these reactions are common to the two processes, however; thus isotopically labelled CO_2 can be shown to appear, at an early stage, in phosphoglyceric acid, which is later reduced to glyceraldehyde phosphate and thence converted to hexose by a reversal of the glycolytic process. In fact it is remarkable that there are only two reactions peculiar to photosynthesis in green plants, both involving the pentose *ribulose*, the diphosphate of this being the molecule that acts as "acceptor" for the atmospheric CO_2 involved in the synthesis.

Assimilation of CO_2. In general, then, we may describe the "assimilation" of CO_2 and its storage as starch by the following series of reactions:

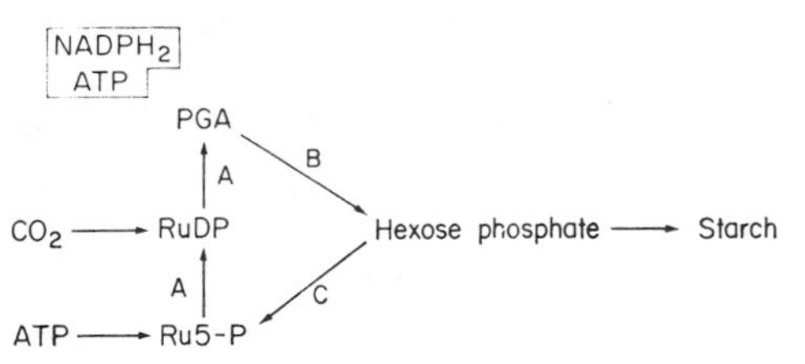

The reactions lead to the formation of hexose phosphates which may be polymerized with loss of H_2O. The processes may be described as the *carboxylative phase* (A), the *reductive phase* B, and the *regenerative phase* C, this last involving resynthesis of ribulose-5-phosphate.
ATP is required to convert ribulose-5 phosphate to the diphosphate, which then allows the acceptance of CO_2 and conversion to phosphoglyceric acid:

$$H_2CO_3 + \begin{array}{c} CH_2OPO_3H_2 \\ | \\ C{=}O \\ | \\ HCOH \\ | \\ HCOH \\ | \\ CH_2\ PO_3\ H_2 \end{array} \longrightarrow 2\ \begin{array}{c} CH_2OPO_3H_2 \\ | \\ HCCH \\ | \\ HO{-}C{=}O \end{array}$$

Ribulose diphosphate

Phosphoglyceric acid

A coenzyme *nicotinamide adenine dinucleotide phosphate*, i.e. the phosphorylated form of NAD (indicated by NADPH), is required for the reductive phase B:

$$NADPH_2 + \text{Substrate} \longrightarrow NADP + \text{Reduced substrate}.$$

Photosynthetic Phosphorylation. We may note that this series of reactions tells us nothing about the role of light. This is absorbed by chlorophyll molecules, which its energy activates, ultimately permitting the breakdown of water molecules with the liberation of O_2, which is a by-product of the photosynthetic process in green plants. During this process ATP is synthesized, the process being described as *photosynthetic phosphorylation* by analogy with the oxidative phosphorylation of carbohydrate metabolism. Under suitable conditions the two processes of oxygen liberation and photosynthetic phosphorylation may be separated. Thus in the *Hill reaction*, isolated chloroplasts can be made to produce O_2 without utilization of CO_2, whilst under other conditions ATP can be synthesized without evolution of O_2 and presumably, therefore, without breakdown of H_2O, the energy of the light being transferred, through appropriate enzymes, to the synthesis of ATP from ADP. Thus, somewhat arbitrarily, we may separate photosynthesis into two processes, namely the *photochemical reaction*, as a result of which the endergonic breakdown of water to O_2 can be achieved, and the *assimilatory reaction*, in which the CO_2 of the atmosphere is converted into carbohydrate, another endergonic reaction. In fact, of course, the processes are intimately linked, so that it is only under artificial conditions that production and breakdown of ATP can be demonstrated.

Sulphur Bacteria. In some bacteria the fundamental energy-giving reaction is also derived from the absorption of light, e.g. the green sulphur bacteria, but here oxygen is not liberated and the reaction may be written:

$$CO_2 + 2H_2S \longrightarrow CH_2O + H_2O + 2S.$$

As with photosynthesis in green plants involving the liberation of O_2, the fundamental process is considered to be the synthesis of ATP and $NADPH_2$ which participate in reactions leading to synthesis of carbohydrate and oxidation of H_2S to S.

Autotrophic Organisms

In other bacteria the energy-giving reaction does not involve utilization of glucose, nor yet the absorption of light energy; instead the basic energy supply is derived from quite different exergonic reactions. For example the sulphur bacterium, *Thiobacillus thioparus*, will grow on thiosulphate solutions, the reaction:

$$2Na_2S_2O_3 + O_2 \longrightarrow 2Na_2SO_4 + 2S$$

taking place within it.

This *autotrophic* organism* carries out a variety of metabolic reactions involving organic matter, but it is unable to utilize organic matter as a primary source of metabolic energy. The energy involved in the oxidation of thiosulphate is, in some way, made available for the

* This is the name given to bacteria that are able to live in a medium completely free from organic matter; the *obligatory* autotroph cannot live in a medium containing organic matter, so that it must be cultured in an inorganic medium, such as silica gel.

synthetic processes involved in growth, and it is interesting that, once again, the production of ATP is concerned in these processes. We may say that the autotrophic bacterium is the most independent of organisms, its synthetic power extending to the synthesis of all its molecules from CO_2 and inorganic sources. In general, we may characterize the development to higher organisms as a progressive loss of synthetic power, so that the animal cell relies on organic compounds for its basic energy-giving reactions as well as for the synthesis of the molecules containing carbon, nitrogen, sulphur, etc.

THE MECHANISM OF CHEMICAL REACTION

Stepwise Conversions

The chains of chemical reactions discussed earlier in this Chapter may be written quite simply when only the initial reactants and end-products are considered, e.g.

$$\text{Glucose} + \text{ADP} = \longrightarrow \text{Lactate} + \text{ATP}$$
$$\text{Glucose} + \text{ADP} = \longrightarrow CO_2 + H_2O + \text{ATP}$$

and we must now consider why so many steps appear to be necessary.

It is clear at once, however, that if glucose were merely burnt to CO_2 and H_2O, as in a crucible, the object of the enterprise, from the point of view of energizing the various processes taking place in the organism, might well be defeated, unless only heat were required. In a steam-engine this *would* be adequate, since the heat is employed in converting water to steam and utilizing its expansive power. In the organism, however, heat must be regarded as a waste product, necessary in the warm-blooded animal, it is true, but still a by-product.

Linking of Reactions

Thus one reason for breaking the reaction down into steps is to permit the involvement of substances whose reactions lead to useful work. For example in the synthesis of protein from amino acids, the exergonic reactions of carbohydrate metabolism must be linked in some way with the condensation of amino acids in the polypeptide chains, so that at some stage amino acids must participate in carbohydrate metabolism. In a similar way, the transport of ions up gradients of electrochemical potential must be linked to the exergonic reactions of metabolism, e.g. anaerobic glycolysis in the accumulation of K^+ by the erythrocyte.

Role of ATP

In all studies of endergonic reactions, e.g. those of protein synthesis, the substance ATP has been shown to play a role; thus we have seen that an amino acid must react with ATP under the influence of the enzyme, aminoacyl synthetase, to permit it to react with transfer RNA (p. 71). Again, the first stage in synthesis of glycogen, an endergonic process, requires phosphorylation of glucose by ATP. Active transport of ions by cells can be accelerated by addition of ATP to the cell, or abolished under conditions where ATP supplies are depleted, whilst contraction of artificial muscle fibres, made up of actin and myosin, may be brought about by ATP. The ubiquitousness of ATP as a participant in metabolic processes, especially those involving endergonic reactions, has suggested that its fundamental role is to provide energy for these endergonic reactions. In a very limited sense this is true, but it is more correct to describe ATP as a compound that acts as a link between reactions. It occupies this role by virtue of its reactivity, being able to transfer a phosphate group to a large variety of compounds. Moreover, it is not only highly reactive but it is also highly specific, in the sense that it has a fairly complex structure and one that allows it to be selectively taken up by a variety of enzymes, each one capable of catalysing a chemical reaction in which ATP is involved.

Thus, the breaking up of the fundamental chemical reactions of glucose oxidation into numerous steps, in many of which ATP is involved, may be seen as the means whereby this fundamental reaction can be linked with those concerned in vital processes, such as protein synthesis.

Activation Energy

A further and closely connected reason for break-up into steps is concerned with the *activation energy* for a given reaction.

The oxidation of glucose to CO_2 and H_2O does not occur spontaneously at a measurable rate, although it is highly exergonic; and this is because the reaction requires the weakening, and ultimate rupture, of highly stable chemical bonds in the glucose and oxygen molecules as a preliminary to the atomic rearrangements that constitute the ultimate reaction. Thus, molecules will react if, by collision with other molecules, they acquire this necessary degree of instability, or in other words, when they have become *activated*. In a photochemical reaction, such as photosynthesis, the *activation energy* is provided by the absorption of light, but the thermal reactions of animal metabolism require activation energy by collision.

Catalysis

There are two main ways of ensuring that the activation energy is available, the first being to employ a catalyst or enzyme. Its action is essentially to reduce the stability of a reacting molecule by attaching this to a specific region on its surface; in this adsorbed or linked state, less energy of collision is required to enable it to react. The second way is to split the reaction into steps, each one involving lower activation energy than the energy required for the direct reaction, and it is here that ATP plays a role.

Reversibility

It will be noted that the energy liberated by a given step varies very considerably, and is always much less than that of the total reaction. The main value of this is to allow the reactions to proceed reversibly under biological conditions. Thus the burning of glucose to CO_2 and H_2O can be achieved directly by providing sufficient activation energy to allow the reaction to begin; then, as each glucose molecule decomposes, sufficiently large amounts of energy are set free to provide activation energy for further molecules, and the reaction proceeds without further help. So great is the change in energy, however, that the reverse reaction, namely the conversion of CO_2 and H_2O, will not proceed to a significant extent. Thus, for reversibility, the changes in energy must not be very large, whilst the activation energies required to initiate the reactions must be such that they can be achieved by thermal collisions within the solution.

Control

The main value of reversibility in the present context is the circumstance that this permits an automatic control over the reaction in accordance with the principle of mass action, e.g. in the reaction:

$$A + B \rightleftharpoons C$$

we may write an *equilibrium constant*, K, given by the equation:

$$\frac{[A] \times [B]}{[C]} = K$$

where the square brackets indicate concentrations. If the reaction is reversible, there will be finite quantities of A, B and C at equilibrium; by reducing the concentration of C—for example, by allowing it to react with another substance—we can make the reaction proceed further "to the right". Alternatively, if, as a result of some other reaction, more C appears, then the reaction will go "to the left".

Furthermore, when the reaction is catalysed by an enzyme, then, although the position of equilibrium is not altered, nevertheless the speed of reaction may be very considerably modified if the product of the reaction, C, tends to inhibit the enzyme—*product inhibition.*

Enzymes

The catalysts of biological reactions are described as enzymes, which have been defined by Dixon and Webb as "proteins with catalytic properties due to their power of specific activation". One of the smallest enzymes is ribonuclease (which breaks down ribonucleic acid) with a molecular weight of 12,700, and one of the biggest is glutamate dehydrogenase (concerned in the reaction of oxoglutarate with ammonia to form glutamate) which has a molecular weight of 1,000,000. Enzymes possess the power of *specific* activation through exposing special groupings of atoms to the substrate molecules, groupings that permit relatively few compounds to react with them. When the grouping of atoms is not a constituent part of the amino acid sequence, as with the haem portion of peroxidase or the flavin of lipoamide dehydrogenase, the grouping is called a *prosthetic group*, similar to the prosthetic group of haemoglobin, *haem*, or of the visual pigment, *retinal*, and so on. Being proteins, their reactivity is sensitive to pH, usually exhibiting an optimum; and experimental studies have shown that these influences of pH, which affect the degree of ionization of the acidic and basic groups of the constituent amino acids, influence both V_{max} and the affinity constant (see p. 185).

Temperature

Temperature can influence enzyme-catalysed reactions in several ways; an extreme effect is through denaturation of the enzyme molecule, a high temperature tending to destroy the tertiary structure, i.e. the manner in which the polypeptide chains are coiled and held together. As a result of this conformational change the active grouping may lose its identity or become unavailable to the substrate. Less extreme effects will be exerted through change in the ionization constants of the acidic and basic groups, whilst the velocities of the component reactions, especially that of the breakdown of the enzyme-substrate complex to products:

$$ES \longrightarrow E + \text{Products},$$

will all be increased by a rise in temperature which increases the average energy of the molecules and so increases the number that are likely to be activated at any moment.

Organization of Reactions

The chain of reactions in the Embden-Meyerhof pathway for glycolysis takes place in ordered succession, in spite of the fact that all the necessary enzymes and substrates are mixed together in solution. The reason why such an ordered progression is possible depends on two main factors, firstly that the product of one reaction is the substrate for the next and secondly that the specificity of the enzymes is such that the product of any given reaction will find only one enzyme for which it can act as substrate. At any rate, this is an ideal situation that would ensure the direction of the chemical process from the moment it began. This arrangement is called *organization by specificity*, which contrasts with the *spatial organization* of enzymes found in the mitochondrion and on the membranes of the endoplasmic reticulum. Here the spatial relations between the enzymes impose a further control on the sequence of reactions, in addition to that imposed by the specificities of the enzymes.

Linking

In a series of reactions:

$$A \xrightarrow[E_1]{} B \xrightarrow[E_2]{} C \xrightarrow[E_3]{} D \xrightarrow[E_4]{} E \text{ - - - and so on,}$$

catalysed by enzymes E_1, E_2, etc., we may regard the two enzymes E_1 and E_2 as being linked by a relationship, namely that the substrate of E_2 is the product of E_1. However, since each of the reactions is to some extent reversible, we may regard A as a product of the reaction of B with enzyme E_1, i.e. B acts as a substrate for both E_1 and E_2 so that we may say that the two reactions are linked by a common substrate. There are two fundamental ways in which this linking by substrate may occur; first we have the above case where E_2 transforms the common substrate B to a new substance, C; secondly we may have the situation where E_2 catalyses the conversion of B back to A. Since the enzymes are not the same, the conversion back to A must be by a different reaction from the conversion of A to B, and this means that a second substance must enter into the reaction.

NAD and NADP

This second type of linkage is exhibited by the *coenzymes*, NAD and NADP. An example is provided by the conversion of a substrate AH_2 to an oxidized form, A, by removal of its two H-atoms and transferring these to oxygen to form hydrogen peroxide.

The oxidized form of NAD reacts with AH_2 to form A and the reduced form of NAD, namely $NADH_2$

$$AH_2 + NAD \xrightarrow[\text{Dehydrogenase}]{} A + NADH_2$$

Subsequently NAD is re-formed by reacting with glyoxalate:

$$NADH_2 + \begin{array}{c} COOH \\ | \\ CHO \end{array} \xrightarrow[\text{Glyoxalate reductase}]{} NAD + \begin{array}{c} COOH \\ | \\ CH_2OH \end{array}$$

to form glycollate, the enzyme being glyoxalate reductase. Thus NAD has acted as the link between the dehydrogenase and the reductase, being reduced by the one and oxidized by the other.

Finally glycollate is oxidized by O_2 to re-form glyoxylate

$$\begin{array}{c} COOH \\ | \\ CH_2OH \end{array} + O_2 \longrightarrow \begin{array}{c} COOH \\ | \\ CHO \end{array} + H_2O_2$$

catalysed by glycollate oxidase.

The net reaction may be written:

$$AH_2 + O_2 = A + H_2O_2$$

and represents the production of peroxide from atmospheric O_2.

NAD in this case acts as part of the catalytic mechanism, and in this context is called a coenzyme, remaining behind as part of the system. The coenzyme, moreover, has operated in a carrier mechanism, transferring two hydrogen atoms from the initial substrate to glyoxylate which then hands them on to O_2.

NAD and Mitochondria. The importance of this linking system in metabolic processes is revealed by the position of NAD in relation to the electron transport chain of the mitochondrion discussed briefly earlier. NAD may act as a link for a large number of dehydrogenases, each capable of removing H-atoms from substrates and transferring them to NAD:

$$AH_2 + NAD \longrightarrow NADH_2$$

The hydrogen from these substrates is then funnelled into the electron transport chain by transfer to flavoprotein:

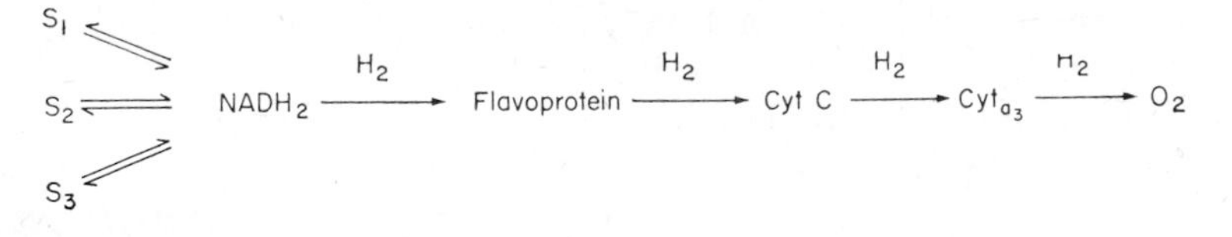

NAD and Glycolysis. In the absence of O_2 the reduced NAD cannot be oxidized by the electron transport chain; it is nevertheless employed as a link in reactions, such as in the anaerobic conversion of glucose to lactic acid or to alcohol, being reduced by one dehydrogenase in a chain and oxidized by the reverse reaction of another dehydrogenase.

Thus we have as a stage in glycolysis (p. 169):

$$\begin{matrix} CH_2OH_2PO_3 \\ | \\ CHOH \\ | \\ CHO \end{matrix} + HS.Enzyme + NAD \longrightarrow \begin{matrix} CH_2OH_3PO_3 \\ | \\ CHOH \\ | \\ CO \\ | \\ S.Enzyme \end{matrix} + NADH_2$$

Glyceraldehyde phosphate + dehydrogenase → Phosphoglyceric acid-enzyme complex

Glyceraldehyde phosphate dehydrogenase

$$\begin{matrix} CH_2OH_2PO_3 \\ | \\ CHOH \\ | \\ CO \\ \\ S.Enzyme \end{matrix} + H_3PO_4 \longrightarrow \begin{matrix} CH_2OH\ \ PO_3 \\ | \\ CHOH \\ | \\ CO.OH_2PO_3 \end{matrix} + Enzyme$$

Phosphoglyceric acid-enzyme complex + phosphate → Diphosphoglycerate + dehydrogenase

Glyceraldehyde phosphate dehydrogenase

$$\begin{matrix} CH_3 \\ | \\ CO \\ | \\ COOH \end{matrix} + NADH_2 \longrightarrow \begin{matrix} CH_3 \\ | \\ CHOH \\ | \\ COOH \end{matrix} + NAD$$

Pyruvate + $NADH_2$ → Lactate + NAD

Lactate dehydrogenase

NAD, therefore, acts as a link between glyceraldehyde phosphate dehydrogenase and lactate dehydrogenase.

NADP. Many dehydrogenases react in conjunction with NADP as coenzyme, and since there is considerable specificity in this respect, it is customary to refer to a dehydrogenase as *NAD-* or *NADP-specific*. Succinic dehydrogenase is an enzyme involved in mitochondrial metabolism that requires neither coenzyme, so that the hydrogen from succinate is transferred directly to flavoprotein.

MICHAELIS-MENTEN KINETICS

The catalysed chemical reaction is fundamental to biological processes, so that the analysis of its kinetics is essential for an adequate understanding of these reactions, and their modifications through inhibition or acceleration. Michaelis and Menten imagined that when a substance, S, was decomposed catalytically by the enzyme, E, to give products, P, the process began by an initial reversible combination of S with the enzyme, followed by the irreversible breakdown to P:

$$E + S \rightleftharpoons ES \longrightarrow \text{Products.}$$

The rate of reaction at any time would be governed by the concentration of the enzyme–substrate complex, ES, and a maximum rate would be achieved when the enzyme was saturated, i.e. all the active groupings on its surface were occupied by reacting molecules. If $[E_0]$ is the total concentration of enzyme and [ES] is the concentration of complex, then the concentration of free enzyme, [E], is:

$$[E] = [E_0] - [ES] \tag{1}$$

The dissociation constant is given by:

$$K_S = \frac{[E] \times [S]}{[ES]} = \frac{[E_0] - [ES] \times [S]}{[ES]} \tag{2}$$

Whence
$$[ES] = \frac{[E_0] \times [S]}{K_S + [S]} \tag{3}$$

The velocity, v, of the reaction giving the products, P, is given by:

$$v = k \times [ES] \text{ or } [ES] = v \div k \tag{4}$$

Substituting for [ES] in (3), we have:

$$v = \frac{k \times [E_0] \times [S]}{K_S + [S]} \tag{5}$$

Maximal Velocity

When the enzyme is saturated, the concentration of ES becomes equal to $[E_0]$; under these conditions the velocity will be maximal, V_{max}:

$$V_{max} = k[E_0].$$

Substituting in equation (5) we get finally:

$$v = \frac{V_{max} \times [S]}{K_S + [S]} \tag{6}$$

Dissociation Constant

K_S, the dissociation constant for the enzyme–substrate complex, may be determined by measuring the velocity of reaction for different values of the substrate concentration, S, and carrying out a Lineweaver–Burk plot, i.e. plotting $1/v$ against $1/[S]$, since it will be seen that, by rearrangement, equation (6) may be written:

$$1/v = \frac{K_S}{V_{max}[S]} + 1/V_{max}$$

i.e. the plot should be a straight line as in Fig. 4.8, with the intercept on the $1/v$ axis equal to $1/V_{max}$ and the intercept on the $1/[S]$ axis equal to $-1/K_S$.

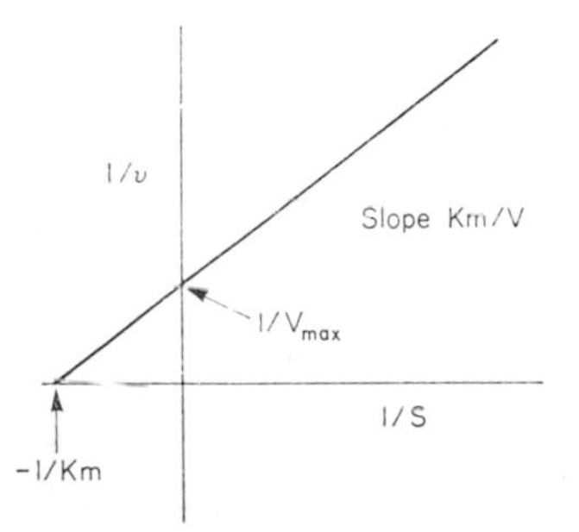

Fig. 4.8. The Lineweaver–Burk plot. The reciprocal of the velocity of the reaction $(1/v)$ is plotted against the reciprocal of the substrate concentration $(1/S)$. If the reaction follows the simple Michaelis-Menten kinetics, the graph should be a straight line with the intercept on the $1/v$ axis equal to $1/V_{max}$ and the intercept on the $1/S$ axis equal to $-1/K_m$.

It will be seen, from its definition, that K_S may be regarded as the reciprocal of the affinity of the enzyme for the substrate, the greater the value of $1/K_S$ the greater will be the concentration of ES. It will also be seen that if, in equation (6), [S] is made equal to K_S, v becomes equal to $V_{max}/2$, i.e. K_S is the concentration of substrate at which the velocity of reaction is half-maximal.

Michaelis Constant

The Michaelis–Menten treatment contains a number of assumptions that are not always applicable, but in fact a great many enzymatically catalysed reactions do follow the equation, giving a good straight line when plotted in the Lineweaver–Burk manner. When K_S has been derived on this basis, however, it is best to describe it as K_m, an *effective equilibrium* or *Michaelis constant*, so that the equation is written:

$$v = \frac{V_{max} \times [S]}{K_m + [S]} \qquad (6')$$

It will be noted that it has been assumed that the maximum velocity of the reaction will be proportional to the concentration of enzyme; this is true if the concentration of enzyme is very small compared with that of the substrate, a condition usually fulfilled in biological systems. This means that a powerful mode of control can be exerted through synthesis of new enzyme.

Modifications of the Reaction Rates

The velocity of the reaction, v, has been assumed to be governed, ultimately, by the rate of decomposition of the enzyme–substrate complex, ES, i.e. by the equation:

$$v = k \times [ES]$$

so that it can be modified by changing the concentration of this complex or by altering k, which is a velocity constant that measures the rate at which a given number of ES molecules decompose in unit time. It is a specific characteristic of the reaction between the substrate and the active grouping on the enzyme, and can be modified by altering the character of the enzyme appropriately. Thus a compound that reacted with the enzyme might either increase k, i.e. make the complex more unstable and promote formation of P, or it might decrease k; in this latter case it would be described as an *inhibitor*.

Enzyme Synthesis and Activation

A further way of modifying the maximum rate of reaction is clearly through altering the concentration of enzyme, E_0, and this is certainly one of the modes of control over biological reactions, i.e. the synthesis of new enzyme or its destruction. In many situations the further quantities of enzyme are synthesized *ab initio*, but in others the enzyme is stored by the cell in an inactive form, and is brought into action by another enzyme, called a *kinase*. Thus trypsin has an inactive form, *trypsinogen*, which, by the action of *enterokinase*, is converted to trypsin and thus becomes capable of hydrolysing proteins in the intestine. The phosphorylating enzyme required in the metabolism of glycogen exists in an inactive form and must be acted on by a phosphorylase-kinase before it can act; and so on. A further method of control over the amount of enzyme is given by sequestering it in granules or vesicles within the cell; if the enzyme is required for use outside the cell, then "packets" of enzyme may be released by exocytosis of individual vesicles on to the scene of action.

Inhibition

The number of reacting groupings on an enzyme molecule may be modified chemically, and this will result in an effective alteration in $[E_0]$; this may be achieved by an *inhibitor* that combines with some or all of them and leaves them unavailable to the substrate, whilst an

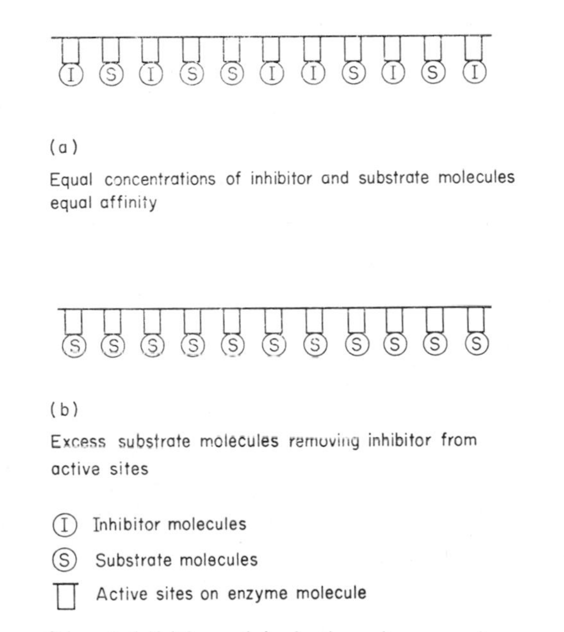

Fig. 4.9. Competitive inhibition. (a) Active sites on an enzyme molecule occupied by equal amounts of inhibitor and substrate molecules both having the same affinity and concentration. (b) The removal of inhibitor molecules by an excess of substrate molecules competing for the active sites.

activator may unmask active groupings that had been combined and were previously unavailable. These influences will affect the velocity of the reaction for a given substrate concentration, but they need not necessarily affect V_{max}.

Competitive Inhibition. Thus, when the inhibitor, for example, is a *competitor* for the active grouping with the substrate, the share it will have of the available groups will depend on its affinity for the groupings and on the relative concentration of the inhibitor compared with the substrate. We may therefore increase the concentration of substrate and ultimately all the inhibitor will be "pushed out of the way", and all the active groupings will be occupied by substrate (Fig. 4.9). Consequently, although over a wide range of concentrations of substrate the inhibitor may reduce the rate of reaction, if the substrate and inhibitor both compete for the enzyme then, ultimately, a concentration of substrate may be reached where the inhibitor exerts no effect, and this is when V_{max} is reached. Thus a competitive inhibitor may be defined operationally as one that does not affect V_{max}. If we return to equation 6′, we see that the rate of reaction is expressible in terms of two parameters—V_{max} and K_m. Therefore, for a given substrate concentration, the inhibition caused by a competitive inhibitor, if it does not affect V_{max}, must affect K_m; i.e. the "apparent K_m", which may be called K_P, will not correspond with the equilibrium constant for the substrate–enzyme complex, but will be composite, involving the equilibrium constants of both substrate and competitor. Thus the kinetics may be treated along the same lines, but we may envisage two reactions of E, the enzyme, as follows:

$$E + S \rightleftharpoons ES \qquad (A)$$

$$E + I \rightleftharpoons EI \qquad (B)$$

whilst we have the single reaction: $ES \longrightarrow E +$ products, as before.

The equation relating velocity with substrate and inhibitor concentrations is:

$$v = \frac{V_{max}}{1 + \frac{K_m}{[S]}\left(1 + \frac{[I]}{K_i}\right)} \qquad (7)$$

Experimentally we may measure the velocity of reaction for different concentrations of substrate and obtain the usual Lineweaver–Burk plot of Fig. 4.8. The process may be repeated in the presence of a known concentration, [I], of the inhibitor, and a new plot may be obtained, as in Fig. 4.10a, with intercepts giving $1/V_{max}$ on the vertical

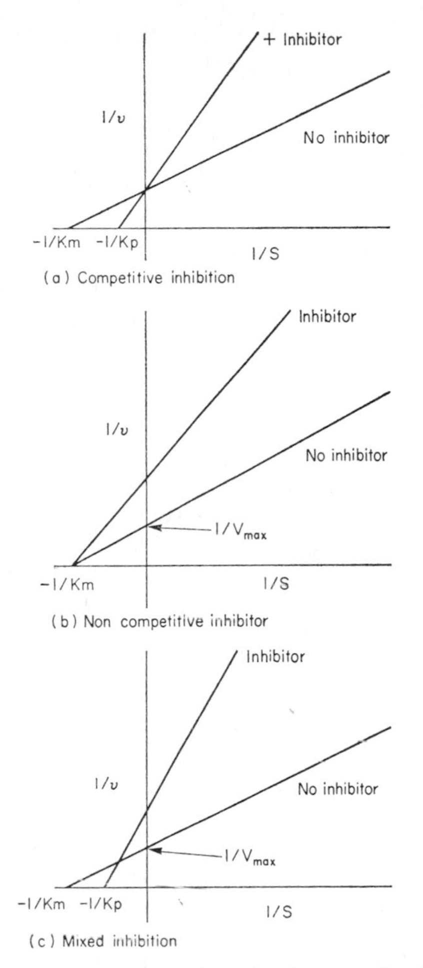

Fig. 4.10. The effect of inhibitors on the Lineweaver–Burk plot for a reaction between substrate and enzyme. (a) The presence of a competitive inhibitor has not changed V_{max} but has increased the slope. (b) Non-competitive inhibition, where the affinity for the substrate (I/K_m) is unchanged. (c) Mixed inhibition where both V_{max} and apparent affinity are altered.

scale and the reciprocal of the apparent K_m, i.e. $1/K_P$, on the horizontal scale, K_P being equal to K_m $(1 + [I]/K_i)$, K_i being defined as the equilibrium constant for reaction (B) above.

Non-competitive Inhibition. In this *competitive type* of inhibition we have modified the affinity of substrate for enzyme only apparently. What we have really done is, by providing competing molecules, to prevent the substrate from exerting its affinity; and it is for this reason that V_{max} remains unchanged. With other ways of modifying affinity (e.g. by chemical reaction with the enzyme molecule), we may alter both V_{max} and apparent K_m together, since in effect we may alter the velocity constant for breakdown of the complex, ES, at the same time as we alter the affinity of the substrate for the enzyme.

Lineweaver–Burk Plots. In general, then, we may expect to find three main types of Lineweaver–Burk plot in the presence of an inhibitor, as indicated in Fig. 4.10; (a) illustrates the purely competitive type with no alteration in V_{max}; Fig. 4.10b illustrates the situation where the apparent affinity of substrate is unchanged, so that the inhibition would be described formally as purely *non-competitive*, whilst in Fig. 4.10c the inhibition would be described as "mixed", with both apparent affinity and V_{max} altered. It should be emphasized that these characterizations of type of inhibition are essentially operational, the classification depending on the type of Lineweaver–Burk plot. To determine the actual mechanism of interaction between the inhibitor and enzyme requires experiments of a different sort.

MECHANISM OF ENZYME ACTION

Internal Strains

A catalyst accelerates a reaction because it reduces the activation energy that is necessary to allow of the rearrangement or rupture of chemical bonds. There have been various hypotheses put forward to explain the mechanism whereby the enzyme is able to reduce the activation energy, and it is likely that the mechanism is not uniform throughout the whole range of catalysed reactions, so that each individual type of reaction must be examined on its own. One general mechanism covering many types of reaction might well be the preliminary adsorption of the substrate molecule on the enzyme through mutual interaction of atomic forces, and as a result, strains might be set up in the molecule thereby reducing its bond-energies and facilitating rupture or rearrangement. Certainly, internal strains in a molecule can increase its reactivity enormously, but it is unlikely that this mechanism operates universally.

Intermediate Reaction

We have seen how the splitting up of a chemical reaction into a number of intermediate stages facilitates its progress by reducing the activation energy required at any given stage; and in many biological instances catalysis may be regarded in this light, the substrate undergoing a chemical reaction with the enzyme to form an intermediate compound which is subsequently decomposed. For this to happen the enzyme must have on its surface, or within its body, an active grouping that is complementary to a reactive grouping on the substrate molecule, and in this way the enzyme can exert its specificity, catalysing only a limited number of reactions. In fact, the reactive grouping would be complementary in chemical structure to the hypothetical intermediate between the reacting substance and the product of reaction, rather than to the reacting substace.

Protein Hydrolysis

For example, the breakdown of proteins or polypeptides is one of hydrolysis of the peptide bond:

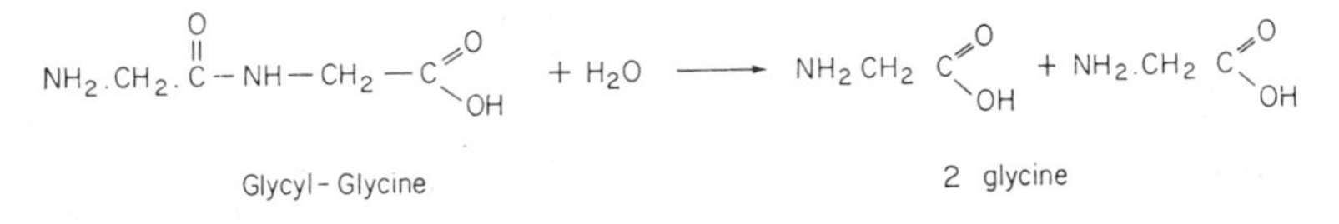

and we may assume that a given proteolytic enzyme has a grouping of atoms that attracts the peptide linkage, which we may write:

$$R-\overset{\displaystyle O}{\overset{\|}{C}}-NH-R'$$

and the hydrolysis:

$$R-\overset{\displaystyle O}{\overset{\|}{C}}-NH-R' + H_2O \longrightarrow R.C\begin{matrix}{=O}\\{\diagdown OH}\end{matrix} + NH_2-R'$$

R′ being the polypeptide contributing the NH_2-group to the peptide bond and R the polypeptide contributing the COOH-group.

Specificity. There is more to it than this, however, since a proteolytic enzyme, like trypsin, will not hydrolyse all peptide bonds indiscriminately, e.g. it will not split off the terminal amino acid residue but must attack a peptide bond within the chain. Other enzymes, called exopeptidases, will only attack terminal peptide bonds. Again, chymotrypsin shows a preference for certain amino acids, such as tyrosine or phenylalanine when these contribute the C=O group in the linkage. Thus, in relating the structure of an enzyme to its chemical activity, it may be necessary to look for at least two sites: the *reactive site* with a structure complementary in some way to the reactive groupings in the substrate; and a "fitting site" that selects only molecules with a specific type of chemical grouping attached to the reactive grouping.

Active Site

As we have indicated, enzymes are proteins, and may have molecular weights as large as a million; however, there is no doubt that their reactivity depends on quite a small portion of the molecule, just as a hormone (like gastrin) can dispense with all but the last five amino acids in its chain (Vol. 3). Nevertheless, the great mass of the protein molecule, besides this reactive portion, is not just a useless appendage; there is little doubt that it contributes to the reactivity of the reactive portion, often by providing a "microenvironment" for it which can be quite different from the watery environment in which the bulk of the reaction takes place. This is certainly true of the catalysis of protein hydrolysis by the enzyme chymotrypsin.

X-ray Analysis

Most protein enzymes are of the "globular" type, their chains being bent into a tertiary structure that approximates to a sphere rather than a long fibre. The amino acid sequence of many enzymes has been determined, and this, together with experimentally induced changes in activity with changes in sequence, have permitted some guesses at the critical amino acid sequence required for the enzyme's activity. A very powerful tool, used in conjunction with amino acid sequence analysis, is the study of X-ray diffraction of the protein crystals which permits the resolution of the secondary and tertiary structures of the amino acid chains, so that in this way we can determine the accessibility, or otherwise, of given amino acid residues. Furthermore, it is usually possible to find a substrate for the enzyme that attaches strongly to the specific reacting site, but reacts very slowly, so that analysis of the X-ray diffraction diagram of this reacted enzyme can often indicate the particular region where the substrate has attached. Additional means of identifying the active site are through the use of inhibitors; thus some enzymes are sensitive to Hg and its derivatives, and this proves to be because of the strong affinity of Hg for SH-groupings derived from cysteine or methionine. Inhibitors of this type are also of great use in X-ray analysis since they provide heavy atoms at known points which show up prominently in the X-ray diagram. In fact it was the application of the preparation of *isomorphous derivatives* by Perutz that made possible the detailed X-ray analysis of protein structure with a resolution of only a few Ångstrom units.

Mechanism of Action of Chymotrypsin

As an example of these applications we may look at Fig. 4.11 which illustrates the tertiary structure of α-chymotrypsin, the intestinal enzyme that hydrolyses proteins or polypeptide chains. The critically

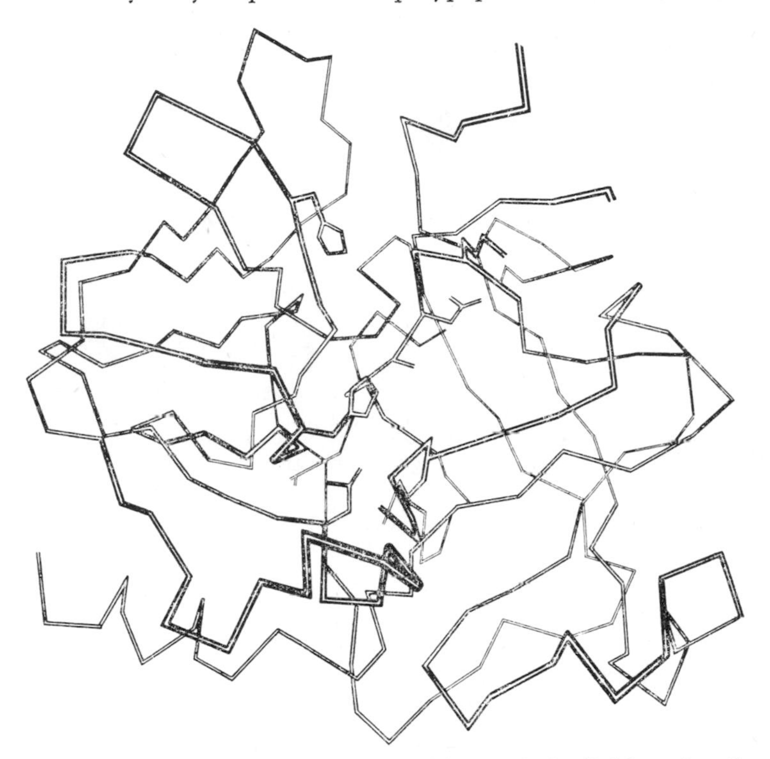

Fig. 4.11. α-chymotrypsin simplified backbone chain linking the C-atoms of each amino acid residue. The side chains of some important amino acid residues are also shown. A portion of the chain, from residues 9 to 13, is not shown, because of the uncertainty in position of these residues. (Frier, Kraut, Robertson, Wright and Xuong, *Biochemistry*, 1920, **9**, 1897.)

important amino acid residues are remarkably few, namely Aspartate 102, Histidine 57 and Serine 195, and the problem is to determine how these three amino acids can co-operate in causing the rupture of a peptide linkage with introduction of a water molecule, as in the above reaction.

Acylation and De-acylation

It seems that the enzyme is able to carry out this hydrolysis by a preliminary reaction with the peptide, described as *acylation*, and illustrated by Fig. 4.12a. As a result of this acylation the $R'NH_2$ group of the polypeptide or protein is split off. In the subsequent *deacylation* step (Fig. 4.12b), the H_2O molecule moves in to take the place of the departing $R'NH_2$, and the remaining half of the peptide is set free, namely $R.CO_2H$.

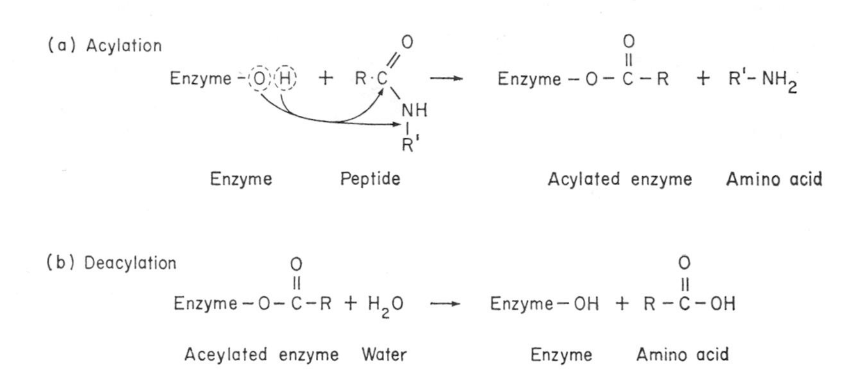

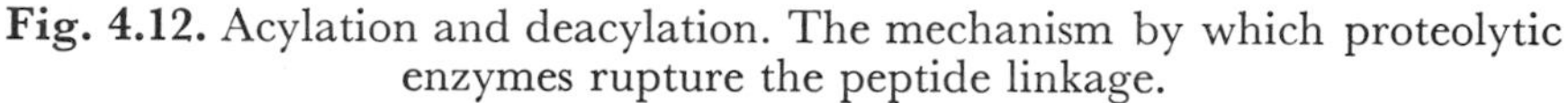

Fig. 4.12. Acylation and deacylation. The mechanism by which proteolytic enzymes rupture the peptide linkage.

Here the enzyme has been represented as *Enzyme-OH*, since the prime requirement is the presence of an OH-group that would be sufficiently reactive to be able to carry out the acylation step. Such a hypothesis, or representation, poses a number of questions, such as which OH-group reacts and why it is able to do so, and why the reactivity is confined to peptide bonds and peptide bonds of a special sort, i.e. why the enzyme should show high specificity. It is considered that the important OH-group in this respect is that of the side-chain of serine:

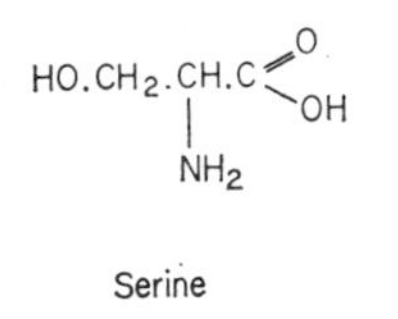

and the main question is to decide how this OH-group becomes so reactive when attached to the serine in the enzyme, since it is known that, without special conditions, the hypothetical acylation step would not proceed to any appreciable extent.

Charge Relay System. The answer is given by the relations of the three important amino acids of the active grouping, namely serine, histidine and aspartic acid:

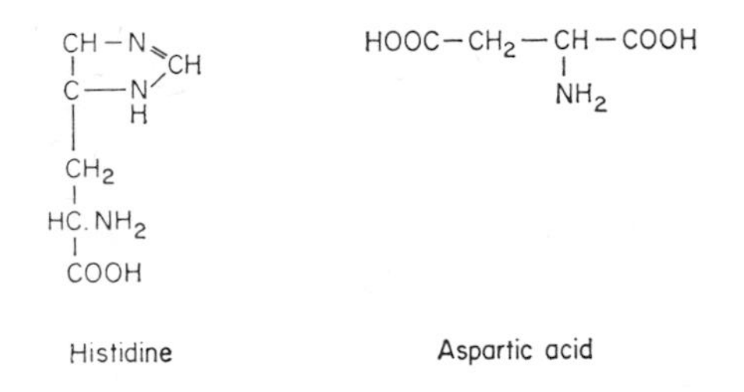

Histidine Aspartic acid

These are linked by hydrogen bonds to form a "charge relay system" that, in effect, allows the shift of an electron over quite a large distance and so permits the OH-group of the aspartate side-chain to co-operate with the OH-group of the serine side-chain, and thereby make this last group sufficiently reactive, or *nucleophilic*, to enable it to attack the peptide bond.

To be able to appreciate the explanation, we must return to our simple description of the acylation step (Fig. 4.12a), and introduce an intermediate stage:

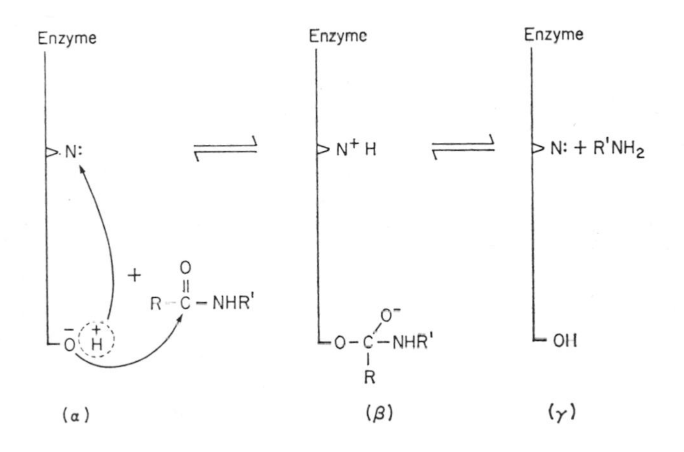

the proton (H^+) of the serine side-chain moving to a nitrogen atom in the enzyme molecule, and back again when the RNH_2 moiety of the polypeptide has been split off.

Similarly the deacylation step requires an intermediate stage with incorporation of a water molecule, and once again the introduction of a nitrogen atom capable of taking on a proton temporarily:

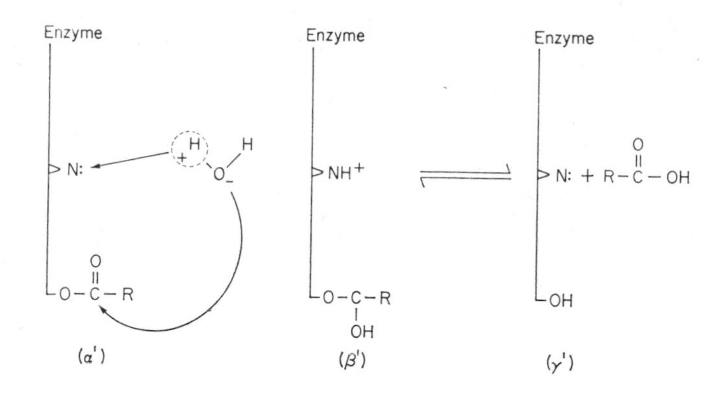

Let us now return to the charge-relay system of the active grouping. The linkage between the three residues is achieved by the formation of hydrogen bonds as indicated below:

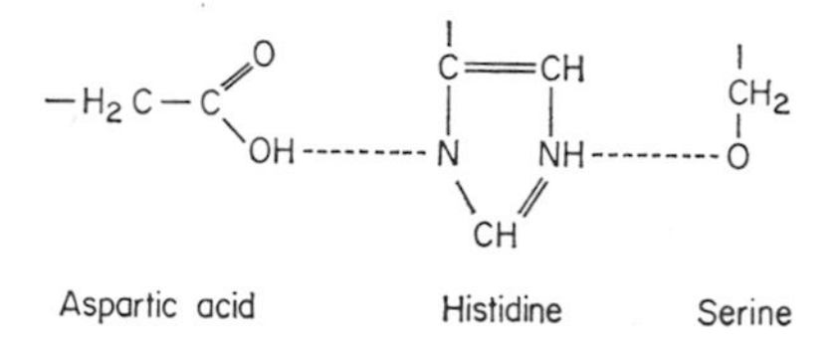

Hydrogen Bonds. It will be recalled that the hydrogen bond is a weak type of semi-covalent link in which a hydrogen atom tends to share its electron partly with one atom, e.g. O, and partly with another, e.g. N. These hydrogen bonds are indicated by dotted lines. In the above arrangement the hydrogen bonds link the H-atom of glutamic acid to an N-atom of the histidine side-chain, and the O-atom of serine to an H-atom of the imidazole NH-group. An alternative linkage is indicated below:

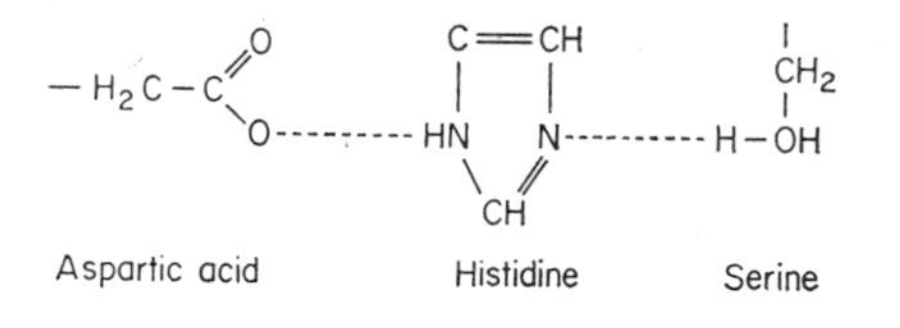

in which now what were formerly hydrogen bonds have become covalent links by, in effect, a shift of electrons along the chain, and we may write an equilibrium formulation:

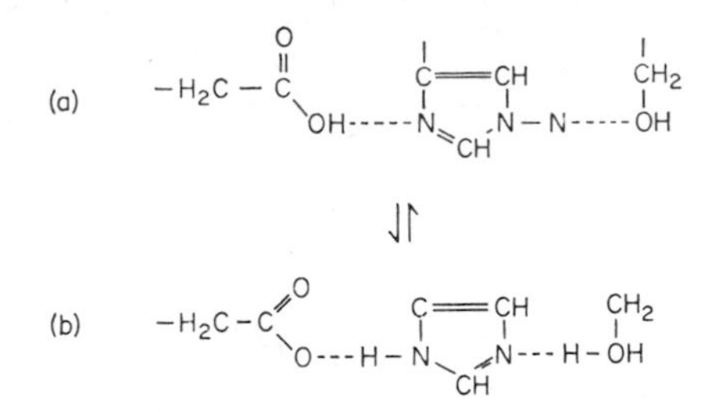

The form favouring reaction with a peptide grouping is form (a) with the hydrogen bond linking the O-atom of the serine OH-group with the histidine hydrogen, a link that generates a negative charge on the serine O-atom, which is the requirement for reaction with the peptide bond (*nucleophyllicity*). It has been concluded that this is the form that would predominate in the "microenvironment" of the active grouping. Thus the intermediate step indicated by (β) on p. 195 becomes possible, with the proton given up by the serine OH-group, being, in effect, passed on to the histidine nitrogen and subsequently returned.

Co-operation between Residues

The important point brought out by this analysis is that the reaction of a single grouping in the enzyme molecule is brought about by co-operation between three amino acid residues in a particular configuration, so that the prime basis of specificity, here, is the ability of the catalysed reactant to fit on to the large enzyme molecule in such a way that the appropriate peptide linkage comes into the correct relation with this configuration.

Action of Lysozyme

Another example is provided by the enzyme *lysozyme*, which exerts its antibacterial action by hydrolysing the mucopolysaccharide coats of certain bacteria. The material susceptible to attack is a copolymer of two sugars, *N*-acetylglucosamine (NAG and *N*-acetylmuramic acid (Fig. 4.13).

X-ray analysis was able to indicate where the important reactive site was and how a sugar, such as *N*-acetylglucosamine (NAM), fitted into it (Fig. 4.14) with the likely hydrogen bonding. It was concluded that specificity of the enzyme was due not only to a configuration of amino acid residues, namely Glutamine 35 and Aspartate 52, but also to the requirement that some six residues had to fit in a cleft in the enzyme molecule. When the six residues have fitted in, then it is residue D

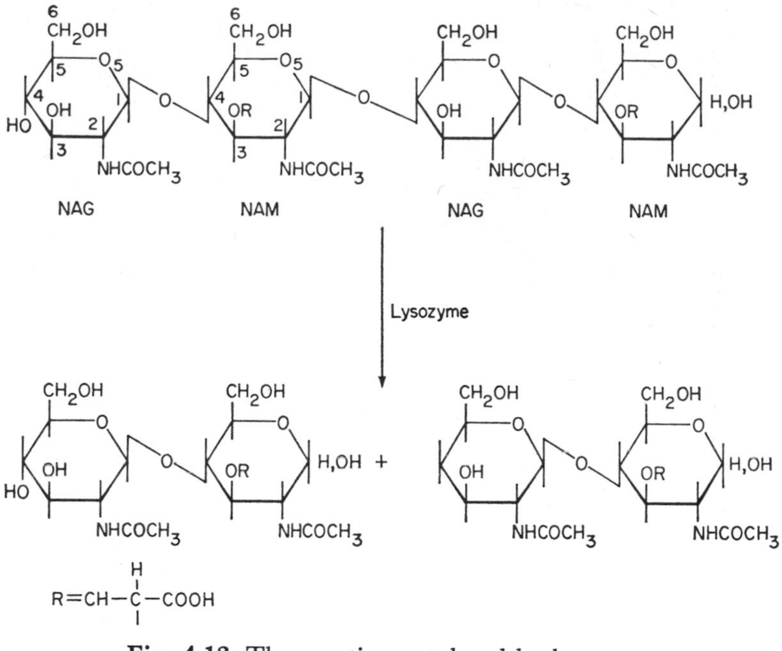

Fig. 4.13. The reaction catalysed by lysozyme. (Philips. *Harvey Lectures*, 1970–71, **66,** 141.)

(Fig. 4.15) that is subjected to strains sufficient to lower the activation energy for the cleavage.

THE ENERGY CHANGES

The activation energy required to make the molecules react determines whether, under given conditions of temperature and pressure, the reaction will proceed. The *extent* to which it will go is determined by quite other factors and is governed essentially by the relative energy-contents of the reacting molecules compared with their products.

Internal Energy

Thus a chemical compound or an atom may be considered to have a certain chemical energy—its *internal energy*—which is potential energy and will only be released during transformation, i.e. as a result of combination with other atoms or molecules or by rearranging its own atoms.

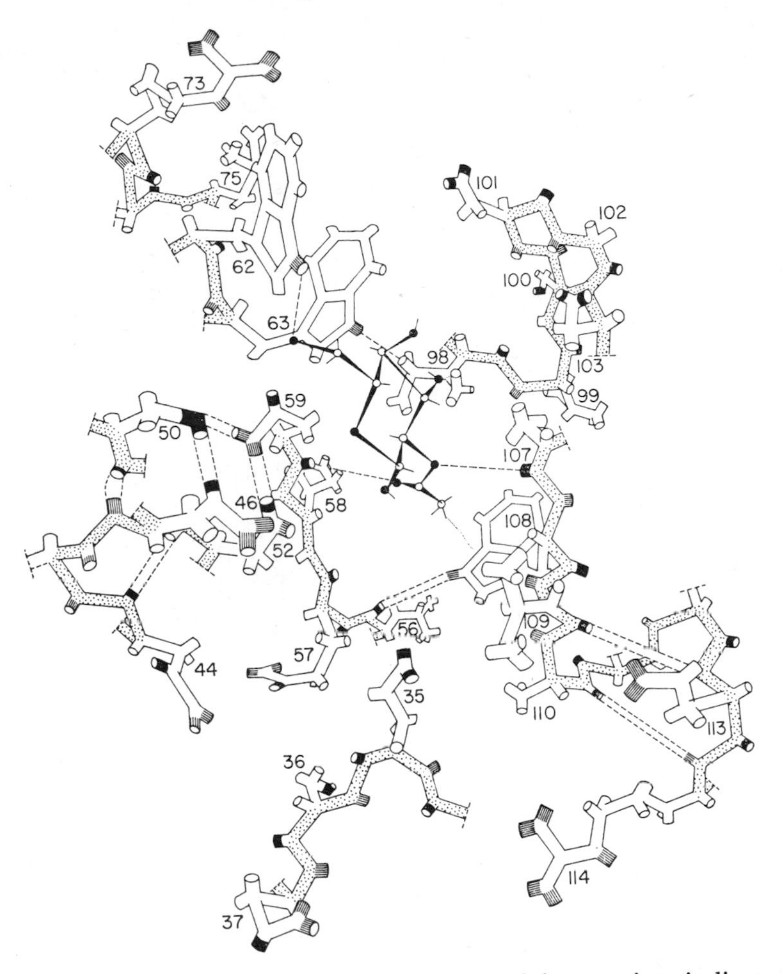

Fig. 4.14. The binding to lysozyme of β-N-acetylglucosamine, indicated by dark lines. (Philips. *Harvey Lectures*, 1970–71, **66,** 141.)

To take a given reaction, say the combination of hydrogen with oxygen to form water:

$$H_2 + O \longrightarrow H_2O$$

the H_2O molecules are said to have less internal energy than that of the H_2 and O_2-molecules, the difference appearing in the form of heat or work during the reaction.

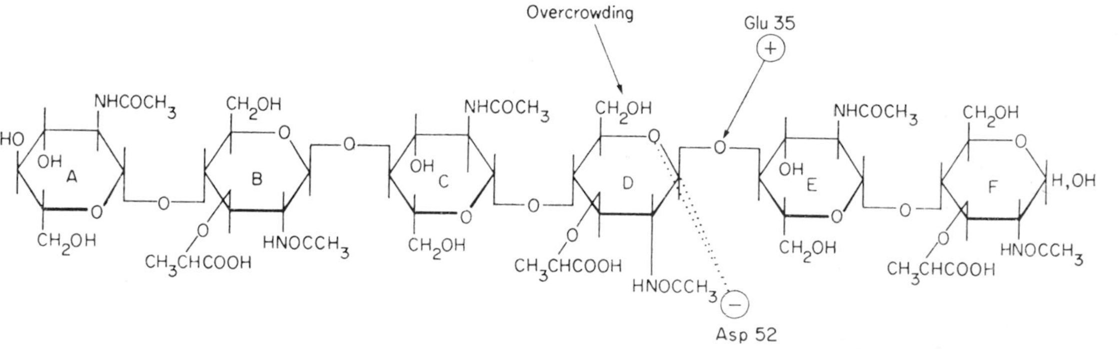

Fig. 4.15. Schematic diagram illustrating the stereospecificity of lysozyme and its proposed mechanism of action. Six sugar residues participate in the formation of the enzyme–substrate complex with groups shown along the upper edge of the molecule directed into the active site cleft of the enzyme. The sugar residue in site D is distorted from the chain formation. This distortion and action of amino acid residue GLU 35 and ASP 52 promote cleavage of the GU-O bond in the glyosidic linkage between sugar residues D and E. (Philips. *Harvey Lectures*, 1970–71, **66,** 141.)

Bond-energy

This concept of internal energy, bound up in a molecule, becomes clearer if we introduce the concept of bond-energy. The energy of a chemical bond, e.g. that connecting H with O in H_2O, may be defined as the amount of energy required to break it, i.e. to set the H-atom free. If, say, 100 kcal per mole are required, then we may say that the total bond-energy of the water molecule is 200 kcal, and by this we mean that 200 kcal are required to split the H-atoms free, or alternatively that 200 kcal will be evolved when two hydrogen atoms react with one atom of oxygen to form a molecule of water. In a similar way we may define the bond-energies of the H_2 and O_2 molecules as the energies required to split the respective atoms apart. Thus in the equation describing the formation of water:

$$2H_2 + O_2 \longrightarrow 2H_2O + \text{Energy.}$$

The amount of energy that will be set free will be determined by the differences in the bond-energies of the H_2 and O_2 molecules on the one hand and those of the water-molecules on the other. The bond-energy of the H—H link is 103 kcal, that of the O—O link is 117 kcal, and that of the two O—H links in water is 222 kcal, so that we may write:

$$\begin{array}{lcl} 2 \times 103 + 117 & \longrightarrow & 2 \times 222 + \text{Energy} \\ 323 & \longrightarrow & 444 - 121 \text{ kcal.} \end{array}$$

Thus the energy set free by what is essentially a rearrangement of chemical bonds follows from the fact that this rearrangement has left us with molecules that have different bond-energies from those of the reactants, and this energy represents the total energy, in the form of heat and work, that can be extracted from the reaction.

Energy of Reaction

In general then, in the reaction describing, for example, the oxidation of glucose to CO_2 and H_2O, if we were to add all the bond-energies of the molecules of the reactants (e.g. of $C_6H_{12}O_6 + 6O_2$), and those of all the products (namely of $6CO_2$ and $6H_2O$) we should find that the difference was the energy set free by the chemical reaction.

Enthalpy Change

Experimentally the heat set free when one mole of glucose is oxidized to CO_2 and H_2O is found to be 673 kcal. If this occurs at constant temperature and pressure the energy set free is called the change in *enthalpy* (ΔH). If no work of any kind is done by the reaction (except to expand the volume of the reaction mixture if there is any change in volume, in which case the value of ΔH includes a term $P\Delta V$) then this is the *heat of reaction* at constant temperature and pressure. It is customary to give the value of the enthalpy change a negative value if heat is liberated, so that the reaction is written:

$$C_6H_{12}O + 6O_2 \longrightarrow 6CO_2 + 6H_2O(\Delta H = -673 \text{ kcal.}).$$

The enthalpy of a reaction is indicated as a difference we have seen that it is essentially the difference between the internal energies of the reactants and products of the reaction. So, just as we can say that the energy liberated by a reaction is given by the difference in the internal energies of the reactant and products, so we can say that the enthalpy liberated, ΔH, is the difference between the enthalpies H_1, $H_2 + H_3$ . . . etc. of the reactants, and the enthalpies, H'_1, H'_2 . . . etc. of the products. The enthalpy is thus one means of defining the nature of the molecules entering into, or resulting from, a reaction, and tells us how much heat can be liberated from it.

"High-energy Bond"

To return to the concept of bond-energy, we must be clear as to what is meant by the bond-energy of one compound being greater, or less, than that of another. If the energy liberated in a reaction is indicated by a negative quantity, as in the equation:

$$AB + CD \longrightarrow AC + BD - x \text{ calories}$$

then in order to make the sum balance, the total bond-energies of AC + BD must be greater than that of AC + CD, in other words the bond-energies of the products of an exothermic reaction are larger than those of the reactants. To revert to the oxidation of glucose, then, we may say that the total bond-energies of the CO_2 and H_2O molecules are greater than those of the glucose and O_2 molecules. What we are saying, in effect, is that more energy must be used to rupture the links in the CO_2 and H_2O molecules than those in the glucose and O_2 molecules, and we are defining our bond-energy as the amount of energy required to *break* the link rather than that which is released when the link is formed. This is the meaning the chemist attaches to a value of bond-energy, so that a "high-energy bond" is one of high stability requiring high energy to rupture it. Unfortunately the term has been applied to just the opposite type of bond, namely the phosphate link in ATP that is broken when ATP becomes ADP; this is highly reactive, in the sense that ATP reacts with many compounds to phosphorylate them, a process that would not take place if a large amount of energy were necessary to break the terminal phosphate link in ATP.

Free Energy

The enthalpy of a chemical reaction tells us how much energy is set free or absorbed. The more important requirement, however, is to know how much of this available chemical or internal energy is available for the performance of work of any sort, and it is here that the development of thermodynamic theory has been of value. The total amount of energy in a chemical reaction that can theoretically be converted to work is the *free energy*, indicated usually by ΔG or ΔF. Thus, under conditions when the maximum work is being obtained from a chemical reaction we have:

$$\text{Total energy set free} = \Delta H = \Delta G + T\Delta S \qquad (8)$$

where ΔS introduces a new concept, namely that of *entropy*, which, multiplied by the absolute temperature, has the dimensions of energy. For the moment we consider that $T\Delta S$ makes up the balance in our energy book-keeping, it represents the theoretically available energy that does not appear as work.

Maximum Work

The concept of free energy is valuable, since it tells us the amount theoretically *available* for useful work, by contrast with the enthalpy, which tells us only the amount of energy set free. As with internal energy, we may speak of the free energy of a given compound, so that when we take the difference between the free energies of the reactants and products of a reaction, the difference indicates the free energy available if the reaction is carried out under the most favourable conditions. In general, because these conditions are "ideal", the amount of free energy liberated will be less. Experimentally, moreover, the free energy of a reaction, ΔF, can be calculated from a knowledge of its equilibrium constant:

$$K = \frac{[A] \times [B]}{[AB]}$$

in the reaction $A + B \rightleftharpoons AB$

$$\Delta F^{\circ} = -RT \ln K \qquad (9)$$

In this case the free energy change is called the *standard free energy change*, ΔF°, defined by fixing the temperature at 25°C, the pressure at 1 atmosphere, and the concentrations of the reactants as 1 M.

If the reaction whose ΔF we are interested in occurs at 37°C, an appropriate correction must be made.

Spontaneous Reaction

From the point of view of deciding whether a given reaction will proceed spontaneously, the Second Law of Thermodynamics tells us that a reaction in which free energy is set free, called *exergonic*, can go spontaneously, whilst an *endergonic* reaction cannot. As to whether an exergonic reaction will go spontaneously depends, as we have seen, on the activation energies. It is conventional to indicate the setting free of energy for work as a negative process hence, if ΔG of a chemical process is negative, it can proceed spontaneously.

Entropy

Let us consider this concept in a little more detail; essentially it takes account of energy changes that are not involved in changes in chemical bonds, or internal energy. Thus we can imagine a gas to expand from a concentration C_1 to a concentration C_2, and in doing so it may perform work, just as work is necessary to compress the gas from C_2 to C_1. Thus the gas has energy available for work independently of the internal energies of its molecules, and if the work is carried out reversibly, i.e. under optimal conditions, we may say that the maximum work done, ΔW, is:

$$-RT \ln C_2/C_1.$$

During expansion heat, ΔQ, is absorbed from the surroundings, so according to the First Law:

$$\Delta Q = \Delta W = -RT \ln C_2/C_1$$

since there is no change in internal energy of the molecules.

If we divide by the absolute temperature, T, we have:

$$\Delta Q/T = -R \ln C_2/C_1 = \Delta S. \tag{10}$$

ΔS is called the change in entropy of the system; in this case it has a positive value, and in general, if, in a reaction like this where there has been no change in the internal energy of the system, the entropy change is positive, the reaction will proceed spontaneously.

Oxidation of Glucose

The significance of entropy changes may be established by the results of calculations of the free energy change in the oxidation of glucose. We have seen that the enthalpy ΔH is -673 kcal; the calculated ΔF is -688 kcal, and this means that the entropy has increased. Thus equation (8) told us:

$$\Delta H = \Delta F + T\Delta S.$$

ΔH is greater than ΔF due to the absorption of heat from the environment during the conversion of sugar to carbon dioxide and water under isothermal and constant pressure conditions, and this heat has been turned into useful work. This portion we ascribe, then, not to the change in internal energy of the molecules but to the change in entropy; this entropy change was provided by the outside environment of the closed system,* and in accordance with our convention, anything received by the reacting system is designated as positive.

The magnitude of the entropy change is given by

$$\frac{688{,}160 - 673{,}000}{298} = 50{\cdot}8 \text{ entropy units (cals mole}^{-1}{}^{\circ}\text{C}^{-1}\text{)}.$$

Functions of State

Just as with free energy and enthalpy, the entropy change during a chemical reaction may be described as the difference between the entropies of the reactants and the products, so that we have three "energetic functions of state" of a system, namely the *enthalpy*, the *free energy*, and the *entropy*. They are defined as functions of state because the magnitude in the change of these functions is independent of the manner in which the change was brought about.

Degree of Disorder

The entropy of a system is often indicated as its degree of disorder; thus we have seen that when a gas does work by expanding, or when a solution is made to do osmotic work by dilution, the work is done, in effect, at the expense of entropy rather than internal energy, and is achieved by the inevitable tendency for a system to pass from a system of lower to one of higher entropy. Thus the solution of higher concentration has more order to it than the one with lower; the transition can be made to do useful work with the appropriate machine, and the work may be said to be the consequence of the increase in entropy. At the absolute zero of temperature the entropies of all molecules are said to be zero, i.e. complete order is established.

* By a "closed" system in this context we mean one that is separated from its surroundings, in so far as exchanges of matter are concerned, but does allow the flow of heat across its boundaries. Thus the expanding gas obtains heat from its surroundings.

Changes in Molecular Shape. Other examples of entropy changes, of significance for the biologist, are those accompanying changes in the shapes of large molecules. Thus when rubber is stretched its entropy is reduced, because its randomly moving chains are constrained to stay in more permanent configurations. When the stretched rubber is released work can be done, and this can be regarded as the result of the increase in entropy of the rubber. In the same way quite large changes in the shapes of protein molecules may be deduced from the changes in entropy.

Coupling of Chemical Reactions

A vast number of chemical reactions taking place in the body appear to be endergonic, and yet they occur spontaneously; the reason for this is that, when any given reaction is examined in isolation, it is found to be exergonic. Thus the straightforward synthesis of glycogen from lactic acid is endergonic; it does not, in fact, proceed in a "straightforward" manner, but with the intervention of ATP, so that the actual reactions that go to the conversion of lactic acid to glycogen are, as we have seen, exergonic. In general, it is customary amongst biochemists to speak of the "coupling" of an exothermic with an endothermic reaction, the energy from the one being employed to "drive" the other, but it must be appreciated that the only way for this to happen is for two reactions to have a common reactant, i.e. the reactions must be connected by *matter* and not by *energy*.

Common Reactant

Thus the conversion of A to B may be endergonic,

$$A \longrightarrow B \qquad (\alpha)$$

with a positive ΔF; for the sake of argument suppose its equilibrium constant $\frac{[B]}{[A]} = 0{\cdot}01$, i.e. that at equilibrium there are 100 molecules of A to one of B. ΔF° is calculated as $-RT \ln K = 2470$ calories. If B can be made to form a compound, C, in accordance with the reaction:

$$B \rightleftarrows C \qquad (\beta)$$

and the equilibrium constant is 1000, i.e. if there are 1000 molecules of C at equilibrium compared with 1 molecule of B, clearly, without argument, the reaction of B to C is going to reduce the concentration of B to the extent that B will be formed from A, i.e. reaction (1) will be forced to go to the right. Ultimately, then, A will be converted to C. The free energy change in the B ⟶ C reaction is given by:

$$-RT \ln 1000, = -4110 \text{ calories.}$$

Thus in a sense we may say that the endergonic reaction, B ⟶ C, is driving (*pulling* is a better word) the reaction A ⟶ B by virtue of its negative free energy change.

The reaction A ⟶ C may be treated as the sum of two separate reactions; the equilibrium constant is given by $K_1 \times K_2 = 0{\cdot}01 \times 1000 = 10$, and the $\Delta F°$ is given either by $-RT \ln 10 = -1640$ calories, or $2470 - 4110 = -1640$ calories. In this case of linked reactions, as with all others, there is a common reactant; moreover, the "driven" reaction must precede the "driving" reaction.

Phosphorylation of Glucose. Another example may be given by the conversion of glucose to glucose 6-phosphate. The biochemist states that the reaction of glucose with phosphate in accordance with equation (γ) is endergonic:

$$\text{Glucose} + \text{Phosphate} \rightleftharpoons \text{Glucose 6-phosphate} \qquad (\gamma)$$

the value of $\Delta F°$ being + 3300 calories/mole. According to his terminology, therefore, the reaction must be driven by an exergonic one, namely the hydrolysis of ATP to ADP and phosphate:

$$\text{ATP} \rightleftharpoons \text{ADP} + \text{Phosphate},$$

the $\Delta F°$ being negative and equal to -7400 calories/mole.

If we add the two reactions we have:

Glucose + ATP ⟶ Glucose 6-phosphate + ADP with a $\Delta F°$ of -4100 calories/mole.

It cannot be too strongly emphasized that this is only a way of looking at the process, and there is no reason to believe that an inorganic phosphate group is ever involved in the reaction, which in fact is catalysed by the single enzyme glucokinase and doubtless represents the direct transfer of phosphate from ATP to glucose. Thus thermodynamics can tell us nothing as to the actual mechanism of the reaction, only whether it is likely to proceed spontaneously or not.

CONTROL MECHANISMS

The metabolism of a complex organism like the mammal is subtly adapted to meet the demands at any given moment, and this clearly needs a highly sophisticated set of controls that modify the rates of the different reactions, and shift the directions in which a given metabolite flows; thus a glucose molecule can be converted to carbon dioxide and water, or alternatively its metabolic processes may be shunted into the formation of lactic acid, or it may be converted into glycogen—*glycogenesis*. The lactic acid produced by glycolysis can only be converted to pyruvic acid, i.e. by reversal of the pathway that produced it, but from this point several pathways are open to it; thus it can enter the tricarboxylic acid cycle through acetyl CoA, or it can be converted to fatty acids or steroids; it can be reconverted to glucose and glycogen (*gluconeogenesis*), or it can react with glutamic acid to give alanine. A variety of methods of control over the numerous possibilities of reaction are available, of which we may indicate a few.

Competition for Substrate

Thus competition for a substrate may well influence the direction in which a reaction goes; the *Pasteur effect* is the name given to the shift from anaerobic glycolysis to oxidative phosphorylation, brought about by the presence of oxygen. A possible explanation for the influence of oxygen is the greater demand for ADP created by the oxidative phosphorylating processes, and this reduces the amount available for the anaerobic mechanism. The *Crabtree effect* is the reverse of the Pasteur effect, consisting in the inhibition of oxidation in the presence of an excess of glucose; this may be due to competition of the glycolytic reactions for the cell's inorganic phosphate and NAD, leaving less for the oxidative respiratory reactions.

Product Inhibition

An important method of built-in control is the inhibition of an enzyme by excess of the product whose formation it catalyses, so that when too much of the substance is produced its production automatically ceases. This inhibition may be brought about by reducing the activity of the enzyme, or it may operate by inhibiting the synthesis of new enzyme.

Enzyme Synthesis

Alternatively the opposite action, namely acceleration of a process already proceeding, or the initiation of one that is in abeyance, may be caused by "de-repression" of the synthesis of an enzyme. Thus the synthetic abilities of a given cell are enormous; theoretically its nucleus has within it the capacity to direct the synthesis of all the enzymes found in the body, and hence to enable the cell to carry out activities that are shared between many different cells. A given cell does not do this, but restricts its activities according to its special degree of differentiation, and this is achieved by the *repression* of the genes that would otherwise induce this behaviour. De-repression represents the release of this inhibition, and seems to be one of the mechanisms for metabolic control.

Blood Glucose

It may well be that the switching of activity from synthesis of glucose (gluconeogenesis) to utilization (glycolysis) or *vice versa* depends on the repression of synthesis of certain key enzymes in the one process, associated with de-repression of key enzymes in the opposing process. Thus the level of sugar in the blood is maintained fairly constant as a result of the interplay between glycolysis and gluconeogenesis, the former being the utilization of sugar by the cells of the body in its multifarious activities, the latter being a process carried out by the liver. The high blood sugar in diabetes is an expression of an imbalance whereby gluconeogenesis dominates; in experimental diabetes in rats the key enzymes for gluconeogenesis in the liver do actually increase whilst those for glycolysis decrease, the situation being reversed by the pancreatic hormone, insulin. Thus insulin may act as a suppressor of the key enzyme for gluconeogenesis and an inducer for the key enzymes of glycolysis. Again, when a tissue is growing rapidly, as in the immature animal, it is found that key enzymes concerned with gluconeogenesis are in high concentration, whilst those for glycolysis are in low concentration. During subsequent development, the relative concentrations alter, with the glycolytic enzymes increasing and the opposing ones decreasing until the adult picture is reached where the opposing enzyme systems reach a balance. With a rapidly proliferating tissue, such as a tumour, it is found that the characteristic feature is a predominance of the glycolytic enzymes, and the ratio to the opposing enzymes, e.g. hexokinase/glucose-6-phosphatase, correlates very closely with tumour growth-rate.

Functional Genic Unit

We may postulate, in accordance with the so-called *Functional Genic Unit Concept*, of Weber, that there are sequential alterations in the read-out of the nuclear genes and in their selective transcription and translation to give the synthesized enzymes. It is these processes that determine the phenotypic pattern of metabolism of cells belonging to different age-groups of animals. The alterations in this read-out are doubtless due to de-repression of the genes.

Nucleic Acid Metabolism

An essentially similar sequence of enzyme activities is found with nucleic acid metabolism, the key enzymes favouring synthesis of nucleic acids or nucleotides being dominant during rapid replication, and in recess in the adult animal; the reverse is found with the key enzymes for degradation. Thus Fig. 4.16 illustrates the synthetic and degradative pathways for a pyrimidine nucleotide, uridine monophosphate (UMP). The rate-limiting enzyme in the degradative

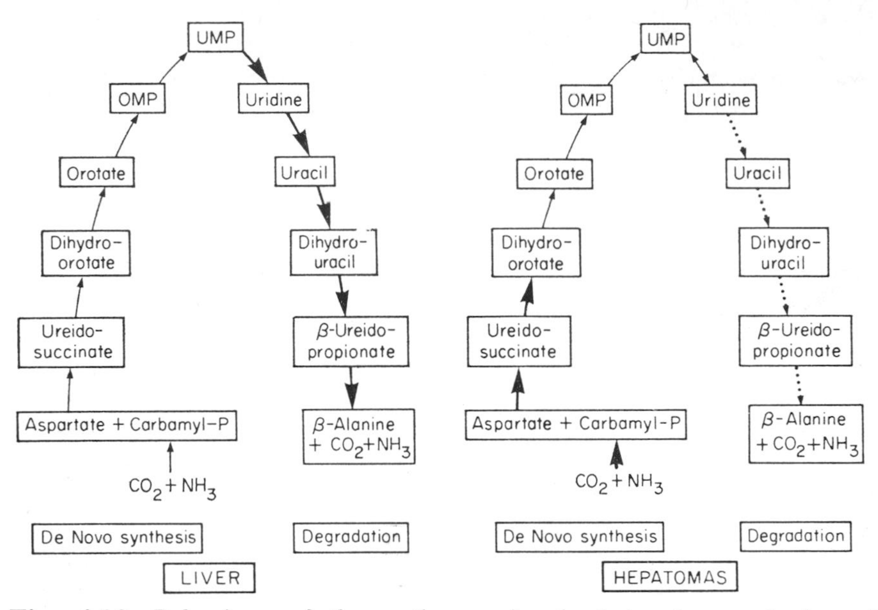

Fig. 4.16. Behaviour of the pathways involved in the synthesis and degradation of uridine monophosphate (UMP) in normal liver and hepatomas. The predominant activities are represented by heavy lines. (Weber *et al.*, *Adv. Enzyme Res.*, Vol. 9, 1971.)

pathway is dihydro-uracil dehydrogenase and it was found that, in tumours, the faster the growth-rate the smaller the activity of this enzyme, whilst that of the enzymes favouring synthesis, namely aspartate transcarbamylase and dihydro-orotase, increased in parallel with tumour growth-rate. Thus tumour growth is essentially a manifestation of an imbalance in the enzymic activities of the cell; so far as DNA synthesis is concerned this is due to a preponderance of key enzymes in synthesis with a repression of those concerned with degradation. As Fig. 4.17 shows, in the normal liver the degradative processes are in control, whereas in hepatoma the synthetic ones are.

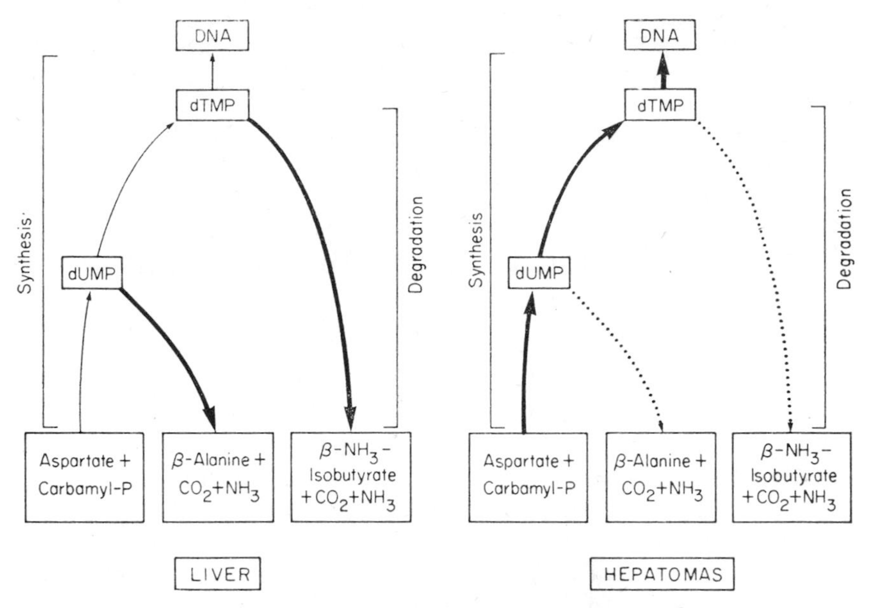

Fig. 4.17. Schematic representation of the metabolic imbalance in pyridine metabolism in hepatomas. The normal liver metabolic conditions favour degradation over synthesis. In the full-blown hepatoma pattern, the synthetic pathways are greatly increased concomitant with a decrease in degradative capacity. (Weber *et al.*, *Adv. Enzyme Res.*, Vol. 9, 1971.)

So much, then, for a very summary review of the energy exchanges in the organism; in succeeding chapters we shall be concerned with the processes for which this energy liberation is required; of fundamental importance, too, is the manner in which the cells are supplied with the components for their chemical reactions and are relieved of those products that are either unnecessary waste, such as urea and CO_2, or are required in other tissues, e.g. the amino acids synthesized by the liver, the hormones secreted by the endocrine glands, and so on. We must pass next, then, to the consideration of the basic mechanisms in the circulation of the body fluids and the relation of the circulation of the blood to the absorption and release of gases in the respiratory system.

CHAPTER 5

Mechanical Aspects of the Blood Circulation

In the complex organism, the sources of the substances necessary for life are usually remote from the cells where they are to be utilized; similarly, the places where the products of metabolism are enabled to escape from the organism are often far from the cells that produced them. Thus the development of the complex organism, as distinct from the unicellular or paucicellular organism, demands the development of a transport system, and this is achieved primarily by the formation of a transporting medium, the blood, and its circulation through a tubular system that enables this blood to come into close relations with every cell of the body.

Basic Principle of the Circulation

In this chapter we shall discuss the basic mechanical principles involved in the circulation of the blood, and in the following three chapters we shall be concerned with the basic mechanism by which fluid and gaseous exchanges between blood and the tissue fluids take place.

In the mammal and many lower vertebrates the basic principle of the circulation consists in the forcing of the blood through elastic tubes, by means of a pump (the heart); the fluid is driven to two primary regions where exchanges of material between it and its surroundings are possible; these regions are (a) the lungs and (b) the rest of the tissues in the body. In the lungs the blood is enabled to take up oxygen from the air and dispose of CO_2 to it, thereby taking on one of the gaseous substrates in the energy-giving chemical reactions of the body and giving up the gaseous product of these reactions, namely carbon dioxide. This, the *pulmonary circulation*, runs in parallel with the *systemic circulation* that transports the blood that has passed through the lungs to the tissues of the body, where the oxygen acquired in the lungs may be given up and the CO_2, produced by the tissues, may be taken away.

General Scheme

Figure 5.1 illustrates the general scheme; thus, to begin at the pulmonary end, the blood is forced out of the right ventricle (R.V.) of the heart along the pulmonary artery into the lungs; during its passage through the lungs it comes into very close relations with the

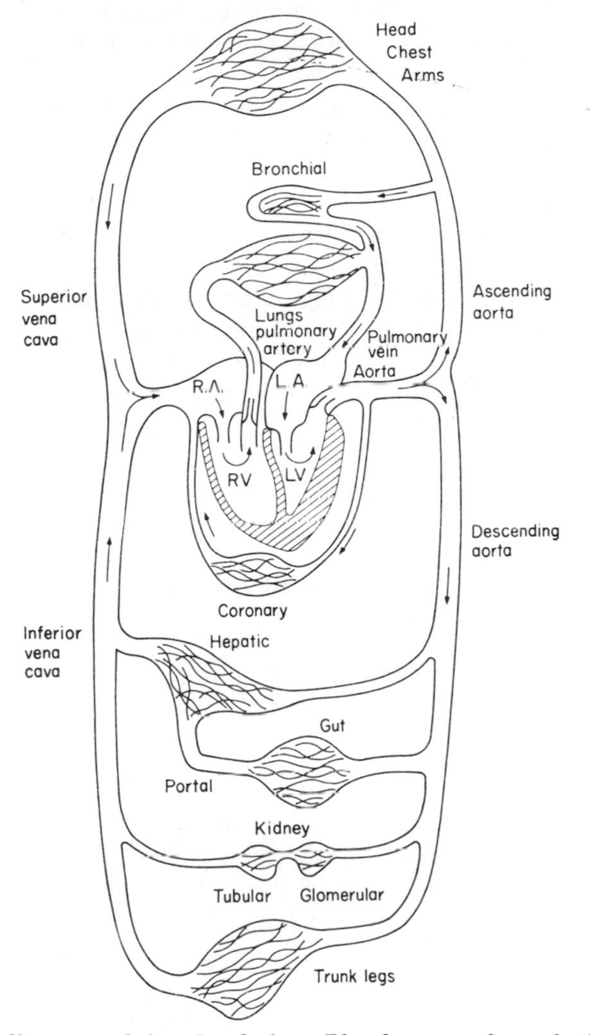

Fig. 5.1. A diagram of the circulation. Blood passes through the capillary beds to each organ by means of a number of parallel routes.

atmospheric air, and the necessary respiratory exchanges of gases take place; the blood returns along the pulmonary veins into the left atrium (L.A.) of the heart, and from this chamber it is forced into the left ventricle (L.V.); from the left ventricle it is expelled at relatively high pressure into the aorta, the single large artery that carries the oxygenated blood from the heart to the tissues of the body. This larger artery bifurcates repeatedly so that each tissue of the body receives its supply; thus the first bifurcation consists in the formation of the ascending and descending aorta, the latter supplying mainly the trunk, lower limbs and the internal organs or viscera. The blood is returned from the tissues to the heart by way of veins, the superior and inferior vena cava, which empty into the right atrium. From here the blood is forced into the right ventricle and the circuit begins again.

Valves

Clearly, if the heart is to function properly, the exits and entries to its chambers, the atria and ventricles, must be governed by valves, whilst the contraction of the walls of each of its chambers must be well co-ordinated in time. Thus contraction of the left ventricle, which expels blood at high pressure into the aorta, must coincide with the opening of the valves controlling flow into and out of the aorta—the semilunar valves—and the closing of the communication between the left atrium and the left ventricle—the mitral valve—otherwise the blood could be forced back into the atrium and along the pulmonary veins.

The Blood

This is the fluid medium through which exchanges between cells and their environment occur. As we shall see, these exchanges are usually not direct, since the blood does not wash the individual cells; instead, the exchanges occur by way of the *interstitial fluid*, which acts as the go-between. The blood is a suspension of cells in a protein-containing fluid, the *plasma;* thus on centrifuging, the cells sediment leaving the yellowish plasma above. The proportion of cells to plasma is indicated by the *haematocrit ratio*, and is normally in the region of 40–50 per cent. The majority of the cells are erythrocytes, some of whose physical characteristics we have discussed earlier; they are red because of the high concentration of haemoglobin within them; this protein is the oxygen-carrier of the body, being capable of reversible combination with atmospheric oxygen as follows:

$$Hb + O_2 \rightleftharpoons HbO_2.$$

Thus, as it passes through the tissues, the reaction goes to the left, i.e. the HbO_2 complex dissociates, and the O_2 set free is taken up by the cells.

Colour

The colour of the haemoglobin in its reduced, or unoxygenated, form is much darker than in its oxygenated form, hence the blood returning in the veins varies in colour through the bright red of arterial blood to a dark brown when the blood has lost most of its oxygen, according as the tissue is using a little or a lot of the oxygen supplied to it. Thus a very sluggish flow of blood to an actively metabolizing tissue will result in a very dark venous blood; by contrast, the very rapid flow of blood through brain means that the venous blood leaving in the internal jugular vein is a fairly bright red, in spite of the high metabolic rate of nervous tissue.

Buffy Coat

Besides the erythrocytes, or red blood cells, the blood contains white cells or *leucocytes;* these, being less dense than the erythrocytes, appear as the top layer of the sedimented cells after centrifugation and are described as the *buffy coat.* These white cells are not concerned with respiration, but with the defence of the body against invasion by foreign material, e.g. bacteria. The buffy coat also contains the *platelets*, protoplasmic fragments derived from *megakaryocytes;* these contain high concentrations of serotonin (Vol. 2) and are intimately concerned with the clotting of the red blood (Vol. 2).

Chemistry

The blood is, as one might expect in view of its function, a highly complex fluid; when the cells have been removed, it remains as a solution containing predominantly salts, of which NaCl and $NaHCO_3$ are the major constituents, but it also contains many organic materials concerned with nutrition, e.g. sugars, amino acids and vitamins; and many other substances, often in minute concentration, concerned with control of the body's mechanisms, namely hormones and humoral transmitters. It is thus pointless to present a detailed table containing the chemical composition of the plasma. In the present chapter we are concerned only with the blood as a circulating fluid which must ultimately be forced through very fine tubes—the capillaries—with diameters of the same size as, or even smaller than, those of the cells it carries in suspension.

THE BLOOD VESSELS

The blood is carried out of the heart by the arteries and returns through the veins; the arteries are the site of the high pressures required to force the fluid through its circuit, and their walls are characteristically thicker and more muscular than those of the veins. The calibres of the arteries become smaller with successive division or ramification, and, associated with this, there is a decrease in thickness of the wall whilst the importance of the muscular elements tends to increase. The smallest arteries are called *arterioles*, with diameters of 20–100 μ, which are to be distinguished from the vessels with which they are in continuity, namely the *capillaries*, largely by virtue of possessing muscular walls, so that the capillaries lack muscular elements and consist of little more than endothelium-lined tubes. From the capillaries the blood, on its return path to the heart, passes through very small veins, or *venules*, which by successive confluence lead to larger and larger veins.

Structure of the Vessels

In general, we can characterize all the blood vessels, with the exception of capillaries, as vessels with three layers; from inwards out: the *intima*, the *media* and *adventitia* (Fig. 5.2).

Intima

The basic feature of the intima is the innermost layer of endothelial cells that forms the cellular lining to the vessel; it is very likely that this acts as a barrier to diffusion of material from the lumen of the vessel to the outer layers and thence into the surrounding tissue spaces, since significant escape of material from arteries and veins does not take place, this being reserved for the capillaries and venules. The endothelial cells lining arteries certainly make the tight junction type of contact, and in the smaller arterioles, of 30 μ diameter or less, the muscle cells make tight junctional contacts with the endothelial cells. The remainder of the intima is made up of connective tissue elements, i.e. fibrillar material described as either elastic and containing elastin, or collagenous; these elements are arranged mostly longitudinally. As its name implies, elastin is a protein with remarkably large rubber-like distensibility; it is chemically different from collagen, which has a much smaller distensibility but contributes greater tensile strength, so that it is a prominent component of cartilage.*

* Reticulin is a form of collagen with different characteristics.

Media

The media contains the muscular elements, together with fibrillar connective tissue material, these being wound in a tight spiral so as to appear circular in arrangement; the circular arrangement of the muscle fibres permits a reduction in diameter of the vessel when these shorten.

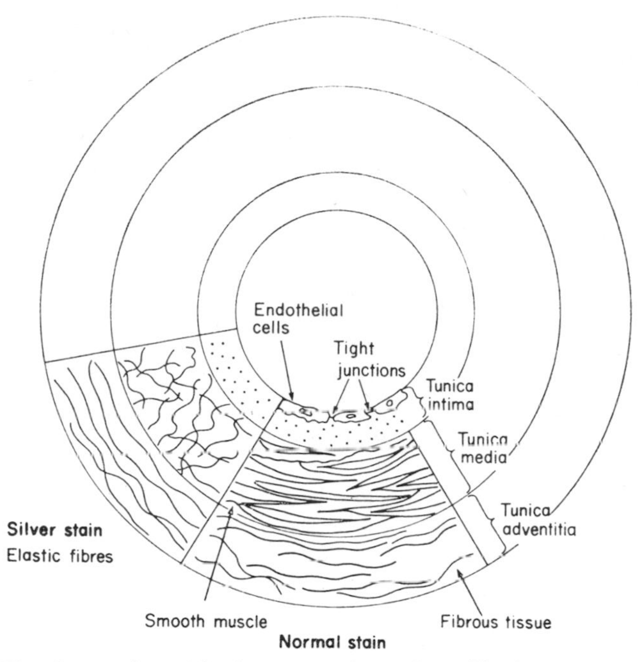

Fig. 5.2. Blood vessels, with the exception of capillaries, are composed of three layers: (a) an inner lining of endothelial cells—the intima; (b) a muscular and elastic tissue layer—the media; (C) an outer layer composed of fibrous tissue—the adventitia.

Adventitia

The adventitia is made up of connective tissue elements arranged mainly as a loose spiral and thus appear to run longitudinally.

Distensibility and Contractility

It is the relative proportions of these layers, and their composition, that determine the character of the vessel so far as its response to changes in internal pressure—distensibility—and its ability to alter its diameter in response to nervous and other stimuli—its contractility—are concerned.

Elastic Arteries

In general, the very large arteries, receiving blood in "boluses" from the heart, are characterized by their distensibility, i.e. they are capable of accommodating relatively large volumes of fluid in response to a given rise in pressure, and they are often referred to as *elastic arteries* on this account; as we shall see, it is this feature that enables the heart to preserve a continuous flow of fluid through the system in spite of the fact that for a large part of its cycle its ventricles are closed off from the arteries. Corresponding with this greater distensibility we find that the tunica media (Fig. 5.3) is composed mostly of elastic tissue in the form of concentric sheets connected by fibres; the spaces

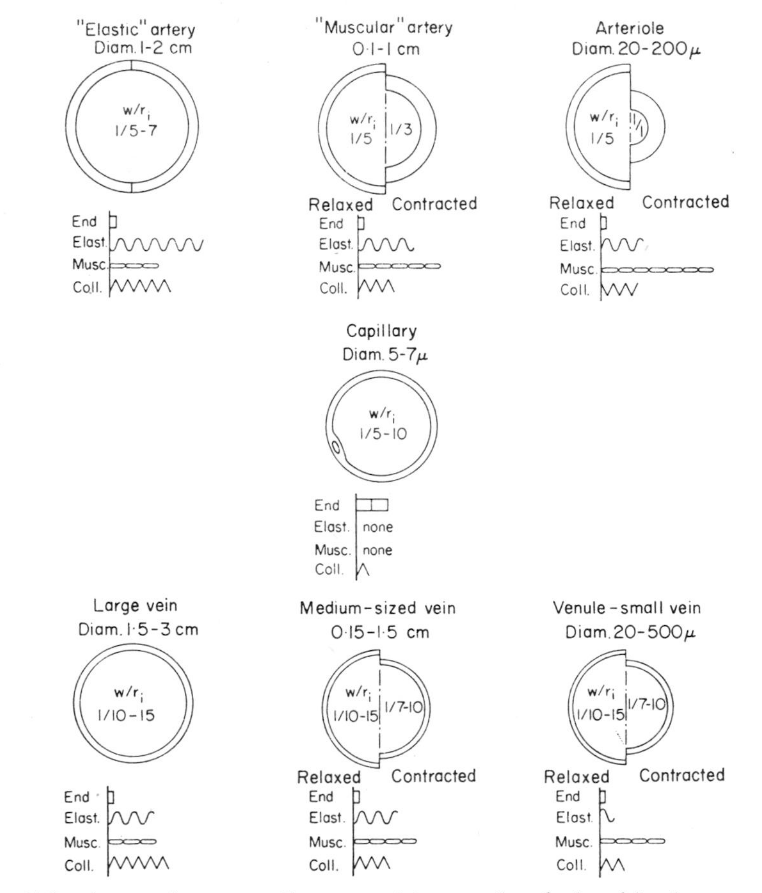

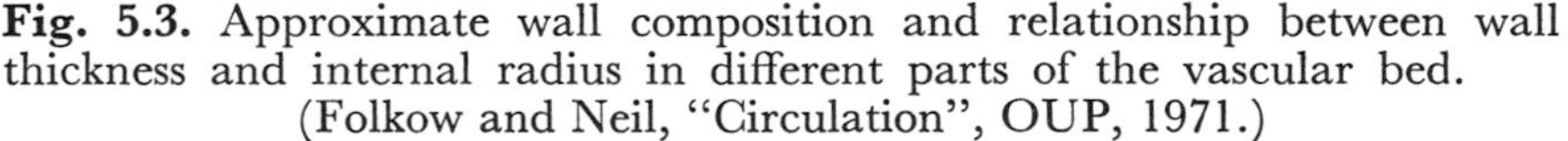

Fig. 5.3. Approximate wall composition and relationship between wall thickness and internal radius in different parts of the vascular bed. (Folkow and Neil, "Circulation", OUP, 1971.)

between the sheets contain thin collagenous and elastic fibres and smooth muscle cells.

Muscular Arteries

By contrast with the very large arteries, in medium arteries, described as *muscular arteries*, smooth muscle makes up a prominent part of the wall, the intima being restricted to the endothelium and an adjacent ring of elastic fibres (the *internal elastic lamina*). The media is thick and consists almost entirely of smooth muscle cells wound in a tight spiral; elastic fibres come into intimate relation with these. The adventitia is made up of longitudinally arranged elastic and collagenous fibres.

Microcirculation

In the finest arteries or arterioles the muscle layer is the predominant feature, an artery of 100 μ diameter having some five layers; this falls to a single layer at 50 μ, and as the capillaries are approached this layer becomes less complete until finally the capillary consists only of a layer of endothelium resting on a minimum of connective tissue elements (p. 277).

Veins

The walls of veins are thinner than those of arteries of comparable diameter; they are less rigid and, up to a point, more distensible. In general there is much greater variability in structure amongst the veins than the arteries, but the general division into intima, media and adventitia is maintained by histologists, although in some veins the tunica media is difficult to distinguish or absent.

Smooth Muscle

The muscle cells of the vascular tree are characterized as unstriated or smooth; they are "involuntary" in the sense that their state is uninfluenced by volition, but they are under control through the autonomic nervous system (Vol. 2) and the endocrine system, so that the diameters of the vessels can be varied, whilst a normal "tone" is maintained by a continuous nervous discharge.

Mechanical Properties of the Tubes

Tension in the Wall

We are primarily interested in the manner in which the tubular system responds to the forcible injection of a "bolus" of blood into it, and also in the power of the muscular system to alter the diameter of a vessel when it has to work against the internal pressure within it. The vessels respond to the injection of fluid by an expansion in diameter, and this generates an elastic tension in the walls. The tension, T, in a cylindrical tube, is related to the pressure within, P, by Laplace's law (Fig. 5.4):

$$T = P \times R.$$

T is the tension measured tangentially and may be considered to represent the tension in individual circularly orientated fibres. Thus the further the vessel expands, increasing its radius, the greater the

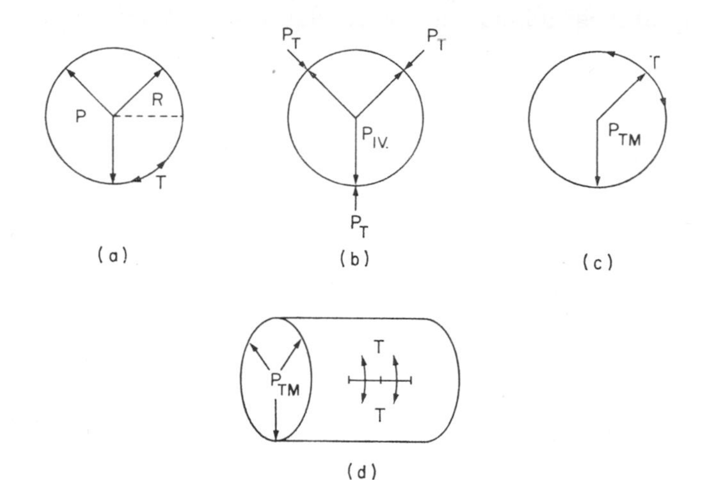

Fig. 5.4. Illustration of the law of Laplace. In (a) T is the tension in the wall balanced by P the intraluminal pressure. R is the radius. $T = P \times R$. In (b) P_T is the tissue pressure acting on the wall which in a vessel may not be zero and therefore the transmural pressure P_{TM} is the balance between the intraluminal pressure P_{IV} and the tissue pressure. $P_{TM} = P_{IV} - P_T$. (c) and (d) illustrate the situation in a vessel $T = P_{TM} \times R$.

elastic tension that develops to resist this expansion. We may note that, for a *given pressure within the vessel*, the tension required to contain this pressure, i.e. to prevent indefinite expansion of the vessel and therefore its rupture, is smaller the smaller the radius. Thus a narrow radius vessel will sustain higher pressures without bursting than will a wider vessel. This means, too, that since the large arteries are subjected to rather larger internal pressures than the smaller ones, the tensions in their walls will be considerably greater. For example, the tension in the aorta is 170,000 dynes/cm compared with only 60,000 dynes/cm in small distributing arteries; in capillaries it is only 16 dynes/cm, and this explains why such fragile vessels are able to withstand quite considerable transmural pressures.

Transmural Pressure. We must note that, here, P is really the *difference in pressure* between the inside and outside of the elastic tube, and for convenience it has been assumed that this is determined entirely by the inside pressure P, i.e. that the outside pressure, P_T, is zero (or rather atmospheric). In the living body the outside pressure, called the *tissue pressure*, P_T, is not necessarily zero, so that more correctly we should say that the tension in the elastic vessel is proportional to the *transmural pressure* (P_{TM}):

$$P_{TM} = P_{IV} - P_T \text{ (Fig. 5.4b, c).}$$

where P_{IV} is the intravascular pressure, i.e. the pressure within the vessel

Distensibility

Practically, the tension in the vessel is not measured directly but rather the pressure, P, within the tube, and this is related experimentally to the volume within the tube, to give what has been called the distensibility, $\Delta V/\Delta P$. Thus the greater the distensibility, the smaller the value of ΔP required to produce a given value of ΔV. Figure 5.5 illustrates the different distensibilities of vein and artery, and shows that a small change of pressure in a vein can cause a very large change in volume, provided we are in the low range of pressures

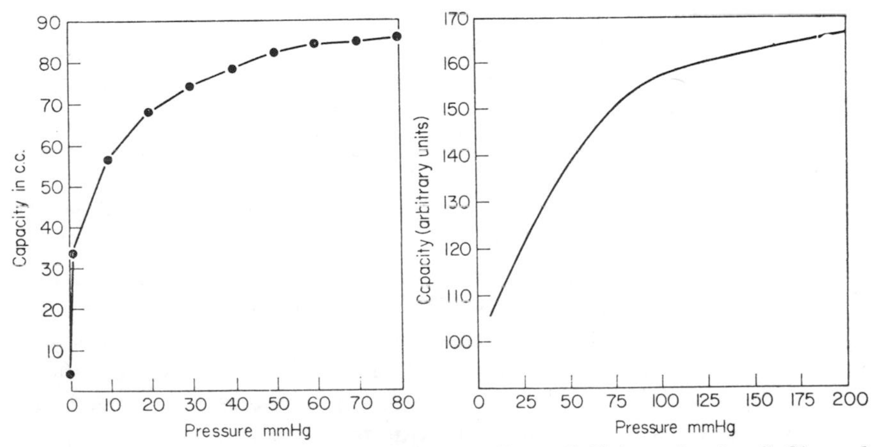

Fig. 5.5. An illustration of the different distensibilities of vein (left) and artery (right). Small changes in venous pressure can cause large changes in volume. At normal arterial pressure there is little "slack" in arteries which operate on the flatter part of the capacity pressure curve. (Wiggers, *Am. J. Physiol*, 1938, **123** 644.)

characteristic of the normal physiological state. We may note with the artery that, at physiologically normal pressures—75–125 mm Hg—the distensibility is much lower than at very low pressures, so that a great deal of the "slack" in an artery has been taken up before it has acquired its appropriate pressure. In the thoracic aorta—a distensible artery—the curve is characteristically different, so that high distensibility is maintained up to 100 mm Hg.

Force Required to Contract Vessel

The relation of T, the tension in the wall, to P the internal pressure, illustrated by Laplace's law, helps to explain why it is easier to cause a small artery to contract against a given pressure than a large one. Thus, for a given arterial pressure, the tension in the walls of the artery, created by the expansive force of this pressure, will be greater the larger the radius. This tension is developed in the elastic structures of the tissue and, in order that the related muscle may cause the vessel to constrict it must develop this tension, i.e. it must take over the load held by the elastic elements of the tissue. Thus the tension required to overcome the expansive force created by the internal pressure increases linearly with the size of the artery. Experimentally it is found, in large arteries, that procedures that cause arterial muscle to contract do not cause measurable changes in calibre at normal arterial pressures. Only when the diameter is in the region of 1 mm does muscular reduction of calibre become feasible, and it is probably for this reason that one may see active contraction in the femoral artery of a rabbit but not in that of a dog or man. When the calibre is of 200–100 μ, i.e. in the region of the arterioles, then it is easy to show, from the known ability of the smooth muscle to develop tension, that contraction of calibre to the point of obliteration of the vessel is feasible.

Critical Closing Pressure. The Laplace relation between pressure and radius also illustrates the possibility of spontaneous closure of a vessel due to muscular action; thus once a muscle has developed tension greater than that in the elastic elements, the radius must decrease. As a result of this constriction, the tension falls (Laplace law) and, unless the pressure rises, this process of constriction may go on until the vessel is completely closed. There are many physiological situations in which complete closure of small vessels is observed, and this may, as Burton has argued, be due to this phenomenon.

Young's Modulus

The elastic properties of a body are described in general terms as the ratio of the applied load (*Stress*) to the change in dimensions (*Strain*):

Young's modulus, E, refers to the change in linear dimension, e.g. of a wire or fibre, in relation to the applied stress.

E = Applied load per unit area/Increase in length per unit length. Thus if a load, F, is applied to a fibre of A cross-sectional area, and L initial length, and if the extension is ΔL, we have:

$$E = \frac{F/A}{\Delta L/L} = \frac{F.L.}{\Delta L.A.}$$

When $\Delta L = L$, the extension is 100 per cent and the dimensions of E are dynes/cm^2 per 100 per cent elongation. Thus Young's modulus for collagen is about 1.10^9 dynes/cm^2 whilst that of elastin is $2\text{–}3.10^6$ dynes/cm^2 very much smaller than that of collagen.* Estimates of Young's modulus for arteries vary from 10^6 to 10^7 dynes/cm^2 and so approximate much more closely to that of elastin than collagen; even if the elastic limit of an artery is approached, when we might expect the collagen fibrils to be taking most of the strain, the estimated modulus is still much less, namely 5.10^7 dynes/cm^2.

Chemical Composition

When arteries from different regions of the vascular tree were analysed for collagen and elastin, it was found that the amount of elastin in the thoracic aorta was 50 per cent of the total, and that this dropped suddenly within quite a short distance from the heart to a proportion of 30 per cent, which remained the same throughout the arterial tree; thus the thoracic aorta is unique in having this high amount of elastin.

THE CARDIAC PUMP

Cardiac Cycle

The mammalian heart consists of two pairs of chambers called *atria* and *ventricles* (Fig. 5.6); blood from all regions of the body except the lungs—the systemic blood—enters the right atrium by way of the superior and inferior venae cavae; from the atrium it passes into the right ventricle, its passage being assisted by the contraction of the muscle of the right atrial wall. At a given moment in the cycle, the right ventricle contracts and expels blood through its emergent—pulmonary—artery, the rise in pressure due to this contraction opening

* In common parlance, elastin would be said to have the higher elasticity, meaning that it is more extensible, than collagen: elasticity is correctly defined in terms of Young's modulus, so that the physicist would describe *collagen* as having the higher elasticity, or elastic modulus.

the semilunar valves in the orifices of the pulmonary artery and right ventricle. Regurgitation of blood back into the right atrium is prevented by closure of the tricuspid valve, so constructed as to close as soon as the pressure within the ventricle exceeds that in the atrium (Fig. 5.6). The blood from the lungs returns in the pulmonary veins to empty into the left atrium; from here it passes into the left ventricle

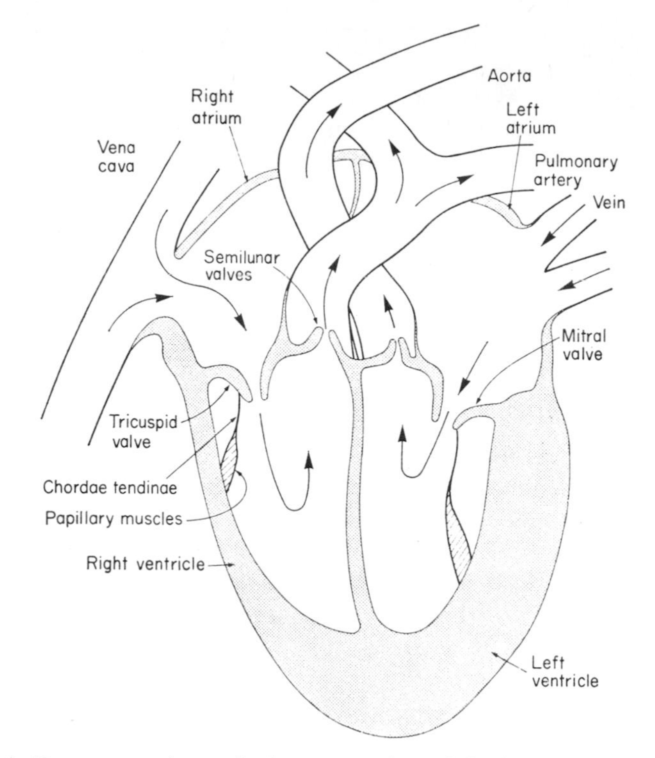

Fig. 5.6. A diagrammatic vertical cross-section of the heart showing direction of blood flow and the major valves.

assisted by contraction of the wall of the atrium, which occurs approximately at the same time as contraction of the right atrium. At the time when the right ventricle contracts, the left contracts also, regurgitation of blood into the left atrium being prevented by closure of the mitral valve. As a result of the contraction of the left ventricle, and the development of pressure within it, the semilunar valves closing the orifice into the aorta are opened and blood is expelled into this large artery. At the end of the contraction-phases of the ventricles, described as *ventricular systole*, the ventricular walls relax, the pressure within them falls, and the semilunar valves close to prevent reflux of blood from the large arterial exits back into the ventricles.

Basic Structure of the Heart

Fibrotendinous Ring

Briefly described, the heart's chambers are made up of muscular tissue; the two atria are separated from the ventricles by a fibrotendinous ring which, in effect, constitutes the main skeleton of the

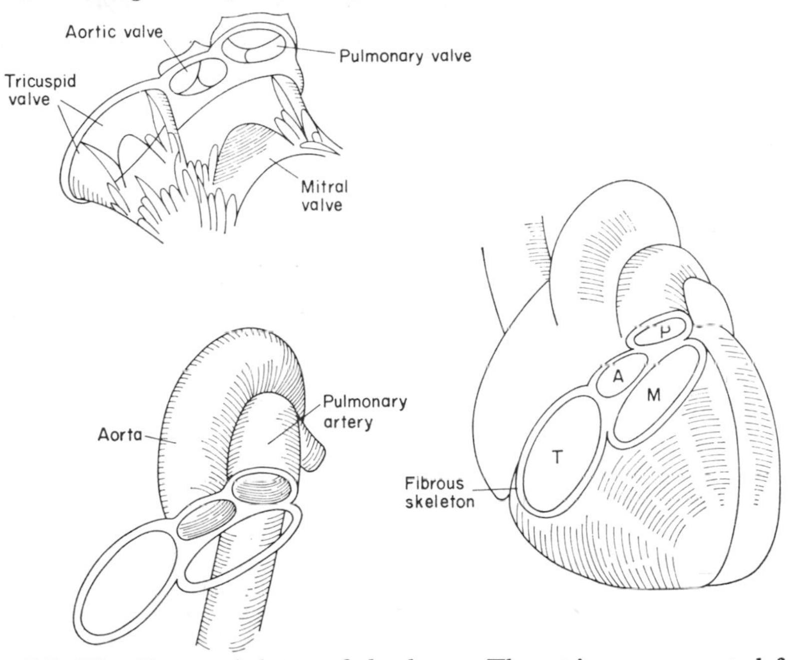

Fig. 5.7. The fibrous skeleton of the heart. The atria are separated from the ventricles by a fibro-tendinous ring which provides the sites of insertion of the ventricular muscle and prevents obliteration of valve orifices during ventricular contraction (systole). There are essentially four interconnected rings. (Rushmer, R. F., "Cardiovascular Dynamics", 2nd Ed., Saunders, Philadelphia, 1961.)

heart, in the sense that it provides the sites of origin and insertion of the ventricular muscular tissue, and imparts to the organ a rigidity that prevents obliteration of the important orifices during the powerful contraction of its muscle. Actually this fibrous skeleton consists of four interconnected rings surrounding the orifices of the great vessels; and the heart valves are attached to these, which also act as anchors to the ventricular muscle (Fig. 5.7).

Ventricular Muscles

The tissue of the ventricles is divided somewhat arbitarily into four separate muscles—the *superficial* and *deep sinospiral bundles*, and the *superficial* and deep *bulbospiral bundles*—according as they lie superficially or deeply, and according to the manner in which their fibres spiral round the chambers. The fibrotendinous ring acts as origin and insertion for the muscles; for example, the superficial bulbospiral muscle, illustrated in Fig. 5.8, originates at the fibrous ring, spirals towards the apex where it forms a vortex and then spirals in the

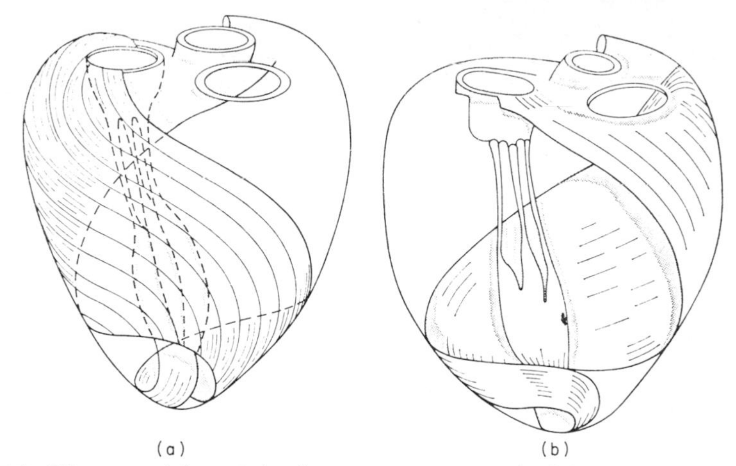

Fig. 5.8. The ventricles of the heart are composed of four muscle bundles. The superficial bulbospiral bundle arises from the mitral ring as shown in (a) The superficial sinospiral bundle is shown in (b). Within these bundles are their deep counterparts, the deep bulbospiral encircling the base of the left ventricle, while the deep sinospiral encircles both ventricles. This division is arbitrary and of no functional importance. (Robb and Robb, *Am. Heart J.*, 1935, **10,** 289.)

opposite direction passing deep and surrounding both ventricular cavities, finally inserting into the fibrous ring around the atrioventricular orifices. The deep muscles behave rather similarly but do not cover the apical region. Because of the much higher pressures that have to be developed by the left ventricle the muscular wall of this chamber is much thicker than that of the right ventricle.

Atria

The atria are contractile, like the ventricles, but their main function is that of reservoirs, which are filled by the great veins and which empty into the ventricles; this emptying is assisted by contraction, but

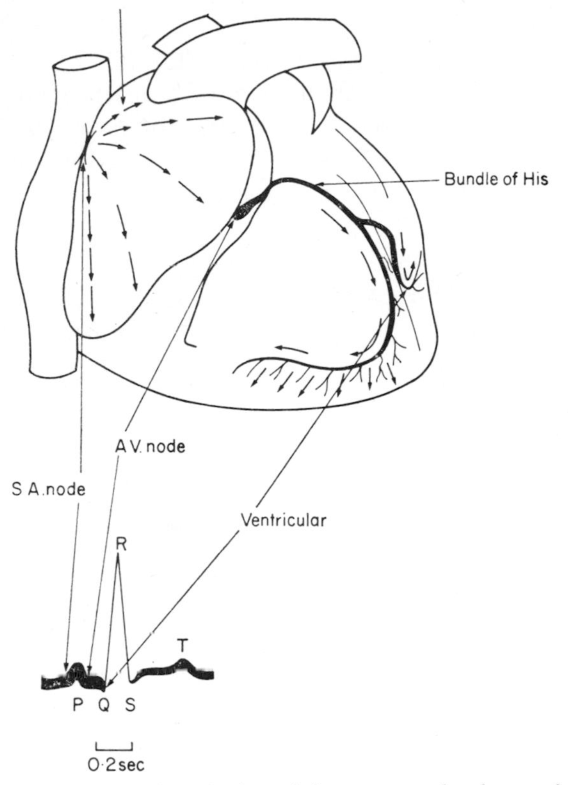

Fig. 5.9. The spread of electrical activity across the heart from the sino-atrial node (SA node). The SA node initiates the wave of depolarization which spreads across the atria until it reaches the atrio-ventricular node (AV node). At the AV node some delay occurs, then there is rapid conduction to the ventricles by the bundle of His. The auricles are separated electrically from the ventricles by a band of fibrous tissue which is non-conducting. The relationship between the spread of depolarization and the electrocardiogram (ECG) is shown below.

relatively little muscular work is required and correspondingly the thickness of muscle is far less than that of the ventricles, which have to develop relatively high pressures, especially the left ventricle. The arrangement in the atria is simpler, the muscle being composed of fibrous sheets made up of two types of fibre, namely those that are confined to one atrium and those that are shared by both; this sharing of fibres means that electrical activity, beginning in one atrium must spread to the other (p. 230). Spread of electrical activity from atria to ventricles is limited, by the fibrotendinous ring separating the tissues, to the specialized conducting tissue of the bundle of His (Fig. 5.9).

Valves

The valves controlling flow of blood through the atrioventricular orifices are membranes attached to the fibrous ring and are divided into flaps that project into the ventricles, the valve on the right side containing three flaps (*tricuspid,* Fig. 5.10) and that on the left two

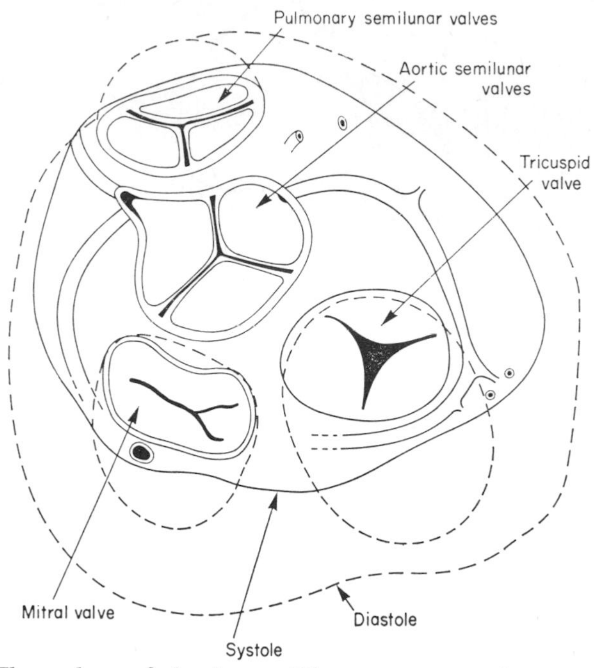

Fig. 5.10. The valves of the heart. There are two valves separating the atria from the ventricles, the mitral on the left and the tricuspid on the right. The pulmonary artery and the aorta are separated from the ventricle by semilunar valves. Each semilunar valve consists of three pocket-like flaps held open by the arterial pressure which seals the exit from the ventricles until the ventricular pressure is greater than arterial pressure. The shape of the orifices alters in contraction, as indicated by the dashed line, being reduced on systole in spite of the support given by the fibrous rings. (Spalteholz, "Hand Atlas and Textbook of Human Anatomy", Little, Brown and Co., Boston, 1954.)

(*mitral,* i.e. looking like a bishop's mitre). They are attached by thin tendinous cords (*chordae tendinae*) to *papillary muscles* (Fig. 5.6) which are projections from the ventricular muscle, and thus, when the ventricle contracts they also do so, preventing the eversion of the valvular flaps that would otherwise occur as pressure developed within the ventricle. The exits to the large arteries—aorta and pulmonary—are each guarded by three semilunar, or crescent-shaped, valves attached to the arterial wall, their free edges projecting into the lumen of the artery (Fig. 5.10). Their structures are such that when the pressure within the artery exceeds that in the ventricle they close.

Results of Muscular Contractions

In general, the directions of the muscular fibres and their mode of insertion into the fibrotendinous ring lead to a narrowing of the chambers when the ventricular muscles contract; this leads to the development of pressure and also to a considerable change in shape, as indicated in Fig. 5.10, where it will be seen that the ostia, i.e. the openings of the large vessels, are considerably constricted during systole in spite of their fibrous support. The change in shape causes a translational movement of the apical region which gives rise to the "apical beat", i.e. the visible and palpable movements of the tissue of the chest immediately above the heart. So far as the atria are concerned, we must note that there are no valves preventing regurgitation up the great veins, so that this undoubtedly occurs; however, the circular arrangement of the atrial muscular fibres gives them a sphincter-like character that may well narrow the venous orifice during contraction. The pressure required to drive blood through the systemic circulation is considerably higher than that required to drive it through the pulmonary circuit, so that it is not surprising that the muscular wall of the left ventricle is some four times thicker than that of the right; since they must expel the same amount of blood, the capacities of the two are about the same.

Coronary System

Branching off the aorta, soon after its emergence from the heart, is the *coronary artery*, supplying the cardiac tissue with blood; the blood is collected into two main venous channels, namely the *coronary sinus*, draining mainly the left ventricular tissue, and the *anterior cardiac veins;* these empty into the right atrium.*

Pericardium

The whole heart is enclosed by a fibrous sac called the pericardium; this is attached to the arterial trunks above, and to the central tendon of the diaphragm below, i.e. to the muscular wall separating the cavity of the thorax from the abdominal cavity, which thus acts as an anchor holding the heart in place. The pericardium is lined by a

* The Thebesian vessels arise from capillaries and veins in the heart tissue and open chiefly into the right ventricle, thus providing an accessory venous return for coronary blood.

layer of mesothelium continuous with a similar layer covering the surface of the heart, which is described as the *visceral pericardium*, by contrast with the *parietal pericardium* (or just plain "pericardium") making up the fibrous sac. The two surfaces are kept moist by the pericardial fluid, which thus acts as a lubricant allowing the heart to move freely within its fibrous container. The volume of the pericardial fluid is normally very small, but it may increase greatly under abnormal conditions; chemical analysis suggests that it is a simple ultrafiltrate of plasma comparable with intercellular fluid of muscle (p. 280). The pericardium is made up of elastic and collagenous fibres and shows a reversible elasticity, developing considerable tension as the heart expands; it thus tends to limit expansion, i.e. over-filling. Not only does the pericardium limit expansion of the heart; it also protects the pulmonary circulation from becoming either engorged or depleted; thus a sudden dilatation of the left ventricle, by increasing intrapericardial pressure, would reduce the filling of the right ventricle and so prevent engorgement of the pulmonary system. Functionally the consequences of engorgement are much more serious in the pulmonary system since here the tissue has to expand and contract to bring air in and out (p. 366). Finally, because during diastole the heart apparently fills the pericardium completely, and expands it a little, the sudden expulsion of blood leads to a negative pressure within the pericardial sac, and this promotes flow of blood into the atria in preparation for the next systole.

Cardiac Rhythm

The contractions of the chambers of the heart occur in a regular sequence of atrial and ventricular *systole* (contraction) and *diastole* (relaxation). This rhythmic behaviour is an intrinsic property of the heart muscle and is thus said to be *myogenic* as opposed to the *neurogenic* type of heart found in insects where the rhythm is governed from without by a cyclical discharge in the nerve supplying the heart.

Spread of Contraction

We shall be discussing the origin of this spontaneous rhythmic type of activity later (Vol. 2), and at present we must accept it for what it is, namely a rhythmically repeated wave of contraction that passes from a *pacemaker zone* at the *sino-atrial node* (S.A.-node), over the atria and thence through another specialized tissue called the *atrio-ventricular* (A.V.) node, along the *bundle of His* and throughout the ventricular walls (Fig. 5.9). This wave of activity is electrical in origin and may be recorded from the surface of the heart with suitable electrodes and a measuring device; in which case it is called the electrocardiogram (ECG). Accompanying the wave of electrical activity, is the wave of muscular contraction, which results in the nearly simultaneous contraction of the two atria, expelling blood into the ventricles, followed by the contraction of the two ventricles.

Asynchrony

The specialized conducting tissue constituting the bundle of His permits a much more nearly synchronous contraction of the ventricular tissue than would be possible were the spread to take place through the individual cardiac muscle fibres, as occurs in the atrium.

However, numerous studies with electrodes placed at different points in the ventricles, both with respect to surface position and depth, indicate some degree of asynchrony; this is indicated graphically by Fig. 5.11 where the different types of shading indicate intervals of

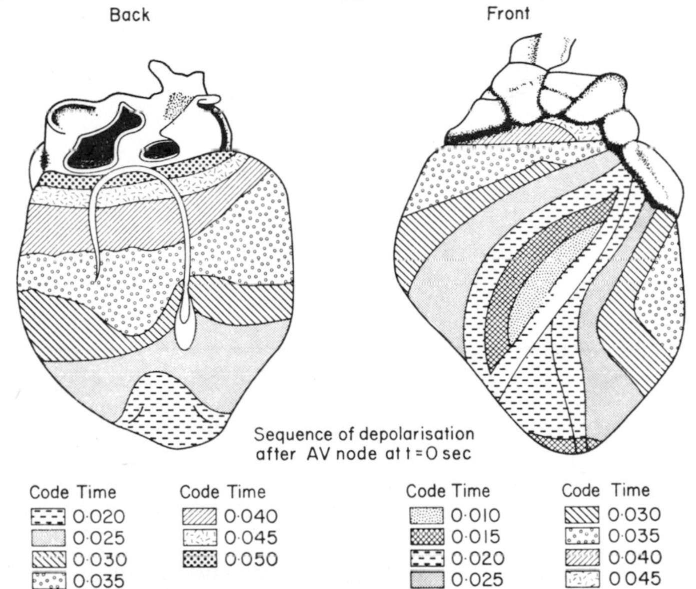

Fig. 5.11. The sequential spread of electrical activity in the ventricles. The different types of shading indicate intervals of 5 msec after the initiation of depolarization at the AV node. The earliest activity is found just to the right of the anterior descending branch of the coronary artery. From this area there is movement to the right. There is also early depolarization in the apical area. On the posterior aspect the earliest activity occurs at between 20 and 25 msec near the apex of the heart while the base of the heart is depolarized last. (Sodi-Pallares and Calder, "New Bases of Electrocardiography", Mosby, St. Louis, 1956.)

5 milliseconds after initiation of the activity at the AV-node. Earliest activity occurs just to the right of the anterior descending branch of the coronary artery; interestingly, the apical region is one of the earliest to receive the conducted electrical change.

Electrocardiogram

In the experimental animal the electrical changes may be recorded from the exposed heart; in man the electrocardiogram (ECG) is

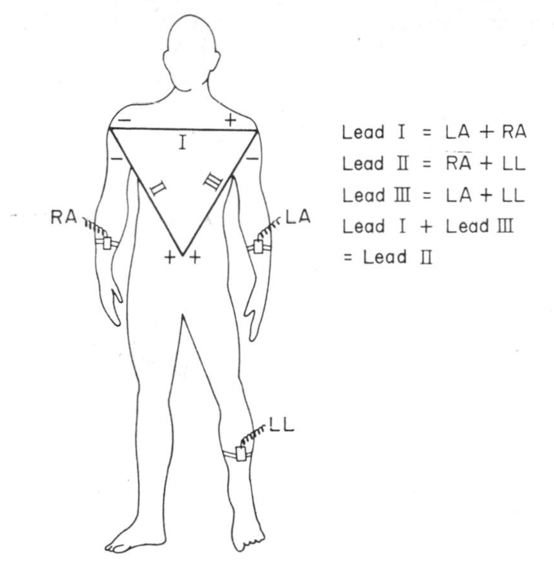

Fig. 5.12. The location of the standard limb leads and their relation to the equilateral triangle of Einthoven. In lead I the deflection is upward when the potential at the left arm lead (LA) exceeds that at the right arm lead (RA) as shown by the plus and minus signs on the line connecting the left and right shoulders. Lead II left leg and right arm, lead III left leg and left arm which also have deflections indicated in a similar manner. Berne and Levy, "Cardiovascular Physiology", Mosby, St. Louis, 1967.)

recorded from electrodes placed at strategic places on the body, e.g. on the left arm and right leg (Fig. 5.12). Obviously a very attenuated and distorted picture of the electrical events taking place on the heart will be obtained by this method, but it is remarkable how useful a diagnostic tool the ECG is for assessing defects in the heart. A typical ECG is shown in Fig. 5.9; it may be divided into a *P-wave*, corresponding with spread of activity over the atria, and a *QRS-complex*, corresponding with ventricular contraction, completed by a T-wave representing subsidence of electrical activity in the ventricles. We may postpone till later any attempt at interpretation of these electrical changes; it is sufficient to know that the duration of the P-wave is about 0·1 sec, and that the interval between the beginning of the P-wave and the beginning of the QRS complex—the P–R interval—is taken as the time for transmission of the electrical change along the bundle of His; it has an average value of 0·16 sec. The upstroke of the R-wave coincides with ventricular systole.

Pacemaker Activity

The rhythm of the heart is a built-in feature of cardiac tissue; as indicated above, it is governed by a pacemaker zone on the right atrium called the S.A.-node, in the sense that it is the electrical activity (depolarization) of this zone that gives the lead to the subsequent spread of activity over the atria and ventricles.

Chronotropic Effects. The frequency of this *pacemaker activity* can be modified by drugs or by the nerves that supply this part of the heart. Thus the right vagus nerve, supplying the S.A.-node, can lengthen the time required for the pacemaker potential to reach its critical level for firing (Vol. 2) and in this way slow the rate of beat. The left vagus, ramifying mainly in the A.V.-node, exerts a slowing effect on transmission of the electrical change in the adjacent atrial tissue through this node. The sympathetic accelerator nerves exert precisely opposite effects, accelerating the beat by shortening the time of development of the pacemaker potential; they also increase the force of the beat. Drugs, such as pilocarpine, imitate the action of the vagus by slowing the heart, whilst adrenaline increases the frequency of beat.

Inotropic Effects. Not only is the frequency of the heart-beat under nervous and chemical control but also the force of the beat, and it is customary to distinguish the *chronotropic* effect, on frequency, from the *inotropic* effect on strength of beat.* In general, the vagus has very little inotropic effect, its influence being only on the frequency of beat; by contrast, the sympathetic nerves, or adrenaline—the adrenal hormone—increase both frequency and strength of contraction, the increased frequency being achieved by shortening the diastolic period.

Pacemaker Substitution. If, for some reason, the S.A.-node ceases to function as a pacemaker, then the heart does not cease beating, because all the cardiac tissue has its own rhythmic capacity to contract; the S.A.-node ceases to act as a pacemaker, but the A.V.-node, at the junction of the atrium and ventricle, may then take over

* For those with a mania for names, the term dromotropic refers to the effect on the capacity of the cardiac tissue to conduct electrical changes.

the task of pacemaker and the heart will continue to beat, but at a slower rate, governed by the intrinsic rhythmicity of the specialized tissue constituting this node. Finally, if this node is blocked too, the ventricles may exert their own rhythmicity giving an even slower beat. The advantage of a beat governed by the S.A.-node is that it induces a smooth contraction of the atria followed, in due course, by smooth contraction of the ventricles.

Vagus Escape. It is probable that the phenomenon of "*vagus escape*" is due to the inherent capacity of the ventricles to contract; thus, when the vagus is stimulated electrically it is possible, by continued stimulation, to cause complete cessation of the beat. After a time, in spite of the maintained stimulation, however, the beat resumes, this time at a slower rate and being confined to the ventricles, the atria being inhibited and remaining relaxed. An alternative explanation is that the slowing of the heart causes an increase in pressure in the great veins entering the auricle, and stretch-receptors here reflexly accelerate the heart via the vasomotor centre (Vol. 3).

Dominance of S.A.-node

The basic feature that allows the well co-ordinated contractions of the various muscular elements of the heart, including those muscles that operate the valves, is primarily the rhythmicity of the S.A.-node, but an additional and necessary feature is the circumstance that the rhythmicities of the other tissues are slower, so that any tendency for them to impose their rhythmicity on the heart is suppressed by the wave of action potential that spreads from the S.A.-node. Thus, while the S.A.-node is developing its "pacemaker potential" preparatory to initiating its wave of propagated action potential, the other tissues (such as the A.V.-node) are also developing their pacemaker potentials, but more slowly, and these potentials are finally suppressed by the passage of the propagated change spreading from the S.A.-node. Thus the S.A.-node dominates the situation by virtue of its more rapid rhythm.

Refractoriness

It is the capacity of a propagated electrical change to suppress activity along its path that permits this domination by the S.A.-node. It may have other consequences; for example, the A.V.-node may become more excitable, for some reason, and its pace-maker activity may increase in rate so that it anticipates the beginning of S.A.-node activity. In this case the heart beat is initiated by the A.V.-node; the electrical change passes along the bundle of His and thus the ventricles contract smoothly, but the atrial contractions will be out of phase. The activity in the A.V.-node may spread backwards along the atria, and if they reach the S.A.-node before it has fired off its pacemaker activity, it will suppress this. Alternatively, we may find the ventricles beating with the rhythm determined by the A.V.-node, and the atria with the rhythm determined by the S.A.-node.

As we shall see, when we discuss electrical activity in excitable tissue, be it muscle or nerve, the suppression of activity resulting from the passage of a wave of action potential over the tissue, discussed above, is a special case of *refractoriness*, the failure for a limited time—called the *absolute refractory period*—of a tissue to respond to any form of stimulus.

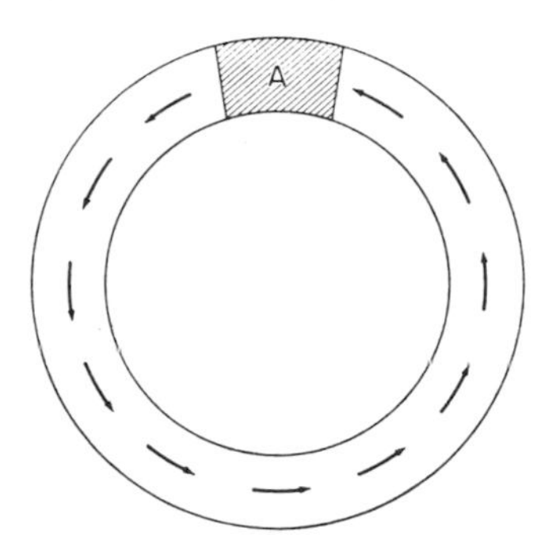

Fig. 5.13. Electrical conduction around the heart. An electrical change initiated at A will travel around the heart in a circle. If this wave of depolarization moves sufficiently slowly, it will arrive back at A when this region has past its refractory period and the wave will continue to travel in a circle indefinitely.

Circular Conduction. This refractoriness serves a useful purpose in a tissue-like cardiac muscle where an electrical change, initiated at one point, may spread long distances, since it prevents the development of circular excitation under normal conditions. Thus in Fig. 5.13 we imagine a ring of atrial tissue, and an electrical change beginning at A and travelling round the ring; if the change made its complete circuit sufficiently slowly that, by the time it returned to A, the refractory period was over, the electrical change would continue to propagate in this circular fashion indefinitely, or until the refractory period was lengthened. Under pathological conditions this circular form of excitation does in fact take place, leading to a fibrillar type of atrial contraction. In this case the ventricular beat occurs independently. Fortunately the filling of the ventricles does not depend critically on atrial contraction, so that this failure of co-ordination does not necessarily seriously impair cardiac function.

One-way Block. It must be appreciated, however, that circus propagation requires something more than a depression of conduction velocity that permits re-entry of the wave of action potential when the refractory period is over. Thus if the spread occurs in both directions

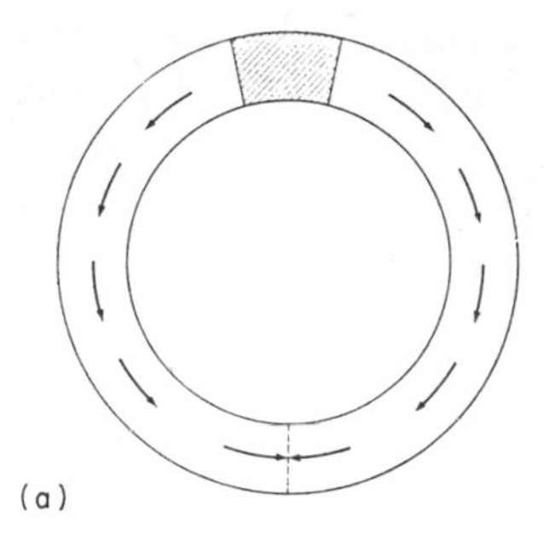

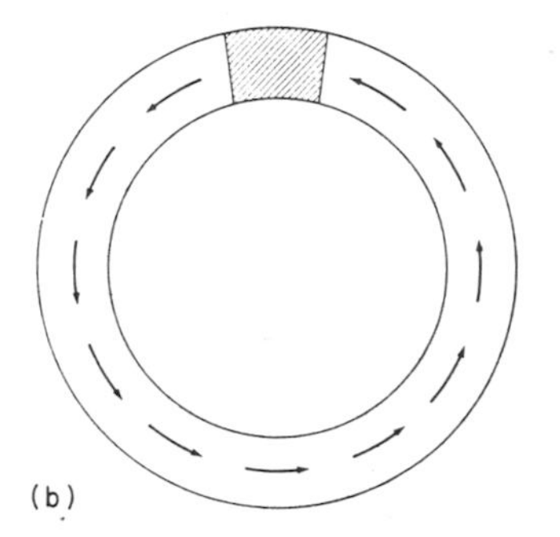

Fig. 5.14. One way block. If a wave of depolarization is initiated in a ring of heart muscle, the wave travels out in both directions as shown in (a), so that when the waves meet they will cancel out. One way propagation can be initiated experimentally as shown in (b) and in this case the wave now travels around the heart in one direction only and continues to propagate in a circular fashion.

round the circle of tissue, the two waves will cancel out when they meet (Fig. 5.14a), hence a second condition is that propagation through a part of the tissue be *one-way* (Fig. 5.14b). Experimentally such a situation can be achieved, in which case circular conduction with re-entry of impulses takes place.

In normal cardiac tissue, such as Purkinje fibres, the path-length through which an impulse must travel in a circle in order to be able to make a second circuit is given by the product of conduction velocity and the duration of the full recovery time or refractory period, i.e. 3 m/sec × 0·3 sec = 900 cm; to achieve a practical path-length the conduction velocity must be reduced very considerably since the refractory period is not, in fact, very variable. Experimentally Wit *et al.* reduced conduction velocity to 0·03 m/sec and obtained circular conduction with a path-length of some 30 mm.

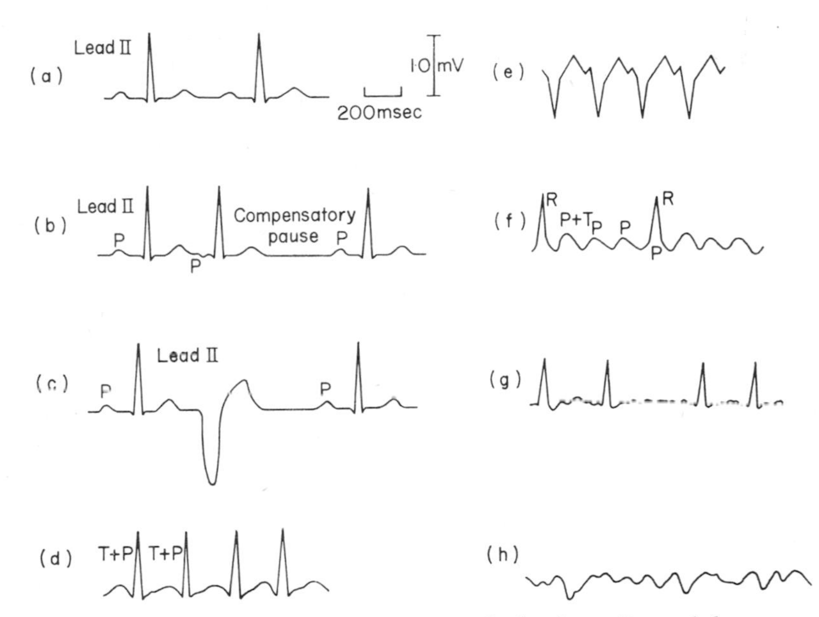

Fig. 5.15. Cardiac arhythmias. A: normal rhythm; B: atrial premature excitation; C: ventricular premature excitation; D: atrial tachycardia (T + P indicates superimposition of T and P waves); E: ventricular tachycardia; F: atrial flutter; G: atrial fibrillation; H: ventricular fibrillation. (Selkurt, "Physiology", Little Brown and Co., Boston, 1966.)

Compensatory Pause. The compensatory pause is another example of refractoriness; let us imagine that the heart is beating normally with a beat governed by the S.A.-node; suddenly during what would be the diastolic pause, a new beat is initiated, say by a focal point of high excitability on the ventricle. This is called an *extra systole*, or more correctly a *premature contraction*, and is revealed in the ECG as an additional QRS complex occurring during what would have been the diastolic period. The next beat does not occur at the moment expected by the previous rhythm; i.e. if the heart had been beating at 60 per min, it does not occur 1 sec later, but there is a *compensatory pause*. This is because the next beat, initiated by the S.A.-node, reaches the ventricular tissue when it is refractory after the extra-systole, so that the heart must wait until a new pacemaker potential has been generated. Figure 5.15 illustrates the ECG's for several abnormal rhythms.

Mechanical Events in the Cycle

Pressures in Left Heart-chambers

The function of the heart is to force fluid through the vascular system, and therefore to generate pressure, which is ultimately dissipated through friction. The mechanical events taking place are therefore best illustrated by measurement of the pressures at strategic points during a cycle. In Fig. 5.16 the changes in pressure in left ventricle, aorta and left atrium are shown during a cycle. Atrial contraction causes a small rise in atrial and ventricular pressures; when the ventricle contracts pressure develops very rapidly in it, but no change in volume occurs since all valves are still closed—the *isovolumetric phase*. At point 2 in Fig. 5.16 the pressure has become equal to the diastolic pressure in the aorta, the valve has opened and blood now flows out of the heart but the ventricle is still developing tension in its walls so that the pressure rises still further, passing through a maximum. Beyond this point blood is being carried away faster than the ventricle can eject it and the pressure tends to fall—a period of reduced ejection. At the point marked 4 the muscle relaxes, giving a steep fall in ventricular pressure bringing it below that in the aorta which has run very close to that in the ventricle. The valves close and the ventricular pressure falls as the muscle continues to relax, actually passing through a negative phase that acts as a suction to bring blood from the venous system into the relaxed atrium and ventricle; during diastole the pressure gradually rises.* The rapid filling of the ventricles at these low pressures is brought about by the elastic fibres intercalated between the muscle layers. When ventricular systole occurs the movement of the layers relative to each other stretches this elastic tissue storing kinetic energy until the active contraction is complete.

Aorta

The pressure in the aorta rises and falls with that in the ventricle while the semilunar valves are open; the sudden relaxation of the ventricle at the beginning of diastole causes a brief fall in aortic

* The negative pressure during diastole is shared by the great veins in the thorax and is due primarily to the negative pressure in the thorax created by the elasticity of the lungs (p. 368). Thus, opening the chest brings the atrial and venous pressures up to atmospheric.

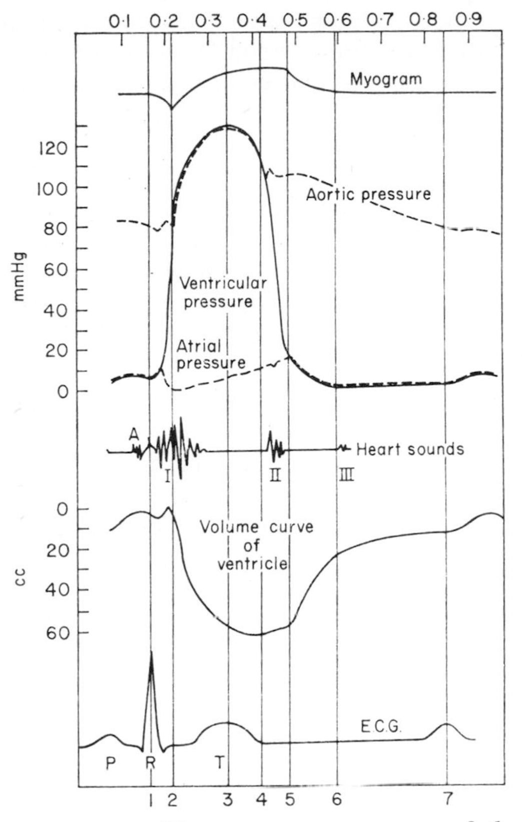

Fig. 5.16. Wiggers curves. The pressure sequences of the cardiac cycle. The ventricular pressure curve shows two steep portions—1 to 2 and 4 to 5—when no change in volume occurs since all valves are closed; these are the isovolumetric contraction and relaxation phases respectively. Once the ventricular pressure exceeds the aortic pressure the valves open and there is a phase of rapid ejection when the main volume charge occurs as can be seen from the volume record in 2 to 3. (Brecher and Galletti, "Handbook of Physiology", Circulation 2, Vol. II. 759, 1953.)

pressure, associated with some regurgitation of blood into the ventricle, to be followed immediately by closure of the semilunar valve. This closure causes a brief rise or *notch* in the aortic pressure and this slowly falls as blood is forced along the vascular tree and the aorta is allowed to reduce in diameter.

As we shall see, it is the forcible expulsion of blood into the aorta that gives the pulsatile character to blood-flow through the arteries, whereas it is the sustained pressure in the aorta, maintained by virtue of the closure of the semilunar valve and the distension of the elastic wall of the vessel, that allows blood to continue to flow during the diastolic pause between contractions.

Heart Sounds

The closure of the atrio-ventricular valves gives rise to sound waves (Fig. 5.16) that may be picked up in the stethoscope as a characteristic "first sound" often described as lubb. The second sound is caused by the vigorous contraction of the ventricles and is synchronous with closure of the arterial valves, being in fact caused by vibrations in these valves; it is often described as "dupp". These sounds thus result from vibrations of solid structures in the heart; the *murmurs* take their origin from the turbulent flow of fluid when it escapes through a narrow orifice into a wider space. Under normal conditions sounds set up in fluid in this way are usually inaudible; in disease conditions, as with, say, mitral insufficiency where the valves are fibrosed and do not close properly (leading to regurgitation of blood back into the atrium) the murmur becomes audible.

The Work of the Heart

Variations in Output

In normal man at rest the output of the heart is some 5–6 litres per minute; in severe exercise it may rise to some 25 litres per minute. Since heart-rate rarely increases beyond 180 beats/minute compared with a normal figure of, say, 75 per minute, this means that the amount expelled per beat (the *stroke volume*) can vary from about 75 ml at rest to some 140 ml.

Starling's Law of the Heart

In his studies on the isolated heart-lung preparation,* Starling found that when the heart was allowed to fill more thoroughly so that the ventricular volume just before systole was larger than normal, the force of contraction increased, thereby helping the over-filled heart to empty itself. Thus this purely physical property of muscle, whereby its development of tension is a function of its initial stretch, can provide a mechanism favouring increased cardiac output during severe exercise. In fact, however, when the heart is studied in the intact animal, the

* In this preparation the heart and lung are effectively separated from the animal; the heart remains *in situ* but blood is driven by it through an artificial oxygenator, and after return to the left side is driven through an artificial resistance that can be varied. In this way it is possible to study the effects of varying venous return and other parameters on the function of the heart.

large increases in output associated with exercise are not found to be accompanied by appreciable increase in pre-systolic size, so that "Starling's law" presumably does not play a dominant role here. In this event the increased stroke volume is not so much due to increased filling of the heart between beats but due to a more efficient emptying of the ventricle.

Systolic Reserve Volume

Thus we may speak of a "systolic reserve volume" meaning the amount of blood usually left behind at the end of each ventricular contraction; increased cardiac output is achieved by encroaching on this, as well as by increasing frequency of beat. The increased strength of beat, required to encroach on this systolic reserve, is provided by the positive inotropic action of nervous and hormonal stimulation, together with the tendency for the muscle to contract more powerfully as soon as the ventricular filling becomes excessive (Starling's law).

Variation in Rate

In this connection it must be emphasized that an increased rate of heart beat, *per se*, is unlikely to increase output; this is because the output depends on the filling of the ventricles and this requires time. Changes in heart rate are made at the expense of the filling time (diastole) so that unless venous return can be augmented as well, increased rate need cause no increase in output.

Equalization of Left and Right Outputs

The left and right ventricles are essentially two pumps in series, so that for efficient working the stroke volume of each should be the same. If, for example, the output of the left ventricle, which expels blood into the general circulation, were less than that of the right, the blood sent by the right ventricle through the lungs would start accumulating in the pulmonary veins, being unable to enter the left atrium and ventricle due to inadequate emptying. Thus the lungs would become congested with blood, and this would impair their capacity to ventilate the blood since this depends on a mechanical expansion of the lung-tissue. It is here that Starling's law is important; the overfilling of the left ventricle, due to the greater output from the right ventricle, causes the left ventricle to contract more powerfully, whilst the underfilling of the right ventricle, due to its putting out more blood than it receives from the systemic circulation, causes it to beat with less force.

THE VASCULAR PRESSURES

The flow of fluid along a system of tubes is governed, at any point, by the pressures up- and down-stream, and we may regard the pressure as the potential energy at this point, potential energy that is converted into the mechanical energy of flow of the fluid and ultimately dissipated as frictional heat. The measurement of the pressures within the vascular tree is therefore of obvious importance.

Measurement

Manometry

In experimental animals the classical method of measurement of pressure within a vessel is to insert a tube, or *cannula*, into the vessel and connect this to a manometer as illustrated in Fig. 5.17a; the manometer is a U-tube containing mercury, and the pressure is registered by the height; if the tube were connected directly between artery and manometer, blood would flow into it until the mercury had been driven up the U-tube to a height that balanced the pressure.

To avoid the loss of blood, a reservoir containing isotonic saline with an anticoagulant to prevent blood-clotting is included in the circuit, so that, before finally connecting the cannula with the system, a pressure approximately equal to that in the vessel may be established. For measurement of venous pressures, which are so much lower than arterial, a water-manometer may be employed, otherwise using the same technique.

Pressure Transducer

The disadvantages of fluid U-tube manometers are numerous, especially their inertia, so that if the pressure is pulsatile the responses are so slow that the pulse is very considerably damped, and what in fact the manometer registers is the *mean pressure* during a cycle. In modern pressure-measuring devices the liquid manometer is replaced by some form of pressure transducer, e.g. the pressure may operate on a chamber containing a thin metal diaphragm as shown in Fig. 5.17b. The diaphragm is connected by a peg to a small cross of metal, each arm of which rests on the wires of a tiny "boxing ring" of strain gauge wire. Small movements of the diaphragm cause the resistance of the wire to change, and by connecting this device to one side of a Wheatstone bridge network, these resistance changes can be converted to voltages which in turn are amplified on to a pen-recorder. The diaphragm is a relatively stiff structure with a high frequency response

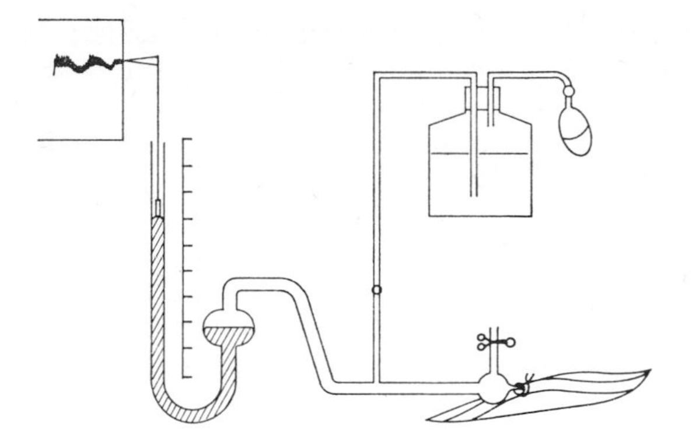

(a) Mercury manometer

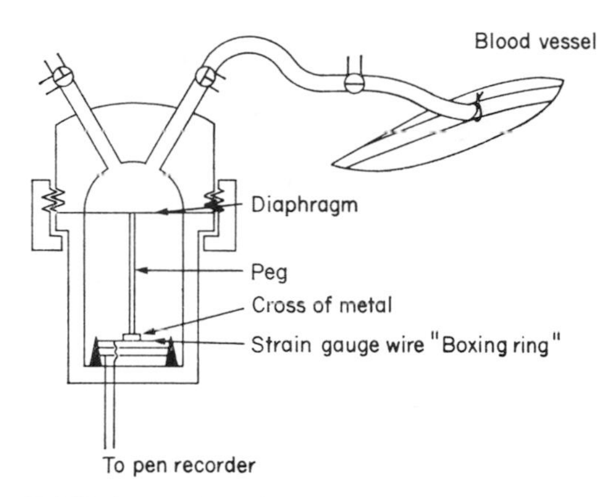

(b) Strain gauge transducer

Fig. 5.17. (a) The mercury manometer. A cannula filled with heparinized saline is inserted into a vessel and then attached to a mercury manometer as shown. The mercury column is initially "blown up" to the expected value of pressure with the pressure bottle to ensure that blood does not enter the cannula. Mercury is heavy and has a high inertia so cannot move fast enough to follow the pulse pressures, so that the value recorded is close to the mean arterial pressure. (b) Pressure transducers. The transducer is connected as before to the vessel and pressures transmitted through the saline cause small displacements of the thin metal diaphragm. The diaphragm is connected by a peg to a cross of metal which rests on a tiny "boxing ring" of strain gauge wire in the base of the transducer. The small movements of the diaphragm thus stretch the wire and cause the resistance of the strain gauge wire to vary. By connection of this wire to one side of a Wheatstone Bridge network these resistance changes can be converted to variations in voltage which in turn can be recorded on a pen recorder or oscilloscope.

and therefore can follow rapid fluctuations in pressure with the cardiac cycle. However, a note of caution must be added since artifacts can be introduced into the system by long tubing between the animal and the transducer, especially if it is of narrow bore and elastic in nature.

Pressure for Collapse

Where a vessel is small but can be easily seen, e.g. a conjunctival vein, the pressure required to make it collapse may often be employed; thus a transparent chamber (Fig. 5.18) covers the tissue forming an air-tight seal, and the pressure inside is raised until the vessel just

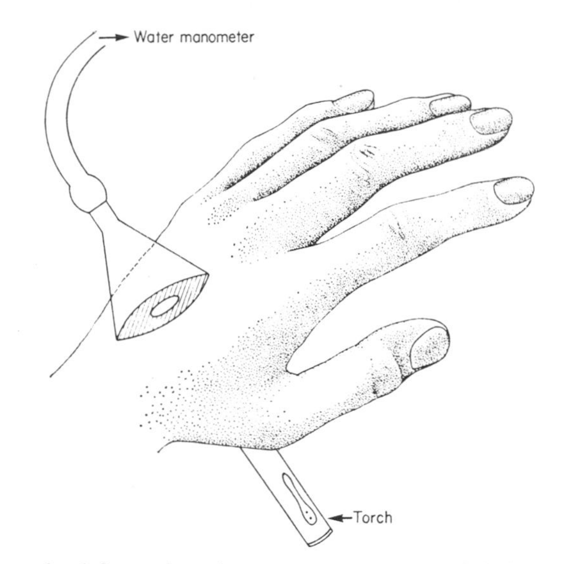

Fig. 5.18. A method for estimating venous pressure. A bright light is placed under the palm to illuminate the veins on the back of the hand. The veins are covered by a glass funnel connected to a water manometer, and by increasing the pressure in this system until the veins just collapse an estimate of venous pressure can be obtained.

collapses. This is, in fact, the basis of the measurement of arterial pressure in man, an inflatable cuff (Riva-Rocci) being placed round the arm, and the pressure required to prevent blood-flow through the artery, as determined by feeling the radial pulse, is determined. The pressure is, of course, pulsatile, so that to prevent all flow, the maximal systolic pressure must be exerted. The experimenter thus increases the pressure in the cuff until pulsation in the wrist can no longer be detected, and this is taken as the systolic pressure. To determine the diastolic pressure, the pressure in the cuff is lowered progressively, and if the experimenter listens to the pulsation above the artery through a stethoscope, a characteristic series of changes will be detected as interference with blood-flow, occurring only during the systolic phase, becomes less and less. When the pressure in the cuff is such that there is no interference with flow, the diastolic pressure has just been reached.

End-pressure

It must be appreciated that all these methods create an artificial system in which flow along the vessel being studied is stopped; this means that the pressure measured—called the end-pressure—is rather

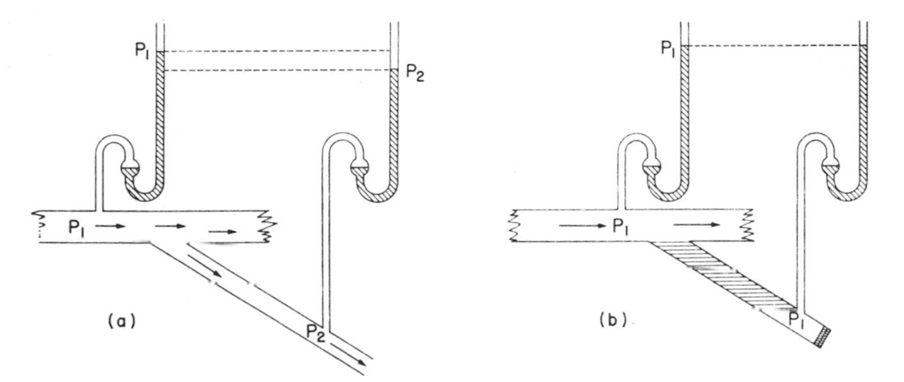

Fig. 5.19. Blood pressure recording. To record a pressure from an artery or vein, the flow is stopped and an end pressure is recorded. This pressure is greater than the true pressure in a vessel since the kinetic energy of blood flow has been converted into potential as pressure energy. In (a) the difference in pressure between the side vessel and main vessel can be seen, since the pressure P_2 is recorded from a smaller side branch and flow is continuing in both vessels. In (b) the side vessel has been blocked off and now there is no difference between P_1 or P_2 as flow has ceased in the side vessel and the pressure recorded is effectively that in the main vessel.

higher than the pressure existing during normal flow, since the kinetic energy of the blood has been converted into potential or pressure-energy. Thus, as Fig. 5.19 illustrates, the pressure actually measured is the pressure in the blood vessel from which the vessel being examined branched, and a better measure would be obtained by inserting a tube into the blood vessel, and measuring the pressure through a side-arm, as in Fig. 5.20.

Measurement of Blood-flow

In experimental animals this may be carried out by some form of *Strohmuhr*, i.e. an instrument that detects the total quantity of blood

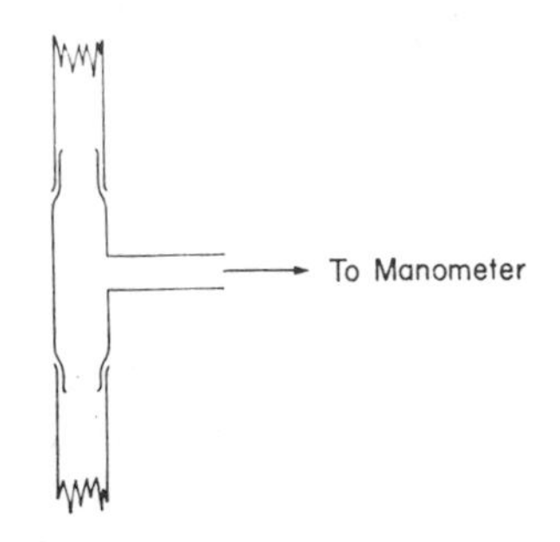

Fig. 5.20. This shows a better method of recording the true pressure in a vessel by means of a T tube so that the kinetic energy of flow is not lost.

flowing through a vessel in unit time. For example, we may add a known quantity of heat to blood in a short section of a vessel by a high-frequency current, and measure the rise in temperature of the blood by thermocouples placed up- and down-stream.

Fick Principle

Most methods employed rely on the Fick principle, i.e. by allowing the blood to carry away a known amount of a test substance in a measured time, and estimating the concentrations in artery and vein entering and leaving the tissue concerned. Thus the blood-flow through the pulmonary system—the *cardiac output*—may be determined by measuring the O_2-concentrations in arterial blood and in mixed venous blood, i.e. the blood in the right ventricle. If the O_2 consumed during a given time is measured, then blood-flow is given by:

$$\text{Blood flow} = \frac{O_2\text{-consumption/min}}{\text{Arterial } O_2 - \text{Venous } O_2}$$

The cardiac output measures the rate of flow of blood through the whole vascular system. The same principle can be applied to measuring the flow through individual parts, e.g. the flow through the kidney may be measured by sampling arterial and renal venous bloods and measuring the rate at which the blood delivers material to this organ (as measured by its rate of accumulation in the urine), employing as a test substance one that is completely absorbed from the circulation during one passage.

Cerebral Blood-flow. The cerebral blood-flow can be measured by employing an inert gas that diffuses readily from blood into the brain. Kety employed nitrous oxide (N_2O) which was inhaled in a mixture with air. The concentrations of the gas in the blood entering and leaving the brain were measured by sampling arterial C_a and jugular venous C_v blood at intervals. After about 10 min the brain becomes saturated with the gas so that the arterial and venous blood concentrations become equal, and this concentration, which may be indicated by C_{eq}, represents the concentration in the brain-tissue or, if the gas is not partitioned equally between brain and blood, the concentration in the tissue is equal to $C_{eq} \times \lambda$, where λ is the partition coefficient. According to the Fick equation the blood-flow is given by:

$$\text{Blood flow (f)} = Q/(A\text{-}V)t$$

where Q is the amount taken up in time t, and (A-V) is the difference between arterial and venous blood concentrations during this time. Q is

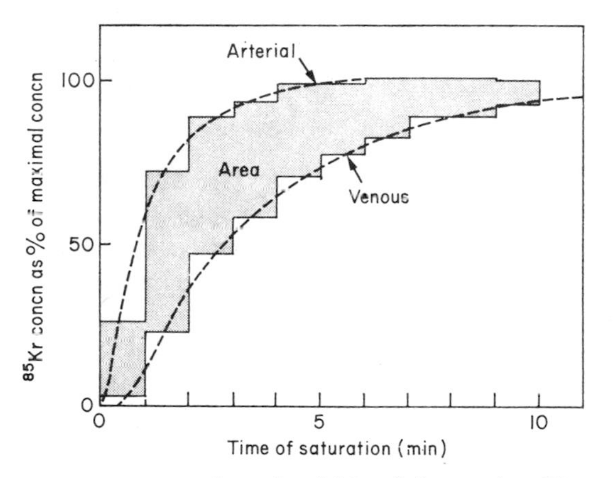

Fig. 5.21. The measurement of cerebral blood flow using Krypton ^{85}Kr and the washout method of Kety and Schmidt. The inert radioactive gas is inhaled and concentrations of blood entering and leaving the brain are measured by sampling arterial and jugular venous blood. After 10 minutes the brain has become saturated with gas so that the arterial and venous concentrations are equal. The cerebral blood flow is derived from the curves of C_v and g against time. (Purves, "Physiology of Cerebral Circulation", C.U.P., 1972.)

given by the weight of brain multiplied by its concentration $Q = W \times C_{eq} \times \lambda$. The average value of (A-V) is given by plotting the two concentrations, C_a and C_v, against time, as in Fig. 5.21, and dividing the area between the two curves by the time. Expressed mathematically it is

$$\frac{1}{t}\int_0^t (C_a - C_v)\, dt$$

Thus the blood-flow (f)

$$= \frac{WC_{eq}\lambda}{\frac{1}{t}\int_0^t (C_a - C_v)\, dt}$$

or the blood-flow per unit weight

$$\frac{f}{W} = \frac{C_{eq}\lambda}{\frac{1}{t}\int_0^t (C_a - C_v)\, dt.}$$

More recently the accuracy of the method has been greatly improved by employing the radioactive inert gas ^{85}Krypton.

Model System

Figure 5.22 illustrates the systemic circuit diagrammatically. The essential feature of the system is the large resistance intercalated between the arterial and venous sides; this resistance is constituted by the very small diameters of the arteriolar and capillary tubes, particu-

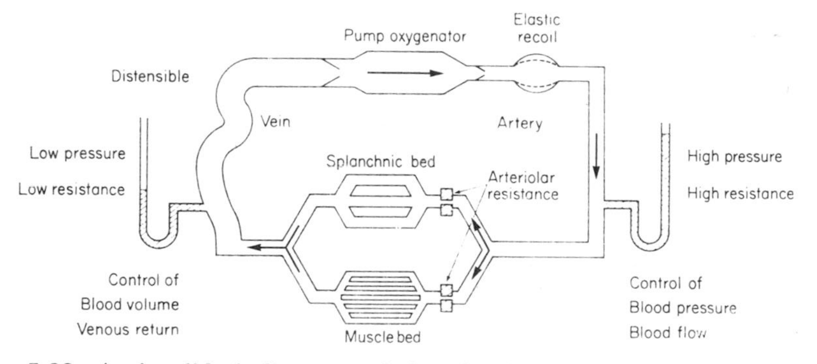

Fig. 5.22. A simplified diagram of the circulation. The essential feature of the system is the large resistance intercalated between the arterial and venous sides of the circulation. The main part of this resistance resides in the arterioles, which are of small diameter, and it is across these structures that the principal pressure-drop occurs in the arterial system. The arteries are mixed elastic and muscular vessels which expand with each systole then recoil sending the blood through the capillaries. If arteries are distended further the pressure rises and there is an increase in the driving force through the resistance vessels into the venous side. The veins are low pressure capacity vessels capable of considerable distension with little increase in pressure.

larly the former. Initially we conceive of the system at uniform pressure; as the pump works, driving fluid into the thick-walled elastic arteries, the pressure rises in these with each systole of the pump, and this causes flow of fluid along the whole system. At first a part of the blood expelled into the arteries will be accommodated by extension of their elastic coats, so that each systole will not cause an equal return of blood to the pump; the accommodation of the blood within the arteries causes the pressure to rise in the arterial system, and this is favoured by the resistance to flow from the arterial to the venous side. The feature of a distensible elastic system, such as that constituted by the blood vessels, is that, as they distend, the pressure required to distend

TABLE I

Average Values of Mean (not Systolic) Pressures in Young Male Adult

	Pressure (mm Hg)
Large arteries (e.g. brachial)	98
Capillaries	10–30
Small veins of arm	9
Portal	10
Inferior vena cava	3
Large veins of neck	0 to −8

them further increases; thus, as more and more blood is forced into the arteries, the arterial pressure rises and this forces blood more and more effectively through the resistance into the venous side. Ultimately a steady state will be reached with the venous return to the pump exactly equalling the systolic output. Under these conditions there will be a characteristic fall in pressure along the vascular tree, the pressure being highest at the aorta and least at the entry of the great veins to the heart. Typical results of measurements are shown in Table I, and we may now proceed to discuss some of the factors determining this gradient.

Poiseuille's Law

Resistance to flow

In general, the flow of fluid along a tube, provided it is not turbulent (p. 273), follows Poiseuille's law (Fig. 5.23):

$$Q = \frac{(P_1 - P_2)\,\pi r^4}{8\eta l}$$

or, if we equate the resistance to flow, R, with $\frac{8\eta l}{\pi r^4}$ we have

$$Q = \frac{(P_1 - P_2)}{R}$$

Q, the volume flow (e.g. in ml/min) is directly proportional to the pressure head driving the fluid along a tube or series of tubes, and inversely proportional to R, the resistance, which we see depends, with a single tube, directly on the length, *l*, of the tube and viscosity, η, of the fluid and inversely on the fourth power of the radius, *r*.

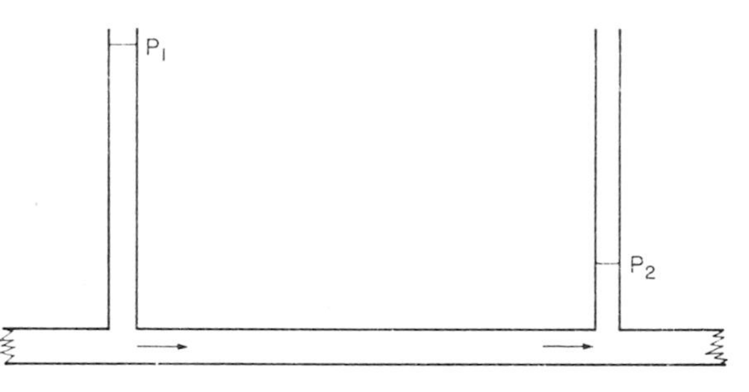

Fig. 5.23. The flow of a fluid through a tube, providing it is not turbulent, depends directly on the pressure difference $(P_1 - P_2)$ and the fourth power of the radius and inversely on the length of tube and viscosity of the fluid.

Variations in Parameters

Thus, with the same volume, Q, flowing through the whole system in unit time (as we would have in any steady state), the change in pressure on passing from one point to the other depends only on the resistance. If we wish to increase the flow through the whole we may increase the pressure-head; e.g. by forcing blood more rapidly and with adequate power into the arterial system; the pressure P_1 will increase and to some extent P_2 will rise, but the pressure-head, P_1–P_2, will increase and this will give us our extra flow. Alternatively, we may reduce the resistance, keeping the pressure-head the same. It is important to appreciate, however, that it is not easy to alter one factor without altering the others; thus reduction in the resistance of the circuit would increase the flow, Q, but it could also reduce the value of P_1, the pressure at the arterial end, unless the heart kept pace with the extra output.

The pressure P_2, at the end of the circuit, is governed by P_1 and R, so that if R is reduced, P_2 will rise, hence the pressure-head will tend to fall as the resistance falls. In order to predict the consequences of a change in a single parameter in this system, governed by three, we must be much more specific in describing the situation.

Distribution of Pressures

We have so far just treated the resistance as the total resistance of the circuit; now this varies as we proceed along the vascular tree, and this means that the drop in pressure from one point to another will vary too; thus the resistance of the large arteries is relatively

TABLE II

Changes which take place with Branching of the Mesenteric Artery of a Dog (Modified from Schleier after Mall)

		Radius (cm)	Total cross-sectional area (cm^2)	Length (cm)	Mean velocity (cm/sec)	Pressure fall (mm/Hg)	Pressure fall 1 cm length (mm Hg)
Mesenteric artery	1	0·15	0·07	6.0	16·80	0·8	0·13
Main branches	15	0·05	0·12	4·5	10·10	3·2	0·72
End branches	45	0·03	0·13	3·91	9·30	7·4	1·88
Short and long intestinal branches	1,899	0·0068	0·20	1·42	5·80	23·5	16·55
Last branches	26,640	0·0025	0·57	0·11	2·10	7·2	65·50
Branches to villi	328,500	0·00155	2·48	0·15	0·48	5·4	36·00
Arteries of villi	1,051,000	0·00122	4·18	0·20	0·28	8·1	40·40
Capillaries of villi	47,300,000	0·00040	23·78	0·04	0·05	2·4	58·80
Total fall in pressure						*58·0*	

small, so that the loss of pressure along these is not large, Table II is especially instructive; here we have the changes in diameter, etc., of the successive ramifications of a large (mesenteric) artery. The precipitate fall is at the arterioles, where the diameters of the vessels become very small whilst they are relatively long, in contrast with the capillaries where the diameters are also small but the vessels are short.

Ramifications of Vascular Tree. These considerations explain in a general way the distribution of pressure along the vascular tree, but they do not tell us the whole story. Thus the resistance of a tube increases conversely as the fourth power of its radius so that for a flow of, say, 10 ml/min along a tube of 1 mm radius and of 10 cm length, a pressure-drop of 417 dynes/cm^2 would be required; with a tube of 10 μ, i.e.

1/1000 the diameter, the pressure-drop would be some million million (10^{12}) times greater to give the same flow. Clearly such large pressure-drops are not practicable, and the factor that we have ignored is the branching of the vessels as they become smaller, so that the actual area exposed to the blood tends to increase as the vessels ramify (Fig. 5.24). As a result, the total resistance, which is made up of the resistances of

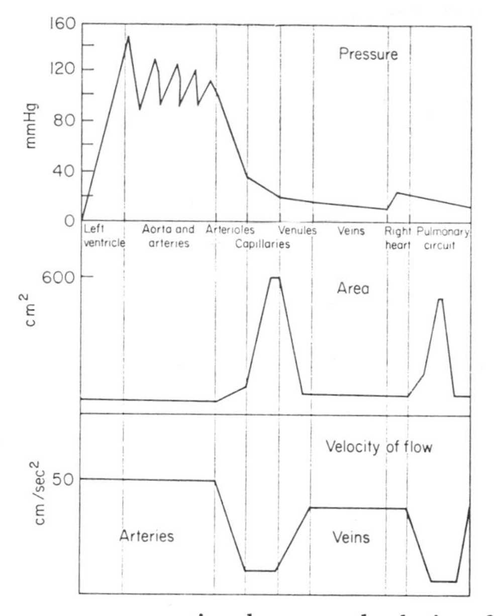

Fig. 5.24. The pressure, cross-sectional area, and velocity of flow throughout the circulatory system. The principal pressure-drop occurs across the arterioles, and although the diameter of each capillary is small, the total cross-sectional surface area is very great. Note the velocity of flow falls as the cross-sectional area increases; also that the velocity on the venous side is nearly as great as that in the arterial side although the pressure gradient is very much lower.

many small tubes *in parallel*, could actually decrease as the vessels became smaller, the actual effect of ramification being determined by the relative areas of the tubes before and after branching. Thus if a tube of 1 cm radius branched to two tubes of 0·5 cm radius (Fig. 5.25), the total area of the branched tubes would be $2 \times \pi \times 0{\cdot}5^2 =$ ca. 1·5 cm^2, compared with the area of the original tube—$\pi \times 1^2 =$ 3·14 cm^2. There has thus been a considerable decrease in area, and a simple computation shows that the resistance per unit length of the system has increased by a factor of 8*. In general, unless the area at a bifurcation increases by a factor of $\sqrt{2} = 1.414$, the combined resistance of the two smaller tubes in parallel will be greater than that of the single tube from which they were derived. Alternatively, it may be shown that the individual *radii* of the two tubes must be some 0·84 times that of the original tube if the combined resistance is not to be greater than that of the original tube. In general, bifurcations lead to radii that are smaller than 0·84 times the original, and so the total resistance per unit length of the vascular system increases as ramification proceeds; and this is illustrated by the figures of Table II relating to the ramifications of the mesenteric artery in the dog. Thus we can account for the large fall in pressure at the arterioles by the

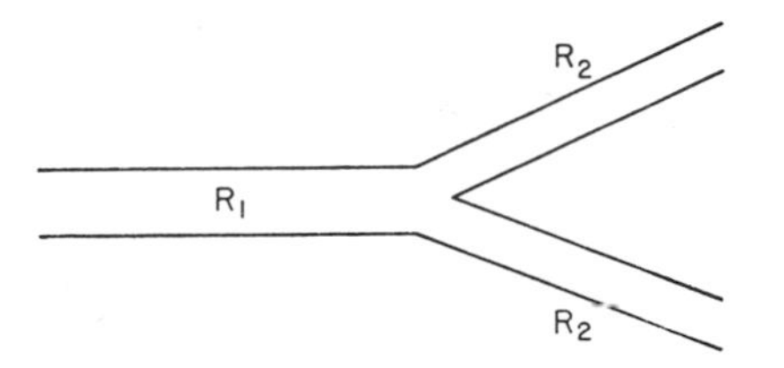

Fig. 5.25. The effect of branching on the resistance to flow. The branching increases the total area of cross-section and on that account decreases the resistance. However, the decrease in radius in the branches causes an increase in resistance, which rises as the inverse fourth power of the radius, so that unless R_2 of each branch is 0·84 times R_1 or more than this, branching will have increased the resistance to flow.

sudden fall in radius which is inadequately compensated by the increase in number of vessels. The fall in pressure as we pass to the capillaries is not so great absolutely, because so much of the pressure has been dissipated already; the capillaries are not so long as the arterioles, whilst the total area of cross-section rises rapidly.

Control of Pressure

Arteriolar Constriction

The arterioles are sited at a convenient place for the control over blood-flow and arterial pressure, and are well suited for the task by virtue of their muscular walls that allow of considerable changes of diameter in response to nerve stimuli. As indicated above, the ultimate

* The resistance per unit length of each of the finer tubes will be $2^4 = 16$ times that of the larger tube. There are two of these finer tubes in parallel, so that the resistance per unit length of the pair is half that of the individuals, and the factor becomes $16 \times \frac{1}{2} = 8$.

effects of a contraction of these vessels cannot be predicted with certainty. If the pump continued to force blood out into the arteries at the same rate, i.e. if Q remained the same, the arterial pressures (Fig. 5.26a) would rise to permit the same flow through the increased arteriolar resistance. Thus, immediately, the response to arteriolar

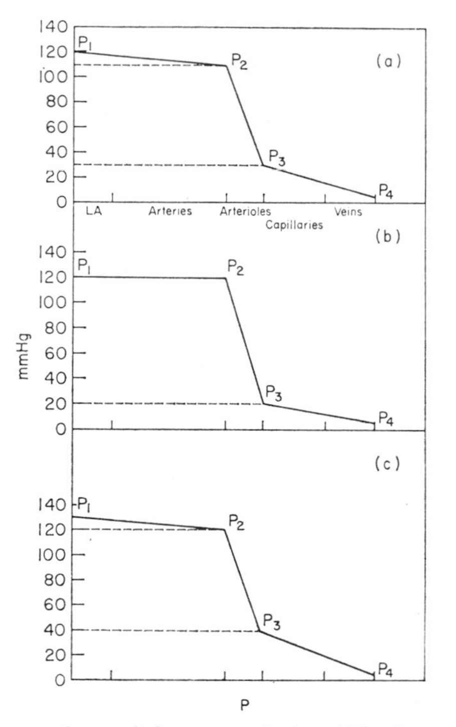

Fig. 5.26. The effects of arteriolar constriction. The immediate effect (b) is a rise in P_2, the pressure in the arteries proximal to the arterioles; P_3, the pressure in the capillaries, will fall. If the same amount of blood-flow is to be maintained, P_1 must rise (c) and ultimately all pressures proximal to the capillaries will remain elevated.

constriction will be a rise in the pressure at the beginning of the arterioles, P_2, because of the damming back of the blood caused by the constriction, which reduces the flow temporarily and thus reduces the dissipation of pressure along the arteries. The pressure at the end of the arterioles, P_3, will fall because of the increased resistance, hence capillary and venous pressures will also fall (Fig. 5.26b). If the pump continues to force blood into the arteries to maintain the original volume-flow, clearly P_1 the pressure in the aorta, must rise to reestablish the gradient P_1–P_2 along the arteries (Fig. 5.26c). Ultimately, all the pressures will be readjusted at generally higher levels, except beyond the arterioles where the same flow could be achieved by the original pressure-drop. Thus the constriction of the arterioles has left the system with a higher arterial pressure; and this is the usual consequence of a generalized constriction of these vessels. Its magnitude in the case illustrated by Fig. 5.26a-c depends on the ability of the pump to work harder to keep the same volume-flow; if this cannot happen, then clearly the system must be content with a lower flow and thus a lower arterial pressure, and the result is different depending on the degree to which the pump responds. The general pattern of pressure along the vascular tree, with its more precipitate drop along the arterioles, will of course remain unaltered.

Shunting of Blood. The control over the flow of blood to different parts of the vascular tree can obviously be exerted through the arterioles; thus we can consider the whole vascular tree as made up of several systems of arteries, arterioles, capillaries and veins in parallel. Alteration of the resistance of one will modify the flow of blood through it in relation to that through the others; thus dilatation of the vessels supplying the limbs during exercise favours flow through these at the expense of that through the viscera; a simultaneous constriction of the arterioles of the viscera (the *splanchnic* arterioles) will assist this diversion of blood to the limbs. Diversion of blood from the viscera during muscular exercise is physiologically feasible, since there is usually no urgency in the matter of digestion; diversion from the brain would, however, be very deleterious and we shall see that the circulation through this organ remains remarkably constant in the face of large variations in arterial pressure and in the relative resistances to flow in the viscera and skeletal musculature.

Capacity

The scheme of the circulation illustrated by Fig. 5.22 envisages a constant volume of fluid, so that at any steady state the volume ejected at the ventricular side of the pump is equal to that returned to the venous side. This is not true in non steady-state conditions, e.g. when the system is starting up from uniform pressures throughout the system, and this is because of the elastic qualities of the tubular system, so that the influx into the aorta, say, is not immediately transmitted into an efflux from the great veins. Thus the *capacity* of the vascular system, and variations in it, are of fundamental significance to the understanding of its mechanics.

Capillary Bed

An increase in capacity is brought about by any factor affecting the distensibility of the vasculature, and in this respect the capillaries play a very important role by virtue of their ability to close down completely, so that in a limb that is not exercising, for example, the actual flow through it may be only 10 per cent of that during maximal exercise. This extra capacity is brought about by the opening up of formerly closed capillaries. The immediate result of an opening up on the main arterial pressure will be to decrease it, unless compensatory mechanisms come into play, since the first effect of the increase in capacity is to reduce the amount of blood passed back into the heart; thus successive heart beats will force smaller amounts of blood into the aorta and distend it less. To prevent this reduced arterial pressure, we may return the capacity of the system as a whole to its original value by causing a constriction of the small vessels in other regions, of which the splanchnic area is the most important. The return to normal arterial pressure, induced in this way, leaves the limb with a greatly increased blood-flow because of its much reduced resistance to flow due to the opening up of new capillaries. During muscular exercise this is accompanied by a dilatation of the arterioles, further reducing resistance.

The Capillary Circulation

At this point a few words on the circulation through this region of the vascular tree are pertinent. The capillaries have no muscular elements in their walls (p. 218) so that their diameters are probably determined exclusively by the pressures in the arterioles and venules. As indicated, the flow of blood through the capillaries may vary enormously, and this seems to be achieved by the diversion of blood through previously empty capillaries or *vice versa*.

Precapillary Sphincters. The mechanism of this diversion is through a system of *precapillary sphincters* that allow the diversion of blood, which would otherwise take a direct course from arteriole to venule, into the "true capillary" circulation. Thus, the microcirculation may be represented schematically by Fig. 5.27. According to this, there is a permanently open thoroughfare channel between arteriole and venule, the arteriole branching to form a *metarteriole* which eventually loses its muscular elements to become a capillary which leads into a prevenule and then into a venule. The capillary portion of this thoroughfare channel is called an *a-v-bridge* (to distinguish it from the "true capillaries" which are off-shoots from the thoroughfare channels), entry from the metarteriole or arteriole being controlled by a muscle, encircling the branching point, called the *precapillary sphincter*. Thus, when the requirements of the tissue are not high, blood is confined mainly to the a-v-bridges through closure of the sphincters, the muscular elements of which are well innervated. By contrast, when the blood requirements are large, as in exercise or when warm blood must be diverted to the skin to cool it, then the sphincters relax and blood fills the previously empty true capillaries.

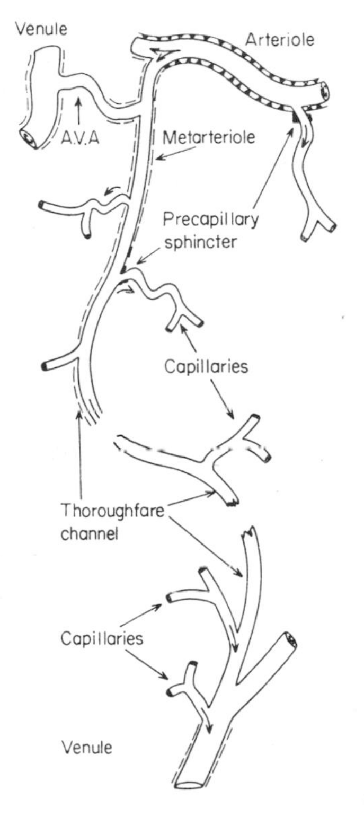

Fig. 5.27. Diagram of a functional unit of the capillary bed, together with a metarteriolar-venular anastomosis (AVA) and a precapillary branching off directly from an arteriole. (Chambers and Zweifach, *Am. J. Anat.*, 1944, **75,** 173.)

Shunts. In Fig. 5.27 it will be seen that the whole capillary system can be shunted through a metarteriolar-venular anastomosis (A.V.A.), thereby reducing blood-flow through the tissue very considerably; such a reduction might be necessary to conserve the body heat, e.g. when a bird's feet rest on ice the loss of heat would be very serious were the capillary circulation not restricted.

Pulsatile Flow

Pulse-pressure

Because the pumping of the heart occurs intermittently, in the sense that blood is forced into the aorta only for short periods of systole (communication with the ventricle being interrupted during diastole by the closure of the semilunar valves), the pressure in the aorta rises and falls with each beat; thus in Fig. 5.24, at the height of systole it is shown as 150 mm Hg and at its bottom, in diastole, it is 80 mm Hg, the difference being described as the *pulse-pressure*. When the pressure is measured at different points along the arterial tree it is found that the pulse-pressure at first increases, but later, as the blood vessels ramify, it becomes very much smaller so that usually in the capillaries and veins the blood-flow has become continuous and non-pulsatile.*

Transmission of the Pulse

The rise and fall in pressure in the aorta constituting the pulse are due to the forcing of a "bolus" of blood into this vessel; the pulse is the response of the elastic arterial wall to the expansion. Because of the rise in pressure, fluid is forced along the arterial tree away from the aorta, and this is accompanied by the diastolic fall in pressure. If the walls of the blood vessels were completely inextensible, the flow of blood would be intermittent, the amount forced in being immediately accompanied by the ejection, at the other end, of the same amount. Thus a pulse of pressure would pass virtually instantaneously along the vascular tree and would be accompanied by a simultaneous flow of fluid at each point. Hence the flow would be *pulsatile*, and between pulses it would drop to very low values and might entirely cease, depending on the kinetic energy imparted at each pulse.

Pulse Wave. In the vascular system, the forcing of blood into the large distensible arteries slows down this transmission and changes its nature; a pulse, consisting of an expansion of the tube and its subsequent return to its original diameter, is transmitted rapidly along the arteries. Thus the fluid first distends that part of the aorta nearest to the ventricle, the tension in its wall rises and hence the pressure within it. This exerts a force on the next segment which itself distends, and so the effects are transmitted, by virtue of this distensibility, as a *wave*. If we measure the pressure in an artery with a sensitive device that indicates rapid variations in it, we find that the oscillation in pressure, constituting the pulse, has a characteristic shape according

* We shall see that there is a pulse in the large veins, but this is not transmitted from the arteries and capillaries; it usually represents events occurring in the atria.

to the distance from the aorta; moreover, it requires a measurable time to reach the given point.

Pulse Wave Velocity. Thus the *pulse wave velocity* can be measured by recording the time at which, say, the foot of the wave reaches two points on an artery; the normal value in the aorta is 4·5–5 m/sec. The formula describing the velocity, c, is given by:

$$c = \frac{Eh}{\rho 2R}$$

where ρ is the density of the blood, h the thickness, R the radius, and E the elastic modulus of the wall of the vessel. Thus we should expect the velocity to be greater in a smaller artery, and in fact it is 8–10 m/sec in the femoral artery.

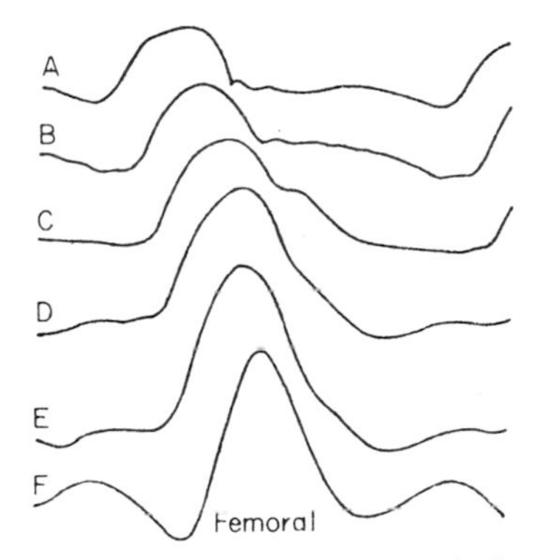

Fig. 5.28. Tracings of the pressure pulses recorded at five sites in the aorta (of a dog) with a long catheter. A is close to the aortic valves and E is close to the origin of the iliac artery. The intermediate records are at 10 cm intervals. F is a femoral record included for comparison but is from another dog. The peak rise is very marked. The pulse/pressure in the proximal aorta was approximately 110/80 mm Hg and in the distal aorta it was 130/75 mm Hg. (MacDonald, "Blood Flow in Arteries," Arnold, 1960.)

Form of Pulse. The form of the pulse-wave, measured as a fluctuation in pressure at a given point in the vascular tree, varies with the position. Figure 5.28 shows the pulse as recorded at different positions in the arterial tree. There is a sharp initial rise followed by a plateau and descending phase followed by a short sharp trough called the *incisura*. The time from the foot of the rise to the incisura measures systole, so that the incisura corresponds to the closure of the aortic valves and is due to the reflex of blood through the aortic valves as they close, the pressure in the ventricle having become temporarily below that in the aorta due to the beginning of relaxation of its muscular wall. The rising limb of the first rise is called the *anacrotic phase.* It will be seen that the height of the pulse in the more peripheral artery is actually higher than in the aorta, whilst the incisura has been damped out. The smoothing out is partly due to the general damping effect, due to the viscosity of the blood. Another factor may well be the effects of reflexion of the pulse-wave back on its course; this would interfere with the oncoming pulse-wave, or, of course, it could add to give peaks or notches. It would seem that the main site for reflexion is the arteriolar bed.

Velocity Pulse. The pulse recorded in Fig. 5.28 is a measure of the changes in pressure. With suitable devices the changes in velocity of

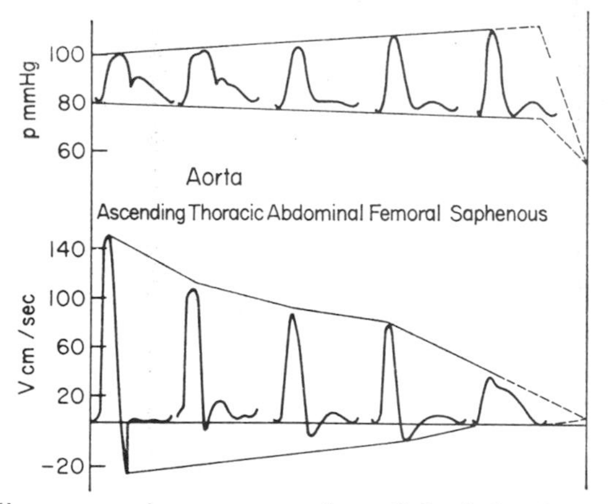

Fig. 5.29. A diagrammatic representation of the behaviour of the pressure and flow pulses in arteries as they travel away from the heart. The upper trace shows how the peak pressure increases with a slow fall in mean pressure. The lower trace shows how the peak velocity progressively declines. (MacDonald, "Blood Flow in Arteries," Arnold, 1960.)

the blood at a given point may be measured; the velocity passes through a cyclic change due to the impetus given to the blood by the pulse-wave, and the shape of this *velocity pulse* is not the same as that of the pressure pulse. Furthermore, the shape of the pulse varies with distance along the vascular tree in a rather different fashion. Thus, instead of the peak velocity increasing as we pass away from the aorta, as the peak pressure does, the peak velocity progressively declines (Fig. 5.29). By the time the arterioles are reached, the pulsatile changes in velocity are very small and none are seen in the capillaries. Ultimately, because of damping and reflexions, the pressure-pulse too must disappear, and this seems to happen in the smallest arteries and proximal arterioles.

Venous Flow

Distensibility of Veins

The flow in the veins obeys the same laws that are followed by flow in the arteries. The essential differences between the two are the generally lower levels of pressure, and the different distensibilities of the vascular tubes. Thus a small change in volume of the artery has to be brought about by a large change of pressure, as manifested by the arterial pulse. The veins will distend with a relatively small change in pressure, and thus their capacities are altogether larger, so that failure to empty the veins back into the heart can lead to a massive engorgement without a very large increase in venous pressure.

Venous Occlusion

When distension has reached a certain point, however, the rise in venous pressure may be much greater, first because the elastic limit has been approximately reached and secondly because the surrounding tissues tend to prevent further expansion. Thus if a cuff is placed on an arm, compressing the large veins but leaving the arteries relatively unaffected, the pressure in the veins may rise to the pressure in the artery supplying the limb, if flow through the limb is prevented entirely. This rise in pressure follows from elementary principles; occluding the end of the tube in Fig. 5.23, p. 250 will result in cessation of flow, and the cause for the loss of pressure, namely viscous resistance or friction, ceases.

Plethysmography. The rapid engorgement of the veins after occlusion of blood-flow provides the basis for the measurement of local blood-flow in a limb or part of it, such as the forearm or a finger. Essentially, the venous outflow is occluded by rapidly inflating a cuff to a pressure sufficient to block venous flow but leaving arterial flow unchanged. A close approximation to the increased volume of the limb is given by measuring the increase in circumference, which can be recorded with a mercury-in-rubber strain-gauge wrapped round the limb. Blood-flow is estimated by measuring the slope of the increase in volume trace.

Effects of Gravity

The large diameters of the veins mean a relatively low resistance, and it is not surprising to find that in the recumbent position the pressure in the large vein of the leg, the femoral vein, is only in the region of 5–10 mm Hg compared with a pressure of about zero or a small negativity in the atrium into which it drains. When the body is upright, it is clear that quite large hydrostatic pressures must be overcome to cause flow back to the heart if the column of blood from the vena cava to the ankle, say, is unbroken (of the order of 85 mm Hg). If a human subject stands for some time in the relaxed state, quite high venous pressures may be recorded, but in general the full potential of pressure is prevented from developing by breaking up the column by virtue of the contractions of the muscles of the limbs and the valvular character of the vessels. It is essentially the pumping action of these muscular contractions that provides the flow of fluid from the limb against the large gravitational pull (Fig. 5.30).

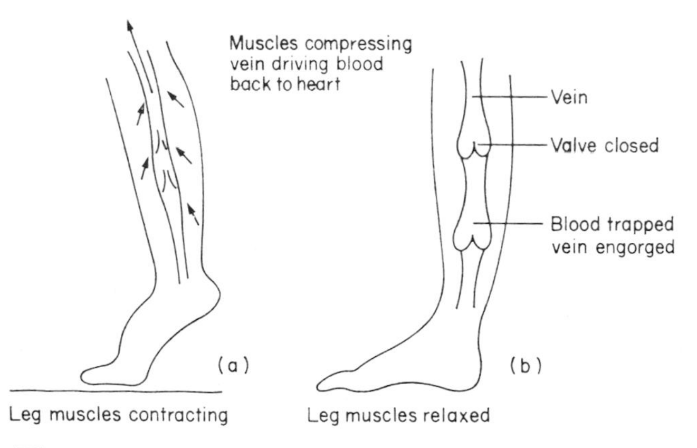

Fig. 5.30. The venous valves of the legs. If an unbroken column of blood extended from the atria to the feet high values of venous pressure would occur in the upright position in the lower limbs. Limiting this rise in pressure, valves are located in the leg veins which reduce the hydrostatic effects by breaking the column of blood into short sections as shown in (b). When the leg muscles contract they effectively squeeze the blood upwards to the next valve acting virtually as a "pump" as shown in (a).

Factors Affecting Venous Pressure

The pressure in a large vein is largely the passive result of many influences, by contrast with that in a large artery where the factors determining it are few. Some of the factors may be summarized here. The pressure will be determined by what has gone on before the blood arrives, namely the pressure-drop from artery to capillaries; if flow through the capillaries is large, the volume of blood carried by the veins will be large; this will result in some venous engorgement with corresponding rise in pressure. The gradient between veins and heart determines rate of flow, and thus the ability of the veins to empty themselves; a rise in the pressure in the atrium is probably the single most important factor in raising venous pressure, and this occurs through the operation of several types of cardiac failure leading to a damming of blood back into the great veins. Finally, an alteration in the circulating blood volume, or an alteration in total capacity of the system, will affect venous pressure.

Venous Pulse

The arterial pulse, transmitted along the arteries, is damped out by the time the capillaries are reached, hence the cyclical fluctuations in pressure observed in the large veins are due to another cause, i.e. the

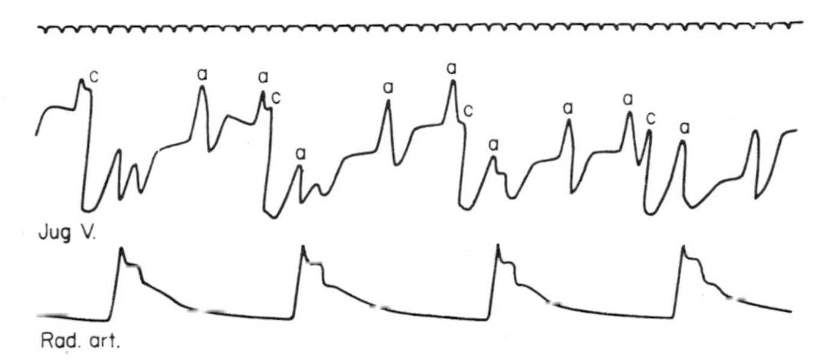

Fig. 5.31. The pulse record from the jugular vein and the radial artery recorded simultaneously. Each heart beat gives rise to two waves of positive pressure (a) is attributed to atrial contraction; (c) is attributed to closure of the atrio-ventricular valve producing a sudden peak as the muscle is still contracting.

transmission backwards along the veins of the alterations in the pressure within the atria. These changes in pressure in the atria are due to contraction and relaxation of their muscular walls, whilst secondary changes are due to the ventricular contractions. The simultaneously recorded pulses in the jugular vein and radial artery are shown in Fig. 5.31. Each heart-beat gives rise to two waves of positive pressure in the great vein, and are similar to those recorded directly from an atrium. Wave *a* is attributed to atrial contraction; *c* is attributed to the closure of the valve separating the atrium from the ventricle and such a closure, with the muscle still contracting, would cause a sudden peak of pressure. A third wave is sometimes seen as the result of the indirect effect of contraction of the ventricle during systole with the atrio-ventricular valves still closed, this latter peak will obviously be greatly increased if there is incompetence of the valve.

The Viscosity of Blood

Resistance to Flow

The viscosity is a feature of blood that receives scant attention in most accounts of the principles governing flow in the vascular system. This is unfortunate both theoretically and practically since the very fact that blood will flow at all through the fine tubes of the blood-vascular system, consisting as it does of some 80 per cent of vessels of diameter less than 180 μ diameter, should be a matter of considerable surprise. We see from Poiseuille's equation governing flow along tubes, that the *coefficient of viscosity*, η, enters in the "resistance" term, and the greater the viscosity of the fluid the greater the resistance to flow under a given pressure-gradient.

Laminar Flow

Mathematically the flow of fluids is treated as the sliding of very thin laminae over each other in response to a force acting tangentially to the laminae, the so-called *shear stress*. It is a shear because the applied force tends to deform the shape of the fluid (see Fig. 5.32), the force being necessary to overcome the friction between the laminae. Thus, because of the friction between the wall of the tube, say, and the lamina immediately adjacent to it, this layer tends to resist the push

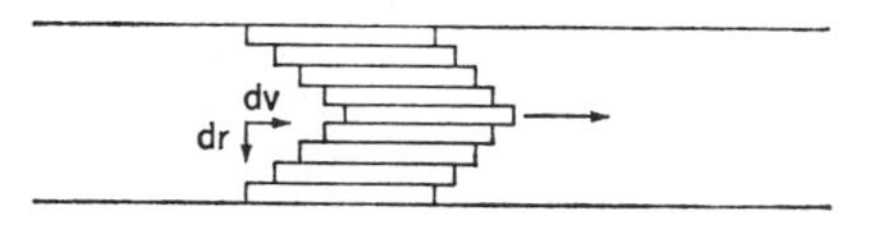

Planar laminar shear

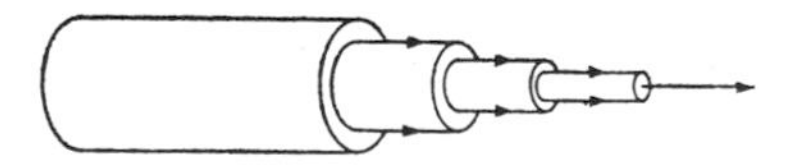

Telescope laminar shear

Fig. 5.32. A representation of laminar shear and deformation in a liquid. The upper figure shows the mathematical visualization of flow in a liquid with thin laminae of liquid slipping over each other in response to a force acting tangentially to the shear stress. The viscosity is the friction between the laminae which must be overcome before the liquid will flow. The lower figure is a representation of the same effect in a tube—telescopic laminar shear.

of the pressure-difference; the immediately adjacent layer moves over this layer, its motion being restrained, however, by the friction between the two; the lamina closer in moves a little faster and so on, so that there is a change of velocity, which reaches its maximum at the centre, falling off in a hyperbolic manner towards each surface of the tube (Fig. 5.33).

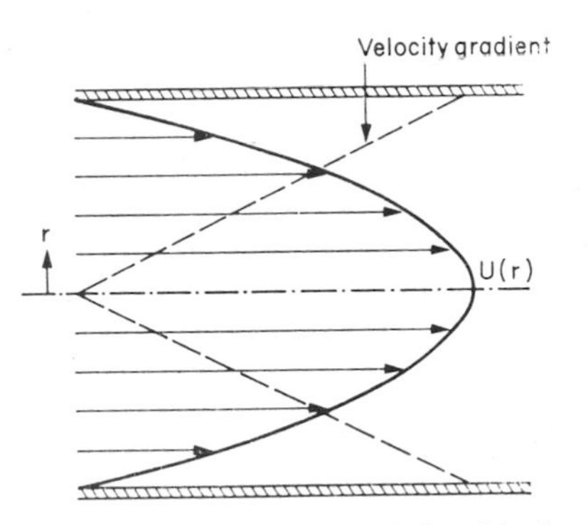

Fig. 5.33. The velocity gradient. Because of the friction between the layers and between the outermost layer and the wall of the tube, the liquid in a tube does not all move at the same velocity, so that there is a gradient of velocities across the tube as shown in the figure. The fluid in contact with the wall is stationary while that in the centre of the tube moves at the greatest velocity.

Rate of Shear

The *velocity gradient* is indicated by the differential, du/dr, and has the dimension of cm/sec ÷ cm = $\sec^{-1}$; and this is called the *rate of shear*, indicated by γ, i.e. the rate at which the column of fluid tends to be deformed. This gradient behaves in the opposite way to velocity, reaching its minimum (zero) value at the centre of the tube and having its maximum at the wall. It is given by the equation:

$$\gamma = \mathrm{du/dt} = \frac{4\mathrm{Q}}{\pi \mathrm{R}^4}\,\mathrm{r}$$

where Q is the rate of volume-flow, R is the radius of the tube and r is the distance from the axis, so that the gradient is zero when $r = 0$.

Shear-stress

The force bringing about this shear is called the *shear-stress*, τ, and is, if the fluid is "Newtonian", directly proportional to the rate of shear:

$$\tau = \eta\gamma$$

the proportionality constant, η, is the *coefficient of viscosity;* a viscosity of one *Poise* is the viscosity of a solution such that a shear-stress of 1 dyne/cm^2 causes a shear-rate of 1 sec^{-1}; its c.g.s. units are dyne.sec cm^{-2}. The viscosity of water at 37°C is 0·007 Poise or 0·7 Centipoise (cP).

Newtonian Fluid

A fluid is said to be Newtonian if this coefficient is independent of shear-stress and since Poiseuille's law is derived on the basis of Newton's law, we can say that the viscosity of a Newtonian fluid should be independent of the pressure-difference along the tube.

Many solutions, especially those of molecules with highly asymmetrical shapes, exhibit "anomalous viscosity", often of the form where the viscosity tends to decrease with increasing shear-rate (shear-thinning), so that their viscous behaviour cannot be described in terms of a true viscosity coefficient, the measured value depending on the conditions of measurement.

Suspension of Particles

Where suspensions of particles are concerned, the viscosity is often Newtonian, in the sense that, for a given concentration of particles in the medium, the viscosity remains independent of the shear-rate, e.g. of the pressure-head driving the fluid along a tube. The actual viscosity of a suspension (η_s) compared with that of the pure fluid (η_0) may be very high indeed, and depends critically on the proportion of particles to suspension medium, or *volume-fraction* as it is called. So high may the viscosity be, indeed, that it can be predicted that a suspension of solid particles with a volume-fraction comparable to that of the blood, i.e. of the order of 0·5, would not be able to flow through the finer arterioles and capillaries of the vascular system. This is because the particles suspended in the liquid phase tend to increase the internal friction (viscosity) by interfering with the flow-lines of the fluid laminae, e.g. the particles rotate thereby dissipating the mechanical energy that should be used to cause the shear-stresses involved in flow. As Fig. 5.34 shows, when the volume-fraction of a suspension of spherical particles becomes high, the viscosity is virtually infinite.

Blood

The viscosity of blood, with a volume-fraction (haematocrit ratio) of about 0·5, is only about twice that of blood plasma, which itself is not greatly different from that of water (1·2 cP for plasma compared with 0·7 cP for water) and, as Fig. 5.34 shows, there is remarkably little increase with increasing volume-fraction even up to a haematocrit-

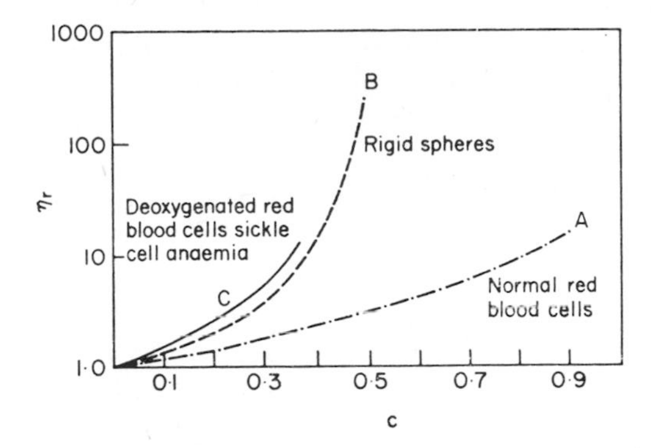

Fig. 5.34. The relative viscosity of a suspension of normal dog red blood cells as a function of the volume–concentration (curve A) compared with corresponding curves for a suspension of small rigid spherical latex particles (curve B) and for a suspension of deoxygenated red blood cells from a patient with sickle-cell anaemia (curve C). With rigid spheres the viscosity becomes very high indeed at a volume–concentration of about 0·5, corresponding to a normal haematocrit value for blood; by contrast, the viscosity of normal blood at this volume–concentration is only a few times that of water ($\eta_r = 1$), and even with a volume–concentration of 0·9 is only about ten times that of water. In sickle-cell anaemia, the suspension of deoxygenated red blood cells behaves as though the cells were rigid spheres. (After Goldsmith, *J. Gen. Physiol.*, 1968, **52** 55.)

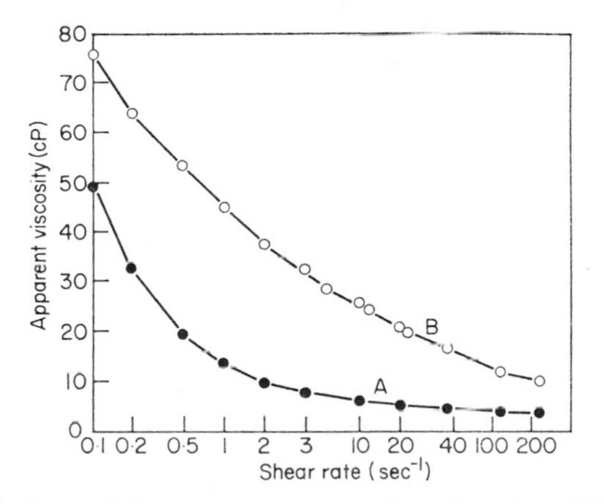

Fig. 5.35. The effect of shear rate on apparent viscosity of blood. Curve A represents normal blood and curve B blood made hypertonic so that the red cells are crenated. Note the higher values of apparent viscosity at all shear rates for the hypertonic blood. (After Schmid-Schönbein and Wells, *Ergebnisse der Physiologie*, 1970, **63,** 146.)

ratio of 0·9. In this sense blood behaves in a highly anomalous manner. Furthermore, when we take a blood of a given haematocrit-ratio, we find that the viscosity varies with the shear-rate, the greater the shear-rate the smaller the viscosity. This is illustrated by Fig. 5.35 where it is seen that the viscosity of a blood, with haematocrit-ratio 0·4, varied from about 50 cP to about 5 cP according as the shear-rate varied from 0·1 to 200 sec^{-1}.

Axial Stream. Thus blood, from a rheological point of view, is a highly anomalous fluid; one aspect of this was recognized by the classical study of Whittaker and Winton in 1933, when they perfused an organ of the dog (e.g. the kidney) and computed the viscosity of the blood flowing through it from the known pressure-head and the measured rate of flow. They found that there was remarkably little

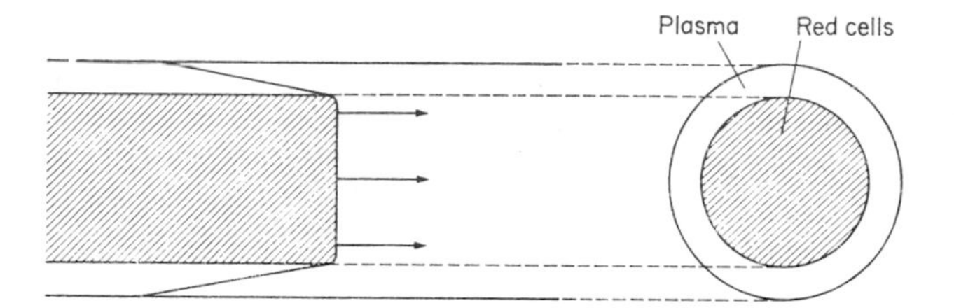

Fig. 5.36. The axial streaming of red cells flowing in a tube. The red cells move *en masse* in a form of "plug flow" with a flat velocity profile. The plasma is in a peripheral annulus and supports a normal velocity gradient. Axial streaming is thought to account for the anomalous behaviour of blood in that the viscosity only marginally increases with increased haematocrit ratio.

increase when the proportion of cells to plasma was increased, and their explanation for this was that the erythrocytes tended to form an "axial stream", being pushed into the centre of the moving column of fluid, so that the effective resistance to flow through the blood vessels was governed by a peripheral annulus of plasma (Fig. 5.36). This explanation of course poses the question as to why the cells should be forced into the axial stream since, with rigid particles, the shear-forces tend to make them rotate about their own axes rather than to move axially. Nevertheless, the explanation for this independence from haematocrit-ratio is probably valid in general terms, although the details are complex and not fully understood. It is unfortunately beyond the scope of this chapter to discuss the details of the flow of particle-suspensions through tubes, a subject that forms a large part of the study of *rheology*, and which has attracted much theoretical and practical study; here we must be content with a description of some of the more important factors.

Deformability of Erythrocytes

In the first place the erythrocyte is not a rigid body, but is highly deformable, and should be compared rather with a liquid droplet so that the flow should correspond more closely with that of an emulsion. In this case, theory indicates that, if the viscosity of the droplet is not greatly different from that of the suspending medium, the viscosity of the emulsion may not be greatly different from that of the medium, the droplet behaving as though it were part of the medium and so contributing little to any disturbance of the lines of flow. Now when the viscosity of blood is measured at high shear-rates, comparable with flow through fine tubes, it behaves in precisely this manner, the viscosity being very low and not greatly different from that of plasma, and the dominant cause is the deformability of the erythrocyte, its membrane presumably slipping over the surface and transmitting the shear-stress from outside to the inside in an even manner.

Influence of Cell Contents. If this is correct, then we may expect the nature of the cell-contents to influence the viscosity of the blood as a whole, a change in the condition of the haemoglobin within them, for example, being reflected in a change in viscosity of the blood. Again, a change in the deformability of the cell should also be reflected in a change in viscosity. Both these predictions are verified; thus in the disease of sickle-cell anaemia the viscosity of the cell contents, which depends on the physical state of the haemoglobin, shows a large increase when the blood is exposed to low O_2-tension; corresponding with this, there is a remarkable increase in the viscosity of the whole blood, i.e. the viscosity of the cell contents is affecting the viscosity of the suspension (Fig. 5.34). Again, in normal blood the internal viscosity of the cells may be increased sharply by making the plasma hypertonic; this withdrawal of water from the cells actually increases the viscosity of the blood as a whole.

Influence of Hypertonicity. The deformability of the blood cells may be modified in several ways, and this can be proved by measuring the ease with which they will pass through a mechanical filter, a deformable cell passing through pores that are considerably smaller than its own diameter. Cells exposed to a hypertonic medium are unable to pass through filters of this sort, the increased concentration of haemoglobin imparting to the cell, as a whole, a rigidity lacking at normal concentrations. As Fig. 5.35 shows, the viscosity of blood made strongly hypertonic is very high and remains high at all shear-rates, so that the blood is now behaving similarly to a suspension of solid particles.

Deformability and the Axial Stream. Let us return for a moment to the concept of the axial stream. The explanation for the anomalous behaviour so far has been on the basis of the deformability of the cells that allows them to behave like liquid droplets of internal viscosity comparable with that of the plasma, and such an explanation would hold independently of any axial stream. Experimentally this axial stream was demonstrated microscopically by Bloch, the annulus representing about one-fifth of the diameter of the vessel and becoming smaller with increasing diameter so that, according to Thomas, with vessels of diameter greater than about 500 μ it is negligible. Thus the peripheral cell-free annulus exists in the smaller vessels, and must contribute to a lowering of viscosity. One might be asked why, if the

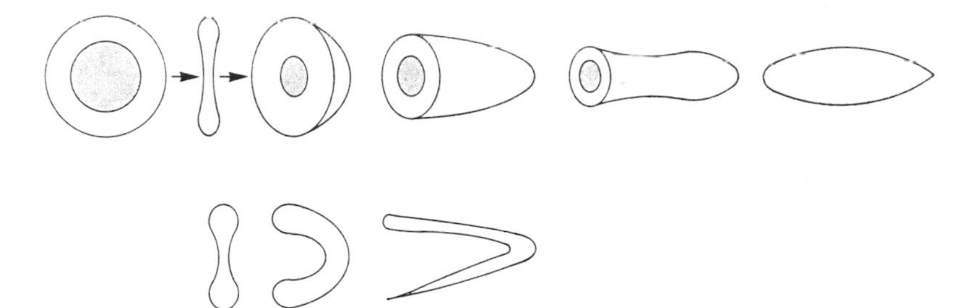

Fig. 5.37. Characteristic changes in shape of a red cell passing through a capillary. (After Branemark and Lindstrom, *Biorheology*, 1962, **1,** 139.)

particles are driven to the axis, should their physical character be important? Part of the answer is that it is because of their deformability that the shear-stresses are able to force them towards the axis of the tube; if they were rigid they would simply rotate. Thus deformability of the cells seems to be the crux to most of the rheological properties of blood; it enables them to behave like fluid droplets, it promotes their entry into the axial stream, and finally it permits the cells to move through channels of diameter less than their own, i.e. under conditions when we cannot speak of an axial stream.

Changes in Cell Shape. This deformability has been demonstrated by high-speed microphotograph of flow of blood in transparent ear-chambers, or in accessible blood vessels such as those in the conjunctiva of the eye. Figure 5.37 illustrates two typical series of changes observed by Branemark and Lindstrom, the cell being either blown out like a jelly-fish or drawn out like a bullet. Both deformations tend to streamline the cell, thereby reducing frictional effects, whilst the deformability *per se* favours the migration of the cells into the axial stream. When we are considering flow in vessels of diameter less than that of the erythrocyte, it is this deformation into a bullet shape that is the main factor permitting any flow at all. The flow consists of individual cells following each other and suffering a degree of distortion depending on the shear velocity; the flow is described as "bolus flow", and the mechanical situation seems to be highly favourable for low viscosity; at any rate it has been found experimentally that, when blood flows through tubes of diameter 5 μ, i.e. less than that of the cells, the viscosity may be only 5 per cent greater than that of plasma.

Rouleaux Formation

The erythrocytes of many species show a tendency to aggregate in a reversible manner to form *rouleaux*, i.e. giving the appearance of a pile of coins; they are able to do this because of their disc-shape, and the extent to which it happens varies with the species, being very manifest in the horse. It also may vary pathologically and physiologically, the tendency being determined by the character of the proteins in the plasma which seem to act as links between the surfaces of adjacent cells. The formation of rouleaux contributes to the non-Newtonian character of blood-flow, since, when the shear-rate is low (i.e., with low-velocity flow in large vessels), it may be seen that the rouleaux tend to build up a three-dimensional framework within the fluid plasma, imparting to the blood a high "structural" viscosity. As the shear-rate increases, this framework tends to break down, and finally with flow in very small vessels rouleaux formation is no longer possible. When red cells are suspended in an isotonic saline medium containing 1 per cent of albumin, rouleaux formation is inhibited, and under these conditions the viscosity remains low over the low shear-rate region.

Sickle-cell Anaemia. In sickle-cell anaemia the cells tend to form more rigid rouleaux than normally, and this contributes to the high viscosity at low shear-rates; since the tendency depends on the plasma rather than the cells, sickle-cells in normal plasma exhibit normal rheological behaviour at these low spear-rates.

Summary

To summarize a very difficult subject, then, we may say that the anomalous viscosity of the blood permits it to carry out its function as a circulating medium. In the small arterioles and capillaries the viscosity of a suspension of particles of volume-fraction about 0·5 would, if these were rigid, be impossibly high to permit flow under the practicable pressure-gradients. Because of the shear-thinning, i.e. the reduction in viscosity with increasing shear-rates, the viscosity in the small vessels, where shear-rates are highest, is sufficiently close to that of plasma to permit adequate flow. Furthermore, quite large variations in the haematocrit-ratio can be tolerated with only slight changes in viscosity. So far as can be ascertained, these remarkable rheological characteristics result from the deformability of the erythrocyte, so that the circulation through the small vessels must be peculiarly susceptible to changes in the physical character of these cells.

Dynamic and Static Haematocrits

The velocity of flow in a tube increases as we approach the centre (Fig. 5.33), consequently the erythrocytes flowing in the axial stream, as predicted by dynamic considerations and as postulated to account for the rheology of blood flow in small vessels, should be moving faster

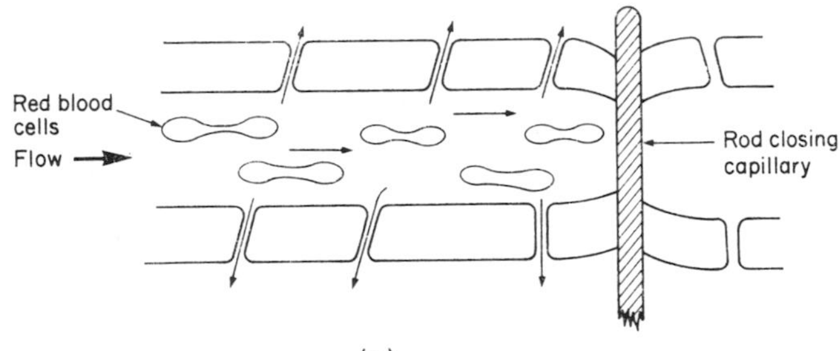

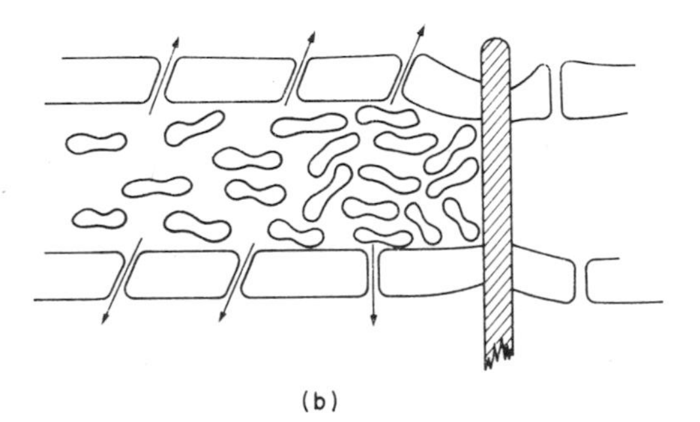

Fig. 5.38. (a) A small venule is compressed downstream and the flow stopped. (b) The escape of plasma through the vessel walls leads to packing of the red cells, and from the measured rate of packing it is possible to estimate the original haematocrit ratio, which is only some 25 per cent compared to the 40–50 per cent of blood from large vessels.

than the plasma. Hence, the "*dynamic haematocrit*" should be different from that of blood drawn from a large vessel, the "*static haematocrit*". Thus, if we have blood flowing from a reservoir A to reservoir B through a fine tube or series of tubes, then to sustain a situation in which cells move faster than plasma, the haematocrit of the blood in the fine tubes must be less than that of the reservoirs. The ability to label plasma and erythrocytes separately with isotopes has permitted the demonstration of this change in haematocrit in blood according as it is in the small vessels or in a large artery. Thus we may excise an organ while the blood stream contains ^{131}I-labelled plasma and ^{51}Cr-labelled cells; the blood remaining in the tissue is largely in the small vessels, so that a comparison of the $^{51}Cr/^{131}I$ ratio in arterial blood and in the tissue will demonstrate the difference. Perhaps this reduction in the haematocrit in small vessels is best demonstrated by compressing a small venule downstream, as in Fig. 5.38. The escape of plasma leads to the packing of the cells, and it is easy to compute, from the measured rate of packing, the original haematocrit-ratio of the blood in this vessel; this is only about 25 per cent compared with some 40–50 per cent in blood from a large vessel.

Turbulence

The laminar flow illustrated by Fig. 5.33 is at the basis of the theoretical treatment of viscous flow, as exemplified, for example, by Poiseuille's law; when certain conditions are not fulfilled the flow becomes turbulent, and in this event the relation between pressure and radius alters, the pressure-gradient being a higher power of the rate of flow than unity, so that a greater pressure-gradient is required to obtain a given rate of flow when this is turbulent. The limiting condition is given by the *Reynolds number*, R_e:

$$R_e = \frac{Vr\rho}{\eta}.$$

Here V is the average velocity; *r* the radius* of the vessel, ρ the density and η the viscosity. The ratio $\frac{\eta}{\rho}$ is called the *kinematic viscosity* of the fluid.

The actual value of the Reynolds number above which turbulence occurs cannot be defined precisely because so much depends on the geometry of the system; thus, with a value of 200, flow along a straight tube will be laminar but turbulence may occur at branches of a system of tubes but, if R_e is less than 1000 it will die away on passing from the branch. With a value of R_e in the region of 1000 turbulence is

* Sometimes the diameter is employed, in which case R_e is twice that as defined above.

likely to occur in straight tubes, but by taking special precautions laminar flow may persist with R_e equal to as much as 10,000.

Recognition by Spleen

We shall see that the spleen "sequestrates" erythrocytes when they are abnormal, either through age or pathology; this is a preliminary to their destruction. The problem is what feature of the ageing or pathologically altered cell is chosen to act as signal for its removal; if the flow-behaviour of the cell depends on the internal viscosity and other geometrical characters it is possible that the altered characteristics of flow are the signal. Thus the rigid abnormal cell might tend to remain at the periphery of the flowing blood-column and so get caught. Certainly when red cells are heated to 50°C they become sequestered in the spleen, and this is accompanied by a marked increase in rigidity. Mere sphericity is not important since cells can be made spherical by treatment with high concentrations of urea, yet they retain their normal mechanical characteristics, as revealed by passage through porous filters.

CHAPTER 6

The Fluid Exchanges between Blood and the Environment. General Aspects of Capillary Permeability

The mechanics of the vascular circulation are of interest to the physiologist only in so far as they lead to an understanding of how the blood reaches the regions where exchanges of matter or heat between blood, on the one hand, and the cells of the body and the atmosphere in the lungs on the other, take place.

These exchanges take place virtually exclusively when the blood is flowing through the capillaries, and in this chapter we shall consider the more basic aspects of the exchange of material across the capillary wall.

Capillary Structure

Endothelial Tube

The capillaries are the smallest blood vessels, consisting of little more than endothelial tubes resting on a basement membrane. They connect the arterial with the venous sides of the circulation and have diameters of the order of 4–13 μ. The structure of the capillary is not the same in all tissues, so for the moment we may consider the capillary of muscle. Figure 6.1 illustrates an electron-micrograph of a cross-section of a muscle capillary, and the essential features are illustrated schematically in Fig. 6.2. The region of the capillary illustrated has been made by the joining together of two flat endothelial cells; the joints are the intercellular clefts, J, and there is reason to believe that the escape of much of the dissolved material from the blood into the surrounding tissue occurs preferentially along these intercellular clefts, rather than across the cytoplasm of the endothelial cells.

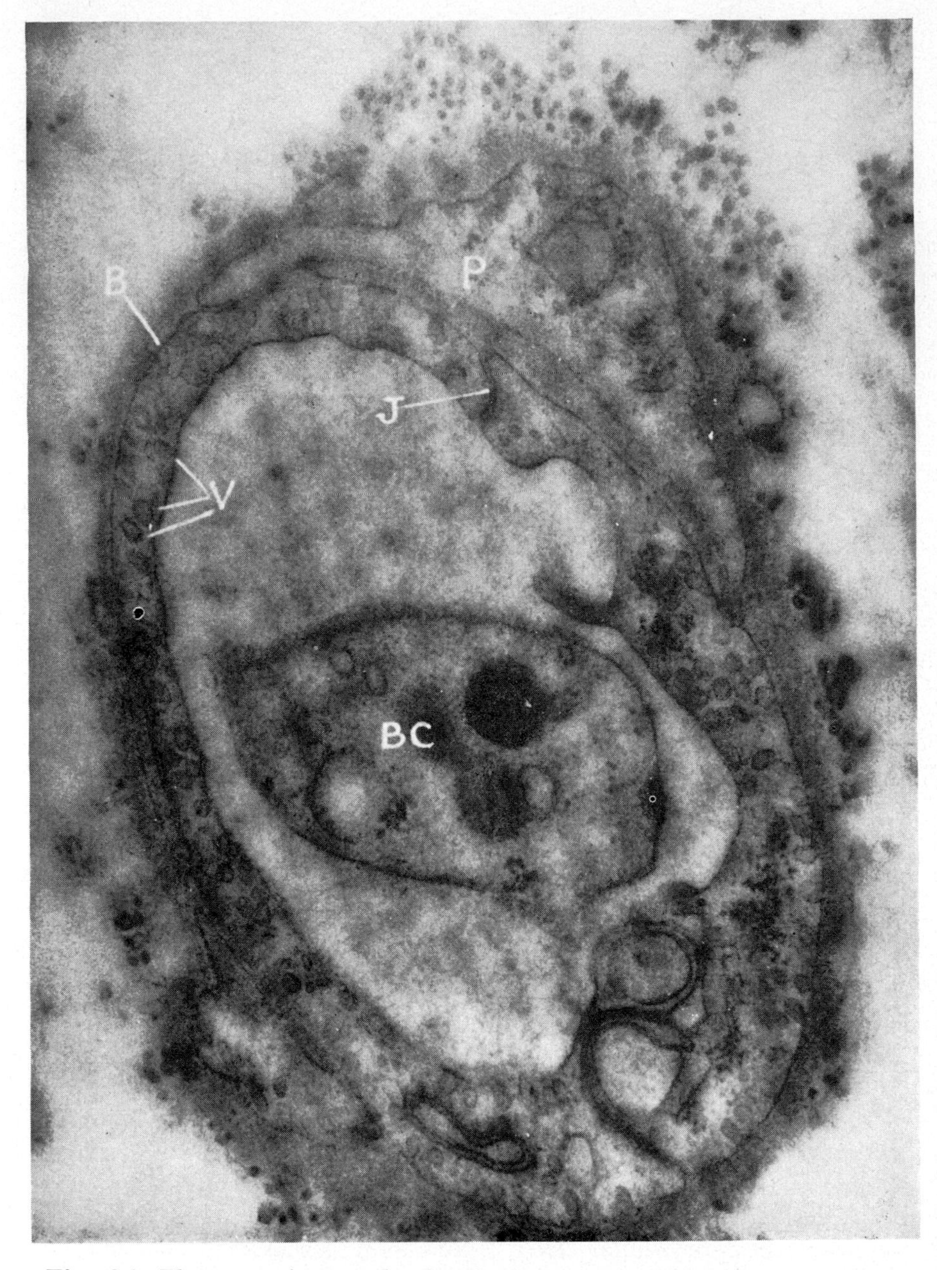

Fig. 6.1. Electron micrograph of rat's skeletal muscle capillary containing a red blood cell (BC). B: basement membrane; V: vesicles; J: junction between two endothelial cells; P: pericyte. Osmium fixation ×42,000. (Courtesy G. E. Palade.)

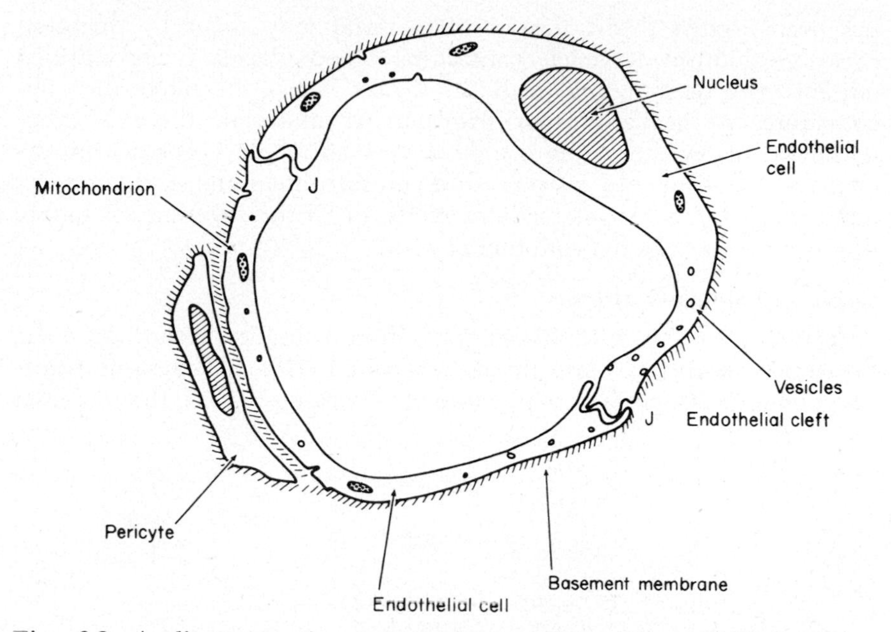

Fig. 6.2. A diagrammatic section of a capillary. The capillary shown consists of two endothelial cells joined by close apposition of the cell membranes at J but there is still a visible pathway between the cells. The basement membrane is shown as a hatched line which divides to enclose a pericyte. Some vesicles are shown present in the cytoplasm, budding off from either membrane.

Basement Membrane

On the outside of the endothelial tube is the basement membrane, and it will be seen that this has split to enclose a *pericyte*, i.e. a cell that characteristically comes into close relation with the capillary tube, its function being by no means clearly understood. The basement membrane at very high magnifications appears as a fibrous pad of otherwise amorphous material; it contributes to the mechanical stability of the vessel, and is analogous with the tunica media of the larger vessels; it is doubtful whether it acts as a restraint to the diffusion of materials out of the capillary, except perhaps for very large particles.

Cytoplasmic Vesicles

Within the cytoplasm of the muscle capillary there are many vesicles, probably formed by endocytosis (p. 48). When large numbers of sections are investigated, some vesicles are found opening into either the lumen of the capillary or the outside medium, consequently it has been argued that transport of material may occur by this *pinocytotic* mechanism, a vesicle containing blood plasma being emptied out into the surrounding medium. Whilst it is quite likely that this constitutes a mechanism for transporting material, there is every reason to believe that this is only a very subsidiary process, the exchanges between the plasma and outside medium occurring by diffusion through the intercellular clefts, or, if the substance is highly lipid-soluble, across the endothelial cells.

Capillary Specializations

As indicated above, capillaries vary in morphology according to the tissue they supply; certain tissues associated with considerable transport of fluid, e.g. the choroid plexuses of the brain (Vol. 3), the proximal

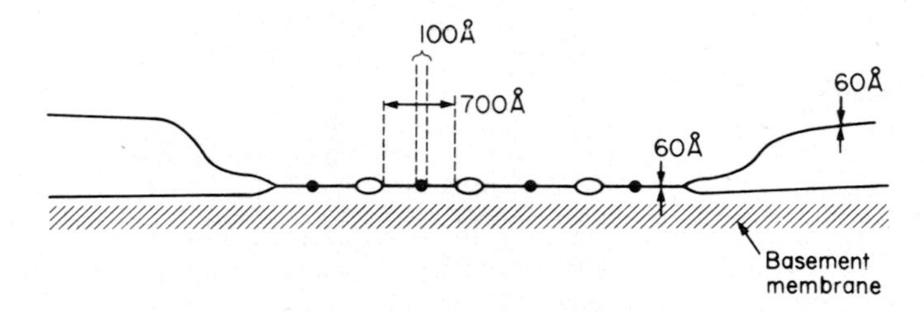

Fig. 6.3. Intercellular fenestrations. In some capillaries attenuation of the cytoplasm occurs in some regions and the two walls of the endothelial cell have come close together to form fenestrations. (Magno, 1965, "Handbook of Physiology 2", Vol. III, 2293.)

tubules of the kidney (p. 488) and the endocrine glands (Vol. 2), have what have been called *intracellular fenestrations*, which consist of an enormous attenuation of the cytoplasmic wall, as illustrated schematically in Fig. 6.3, where it is seen that the fenestration consists of a mere apposition of the plasma membranes of the endothelial cell, the intervening cytoplasm having been squeezed away. So thin is the capillary membrane here that it may well be that transport of fluid and solutes occurs in significant amounts across the fenestration, in addition to that carried across the intercellular clefts. A third category of capillary exhibits large *inter*cellular gaps; here, restraints on the escape of material will probably be governed only by the basement membrane, and, as such, the restraints will not be great. These vessels occur in the sinusoids of liver, spleen and bone-marrow, vessels that allow relatively unimpeded passage of large protein molecules, and even cells (Chapter 9).

Capillary Permeability

Permeability Coefficients

We have seen that the ease or difficulty with which a substance may penetrate or leave a cell can be measured by a permeability coefficient, the amount crossing unit area of the membrane in unit time when unit difference of concentration is present. In the same way we may study the permeability of a more complex structure like the capillary, measuring the amount escaping from the blood plasma into the surrounding tissue in unit time. Permeability coefficients measured in this way are very much higher than those for cells; e.g. the permeability coefficient for glycerol penetrating the red blood cell is about $1{\cdot}7 . 10^{-8}$ cm/sec compared with about $1{\cdot}5 . 10^{-4}$ for a muscle capillary. Associated with this much greater permeability there is a lack of selectivity, so that the difference between sucrose and glycerol, for example, is only a factor of about 2 compared with one of hundreds or even thousands when single cells are examined. In fact, for most mammalian cells the permeability coefficient for sucrose is infinitesimally small, and the molecule is said to be excluded from the cells of the body.

Intercellular Clefts

This high permeability, and lack of discrimination, could be accounted for if transport was largely governed by diffusion, or flow, through large intercellular clefts. If these had a diameter of about 80 Å the capillaries would, indeed, have high and indiscriminate permeability, and would only exert significant restraint on large molecules such as the plasma proteins, e.g. serum albumin with a molecular diameter of about 72 Å.*

Gap Junctions. In discussing the permeability of epithelia it was shown that their low permeability and high selectivity could be due to the occluding type of junction that debarred passage through the intercellular clefts, so that the permeability of the epithelium represented the permeabilities of the plasma membranes of the epithelial cells. The intercellular clefts of the capillary endothelium were for some time considered to be occluded in the same way, so that the high permeability of the capillary was something of a puzzle. However, more recent studies, employing lanthanum fixation, have shown that the junctions between endothelial cells are not completely occluded;

* The pores may, on average, be of about the same diameter as serum albumin, and therefore would allow of some escape of this protein; they would, however, exert a considerable restraint for geometrical reasons, since a protein molecule would have to approach the pore at a definite angle if it were to penetrate it.

the space narrows to about 80 Å or less and is described as a gap-junction.

Flow-limited Exchanges. When the solute considered is highly lipid-soluble, e.g. an anaesthetic, then it may diffuse rapidly across the endothelial cell membranes as well as through the intercellular clefts, and this means that exchanges may be very rapid indeed, in fact too rapid to measure quantitatively by the usual techniques. Under these conditions, moreover, the limiting factor is not the permeability of the capillary membrane but the rate of flow of blood through the capillary. It is for this reason that the rate of flow of blood through a limb, or through the brain, can be measured by estimating the time required for a solute like radioactive krypton to escape from the blood and accumulate in the tissue.

This rapid permeability extends to the respiratory gases, CO_2 and O_2, so that exchanges of these important solutes are governed primarily by the rate of blood-flow through a tissue. As we shall see, the actual flow through an organ, such as the coronary circulation to the heart, is delicately adjusted to the O_2-requirement, so that an increased utilization, which causes the partial pressure of O_2 in the tissue to fall, results in an immediate increase in the blood-flow.

Plasma–Interstitial Fluid Relations

Interstitial Fluid

The nutrition of the cells of the body occurs through a thin film of fluid surrounding the cells, called the *interstitial fluid;* thus material escapes first from the capillary into this interstitial fluid and thence into the cells. The volume of this interstitial fluid may change very considerably in accordance with the state of hydration of the animal as a whole, and it is therefore of considerable interest to examine the relation between the fluid and the capillary blood. Because of the low permeability of the capillaries to proteins, the interstitial fluid normally has a low concentration of these plasma constituents, which are predominantly (67 per cent) serum albumin, with a molecular weight of 67,000, and serum globulins (33 per cent) with considerably greater molecular weights most of which are above 100,000. The rapid exchanges of smaller molecules and ions ensure that the fluid has about the same concentrations of these as in plasma, so that in discussing the fluid-relations between the plasma and interstitial fluid we are concerned with the relations between two salt solutions, the one containing protein and the other none or very little.

Colloid Osmotic Pressure

This relation was first examined by Starling in 1896, who pointed out that the interstitial fluid could be considered as an ultrafiltrate of plasma, and consequently the Gibbs–Donnan equilibrium should apply to this situation. Thus we may place plasma in a vessel closed at the end by a membrane permeable to the salts, etc. of the plasma, but impermeable to the proteins (Fig. 6.4); then, on exerting a suitable pressure on the vessel, we may extrude across the membrane an ultrafiltrate, containing the permeable solutes in about the same concentrations as in plasma but being protein-free. If the pressure were to be

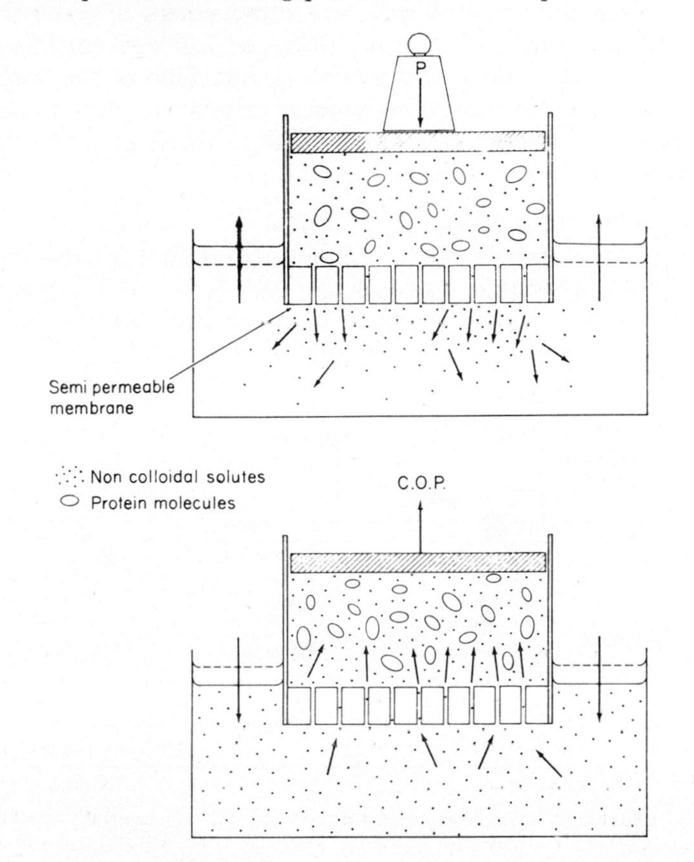

Fig. 6.4. The ultrafiltration of plasma. *Top:* the hydrostatic pressure represented by P forces water and salts through the pores of the semi-permeable membrane. The pores of the membrane are too small to permit the passage of the large protein molecules. *Bottom:* the hydrostatic pressure has been removed and now the colloid osmotic pressure of the protein (COP) draws salts and water back through the membrane into the chamber.

released, the opposite process of *osmosis* would now occur, because the osmotic pressure of the plasma would be greater than that of its filtrate by virtue of the presence of the impermeable protein molecules. Only if a pressure equal to this difference of osmotic pressure were exerted on the plasma would the osmosis cease. This pressure is called the *colloid osmotic pressure* of the plasma, and amounts to about 25 mm Hg in the mammal.

Flow of Filtrate. It is important to appreciate that, when the osmotic process is allowed to occur, the result is the movement of the whole plasma ultrafiltrate back into the plasma, not just a movement of water. Thus the first step will be the movement of water from the filtrate into the plasma, but this causes the salt concentration in the filtrate to rise and some movement of salt follows the water. The movement of the salt causes the osmotic pressure to rise in the plasma and so more water enters; in fact, the final result could be the complete reabsorption of the filtrate.

Filtration and Absorption

To apply this consideration to the interstitial fluid, we may represent by Fig. 6.5 the plasma in its capillary surrounded by interstitial fluid. The pressure within the capillary at its arterial end is, let us say, 35 mm Hg; this pressure is laterally directed and tends to squeeze

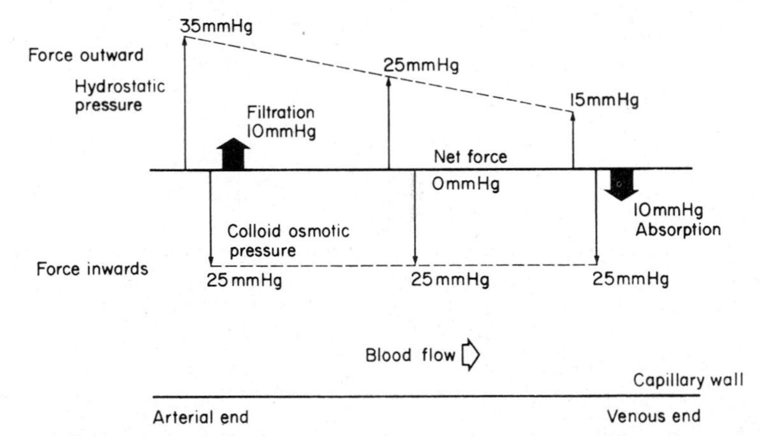

Fig. 6.5. Filtration and absorption forces along a capillary bed. At the arteriolar end the hydrostatic pressure exceeds the colloid osmotic pressure by some 10 mm Hg and there is a net filtration of salts and water outward. This net outward force gradually declines until at a certain point the outward force exactly balances the inward force and no fluid movement occurs. At the venular end the hydrostatic pressure is less than the colloid osmotic pressure so that there is a net inward force causing absorption of fluid back into the capillary.

plasma out of the vessel through the pores, in the same way that water leaks out of a leaky hose if it is under pressure. Thus there is a tendency for the capillary to produce an ultrafiltrate into its surroundings at its arterial end, the actual tendency being measured by the difference between two opposing influences, namely the hydrostatic pressure causing the filtration, and the difference of osmotic pressure which favours reabsorption of the filtrate.

The colloid osmotic pressure is, as we have said, about 25 mm Hg, so that there is a net tendency for filtration to occur. At the venous end of the capillary the situation is reversed, the hydrostatic pressure being lower, say 15 mm Hg, whilst the colloid osmotic pressure is still 25 mm Hg.* Thus the interstitial fluid tends to be reabsorbed at the venous end.

Fluid Pressure. Since the capillary pressure decreases continuously from arteriolar to venular end of the capillary, a point will be reached when the tendency to filter is zero, being just balanced by the opposing forces. At this point the capillary pressure and colloid osmotic pressure are not necessarily equal, since the fluid in the tissue will be under some pressure, which we may call the *interstitial fluid pressure.* If this were positive, it would oppose the filtration, and we should have equilibrium with the capillary pressure greater than the colloid osmotic pressure difference: e.g. if the tissue fluid pressure were +5 mm Hg, and the colloid osmotic pressure difference were 25 mm Hg, we should have:

$$\underset{\text{C.P.}}{30\ \text{mm Hg}} - \underset{\text{C.O.P.}}{25\ \text{mm Hg}} - \underset{\text{I.F.P.}}{5\ \text{mm Hg}} = 0$$

Thus, in general terms we may say that to have no net flow of fluid into or out of the tissue we must have, on average across the capillary, the relationship:

Capillary pressure = Colloid osmotic pressure + Tissue fluid pressure.

At the arterial end, the capillary pressure must, therefore, be greater than the colloid osmotic pressure + tissue fluid pressure, if filtrate is to be formed; and in fact there is little doubt that there is normally a continuous escape of fluid at the arterial end with reabsorption at the venous end.

* Actually we may expect the colloid osmotic pressure difference to be rather greater because the plasma will have become concentrated as a result of the ultrafiltration taking place at the arterial end. This difference allows of some homeostasis so far as fluid exchanges between plasma and interstitial fluid are concerned, the greater the loss at the arterial end the more favourable the conditions for recovery of fluid at the venous end. However, it is now well established that the venules are much more "leaky" in respect to proteins than the capillaries, so that at the venous end the concentration of protein in the tissue fluid will be larger, thereby reducing the colloid osmotic pressure difference.

Negative Interstitial Fluid Pressure

For a long time it was considered that the tissue fluid, such as that in muscle or skin, was under a positive pressure. Experiments, in which a needle was inserted into the tissue and the pressure required to force fluid into this tissue measured, indicated a positive value. However, Guyton emphasized that this technique would be most unlikely to measure the actual tendency for fluid to pass into, or out of, the capillary, which is what is really required. He implanted into the tissue a small hollow plastic capsule punctured by numerous holes; the fluid accumulating within this capsule was apparently representative of

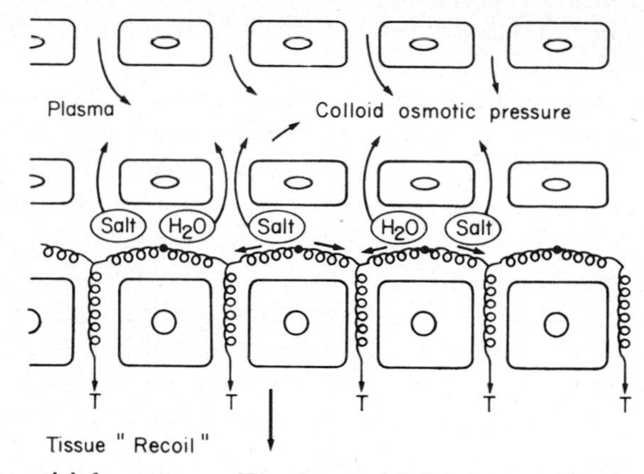

Fig. 6.6. Interstitial pressure. The interstitial pressure has been shown to be most probably some 6 mm Hg negative so in fact acts as force opposing the movement of fluid back into vessels. The diagram represents this negative pressure as springs opposing the fluid movement, and is the elastic resistance of tissue balancing the colloid osmotic forces.

tissue fluid so that its pressure could be measured by inserting a needle connected to a manometer. Alterations in the colloid osmotic pressure of the blood were reflected in alterations in the pressure of the fluid, confirming this postulate regarding the nature of the fluid. Surprisingly, the normal interstitial fluid pressure measured in this way was negative —about −6 mm Hg—indicating a net tendency for fluid to be absorbed from the tissue into the blood. Only under pathological conditions, when fluid tended to accumulate in the tissue (*oedema*, p. 290), did the pressure become positive.

Tissue Recoil. It might be argued that a negative interstitial fluid pressure means a tendency for absorption of the fluid, and therefore absorption should occur indefinitely; however, this is to ignore the elastic resistance of the tissue opposing the removal. Thus the negative pressure represents the balance between the escaping tendency of the fluid under a difference of colloid osmotic pressure and the "retaining tendency" of the vacuum that would be created were the fluid to leave (Fig. 6.6). As we shall see, similar negative pressures can be created in the pleural cavity through the opposing tendency of the lungs to open the cavity through their elastic recoil and the vacuum that would be created were the cavity to expand (p. 367).

Mucopolysaccharide Gel. We should note that Guyton's capsule presents an artificial situation which permits the measurement of the escaping tendencies of fluid in capillary and the capsule, and this is not necessarily applicable to the tissue fluid, which is retained within the meshes of a hyaluronic acid gel which exerts its own colloid osmotic pressure, called in this context its *swelling pressure*. This pressure is probably only one or two mm Hg, and tends to attract water in the spaces and therefore opposes the colloid osmotic pressure of the plasma and the tissue-fluid pressure as measured in the capsule.

Mean Capillary Pressure

A way of estimating the average capillary pressure was described by Pappenheimer and Soto-Rivera, who perfused the hindlimb of a cat through an artery, collecting the blood from the vein. The leg was held on a balance so that any change in weight, due to absorption or loss of fluid from the blood, could be determined. When arterial pressure was varied a venous pressure could be found such that there was no change in weight—the *isogravimetric state*—and in fact there must exist an infinite number of pairs of arterial and venous pressures conforming to this condition where the outflow of fluid is just balanced by the inflow. As the arterial pressure is progressively lowered and venous pressure is raised, a point will be reached at which venous pressure for the isogravimetric state is just equal to the arterial pressure. Under these conditions, of course, flow of blood would cease and therefore the arterial and venous pressures would equal the capillary pressure. Since this capillary pressure corresponds to the isogravimetric state we may say that it just balances the tendency for fluid to enter the capillaries from the tissue and is, in fact, the mean capillary pressure. The situation with zero blood-flow when arterial and venous pressures are equal is too unphysiological to permit determination of a capillary pressure with any meaning. However, by extrapolation of the curve of arterial versus venous pressure for the isogravimetric state, to the situation of equal arterial and venous pressures an estimate may be obtained. Since the relation between the two parameters is not linear, however, the extrapolation is not very accurate and a more accurate one is given by plotting blood-flow against venous pressure for the isogravimetric state, and extrapolating to zero blood-flow, when venous pressure is equal to capillary pressure. This has been done in Fig. 6.7, where the extrapolated venous pressure was 13 mm Hg, which is thus the mean level of capillary pressure.

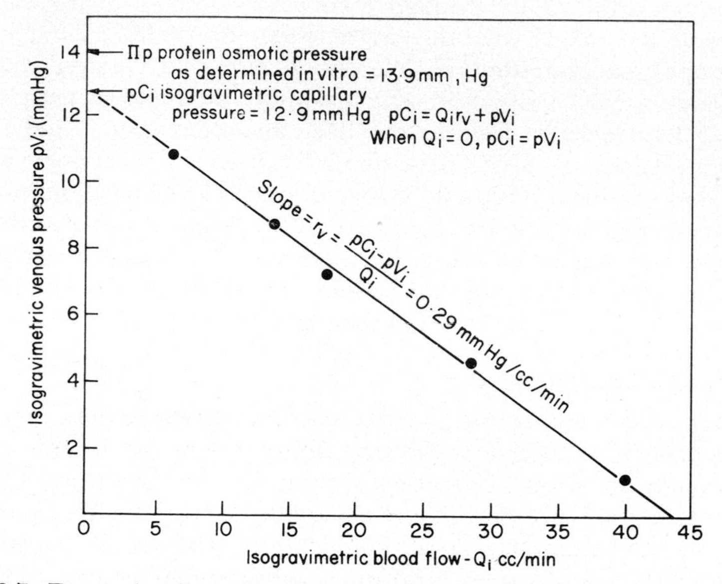

Fig. 6.7. Determination of isogravimetric capillary pressure. By perfusing an isolated cat hind limb with blood at varying arterial and venous pressures a series of pressure pairs can be found when there is no weight change in the limb—the *isogravimetric state*. This isogravimetric state represents the situation when loss of fluid from the capillaries just balances the gain of fluid by absorption. The values obtained are not physiological but by plotting isogravimetric venous pressure against flow and extrapolating back to zero flow, an estimate of the mean capillary pressure can be made. (Pappenheimer and Soto-Rivera, *Am. J. Physiol.* 1948, **152,** 471.)

Lymph

Absorption of Large Molecules from Tissue

When a dye-stuff is injected into a tissue, e.g. into the skin, in a short time a series of vessels will be seen under the surface filled with the dye. These are lymphatic channels carrying fluid from the interstitial spaces back into the blood stream; the small vessels may be followed centrally where they are in confluence with larger and larger vessels until finally large lymphatic ducts empty their contents into the large veins. These vessels may be regarded as a subsidiary means of exchange between the tissue and the blood stream; thus the most obvious and most direct way of clearing the injected dye would be for the dye to diffuse into the capillaries of the tissue and be carried away. The alternative or additional route is by way of these lymphatic channels. If the material injected consists of large molecules, such as ferritin,* the capillary membrane may be incapable of allowing passage from interstitial fluid into the blood, and the lymphatic capillary, being much more permeable, then becomes the sole route.

Reabsorption of Plasma Proteins. If plasma proteins escape from the capillary into the interstitial fluid, and there is indeed normally a continuous but slow leak, then the way back to the blood for these molecules must be by way of the lymphatic channels because the concentration gradients are unfavourable for direct diffusion into the blood capillaries. Thus the normal concentration of proteins in the plasma is some 7 g/100 ml, whilst the concentration in the interstitial fluid is probably only of the order of 1–2 g/100 ml. Return of these proteins to the blood, which is physiologically not only desirable but absolutely necessary, relies, therefore, on the lymphatic vessels. The proteins diffuse into the lymphatic capillaries, and the fluid within these vessels is carried proximally towards the larger lymph vessels by virtue of valves which permit proximal movement; as with the large veins, this proximal movement is assisted by contraction of the skeletal muscles.

Valvular Action

This valvular action, whereby movement of fluid away from the limb towards the large blood vessels is favoured, is probably very important. It seems likely that even in the smallest lymph capillaries the arrangement of the endothelial cells is such that the intercellular clefts have a valvular character, so that when the tissue is compressed the overlapping flaps of endothelial cells close, preventing fluid from emerging from the lymph capillary and forcing it to a region of lower pressure (Fig. 6.8). In this context we may note that the pulsatile nature of the blood-flow seems to be important for lymph clearance, the changes in pressure in the blood vessels being transmitted to the lymph vessels and causing a pumping action. Thus Parsons and McMaster, many years ago, showed that fluid tended to collect in the tissue of the isolated rabbit's ear if this were perfused through the artery by blood at a steady pressure, whereas if a pulsatile pressure was employed there was no accumulation.

* This is a naturally occurring particle of molecular weight 500,000, and diameter of about 100 Å, containing an electron-dense core of iron.

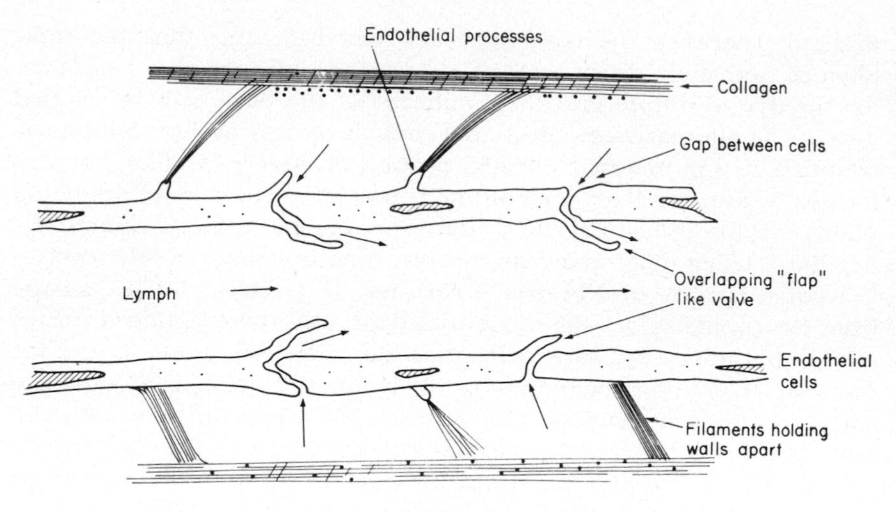

Fig. 6.8. Lymphatic wall and vessel valves. The endothelial cells of lymphatics are held apart by filaments attached to the collagen so that they can act as drainage canals. There appear to be two sorts of valves aiding drainage of interstitial fluid; between the endothelial cells there are valve flaps which permit fluid to enter the lymphatics but close when muscular contraction occurs, and luminal valves which direct lymph towards the blood stream like those found in the venous system.

Lymph Composition

By placing a cannula in a large lymphatic duct the lymph may be collected and analysed. Its protein content is the feature that distinguishes it from a pure filtrate of plasma, so that we may regard it as a sample of interstitial fluid with, however, probably a rather higher concentration of protein than normally prevails in the interstitial fluid, as we shall see later. When different tissues are examined, the percentage of proteins is found to vary, the liver having a high concentration, for example, compared with a limb. In general, the content will depend not only on the tissue but on its activity; the high content of the liver depends on the very leaky character of its blood vessels, and probably too on the synthesis of proteins by it.

Capillary Leaks

The electron microscopical evidence suggests that molecules with diameters much larger than 80 Å will be excluded from the interstitial spaces of tissues such as muscle or skin. The fact that only small leaks of serum albumin, with a diameter of this order, occur is consistent with this view. However the blood globulins, much larger molecules, are also found in lymph. If artificial materials of varying molecular weight, such as dextrans, were injected into the blood, it was found that, when relatively small molecular weight dextrans were employed, in the region of 5000–9000, there was some selectivity, the smaller molecules escaping in preference to the larger. When much larger molecules were studied—100,000 to 350,000—there was still some

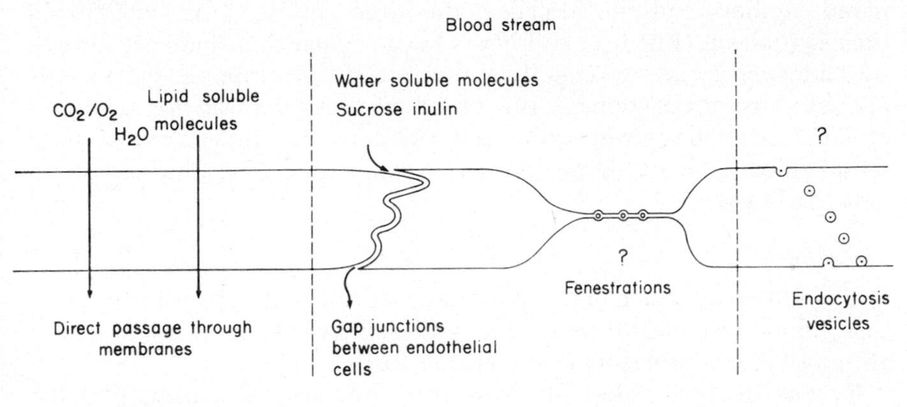

Fig. 6.9. Routes of passage through capillary walls. The figure illustrates the three principal routes by which substances can cross the endothelium of capillaries. Lipid-soluble substances, respiratory gases and water can cross directly through the endothelial membranes. Lipid insoluble molecules including small proteins pass through the gaps between endothelial cell junctions. Large macromolecules probably pass through the vesicles that form and empty at the membrane surfaces. The barrier provided by the fenestrations is not certain; if it were indeed open very large protein molecules could pass through as in the liver sinusoids. If the fenestrations were closed by a membrane, protein molecules would be unlikely to pass through.

escape, but this was small and was now unselective, all molecules escaping at about the same rate. It seemed that these large molecules escaped by a different route, namely through a few very large pores, so large (500 Å diameter) that they could all pass through without hindrance.

Categories of Capillary Permeability. Thus capillary permeability could be considered to fall into three main categories: (a) Substances passing easily through cell membranes, such as CO_2, O_2, H_2O, and lipid-soluble materials, would pass readily through the cells as well as through the intercellular gaps and the few large pores; (b) relatively large lipid-insoluble substances, like sucrose, inulin, and small proteins would be largely confined to the gap-junctions between endothelial cells and to the large pores; (c) the very large molecules, like globulins, would be confined to the large pores (Fig. 6.9). As to the nature of these pores, electron microscopical studies, using as test molecules ferritin (diameter 100 Å) and glycogen (300 Å) suggested that, where the capillary is fenestrated, as in the intestine, the large molecules escape through the fenestrations, the intercellular clefts being at all times free from these molecules. A prominent feature of many capillary endothelial cells is the large number of vesicles within their cytoplasm (Fig. 6.1) and it has been argued that these are formed by endocytosis, i.e. by engulfing the outside medium. If the vesicles also emptied their contents this could be a mechanism for transport of large particles across the capillary; certainly these vesicles were found filled with the large molecules some time after their injection into the blood.

Oedema

The full significance of the lymphatic drainage mechanism becomes clear when we consider the phenomenon of *oedema*, by which is meant an unusual accumulation of interstitial fluid.

Excess Fluid Intake. This can arise from several causes; thus the ingestion of large amounts of fluid tends to dilute the plasma proteins and to raise the general level of vascular pressures; the effect then is to favour the ultrafiltration process and impair the osmotic reabsorption at the venous end of the capillary, with the result that the tissue tends to gain fluid. As we shall see, this "physiological oedema" constitutes a means whereby the interstitial spaces of the body may act as a temporary reservoir for fluid ingested in large quantities as, for example, when a man drinks ten or more pints of beer in a few hours.

Lowered Colloid Osmotic Pressure. Another cause of oedema is the loss of plasma proteins, as in severe protein deficiency—the so-called *hunger oedema*—which is accentuated by the tendency to drink large amounts of fluid to stay the cravings of hunger.

Venous Congestion. Yet another cause is the increased venous capillary pressure associated with venous congestion as in heart failure, although other factors are doubtless concerned (Vol: 3).

Increased Protein in Tissue Fluid. A very potent cause of oedema is the accumulation of protein in the interstitial fluid, since if the concentration approximates to that in the plasma there is no force favouring absorption of filtrate except the positive tissue pressure which now builds up. The build-up of protein can be due to an increased capillary permeability, such as occurs in certain shock conditions, or more locally, in a blister due to a burn. Alternatively, the concentration can build up by virtue of defective lymphatic clearance, the extreme form of this is the *elephantiasis* associated with blockage of a main lymph trunk. It is this condition that emphasizes the value of the lymphatic system in the normal physiology of the tissue. Capillaries are never "perfect" but continuously leak proteins into the interstitial fluid; this protein cannot return by the way it came because the concentration gradient is in the wrong direction; hence the only way out is by a flow along the lymphatic vessels, being first entrapped into the lymphatic capillaries and drawn by the valvular suction developed in the larger channels.

Standing Oedema. Physiologically a defective lymphatic drainage may occur through standing for a period of time, the absence of muscular movements restricting the forced flow of lymph; the absence of muscular movements also militates against adequate venous return from the limb so that the capillary pressure rises and favours retention. Under these conditions, then, the interstitial fluid pressure, measured with Guyton's capsule, becomes positive.

Lymph as Tissue Fluid

The lymph is a sample of interstitial fluid. This need not mean that the protein concentration in the lymph obtained from a large duct is the same as in the interstitial fluid since it may well be that lymph is only formed when the protein concentration has increased in the tissue for some reason. In general, the flow of lymph from a resting limb is barely measurable, and this may be because there is no accumulation of fluid making it necessary. With exercise, flow of lymph occurs, and this may be because the capillary dilatation has allowed escape of proteins into the interstitial fluid in unusually large amounts. Thus lymph flows from a tissue when proteins tend to accumulate, so that it can well be that the protein concentration in the interstitial fluid of a resting limb is much lower than that in the lymph collected from it during exercise.

The Blood Volume

The volume of blood in the circulatory system, and the obverse of this, namely the *capacity* of the vascular system, are important parameters in determining the efficiency of the circulation as a whole. Thus an abrupt diminution in the blood volume, say by haemorrhage, may leave the heart with an inadequate venous return and thus lead to a drastic fall of arterial blood pressure, in the same way that an abrupt increase in capacity will do so (Vol. 3). The volume of the blood can alter physiologically by a change in the volume of its plasma brought about either by an increased ultrafiltration into the interstitial spaces, or by an increased reabsorption of interstitial fluid. It is for this reason that an adequate understanding of the fluid relations between blood and interstitial spaces is so important. In general, factors favouring oedema, such as increased capillary permeability to proteins, cause loss of blood volume, and this accounts for the low blood pressure in conditions of anaphylactic shock when capillary permeability to proteins has increased. The loss of blood volume caused by haemorrhage tends to be compensated by the low arterial pressure; this leads to lowered capillary pressure and so favours reabsorption of interstitial fluid. Thus the blood draws on the reservoir of the interstitial fluid. By contrast, when the blood volumc has increased abnormally, say by the ingestion of large amounts of fluid, first into the gastro-intestinal tract and subsequently into the blood, the increased capillary pressure favours ultrafiltration, so that the interstitial spaces act as a reservoir for the extra fluid intake until renal mechanisms can rid the body of it.

Compensations in Haemorrhage. However, at any rate where haemorrhage is concerned, physiological mechanisms also are brought into play to ameliorate the effects of lost blood; there is a constriction of the resistance vessels—arterioles—that helps to raise arterial pressure, and a constriction of capacitance vessels that tends to reduce the capacity of the vascular system, thus making more blood available for filling the heart. The construction of the arterioles, brought about reflexly and through the liberation of renin into the blood (Vol. 3), causes a drop in capillary pressure, and favours reabsorption of tissue fluid. This may be demonstrated experimentally in man by "trapping" a large volume of his blood in his extremities by inflating a cuff round the thighs; it is found that the volume of the arm progressively decreases due to absorption of tissue fluid.

Measurement of Fluid Compartments

The fluid compartments of the human body can be represented diagramatically by Fig. 6.10. The blood consists of cells and plasma whose water constitutes some 9·7 per cent of the whole body-water; the interstitial fluid is much larger, namely 17 per cent of the whole, and the intracellular fluid is the largest water compartment of all, representing some 73 per cent of the whole.

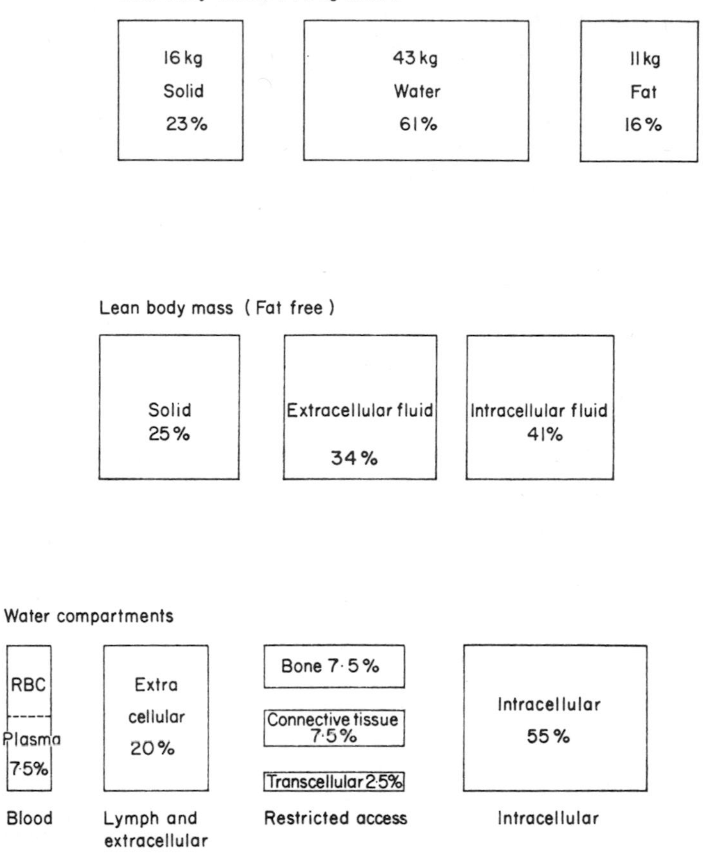

Fig. 6.10. The fluid compartments of the body. The upper diagram gives the distribution of solid water and fat in a 70 kg male. It is usual to correct for the fat, the so-called lean body mass as shown in the middle diagram. The water compartment has been divided into extracellular fluid and cell water. In the lower diagram the subcompartments of fluid distribution are shown as percentages of total water. Connective tissue water, bone water and specialized extracellular fluids (such as cerebrospinal fluid) are compartments of restricted access to marker compounds, unless highly lipid soluble.

Dilution Principle

To measure these compartments individually we make use of the dilution principle. If we have a volume, V, of water and we cannot measure it directly, we add a known quantity, M, of a tag, allow it to mix with this water, and remove a sample and measure the concentration, C, of the tag in it. $M/V = C$. Knowing M and C, we can obtain V. The experimenter's problem is to find tags that are confined to certain compartments.

Blood Volume

Thus if we add a known mass, M, of labelled red cells* to the animal's blood-stream and allow them to mix, we may take a sample of its blood and measure the concentration, *C*, of the labelled cells in the blood. Hence we can obtain *V*, the volume of the circulating blood with which these cells have mixed. We may measure the volume of the plasma if we inject a known mass of labelled plasma protein.

Interstitial Fluid Volume

To find the volume of interstitial fluid we must inject a tag that passes out of the blood into the interstitial fluid but does not pass into the cells. If complete mixing in the blood and interstitial fluid has occurred, the volume, *V*, estimated by applying the equation, $M/V=C$, will be the combined water of the blood and interstitial fluid, or more usually of the plasma and interstitial fluid, because the tag usually does not penetrate the blood cells. A typical tag is sucrose or mannitol, substances that do not penetrate cells with ease, if at all. By subtraction, then, the volume of the interstitial fluid is obtained.

Intracellular Volume

Finally, we may measure the total body-water by injecting a known amount of water labelled with tritium, ^{3}H, a radioactive isotope of hydrogen; this mixes very rapidly with all the water in the body, so that a sample of blood plasma indicates the concentration everywhere in the body-water. *V*, in the equation, thus indicates the total body-water, whence the intracellular water may be obtained by subtraction.

Changes in the Compartments following Ingestion of Fluid

Absorption of Water

If a human subject absorbs, say, 6 pints (ca. 3 litres) of water by way of his gastrointestinal tract, we shall see that this water will be absorbed osmotically into the blood thereby diluting the salts and other solutes within it. If this occurred only into the blood-plasma the dilution would be considerable; thus if we take the total plasma volume

* The cells are best labelled by mixing them with radioactive chromium (^{51}Cr); this attaches very firmly to the surface, so that radioactivity emitted by a given volume is a measure of the number of cells in this volume.

as 2·5 litres with a total solute concentration of, say, 300 milliosmoles/litre, the concentrations would become

$$\frac{300 \times 2{\cdot}5}{5{\cdot}5} = 140 \text{ milliosmoles/litre}$$

a reduction to 47 per cent of normal and one that would be physiologically intolerable. If the water distributed by osmosis is divided equally between cells and plasma, the available fluid for dilution becomes 5, and the concentration becomes

$$\frac{300 \times 5}{8} = 190 \text{ milliosmoles/litre}$$

or 63 per cent of normal, still a serious dilution. The absorbed fluid would not stay in the blood, however, and would tend to distribute itself between plasma and interstitial fluid, so that the effective dilution is given by:

$$\frac{300 \times 11{\cdot}5}{14{\cdot}5} = 230 \text{ milliosmoles/litre}$$

or a dilution to 70 per cent. The reduction in the osmolality of the interstitial fluid would lead to an osmotic influx into all the cells of the body, so that the effective dilution would be given by

$$\frac{300 \times 43}{46} = 280 \text{ milliosmoles/litre}$$

or 93 per cent of normal.

Osmotic Buffering

This, in fact, would happen and illustrates the great osmotic buffering power of the body, a buffering power that mitigates the large fluctuations in plasma osmolality that would occur were the ingested water to be confined to the blood plasma. We shall see that the ultimate elimination of the water is effected by the kidneys, but this is a process that is not initiated instantaneously, hence the value of the buffering process, especially that of the interstitial fluid, since changes in the volume of this compartment do not involve changes in cell volume, large fluctuations in which are undesirable.

Fluid Shifts

The mechanisms of these water shifts are easy to understand in the light of what has been discussed before. The dilution of the plasma is rapidly reflected in a dilution of the interstitial fluid because of the

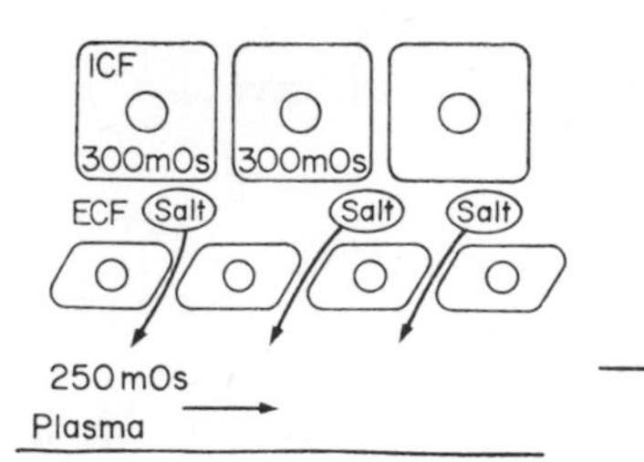

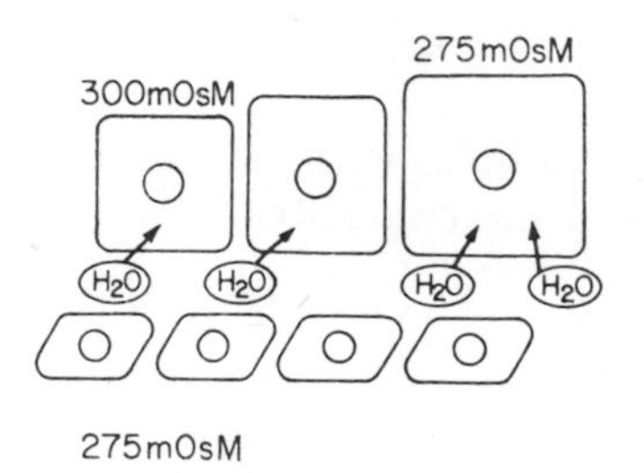

(a) Salt diffuses from ECF to plasma

ECF now at 275 mOsm cells swell

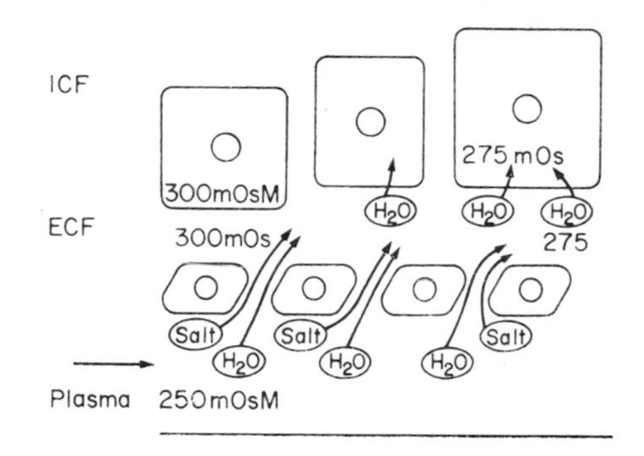

(b) Bulk flow of fluid occurs ECF space expands, cells expand.

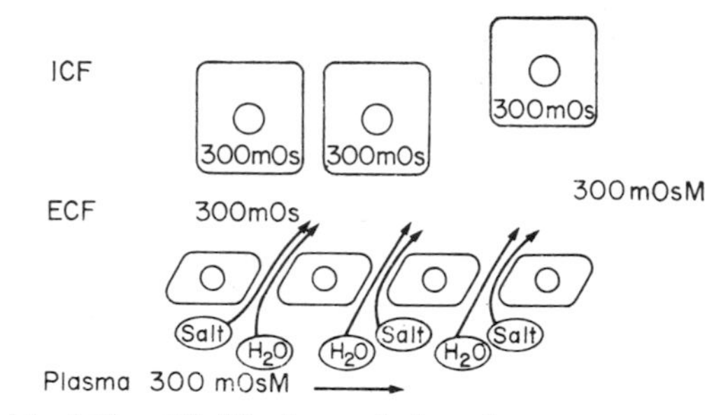

(c)

Fig. 6.11. Fluid shifts. If dilution of the plasma occurs as in (a) solutes move rapidly into the circulation through the endothelial gaps and the extracellular fluid becomes hypotonic. This fall in osmotic pressure now leads to entry of water into cells causing swelling, as the cell membrane is effectively impermeable to solutes. (a) only applies to an isolated system and in the body, as shown in (b), bulk flow of fluid also occurs so that an increase in extracellular fluid volume also occurs. The driving forces for this fluid movement are the increased capillary pressure and the decreased colloid osmotic pressure. If, as in (c), saline, isotonic with plasma, is ingested, the increased blood volume increases capillary pressure and decreases the colloid osmotic pressure as before, but now no water enters cells so that the extracellular fluid must act as the only reservoir for this excess fluid.

very high permeability of the capillary membrane to all the solutes of blood except proteins (Fig. 6.11a). This dilution need not necessarily be associated with significant osmotic flow, owing to the high capillary permeability to the solutes, so that it could be largely achieved by diffusion of solutes from interstitial spaces to the plasma, as illustrated. The cells of the tissue are thus exposed to an interstitial fluid of lower osmolality than their own, and swell osmotically owing to their high permeability to water and virtual impermeability to most of the solutes of the tissue fluid.

Enlargement of Tissue Fluid Compartment. However, the mechanism illustrated by Fig. 6.11a, which leads to no net increase in interstitial fluid volume, is only applicable to a simple isolated system, and in the living body it is found that ingestion of water not only leads to dilution of the plasma, interstitial fluid, and cells, but also to a *net increase in the volume of the interstitial fluid compartment*, in fact this compartment may be regarded as a reservoir for extra fluid absorbed beyond the body's requirements. The basis of this net flow of fluid into the extracellular compartment is the increased capillary pressure that favours filtration, together with the reduced colloid osmotic pressure. Both factors favour flow out of the capillary so that not only is the interstitial fluid diluted by rapid diffusive exchanges of solutes, as illustrated in Fig. 6.11a, but also by a net flow of hypotonic fluid by filtration as in Fig. 6.11b.

Ingestion of Saline Solution. The importance of this mechanism for transfer of absorbed fluid into the interstitial fluid compartment is shown by the effects of ingestion of isotonic saline. The saline solution is absorbed into the blood by active processes in the small intestine, but there is now no difference in osmolality between plasma and interstitial fluid, and the extra fluid is largely transferred to the interstitial fluid compartment by virtue of the effects of the increased blood-volume and reduced colloid osmotic pressure (Fig. 6.11c). In this instance, then, the absorption of large volumes of isotonic salt solution results in a partition of this between plasma and interstitial spaces—the cells are unable to take up any of the fluid, so the interstitial space acts as the exclusive reservoir for the unwanted fluid.

Tissue Compliance. This reservoir is a very significant quantity owing to the remarkable "compliance" of the tissue spaces, i.e. their ability to take up fluid without a large increase in pressure. This may be illustrated by Fig. 6.12 which shows the interstitial fluid pressure of a dog's limb perfused with fluids of low colloid osmotic pressure and therefore favouring oedema. The interstitial fluid pressure very soon rises to about zero (atmospheric) after only a small increase in weight

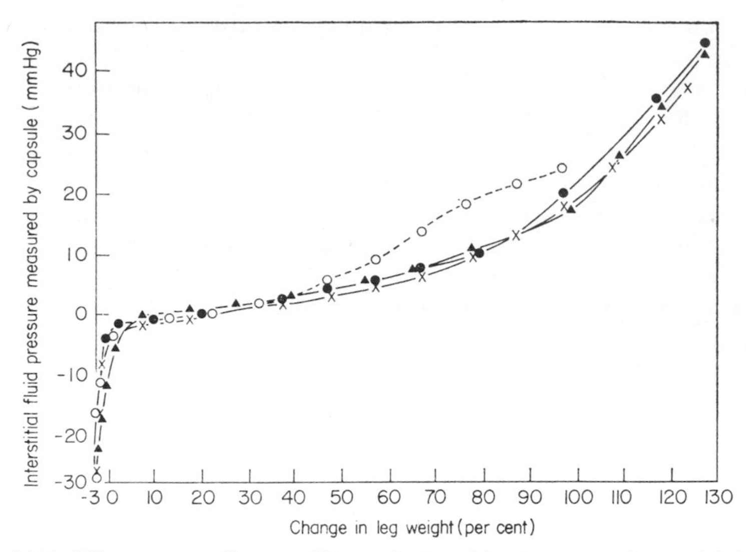

Fig. 6.12. Tissue compliance. The relationship between interstitial fluid pressure and change in leg weight during progressive increase in interstitial pressure. Note that after the initial steep rise of interstitial pressure to 0 mm Hg no further increase occurs until the limb weight has increased by 100 per cent indicating a large tissue compliance. Each curve represents results from a separate isolated perfused leg with an implanted capsule in the muscle to measure interstitial pressure. (Guyton *et al.*, *Physiol. Rev.* 1971, **51,** 551.)

of the limb; when this change of pressure has occurred, however, the limb may increase in weight by nearly 100 per cent with only a small change in interstitial fluid pressure, indicating a high compliance of the tissue. Only when the limb has taken up this amount of extra fluid does it show a steep rise in pressure with further uptake, and now the tissue appears tightly swollen.

The Gibbs–Donnan Equilibrium

In discussing the distribution of ions between the inside and outside of the muscle or nerve cell we have referred to the Gibbs–Donnan equilibrium, which applies to the situation when a membrane separates a solution of electrolyte containing an impermeable ion, from another solution of electrolyte not containing this ion. With muscle or nerve the impermeant ions were anions, being a mixture of proteins and probably organic phosphates, etc. The Gibbs–Donnan equilibrium states that, when two solutions are separated by a membrane, the concentrations of the permeable ions are related by equations of the type:

$$[Na^+]_1 \times [Cl^-]_1 = [Na^+]_2 \times [Cl^-]_2 = [K^+]_1 \times [Cl^-]_1 = [K^+]_2 \times [Cl^-]_2 \quad (1)$$

i.e. the products of the concentrations (activities) of positive and negative univalent ions are equal, provided these ions can diffuse across the membrane.

Gibbs–Donnan Ratio

Thus, expressing the equations in the form of ratios we have:

$$[Na^+]_1/[Na^+]_2 = [K^+]_1/[K^+]_2 = [Cl^-]_2/[Cl^-]_1 = \ldots . = r \quad (2)$$

where r is the Gibbs–Donnan ratio.

With a divalent cation, such as Ca^{2+}, we should have:

$$[Na^+]_1 \times [Cl^-]_1 = [Ca^{2+}]_1 \times [Cl^-]_1^2 = [Ca^{2+}]_2 \times [Cl^-]_2^2$$

whence $$[Ca^{2+}]_1/[Ca^{2+}]_2 = [Cl^-]_2/[Cl^-]_1^2 = r^2 \quad (3)$$

whence $$\sqrt{\frac{[Ca^{2+}]_1}{[Ca^{2+}]_2}} = r$$

Plasma–Interstitial Fluid Relation. As indicated earlier (p. 280), the plasma in the capillary separated from the interstitial fluid by the plasma membrane, impermeable to the plasma proteins, provides just such a condition as is covered by this equilibrium, and in fact the equation relating the concentrations of ions was deduced by Donnan in response to Starling's query. Thus the plasma proteins are in the form of anions at physiological pH, so that the ionic relations may be represented by Fig. 6.13a, where the cations are Na^+, K^+, Ca^{2+}, etc., and the anions Cl^-, HCO_3^-, etc.

For simplicity's sake, let us imagine that the only salt present is NaCl and that the protein may be represented by NaP (Fig. 6.13b). The Gibbs–Donnan equation tells us that the products of the diffusible ions on both sides of the membrane are equal at equilibrium:

$$[Na^+]_{Int} \times [Cl^-]_{Int} = [Na^+]_{Plasma} \times [Cl^-]_{Plasma} \quad (4)$$

Since there are equal concentrations of positive and negative ions in each compartment, we may put $[Na^+]_{Int} = [Cl^-]_{Int}$

whence we have:

$$[Cl^-]^2_{Int} = [Na^+]_{Plasma} \times [Cl^-]_{Plasma} \quad (5)$$

From Fig. 6.13b it is clear that $[Na^+]_{Plasma}$ is greater than $[Cl^-]_{Plasma}$, since some of the Na^+-ions in plasma are associated with the protein

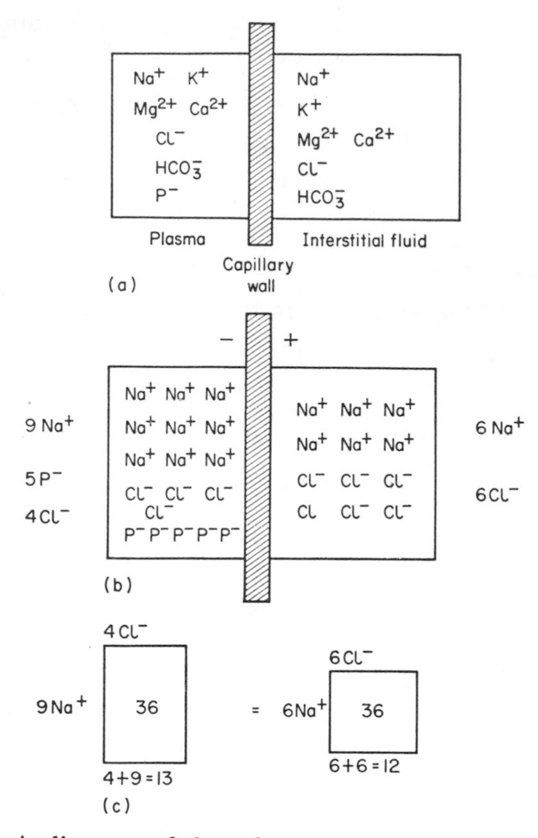

Fig. 6.13. (a) A diagram of the solute constituents on either side of the capillary endothelium which is permeable to all solutes but protein (P^-). (b) A simplification of the solute balance across the capillary wall. In this diagram the plasma sodium concentration is represented by nine Na^+ ions which balance the negative charge of four Cl^- ions and the five protein negative groups P^-. In the interstitial fluid the ionic composition is represented by six Na^+ ions and six Cl^- ions. The Gibbs-Donnan equilibrium states that:

$$Na^+_{\text{plasma}} \times Cl^-_{\text{plasma}} = Na^+_{\text{Int}} \times Cl^-_{\text{Int}}$$

that is the product of the concentrations of ions must balance

$$9 \times 4 = 6 \times 6.$$

A potential is established across the membrane because of the tendency for Na^+-ions to pass from left to right down their concentration gradient and for Cl^--ions to pass from right to left. (c) Illustrates that the products of the sides can be equal but the sums of the sides not.

anions, P^-. Thus the right-hand-side of equation (5) is the product of two unequal quantities whilst the left-hand side is a square, and since the sum of the sides of a rectangle is greater than the sum of the sides of a square of equal area, it follows that the sum of the concentrations of Na^+ and Cl^- in the plasma is greater than the sum of the concentrations of Na^+ and Cl^- in the interstitial fluid, i.e.:

$$[Na^+]_{Plasma} + [Cl^-]_{Plasma} > [Na^+]_{Int} + [Cl^-]_{Int}.$$

Colloid Osmotic Pressure. Thus the diffusible ions Na^+ and Cl^- distribute themselves in such a way that the total concentration is greater on the side of the plasma. If we add to this the protein anions, it is clear that there is an excess of osmolality on the side of the plasma, and this is the basis of the *colloid osmotic pressure*, discussed earlier. It means that there is a tendency for water to pass into the plasma from the extracellular fluid unless this is opposed by the exertion of a counter-pressure. If the two solutions were contained in inexpansible chambers, as in Fig. 6.13, net movement of fluid would be impossible, but insertion of a pressure-measuring device into the plasma compartment would reveal a pressure of about 25 mm Hg, the colloid osmotic pressure. As indicated earlier (p. 295) if the volumes of the compartments could change, then the interstitial fluid would pass into the plasma compartment. The initial process would be a movement of water, but this movement would raise the concentrations of ions in the interstitial fluid and reduce those in the plasma, and because the membrane is permeable to these ions, there would be net movements from interstitial fluid into plasma, re-establishing the higher osmolality in the plasma; water would move again, salts would follow, and so on.

If we apply the Gibbs–Donnan equation to the concentrations of all the permeable ions of plasma, we have:

$$[Na^+]_{Int}/[Na^+]_{Plasma} = [K^+]_{Int}/[K^+]_{Plasma} = [Cl^-]_{Plasma}/[Cl^-]_{Int} = r$$

From the concentrations of plasma proteins and their degrees of dissociation, the ratio can be calculated theoretically, and it comes out at 0·96, i.e. the concentrations of Na^+ and K^+ are some 4 per cent less in the filtrate than in plasma, and the concentrations of Cl^- and HCO_3^- some 4 per cent more. With a divalent ion like Ca^{2+} the ratio of concentrations is equal to r^2, i.e. $0{\cdot}96^2 = 0{\cdot}92$, so that the concentration in plasma is some 8 per cent higher than in the interstitial fluid.

It is quite simple, experimentally, to place plasma in a sac composed of a membrane impermeable to proteins but permeable to the non-colloidal constituents, and by exerting a pressure on the plasma to produce an ultrafiltrate; analysis of this shows that the relative concentrations of, say, Na^+ in plasma and the filtrate conform reasonably well with the calculated value.

Aqueous Humour and Cerebrospinal Fluid

Physiologically it is important to know just what the expected concentrations of ions in a plasma filtrate are, since some body-fluids, in addition to the interstitial fluid of tissues, have ionic concentrations that are similar to those of a plasma filtrate (e.g. the aqueous humour of the eye, Vol. 4; and cerebrospinal fluid of the brain, Vol. 3), and it might be argued that they were formed by filtration from the blood capillaries of the adjacent tissue. In fact, however, as Table 1 shows, the ratios of the concentrations in the fluids: C_{Aq}/C_{Pl} and C_{CSF}/C_{Pl} are significantly different from the corresponding ratios: C_{Fil}/C_{Pl} where C_{Aq}, C_{CSF}, C_{Fil}, and C_{Pl} are the concentrations in aqueous humour, cerebrospinal fluid, plasma filtrate and plasma respectively. Thus something more than filtration is necessary to account for the formation of the fluids.

TABLE I

Distributions of certain Ions between Plasma on the one hand and Aqueous Humour Cerebrospinal Fluid, and Plasma Filtrate on the other

Ion	C_{Aq}/C_{Pl}	C_{CSF}/C_{Pl}	C_{Fil}/C_{Pl}
Na^+	0·96	1·03	0·945
K^+	0·955	0·52	0·96
Cl^-	1·015	1·21	1·04

Gibbs–Donnan Potential

We may ask how it is that the presence of the protein anions is able to prevent the concentrations of diffusible cations and anions from becoming equal on each side of the membrane; the ions are free to move as individuals and we may ask what is preventing the Na^+ in the plasma from diffusing down its concentration gradient to give equal concentrations in filtrate and plasma, or why Cl^- does not diffuse down its concentration gradient from filtrate into plasma. A little consideration will show that, were such processes to occur, the bulk of the plasma would have more negative ions in it than positive ions and so would be negatively charged, and the bulk of the filtrate would have more positive ions than negative ions and become positively charged. In fact the Gibbs–Donnan expression shows us that in the bulk of the two fluids there will be equal numbers of cations and anions, which is a general requirement for all solutions.

Membrane Potential. However, there is undoubtedly a *tendency* for the positive Na^+-ions to pass down their concentration gradient into the filtrate, and for Cl^--ions to pass in the other direction. This actually happens just across the membrane, but the resulting separation of charges establishes a potential across the membranc that effectively prevents more ions going down the gradients. Thus the small excess of Na^+-ions on the filtrate side of the membrane and excess of Cl^--ions on the plasma side make the membrane positive on the filtrate side and negative on the plasma side, as illustrated in Fig. 6.13b. It is this difference of potential—the *membrane potential*—that by slowing down the migration of Na^+-ions out of the plasma and accelerating them out of the filtrate, permits the maintenance of higher concentration of Na^+-ions in the plasma. Essentially similar types of potential have already been described in relation to the membrane-potentials across cells; they are all due to the restraint on movement of one species of ion; in the present instance the restraint is on the negative protein ion. The magnitude of the potential in this case is small, being given by the Nernst equation:

$$E = RT/nF \ln \frac{[C]_{Pl}}{[C]_{Fl}}$$

Thus with the ratio of $[Na]_{Pl}/[Na^+{}_{Fl}] = 1{\cdot}04$ the computed potential is $58 \log_{10} 1{\cdot}04 = 0{\cdot}98$ millivolts.

CHAPTER 7

The Carriage and Release of the Blood Gases

GAS TENSION OR PRESSURE

In this chapter we shall be concerned with those characteristics of the blood that are important for the carriage and release of respiratory blood-gases, O_2 and CO_2; thus we are concerned with the uptake of O_2 and release of CO_2 in the lungs, and the release of O_2 and uptake of CO_2 in the tissues. If the gases were simply dissolved in the water of the blood, as with N_2, the physical problems would be simple to describe, the uptake and release being governed by the partial pressures of the gas in blood, lungs and tissue. In fact, however, the situation is more complex, since the O_2 is reversibly combined with haemoglobin in the red blood cells, whilst the CO_2 dissolves to form an acid, H_2CO_3, which participates in the complex acid-base equilibrium of plasma and erythrocytes. Nevertheless, the partial pressures of the gases in the system do, in fact, play the dominant role, and we may profitably begin with a statement of fundamental principles governing equilibria of gases in aqueous and gaseous phases.

Concentration

Boyle's and Charles' Laws

The concentration of a gas in a mixture of gases may be indicated in a variety of ways, e.g. moles per litre, volumes per cent, and so on, but for purposes of comparing concentrations in the gas and liquid phases, when the gas has access to both, it is necessary to use the concepts of *gas tension*, or *pressure*, and *partial pressure*. Thus, because of the relations between pressure of a gas and its volume, and between its temperature and volume, known as Boyle's and Charles' laws respectively, the pressure of a gas, at a given temperature, defines its concentration for us:

$$PV = nRT \tag{1}$$

where n is the number of moles of the gas in the volume V; hence the concentration, n/V, is equal to P/RT where R is the gas constant and T the absolute temperature.

Partial Pressure

If we are dealing with a system containing a single gas, we may separate two chambers by a permeable membrane, and we may say that the gas will pass from the region of higher tension, or pressure,

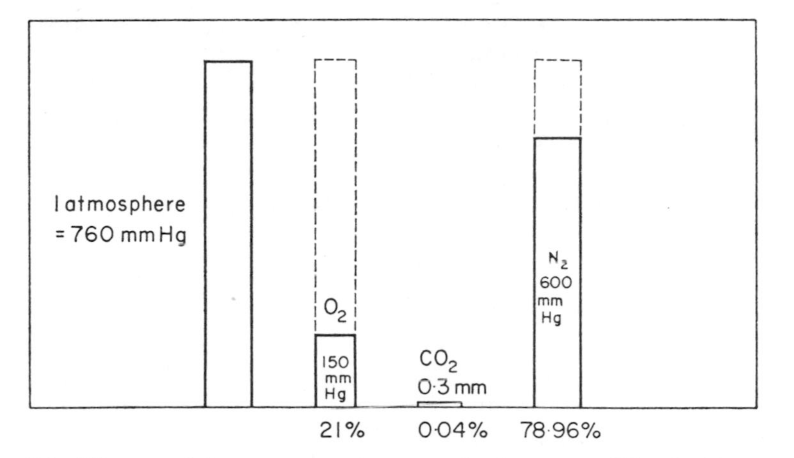

Fig. 7.1. The partial pressures of atmospheric gases. The height of the column represents the contribution the gas makes to the total atmospheric pressure. The percentage composition of each gas is indicated.

to one of lower (or from one of higher concentration to one of lower) indifferently. When we are dealing with a mixture of gases, then the *partial pressure* substitutes for the pressure of the pure gas, the partial pressure being defined as the contribution of a gas to the total pressure. The composition of dry air at standard temperature (273°K or 0°C) and pressure (760 mm Hg) is:

O_2	20·96 per cent
N_2	70·00 per cent
CO_2	0·04 per cent. (Fig. 7.1)

Thus the mixture of these gases exerts a pressure of 760 mm Hg at 273°K. The partial pressures of the gases, indicated by P_{O_2}, P_{N_2} and

P_{CO_2}, are derived from these proportions. Thus for O_2 it will be given by:

$$\frac{20{\cdot}96}{100} \times 760 = 159{\cdot}2 \text{ mm Hg};$$

the partial pressures of N_2 and CO_2 are 600·5 and 0·30 mm Hg respectively.

Water Vapour Pressure. When the air is moist, then the water vapour contributes its partial pressure to the whole. When the gas mixture is saturated with water vapour this contribution is obtainable from an appropriate Table, giving the partial pressure of water vapour at the particular temperature. At 37°C the partial pressure of water vapour in saturated air is 47 mm Hg, so that the air entering the alveoli of the lung, which becomes saturated with water during its passage through the conducting pathways, has a lower partial pressure of O_2 than that in the relatively dry atmospheric air. Thus if the inspired air is dry, at an atmospheric pressure of 760 mm Hg, the partial pressure of O_2 becomes:

$$(760 - 47) \times \frac{20{\cdot}99}{100} = 149{\cdot}6 \text{ mm Hg}.$$

As we shall see, the actual partial pressure of O_2 in the alveolar air which exchanges with the blood in the lung is considerably less than that of the inspired air even when allowance is made for saturation with water vapour; this is because O_2 is being removed from the alveolar air by the blood and a steady state is reached with a partial pressure, P_{O_2}, of about 100 mm Hg.

Molar Concentrations

From the partial pressures we may, if we wish, calculate the molar concentrations, making use of equation (1) and the gas constant, R. Thus, if the pressure is expressed in atmospheres and the volume in litres, R has a value of 0·08 litre-atm. per °C; a partial pressure of, say, 100 mm Hg or 100/760 atmosphere would have a concentration of

$$\frac{100}{760 \times 0{\cdot}08 \times 273} = 0{\cdot}006 \text{ moles/litre}$$

Liquid Phase

If we pass, now, to the concentration of a gas in the liquid phase we may, if we wish, express it as a molar concentration, just as with non-gaseous solutes, so that if we separate two solutions of different concentration from each other by a membrane permeable to the gas molecules, we may say that the gas will pass from the higher concentration to the lower, as before. However, when dealing with the relations between two separate phases, namely a gaseous phase, such as the atmosphere, and a liquid phase, such as the blood, we cannot use this simple mode of expression to tell us the direction in which the gas will pass to achieve equilibrium, since the concentration of a molecule in the gas phase does not tell us enough about its escaping tendency, or chemical potential, when we wish to compare this with its escaping tendency in a liquid phase.

Henry's Law. The difficulty is avoided if we use, instead of the concentration, the partial pressure of the gas as a measure of its chemical potential, or escaping tendency. So far as the gas phase is concerned, the partial pressure can be converted into a concentration by application of Boyle's and Charles' laws. The partial pressure of a gas in the liquid phase is defined as the partial pressure of gas in the gaseous phase with which the liquid would be equilibrium. Henry's law tells us that the concentration in the solution, at equilibrium, is proportional to the partial pressure:

$$C = qP$$

where q/ is the *solubility coefficient.* Thus we may shake water in a vessel, containing a gas at a known partial pressure, and the amount dissolved will be uniquely determined, at a given temperature, by this partial pressure.

Escaping Tendency of Gas. By definition, then, a solution of a gas in a liquid will be in equilibrium with the gaseous phase when the partial pressures in liquid and gaseous phase are equal and there will be no net flow from one to the other. When the partial pressure is greater in the solution, gas will flow into the gaseous phase and *vice versa.* Furthermore, if we consider two aqueous solutions of gas, separated by a membrane permeable to this gas, we can state at once that, if their partial pressures are different, net movement of the gas will be from the solution with the higher partial pressure to the one with the lower. Thus the partial pressure of a gas provides a much more general expression of the escaping tendency of the solute than does the concentration, in moles/litre, which does not allow of comparisons between different phases.

Reversible Complex Formation. This value of the partial pressure, as a unit defining escaping tendency, extends to the comparison of two aqueous solutions when there has been some reversible chemical combination of the gas with the solvent or other molecules in the solution. This is typically seen when comparing, say, a saline medium with blood; one may prepare two solutions of O_2 by equilibrating (a) a sample of blood with the gas at a given partial pressure, and (b) a sample of isotonic saline. The escaping tendencies of the gas in the two fluids are the same, but the actual concentrations, in moles/litre

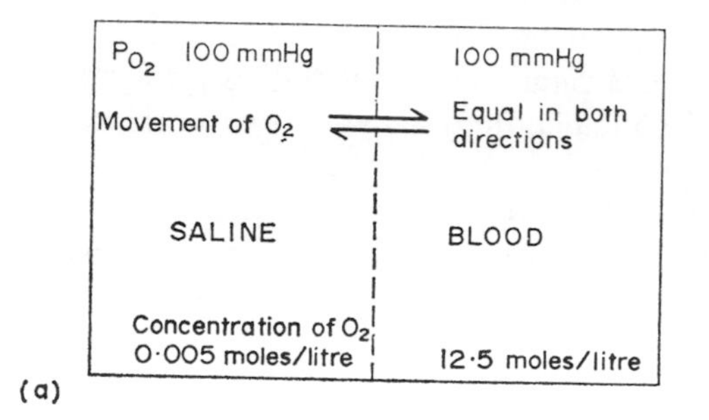

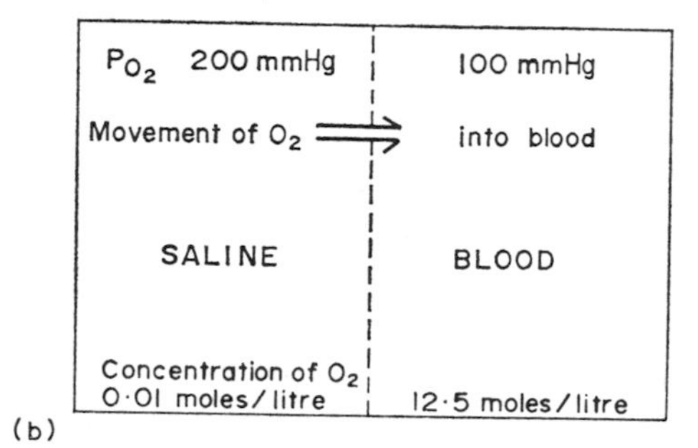

Fig. 7.2. In the diagram a saline-filled compartment is separated from a blood-filled compartment by a membrane permeable to oxygen. In (*a*) the concentration of oxygen in the blood is some 2,000 times greater than in the saline yet, since the partial pressures are the same in both compartments, there is no net movement of oxygen in either direction. In (*b*) the concentration of oxygen in the blood is the same as in (*a*) but now the partial pressure of oxygen in the saline is twice that of the blood and oxygen moves from the saline, at a lower concentration, into blood at higher concentration.

(Fig. 7.2a), are vastly different because O_2 combines reversibly with haemoglobin to form a complex. When the blood is juxtaposed with an isotonic saline of, say, twice the partial pressure, O_2 will pass into the blood, although the concentration in the saline medium may be only a small fraction of that in the blood (Fig. 7.2b). Similarly, if we take two samples of blood equilibrated with O_2 at the same partial pressure but in equilibrating one sample we have included carbon monoxide in the gas mixture, then the two bloods are in equilibrium in respect to O_2, but the sample containing the carbon monoxide has by far the lower concentration of O_2 because the carbon monoxide has competed for the oxygen-sites on the haemoglobin and, having a very much higher affinity for them, has almost entirely monopolized them.

Summary

To summarize, then, the escaping tendency or chemical potential of a gas, by which we mean its tendency to pass from one region to another, can only be indicated by its concentration when we are dealing with the same phase. By employing the partial pressure of the gas with which the solution is in equilibrium, we have a more general method of indicating escaping tendency, or chemical potential, so that at a given temperature we may say that a gas will pass from a higher partial pressure to one of lower. If we wish to know the actual concentration of the gas in the liquid, then, if there is no chemical combination, we may employ the solubility coefficient; if chemical combination has occurred, we must study the equilibrium between the gas and the solution empirically, measuring the amounts of gas taken up under varying partial pressures.

CARRIAGE OF CARBON DIOXIDE

The carriage of O_2 by the blood is largely understandable in terms of its combination with haemoglobin within its erythrocytes, and its reversible dissociation. The carriage of CO_2 is more complex, being linked with the formation of bicarbonate-ions and exchanges of this ion with Cl^- across the erythrocyte membrane. To understand the basic principles of carriage of the respiratory gases, then, it is important that we examine some of the ionic relations between the erythrocyte and plasma.

Erythrocyte–Plasma Ionic Relations

The internal and external media of the erythrocyte may be indicated by Table I; the high internal concentration of K^+ and low concen- of Na^+ have already been remarked on, together with the fact that the total osmolar concentrations inside the cell and outside (i.e. in the plasma) are the same. It will be seen that the concentrations of Cl^- and HCO_3^- are lower inside the erythrocyte than in the plasma, and this is because of the presence of organic anions in the erythrocyte, which are largely constituted by the haemoglobin, which is present in very high concentration (about 30 per cent). The concentration of protein in the plasma is much less, namely about 7 per cent, and for the purposes of the present argument, which discusses the relations between the erythrocyte and the medium surrounding it, we may ignore these. As we have seen, when we consider the relations between the plasma and the interstitial fluid surrounding cells, on the other hand, these proteins are of the utmost significance.

TABLE I

Approximate Concentrations (meq/kg H_2O) of certain Solutes in Human Erythrocytes and Plasma

	Erythrocytes	Plasma
K^+	150	5
Na^+	26	144
Cl^-	74	111
HCO_3^-	24	35

Anion Exchanges

When the erythrocyte is placed in a solution containing, say $NaNO_3$ of the same osmolality as that of the plasma, the Cl^- and HCO_3^- ions diffuse out in exchange for NO_3^- ions in the medium (Fig. 7.3); as explained in Chapter 2 this is a very rapid process indicating a very high permeability of the erythrocyte membrane to anions. By contrast, the permeability to cations is so slow that for the purposes of the present argument we may treat the membrane as impermeable to these. Thus ionic exchanges between the erythrocyte and its environment, the plasma, are essentially those of negative ions, and this includes the OH^--ion, the movements of which are very important physiologically, since they influence directly the acidity or alkalinity of the medium, i.e. the pH.

H^+- and OH^--Ions. In general, it is customary to consider the erythrocyte membrane as rapidly permeable to the H^+-ion, but this need not necessarily be true, since if, say, the concentration of H^+-ions is raised outside the cell, equilibrium could be achieved either by the movement of H^+-ions into the cell or by the movement of OH^--ions out; the result would be the same, namely an acidification of the cell contents, and this follows from the constancy of the product:

$$[H^+] \times [OH^-] = K_W = 10^{-14} \qquad (2)$$

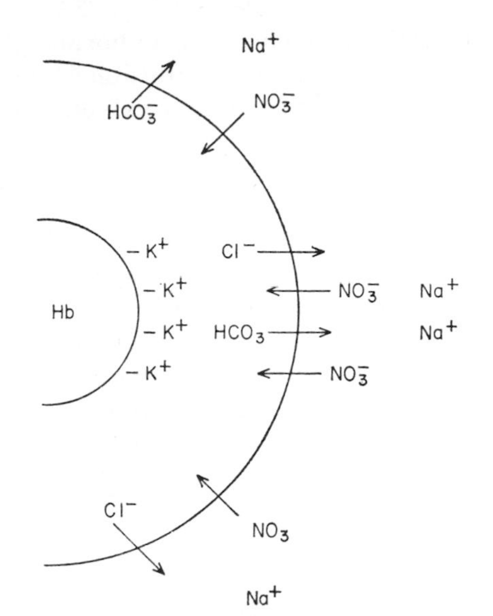

Fig. 7.3. An illustration of anionic permeability of the red cell membrane. The erythrocyte in the diagram has been placed in isotonic sodium nitrate and the Cl^- and HCO_3^- ions inside the cell are exchanging with the NO_3^- ions outside. Na^+ and K^+ do not exchange as the red cell membrane is effectively impermeable to cations.

Thus, removing OH^- from the solution causes more H^+-ions to be formed by the dissociation of water:

$$H_2O \rightleftharpoons H^+ + OH^-.$$

Gibbs–Donnan Equilibrium

The fact that the cell membrane is permeable to negative ions except for the large organic anions (chiefly haemoglobin), means that a special ionic relationship between the concentrations of ions inside and outside the cell will hold, namely that the ratios of these concentrations will be the same:

$$\frac{[Cl^-]_{in}}{[Cl^-]_{out}} = \frac{[HCO_3^-]_{in}}{[HCO_3^-]_{out}} = \frac{[OH^-]_{in}}{[OH^-]_{out}} = \frac{[H^+]_{out}}{[H^+]_{in}} = \mathbf{r} \qquad (3)$$

This is an expression of the *Gibbs–Donnan Equilibrium*, which was discussed earlier (Chapter 6). We may recall that it is a relationship that applies to the ions that will diffuse readily across the membrane; it applies therefore to the negative ions, but also to H^+ because of the special relation of this to the OH^--ion indicated above.

Gibbs–Donnan Ratio. The value of r in normal blood is about 0·6, indicating that the concentrations of diffusible anions inside the cell are only some 60 per cent of the plasma values. The magnitude of r depends on the concentration of haemoglobin ions, so that any alteration in this will influence the concentrations of diffusible anions; thus if the concentration of haemoglobin ions became zero, r would be unity and the concentrations of Cl^- and HCO_3^- inside the cell would rise at the expense of the concentrations in the plasma.

Effects of pH and Oxygenation. Haemoglobin is a protein, and the degree of dissociation, and therefore the concentration of Hb^--anions, depends on the acidity or alkalinity of the medium, i.e. on its pH (p. 322). Hence, alterations in the pH of the cell can alter the concentrations of HCO_3^- and Cl^- within it. Haemoglobin becomes a stronger acid when it is in the oxygenated, rather than in the reduced, form (the Bohr effect), and this means that oxygenation of blood causes the degree of dissociation to become greater and thus increases the concentration of Hb^--anions. Hence oxygenation of the cells should decrease r and therefore decrease the concentrations of diffusible anions in the cells; loss of oxygen should have the reverse effect. These effects are illustrated schematically in Fig. 7.4. An interesting corollary of these shifts of anions in accordance with the oxygenation of the cells is that the total osmolality within the cells changes; thus when anions leave, as in oxygenation, the osmolality falls and, in consequence, the cells lose water to the plasma; when the blood is deoxygenated the cells swell. These changes are sufficiently large to be measurable. It might be argued that the lost anions—Cl^- and HCO_3^-—are replaced by appearance of haemoglobin anions; this is true electrically, but of course these haemoglobin ions result from dissociation of molecules that were already present, namely haemoglobin molecules, and so there has been no increase in osmolality to compensate for the lost mobile anions.

Hamburger Shifts

A related movement of anions occurs when the concentration of CO_2 is altered in the blood plasma; thus suppose we expose blood to a high partial pressure of CO_2; this gas diffuses very rapidly across cell membranes and therefore the concentration increases in the cells at a parallel rate. The CO_2 within the cells becomes transformed mainly into H_2CO_3, largely by virtue of the enzyme, *carbonic anhydrase*, which accelerates the reaction:

$$CO_2 + H_2O \rightleftharpoons H_2CO_3$$

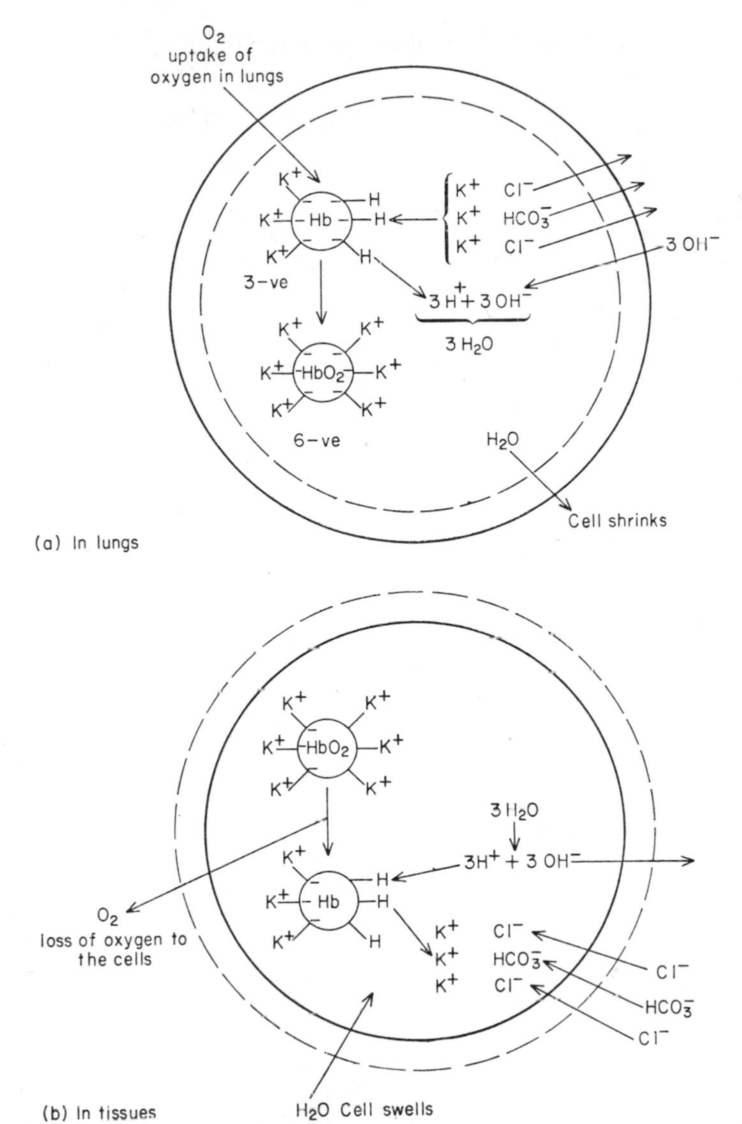

Fig. 7.4. (*a*) The effect of oxygenation of haemoglobin on the osmolality and size of the red blood cell. When haemoglobin is oxygenated it becomes a stronger acid and thus more of its acidic groups become ionized. This increase in negative charges causes the diffusible anions to be in excess and so Cl^- and HCO_3^- ions move out of the cell reducing the total internal osmolality. The reduction of cellular osmolality causes water to move out of the red cell with consequent shrinking. The converse occurs when haemoglobin is deoxygenated, as illustrated in (*b*), and in this case the cells increase in osmolality and swell.

The H_2CO_3 is a stronger acid than the haemoglobin of the cells and therefore the reaction:

$$KHb + H_2CO_3 \rightleftharpoons KHCO_3 + HHb$$

proceeds to the right. Thus the concentration of HCO_3^- anions in the cell has increased, and the equilibrium across the membrane is disturbed, so that HCO_3^- diffuses out of the cell into the plasma. Because of the essential impermeability of the membrane to cations, electric

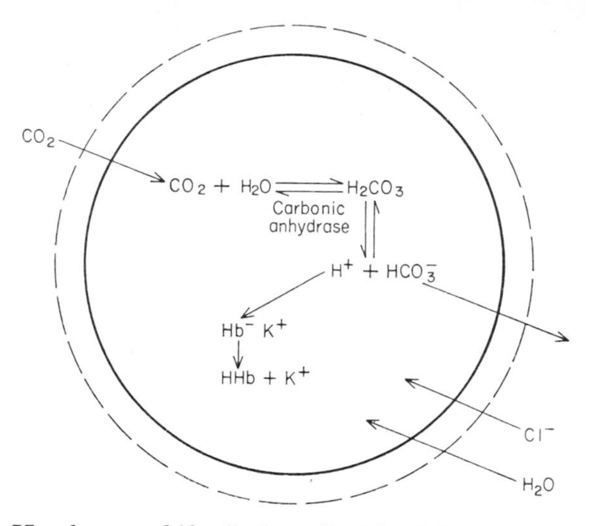

Fig. 7.5. The Hamburger shift. Carbon dioxide diffuses into the red blood cells where its conversion to carbonic acid is accelerated by the enzyme carbonic anhydrase. The carbonic acid reacts with haemoglobin, and the hydrogen ion is effectively buffered. The bicarbonate anion produced by this reaction diffuses out of the cell into the plasma and chloride diffuses in to maintain electrical neutrality. The extra ions in the red cell increase the internal osmotic pressure and hence water enters and the cell swells.

neutrality must be maintained by movement inwards of another anion; and this is achieved by the inward movement of Cl^-. Thus, as CO_2 continues to dissolve in the plasma, HCO_3^- is formed with the cell, part of which diffuses into the plasma, raising its concentration, whilst Cl^- moves into the cells (Fig. 7.5). This is called the *Hamburger shift*, and accounts for the fact that the concentration of Cl^- in venous plasma, i.e. in blood with a high partial pressure of CO_2, is lower than in arterial plasma. An additional cause is the lower oxygenation of the haemoglobin in venous blood which, as we have seen above, causes anions to enter the cells.

Change in "r" We may ask how it is that a new ionic equilibrium is achieved when HCO_3^- diffuses out of the cell and Cl^- diffuses in. Surely, because the ratio:

$$[Cl^-]_{in}/[Cl]_{out} = [HCO_3^-]_{in}/[HCO_3^-]_{out} = r$$

must be maintained, decreasing HCO_3^- in the cell by its outward movement might be expected to be accompanied by movement of Cl^- out too. In fact, however, the prime event is an increase of $[HCO_3^-]_{in}$ above its equilibrium value, so that the ratio has become too large; HCO_3^- must move out. Cl^- moves in to preserve electrical neutrality, and this does, indeed, raise the value of r to above its previous value. In effect, however, this is a movement towards equilibrium because the equilibrium value of r has changed as a result of the acidification of the medium by the CO_2. Thus r for equilibrium has risen, and this means that the value of $[Cl^-]_{in}$ must rise whilst that for $[Cl^-]_{out}$ falls. It will be clear that removing CO_2 from the blood, by exposing it to a low partial pressure of the gas, will have the opposite effects, causing the concentration of HCO_3^- in the cells to fall; this is followed by passage of HCO_3^- into the cells and movement of Cl^- out.

Exchanges in Lungs and Tissue

Thus, in essence, because the cells are the home of the carbonic anhydrase, the shifts in anion concentration reflect the movements of CO_2; when the blood is exposed to the tissues it takes up CO_2 and this must pass into the cells to be converted into HCO_3^- whence it subsequently moves out in exchange for Cl^-. In the lungs, the partial pressure of CO_2 is low, the CO_2 escapes out of the plasma into the air in the lungs. This causes CO_2 to pass out of the cells and causes the reaction:

$$KHb + H_2CO_3 \rightleftharpoons KHCO_3 + HHb$$

to go to the left causing the decomposition of bicarbonate within the cell, producing more CO_2, which passes into the plasma and thence into the lung gases. The lowered HCO_3^- concentration in the cells causes HCO_3^- to pass out of the plasma into the cells in exchange for Cl^-. Thus, in effect, the plasma bicarbonate is being dissociated within the cells, and this means, as we shall see, that far more CO_2 may be removed from whole blood than from the plasma alone.

Carbon Dioxide Dissociation Curve

The blood carries CO_2 away from the tissues and gives it up to the lungs; its power to carry the gas is therefore of great significance. We have seen that when CO_2 is pumped into the blood it appears in the plasma as the bicarbonate ion; and it is because of this process, which occurs by the intervention of the red cells, that the blood is an efficient carrier of CO_2. Thus, if we bubbled CO_2 into plasma alone, very little extra CO_2 would be taken up; similarly if we exposed plasma to a reduced partial pressure, very little would be given up compared with the amount given up by exposing whole blood.

Plasma and Whole Blood

These facts are illustrated when we measure and plot the *carbon dioxide dissociation curves* of plasma and whole blood separately. Let us imagine that we expose a solution of 0·2 M $NaHCO_3$ in water to different partial pressures of CO_2, and measure the amounts dissolved at each partial pressure. If the gas did not react with water, e.g. as with O_2 or N_2, the amount taken up would rise in simple proportion to the partial pressure in accordance with Henry's law. In fact, because the CO_2 combines with water to form carbonic acid:

$$CO_2 + H_2O \rightleftharpoons H_2CO_3$$

the amount dissolved will increase to a greater extent as the CO_2 pressure rises, the above reaction being pushed over to the right. Nevertheless, as the partial pressure is increased, the total amount of CO_2 taken up, in the form of dissolved gas and carbonic acid, does not rise rapidly, and the carbon dioxide dissociation curve is that shown in Fig. 7.6a, where the total carbon dioxide in the solution is plotted against the partial pressure of CO_2 in the atmosphere with which it is equilibrated. Here the CO_2 in the form of bicarbonate has been included in the total-CO_2.

Inefficiency of Plasma. It will be seen that the solution takes up barely a millimole of CO_2 on changing the partial pressure from 30 mm Hg to one of 60 mm Hg. This represents a very small carrying power when the actual requirements are considered. Thus the blood exposed to an actively metabolizing tissue might leave this tissue with a partial pressure of, say, 50 mm Hg; on passing through the lungs it is exposed to a gas mixture, called alveolar air, with a partial pressure of about 40 mm Hg, and CO_2 therefore passes into the alveolar air out of the blood. If we assume that it has time to come into complete equilibrium with this alveolar air, its partial pressure drops from 50 mm Hg to 40 mm Hg, a difference of 10 mm Hg. Examination of the dissociation curve for plasma shows that barely a millimole per litre of blood would be given up, and this would be quite inadequate to deal with the production of CO_2 by the tissues. In fact, however,

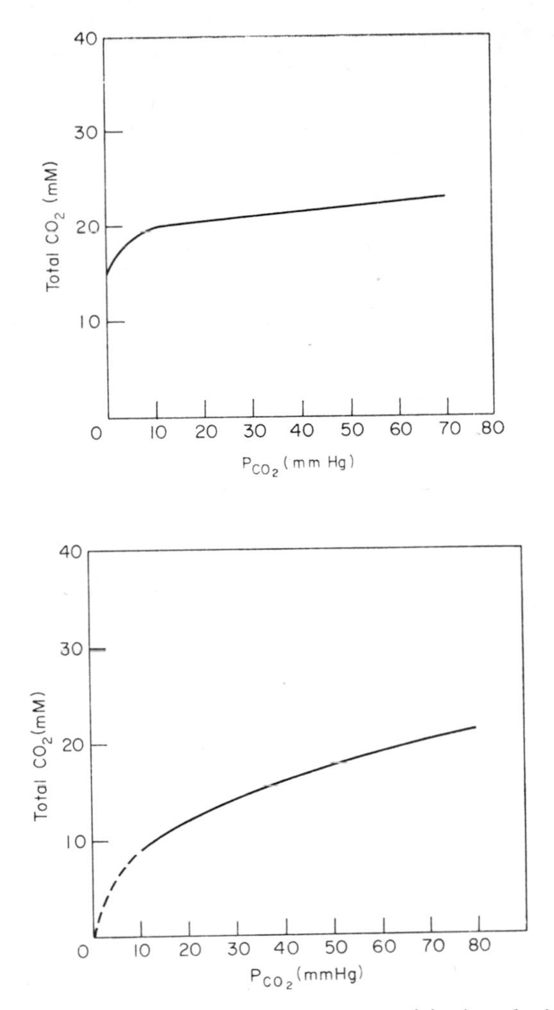

Fig. 7.6. Carbon dioxide dissociation curves. (*a*) A solution of sodium bicarbonate. The changes in total CO_2, consisting of bicarbonate ion plus carbonic acid, due to altered CO_2-tension are not large; moreover, at zero tension a large proportion is retained. (*b*) Whole blood. Here the changes in total CO_2 with altered CO_2-tension are larger because of the buffering action of the haemoglobin in the cells which favours breakdown of bicarbonate to CO_2 and OH^- at low CO_2 tensions, and the reverse at high tensions.

whole blood takes up and gives off considerably more CO_2 when exposed to changes of 10 mm Hg partial pressure, and this is shown by the curve for the whole blood in Fig. 7.6b.

Importance of Haemoglobin

The reason for this greater efficiency of whole blood as a carrier resides entirely in the presence of the haemoglobin in the erythrocytes which assists in converting the absorbed CO_2 to HCO_3^-. Thus, if we look back to the phenomenon of the Hamburger shift (p. 312), we see that the CO_2 entering the plasma passes into the cells where it is converted into H_2CO_3. If it merely remained as H_2CO_3, the uptake of the blood would be similar to that of our solution of $NaHCO_3$, i.e. limited by the solubilities of CO_2 in water and the extent to which the reaction:

$$CO_2 + H_2O \rightleftharpoons H_2CO_3$$

goes to the right.

However, the next step is to convert the H_2CO_3 into HCO_3^- through the intervention of the potassium–haemoglobinate salt in the cells, haemoglobin behaving as a weaker acid than H_2CO_3:

$$H_2CO_3 + KHb \rightleftharpoons KHCO_3 + HHb$$

The bicarbonate diffuses out of the cell in exchange for Cl^-, and this happens rapidly because of the extremely high permeability of the red cell membrane to these anions. As indicated in Chapter 2 this is an example of membrane specialization that enables the cell to carry out this function, no other cell of the body having such high anion-permeability.

Release of CO_2

In a similar way we can explain the reverse process, namely the giving up of a large amount of CO_2 when we reduce the partial pressure of CO_2 in the atmosphere to which the whole blood is exposed. Thus let us consider the extreme case when either plasma containing 20 millimoles/litre or whole blood is exposed to zero partial pressure of CO_2. The CO_2-dissociation curve for plasma shows us that, in spite of this low partial pressure, the solution contains nearly 20 millimoles/litre, and the essential reason for this is the limitation put on the reaction:

$$NaHCO_3 \rightleftharpoons CO_2 + Na^+ \, OH^-.$$

If this could go indefinitely to the right, by removing CO_2 as fast as formed, we would expect complete dissociation of the bicarbonate and complete loss of all the CO_2 in the blood plasma. However, this reaction to the right is strongly limited by the formation of the strong base, NaOH. This reacts with the bicarbonate-ion as follows:

$$NaHCO_3 + NaOH \rightleftharpoons Na_2CO_3 + H_2O$$

and the carbonate so formed does not give up its CO_2 on further reducing the pressure. Thus it is essentially this formation of strong base, NaOH, that prevents the bicarbonate in plasma from giving up all its CO_2, but if this base can react with an acid the formation of carbonate is prevented, and the dissociation of $NaHCO_3$ can proceed.

Neutralization of Base. It is the haemoglobin in the cells that provides the acid with which to neutralize, or buffer, the base:

$$HHb + OH^- \rightleftharpoons H_2O + Hb^-$$

the mechanism being the ionic shifts described above.

Thus the ionic shifts, when the tension of CO_2 of blood is held at a low value, may be represented by Fig. 7.7. CO_2 diffuses out of the

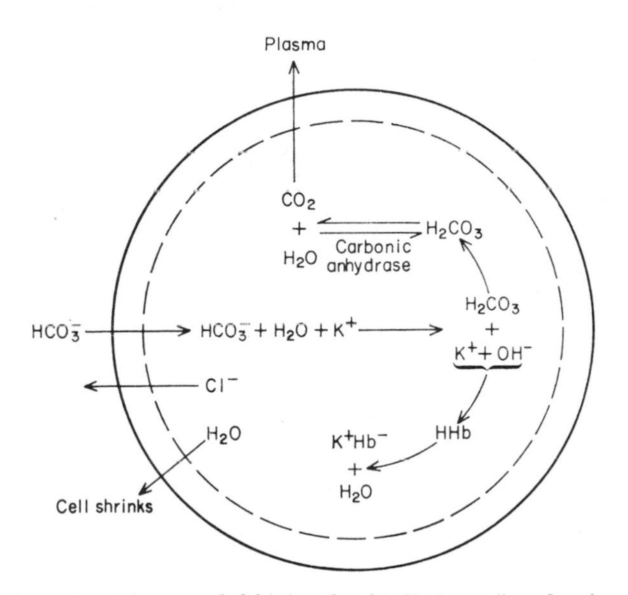

Fig. 7.7. The role of haemoglobin in the buffering of carbonic acid in the red cell. When the P_{CO_2} in plasma falls, CO_2 diffuses out of the cell; this favours decomposition of H_2CO_3, thereby leading to the breakdown of bicarbonate; the OH ions formed in this breakdown are buffered by the reaction with un-ionized haemoglobin, HHb. The fall in internal bicarbonate concentration creates a concentration gradient so that more diffuses in from outside, exchanging with internal chloride in the interests of electrical neutrality. There is a net loss of osmolality because the appearance of the Hb^+ ion does not increase the number of diffusing particles, but it reduces the number of Cl^- or HCO_3^- ions required to neutralize the cations within the cell.

plasma and CO_2 passes from the cells to the plasma because of the favourable concentration gradient. This causes the reaction:

$$H_2CO_3 \rightleftharpoons H_2O + CO_2,$$

catalysed by carbonic anhydrase, to go to the right, causing more CO_2 to pass into the plasma. The decomposition of H_2CO_3 in the cell promotes the decomposition of ionized bicarbonate

$$KHCO_3 + H_2O \rightleftharpoons H_2CO_3 + KOH,$$

so that the concentration of the ion falls and HCO_3^- from the plasma passes into the cells in exchange for Cl^-—the Hamburger Shift. The KOH formed in the cell reacts with un-ionized haemoglobin, HHb:

$$KOH + HHb \rightleftharpoons K^+Hb^- + H_2O.$$

If the CO_2-tension is held at zero the processes envisaged above will continue so that ultimately, if there is sufficient HHb present in the cells to react with the OH^--ions formed by the dissociation of $NaHCO_3$, all the CO_2 of the blood should be removed.

Water Shifts. The increased formation of K^+Hb^-, in place of the undissociated HHb, leads to a reduction in the total number of osmotically active ions in the cell so that there is a movement of water outwards into the plasma. Thus exposure of the red cells to a high partial pressure of O_2 and a low partial pressure of CO_2 both cause the cells to shrink, the situation encountered in the lungs. The reverse conditions, encountered in the tissues, namely low partial pressure of O_2 and high partial pressure of CO_2, cause the cells to swell.

Lungs and Tissues. The shifts of anions taking place in lungs and tissues are illustrated schematically in Fig. 7.8, which is, of course, a composite diagram combining the features of Fig. 7.4 showing the effects of changed P_{O_2} and of Figs. 7.5 and 7.7 showing the effects of changed P_{CO_2}.

BUFFERS AND pH

We may note that the haemoglobin is behaving as a *buffer*, preventing the solution from becoming very alkaline when the partial pressure of CO_2 is reduced, or too acid when it is increased. This is a very important function of the blood, quite apart from this particular aspect of the carriage of CO_2, and it is worth digressing to consider the physicochemical basis of buffers.

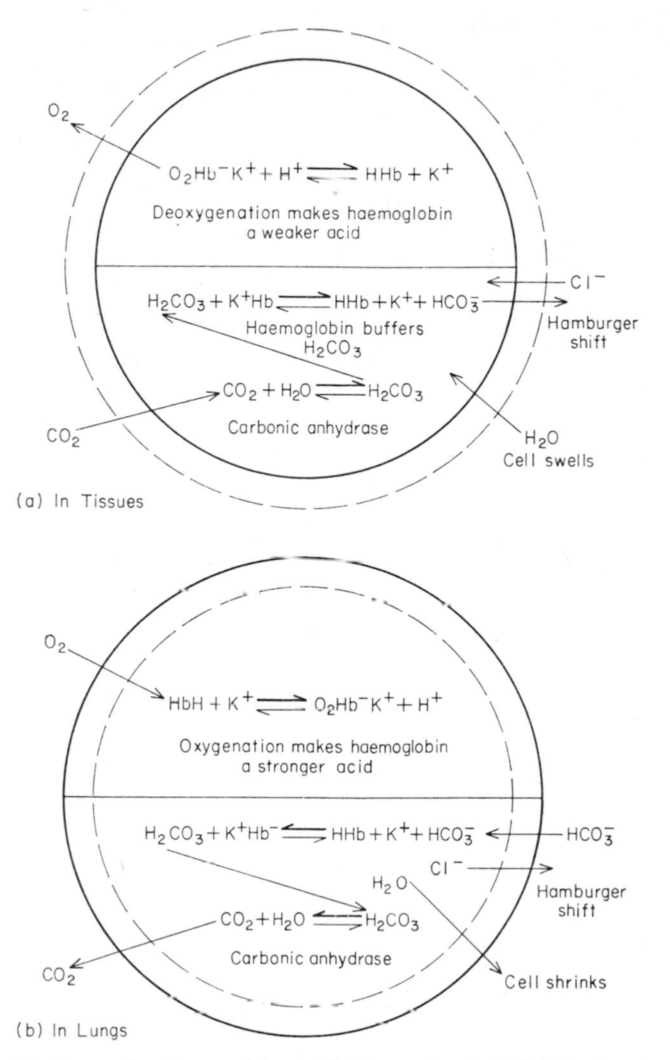

Fig. 7.8. (*a*) Overall effects of loss of O_2 and gain of CO_2 in the tissues. Deoxygenation of haemoglobin makes it a weaker acid and increases formation of HHb. Uptake of CO_2 causes increased formation of H_2CO_3 which is buffered by KHb to give rise to HCO_3^- ions which exchange with Cl^- ions in the plasma (Hamburger shift). (*b*) Overall effects of uptake of O_2 and loss of CO_2 in the lungs. These are the reverse of those shown in (*a*).

Strong and Weak Acids

Strong acids and bases are such that in aqueous solution they are almost completely dissociated to give H^+- or OH^--ions respectively, e.g.

$$HCl \rightarrow H^+ + Cl^-$$
$$NaOH \rightarrow Na^+ + OH^-$$

whereas weak acids are, by definition, those that are not completely dissociated into H^+-ions and their corresponding anion. Hence addition of a strong acid to water will make the concentration of H^+-ions, or its acidity, greater than the addition of a corresponding quantity of the weak acid.

pH

For many reasons it is preferable to refer to the H^+-ion concentration in terms of a logarithmic scale—the pH—which is the negative logarithm to base 10 of the H^+-ion concentration. The concentration of H^+-ions in pure water is 10^{-7} molar, hence the pH of this, neutral medium, is 7·0. Because of the reciprocal relation between the H^+- and OH^--ions in water, indicated by the relation:

$$[H^+] \times [OH^-] = K_W = 10^{-14}$$

we may deduce the concentration of OH^--ions from that of H^+-ions, or *vice versa*. Thus in pure water the concentrations of OH^--ions is 10^{-7} M.

Additions to Pure H_2O. The effects of adding strong acids or bases to pure water are large from a biological point of view where alterations in pH of more than one or two units are rare, whilst the pH of blood is usually held within quite narrow limits of between 7·2 and 7·4. Thus addition of pure HCl to water to make its concentration 0·1 Molal would lead to a concentration of H^+-ions of approximately 10^{-1} M, i.e. a pH of 1, a shift of some six units and an enormous shift in biological terms, compatible only with continued life of the cells in specialized situations, such as in the stomach (Chapter 9). The quite large shifts in blood-pH that might be produced by the formation of acids, e.g. of lactic acid during exercise, are held to a low level by the presence of buffers in the blood as well as by the operation of special physiological adaptations, such as the selective removal of acid from the blood by the kidney and by the selective removal or retention of CO_2 through the lungs.

Buffer Mixtures

Buffers are the systems that damp down the effects of added strong acids and bases; they consist in a mixture of a weak acid and its salt with a strong (conjugate) base, or a weak base and its conjugate strong acid salt. Thus acetic acid is a weak acid, not being completely dissociated into H^+ and acetate-ions in aqueous solution; this, combined with sodium acetate, constitutes a buffer system. The mechanism is as follows:

We imagine that some strong acid has been added to the mixture:

$$NaAc + HCl \rightleftharpoons HAc + NaCl$$

the strong acid reacts with the salt of the weak acid forming a weak acid and a salt, NaCl; now the weak acid is only partially dissociated, and hence the concentration of H^+-ions in the mixture is far less than if the strong acid had been added to water. In this way the salt has *buffered* the strong acid. If strong alkali is added to the mixture, the weak acid component does the buffering:

$$NaOH + HAc \rightleftharpoons NaAc + H_2O.$$

Optimum Buffering Range

The effectiveness of a given buffer mixture depends on the range over which it is required to work, e.g. if we wish to maintain the pH constant in the range of 4, a different mixture would be required from that required to maintain constancy in the region of neutrality, i.e. pH 7·0. The range of effectiveness is governed by the degree of dissociation of the particular weak acid or base employed. Thus the dissociation of a weak acid in solution is governed by the Law of Mass Action:

$$HA \rightleftharpoons A^- + H^+$$

$$K' = \frac{[H^+] \times [A^-]}{[HA]}$$

where the square brackets indicate concentrations. K′ is called the dissociation constant,* and it is clear that the greater the degree of dissociation, the larger will be the constant, so that acetic acid with a K′ of $2{\cdot}24 . 10^{-5}$ is a stronger acid than $H_2PO_4^-$ with a K′ of $1{\cdot}58 . 10^{-7}$.

pK

As with H^+-ion concentrations, it is convenient to use a negative logarithmic scale of notation, so that the pK of an acid is the negative

* The dissociation constant is indicated by K′ when *concentrations* of the reactants are used as opposed to K, when *activities* are employed.

logarithm of its dissociation constant. Thus the logarithm, to base 10, of $2{\cdot}24 . 10^{-5}$ is $\bar{5}{\cdot}35$ or $-4{\cdot}65$, whence the pK is 4·65. A similar calculation gives a value of 6·80 for the phosphoric acid pK.

In general, then, the smaller the pK the stronger the acid, just as with pH where the smaller the pH the greater the acidity.

Henderson–Hasselbalch Relation. An important relation between pK and pH is given by the Henderson–Hasselbalch equation derived quite simply from the equation defining the dissociation constant by taking logarithms. Thus from:

$$K' = \frac{[H^+] \times [A^-]}{[HA]}$$

$$\text{Log } K' = \log [H^+] + \log \frac{[A^-]}{[HA]}$$

$$-\log K' = -\log [H^+] - \log \frac{[A^-]}{[HA]}$$

$$pK' = pH - \log \frac{[A^-]}{[HA]}$$

or

$$pH = pK' + \log \frac{[A^-]}{[HA]}$$

In a mixture of a salt and a weak acid, the concentration of anions, A^-, is approximately equal to the concentration of the salt since this is almost completely dissociated into ions: hence we may write

$$pH = pK' + \log_{10} \frac{[\text{conjugate base}]}{[\text{acid}]} \qquad (4)$$

Titration Curve. Equation (4) can be used to predict the general nature of the "titration curve" of a buffer mixture, i.e. the effects of progressively changing the proportions of acid to base on the pH of the mixture, as in Fig. 7.9. It will be seen that there is a flat region, between pH 4 and about 5, where quite large changes in these proportions are reflected in relatively small changes in pH, and it is in this range that the particular mixture—acetic acid/acetate—is most effective as a buffer. The minimum sensitivity to changed proportions is obtained, generally, when the concentrations of acid and conjugate base are equal, whence [Acid]/[Base] = 1 and its logarithm is zero, and the pH for least sensitivity, or greatest buffering power, is thus the pK′ of the acid.

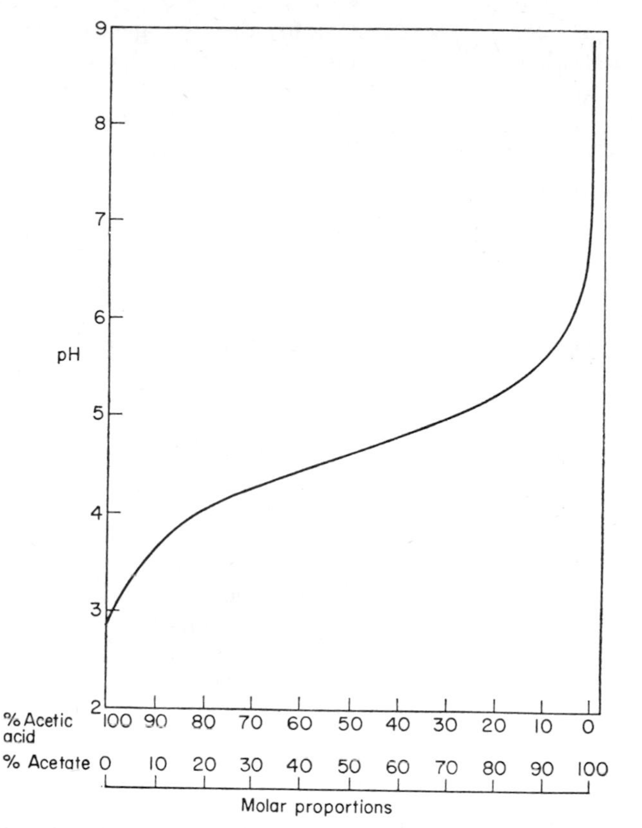

Fig. 7.9. The titration curve of acetic acid and sodium acetate, a buffer mixture of a weak acid and a strong base. Note that the curve is relatively flat over the range pH 4 to 5. This flat portion indicates that quite large quantities of acid or alkali can be added to this mixture without producing much change in the pH or of the solution.

Buffering of Blood

Haemoglobin as Buffer

The pH of blood is kept within fairly narrow limits; and this is achieved by the use of two buffers, the haemoglobin within the cells and the bicarbonate–carbonic acid system. Thus the ionic exchanges making up the Hamburger shifts result in two basic reactions, according as the CO_2-tension in the plasma is increased or decreased, namely:

$$H_2CO_3 + KHb \longrightarrow HHb + KHCO_3$$

and

$$HHb + KHCO_3 \longrightarrow KHb + H_2CO_3$$

as a result of both of these reactions, the changes in pH that would occur as a result of the increased or decreased H_2CO_3 concentrations in plasma, are reduced, the haemoglobin being a weaker acid than the H_2CO_3. Thus the mixture of HHb and KHb in the cell constitutes a buffering system that becomes available to the plasma by virtue of the ionic shifts across the erythrocyte membrane.

Bicarbonate–CO_2 Buffer

Carbonic acid functions as a weak acid, stronger indeed than haemoglobin, but weaker than metabolically produced lactic and pyruvic acids, so that it, together with $NaHCO_3$ in the plasma, constitutes a second buffer system. The buffering may be described by the Henderson–Hasselblach equation:

$$\text{pH} = \text{pK}' + \log \frac{[\text{HCO}_3^-]}{[\text{CO}_2 + \text{H}_2\text{CO}_3]}$$

where it will be seen that the concentration of the acid is indicated by the concentration of dissolved CO_2 plus the carbonic acid, H_2CO_3. This is not a strictly correct formulation, so that the value of K′ employed is not the true equilibrium constant for dissociation of carbonic acid. The value of $[CO_2 + H_2CO_3]$ may be estimated chemically, e.g. by precipitating all as Ba_2CO_3, or it may be computed from the known solubility coefficient, q′, based on Henry's law:

$$q'P_{CO_2} = CO_2 + H_2CO_3.$$

At body temperature this "effective pK′" is 6·1; in plasma the ratio of $[HCO_3^-]$ to $[CO_2 + H_2CO_3]$ is usually approximately 20:1 whence the plasma pH is given by:

$$\text{plasma pH} = 6{\cdot}1 + \log 20 = 7{\cdot}4 \text{ approx.}$$

Series Action

The two buffering systems act in series; thus, when a strong acid such as lactic, enters the blood it is buffered by the CO_2/HCO_3^- system. The result is, temporarily at any rate, an increased concentration of H_2CO_3 in the plasma, but the effects of this are themselves buffered by the haemoglobin system in the erythrocytes.

Co-operation of Respiratory System

Carbon dioxide, and hence carbonic acid, are continuously produced in large quantities by the aerobic metabolism of the cells, and this production can vary in magnitude very considerably during rest and activity. The CO_2/HCO_3^- system cannot, of course, buffer against the effects of these changes in carbonic acid production, so that the haemoglobin system must do this alone. However, in the intact animal, the ventilation of the blood through the respiratory system is very delicately adjusted to the amount of carbonic acid in the plasma, i.e. with the P_{CO_2}, so that if this increases, as a result of greater cellular activity, the tendency for CO_2 to be washed out of the blood in the lungs will be augmented. Thus the respiratory system, acting together with the erythrocyte's ionic exchange mechanisms, provides a highly efficient mechanism for maintaining the pH of the blood within a narrow range, the pH of the CO_2-laden blood arriving in the lungs being about 7·2 compared with 7·4 when the excess CO_2 has been removed in this tissue.

Carbonate Mechanism

This digression on pH has taken us away from the main thesis, namely the carriage of gases by the blood. We have seen that CO_2, formed in the tissues, is carried largely by virtue of its conversion to bicarbonate. An additional mechanism is provided by the formation of carbamates, the products of reaction of CO_2 with amino groups in haemoglobin. This reaction is reversible, and the binding is favoured when the haemoglobin in the blood is in the reduced condition, whilst dissociation of the complex is favoured when the blood is fully oxygenated. This is an ideal arrangement for the respiratory requirements, since it is precisely when the blood is being well oxygenated (in the lungs) that it is necessary to allow the CO_2, bound as carbamate, to escape.

Importance of Carbamate. It has been estimated, in fact, that some 50 per cent of the CO_2 taken up from the tissues, and released in the lung, is carried as haemoglobin–carbamate.* This does not mean, of course, that 50 per cent of the CO_2 in blood is carbamate, since the great bulk is in the form of inorganic bicarbonate, and the amount taken up or released in a cycle represents quite a small percentage of the total CO_2-content. It is because of this that large *percentage* changes in the carbamate content of blood are reflected in large changes in the *amount taken up or released.*

Bohr Effect

An essentially similar change in the power of the blood to hold on to its CO_2 is provided by the change in the strength of haemoglobin as an acid according as it is well oxygenated or not. In the lungs it is necessary to lose CO_2. We have seen that the power of the blood to

* Also called carbomino-haemoglobin.

take up CO_2 depends mainly on the behaviour of HHb as a weak acid; clearly, then, if the strength of HHb as an acid is increased while the blood is in the lungs, i.e. when the haemoglobin is being converted from its reduced to its oxygenated form, then the power to hold on to the CO_2 is decreased. The alternative situation occurs in the tissues when the haemoglobin loses its oxygen; it becomes a weaker acid and hence its power of taking up CO_2 is increased. Thus the Bohr effect, whereby the dissociation constant of haemoglobin varies according to the degree of oxygenation of haemoglobin, is an important factor in the efficient carriage of CO_2.

The pH Inside the Erythrocyte

The uneven distribution of diffusible anions across the erythrocyte membrane occasioned by the presence of ionized haemoglobin in the erythrocyte extends to the OH^--ion, so that the inside of the cell should be less alkaline—or more acid—than the plasma. In fact, when the pH of the plasma was 7·4 the pH inside the cells was found to be 7·2, or more generally, the pH in the cells, pH_C, was related to the pH of the plasma, pH_{Pl}, by:

$$pH_C = 0{\cdot}796\,(pH_{Pl} + 1{\cdot}644).$$

Intracellular Location of Haemoglobin

It should be clear that haemoglobin could perfectly well exert its buffering action, and, indeed, its carriage of oxygen, if it were dissolved in plasma and not sequestered in erythrocytes. Thus the CO_2-dissociation curve of haemolysed whole blood is not significantly different from that of blood with its erythrocytes intact. The exchanges of anions involved in the buffering action of haemoglobin are, therefore, a complication introduced through the requirement to segregate the haemoglobin from plasma, and an important reason for this requirement is the high viscosity of concentrated haemoglobin solutions. We shall see that the viscosity of whole blood is very little different from that of plasma, which itself is little different from that of water; and this is achieved by keeping the haemoglobin separate from the fluid medium. Additional benefits obtained by this separation are the employment of an oxygen-carrying molecule of relatively low molecular weight, which would otherwise escape from the capillaries; and the restriction of the colloid osmotic pressure of the blood to that due to the plasma proteins. If the haemoglobin were free in the plasma the total colloid osmotic pressure would be several times greater, requiring very high capillary pressures to maintain fluid balance in the tissues.

CARRIAGE OF OXYGEN

Plasma as Whole Blood

The haemoglobin in the red cells is the main carrier of O_2. Thus the amount of O_2 dissolved in plasma at a pressure of 760 mm Hg is 0·023 ml per ml plasma. Since the partial pressure of O_2 in arterial blood P_{O_2}, is about 100 mm Hg, the amount dissolved in plasma is $0{\cdot}023 \times 100/760 = 0{\cdot}0032$ ml/ml. The volume of a gas dissolved in blood is usually expressed in terms of "volumes per cent", i.e. ml/100 ml, hence the amount dissolved in plasma is $100 \times 0{\cdot}003 = 0{\cdot}3$ ml/100 ml. This compares with a total of 20 ml/100 ml for whole blood exposed to a P_{O_2} of about 100 mm Hg.

O_2-Dissociation Curve

The combination between Hb and O_2 is reversible and depends on the partial pressure of O_2 in the solution to which the haemoglobin is exposed. Hence we may plot an *oxygen-dissociation curve*, similar to that for CO_2. The technique for measurement is to allow a sample of blood to come into equilibrium with an atmosphere containing a measured partial pressure of O_2; when equilibrium is achieved the concentration of O_2 in the blood is measured. Figure 7.10 shows a typical curve; here, as ordinates, we have plotted *not* the actual oxygen concentration but the *percentage saturation.* 100 per cent saturation corresponds to about 20 ml/100 ml, but the figure depends on the amount of haemoglobin in the blood, so that an anaemic blood would give a much smaller value than a polycythaemic one. However, if the amount is expressed as a percentage of the maximum that can be taken up, the shape of the curve remains characteristic of the behaviour of haemoglobin.

Uptake in Lungs

The curve shows that, when the blood is exposed to partial pressures of about 80 mm Hg or more it is very nearly saturated; the partial pressure of O_2 in the alveoli of the lungs is about 100 mm Hg, hence we may expect the blood passing through the lungs to become 100 per cent saturated provided it has time to equilibrate.

Release in Tissues

The partial pressure of O_2 in the tissues will vary according to the degree of oxygen utilization; if utilization is high, then the partial pressure will be low and the ability of the blood to give up oxygen will be correspondingly high. Thus to look at the graph, the steep portion occurs over the ranges of oxygen partial pressures likely to occur in actively metabolizing tissues; if the partial pressure in the tissue is 40 mm Hg, the blood can give up rather less than 30 per cent of its

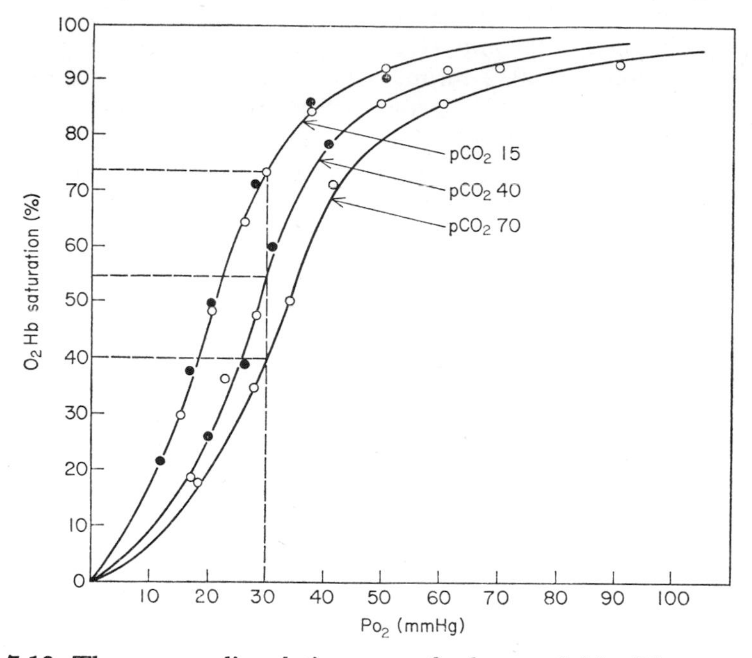

Fig. 7.10. The oxygen dissociation curve for haemoglobin. The percentage saturation of haemoglobin with oxygen is plotted against oxygen tension (P_{O_2}). At a P_{O_2} of 80 mmHg, i.e. below that in the lungs, the haemoglobin is almost completely saturated. In the tissues, where the P_{O_2} may be 40 mmHg or less, dissociation of the hacmoglobin-oxygen complex is favoured, and the steepness of the curve in this range indicates that the release of oxygen to the tissues will be well adjusted to the requirements. The curve is shifted to the left at low CO_2 tension so that in the lungs uptake of O_2 is favoured, whilst in the tissues the shift to the right favours release of O_2 (the Bohr effect). (Joels and Pugh, *J. Physiol.* 1958, **52,** 63.)

contents, if we assume that it was 100 per cent saturated on arrival at the tissue. If the partial pressure were only 20 mm Hg, indicating a greater utilization of O_2 by the tissue, then the blood would be able to give up nearly 70 per cent of its oxygen content. In general, the shape of the oxygen-dissociation curve of blood is well adapted to suit the oxygen carrying function of the blood.

Bohr Effect

An additional factor that improves the function of haemoglobin as a carrier of O_2 is the circumstance that the tendency to give up O_2 is increased when the CO_2 tension is high, and reduced when it is low. This is revealed by the shift in the dissociation curve to the right (Fig. 7.10), which means that, at a given partial pressure of oxygen, the haemoglobin is less saturated when the partial pressure of CO_2 is high than when it is low. This, described as the *Bohr effect*, is a necessary thermodynamic consequence of the increased strength of haemoglobin as an acid when oxygenated, which is often described as the "reversed Bohr effect". Thus, when the blood reaches an actively metabolizing tissue it is exposed to a high partial pressure of CO_2; this reduces the O_2-binding capacity of the blood, and so favours loss of O_2 from the blood. Hence, two factors contribute to the giving up of O_2 from the blood where it is needed: first the low partial pressure of O_2 and second, the high partial pressure of CO_2.

Actual Values of Blood Parameters

For convenience we may summarize mean values for blood O_2, CO_2 and pH in Table II.

TABLE II

Mean Values for Blood O_2, CO_2 and pH in Healthy Young Men (Comroe *et al.*, after Albritton)

	Arterial Blood	Mixed Venous Blood
1. O_2 pressure (mm Hg)	95	40
2. Dissolved O_2 (ml O_2/100 ml W.B.*)	0·29	0·12
3. O_2 content (ml O_2/100 ml W.B.)	20·3	15·5
4. O_2 combined with Hb (ml O_2/100 ml W.B.)	20·0	15·4
5. O_2 capacity of Hb (ml O_2/100 ml W.B.)	20·6	20·6
6. % saturation of Hb with O_2	97·1	75·0
7. Total CO_2 (ml CO_2/100 ml W.B.)	49·0	53·1
(mM/L)	21·9	23·8
8. Plasma CO_2 (ml CO_2/100 ml plasma)	59·6	63·8
(*a*) Dissolved CO_2 (ml CO_2/100 ml)	2·84	3·2
(*b*) Combined CO_2 (ml CO_2/100 ml)	56·8	60·5
(*c*) Combined CO_2/dissolved CO_2	20/1	18·9/1
(*d*) CO_2 pressure (mm Hg)	41	46·5
Plasma pH	7·40	7·376

* Whole blood.

Physical Basis of the O_2-Dissociation Curve

The S-shape of the O_2-dissociation curve of mammalian haemoglobin, exhibiting as it does an ability of the molecule to take up or release large amounts of O_2 when the surrounding O_2-tension is altered over a small and physiologically important range, represents a fundamental development in evolution that has made possible the maintenance of a high metabolic rate in the complex organism. Thus, as Haldane pointed out a long time ago, a change in the O_2-dissociation curve of human blood to the hyperbolic curve of, say, myoglobin, would result in death.

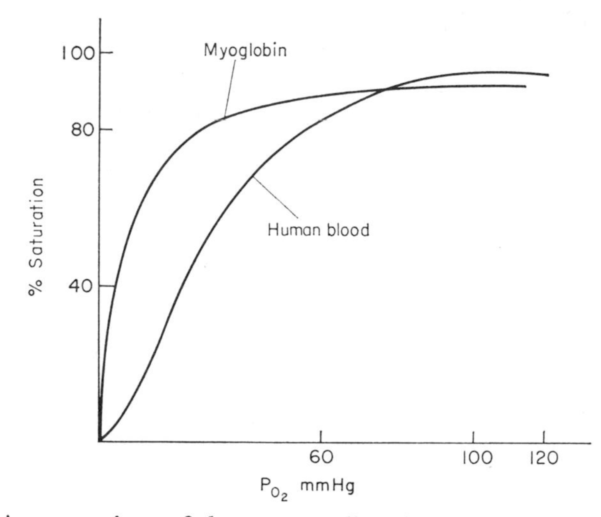

Fig. 7.11. A comparison of the oxygen dissociation curves of haemoglobin and myoglobin. At a tissue P_{O_2} of 40 mmHg myoglobin is still almost 90 per cent saturated so that it can release very little oxygen. Haemoglobin is only 60 per cent saturated at 40 mmHg and so can release much more O_2. (Roughton, "Handbook of Physiology", Respiration 3, Vol. I, **775,** 1964.)

Myoglobin and Haemoglobin

This becomes clear from Fig. 7.11 in which the dissociation curves for haemoglobin and myoglobin are compared; it is clear that a given change in oxygen tension, e.g. from 60 to 20 mm Hg, produces a much greater release, or uptake, of O_2 with haemoglobin than with myoglobin. This difference in behaviour between the two respiratory pigments, both made up on the same chemical basis of a hacm prosthetic group and globin protein molecule, must be related to a subtle difference in their molecular structures; and the demonstration of the atomic basis for this difference by Perutz represents one of the most striking successes in the application of X-ray analysis of protein structure to the interpretation of physiological phenomena.

Haemoglobin–O_2 Reaction

Human haemoglobin has a molecular weight of approximately 64,000. It contains four atoms of ferrous iron per molecule, and it is with this Fe that the O_2 molecule reversibly combines, so that the reaction may be written:

$$Hb + 4\,O_2 \rightleftharpoons HbO_8$$

and we may write an equilibrium equation:

$$\frac{[Hb]\,[O_2]^4}{[HHbO_8]} = K_{Eq}$$

with a dissociation constant, given by K_{Eq}, the reciprocal of which would be equated with the affinity of the haemoglobin for O_2 (p. 185). Application of simple equilibrium kinetics to the system would indicate a hyperbolic type of dissociation curve, similar to that of myoglobin, but possibilities for different shapes of this curve would arise if the uptake and release were governed by a set of reactions:

$$\begin{aligned} Hb \quad &+ O_2 \rightleftharpoons HbO_2\ (K_1) \\ HbO_2 &+ O_2 \rightleftharpoons HbO_4\ (K_2) \\ HbO_4 &+ O_2 \rightleftharpoons HbO_6\ (K_3) \\ HbO_6 &+ O_2 \rightleftharpoons HbO_8\ (K_4) \end{aligned}$$

each with its own equilibrium constant.

Varied Affinity for O_2. Thus the affinity for the initial uptake at low tensions would be largely governed by K_1, but as more and more O_2 was taken up the other equilibrium constants would become more and more important. It is not difficult to envisage the situation in which small changes in O_2-tension, by causing changes in the proportions of HbO_2, HbO_4, etc., would cause large changes in the average affinity of the solution of haemoglobin. Stated in kinetic terms, this is essentially what is found—the attachment of one molecule of O_2 influencing the affinity of the haemoglobin molecule to O_2 thereby modifying the ease with which it would take up a second, third and fourth molecule.

Co-operative Action of O_2 Molecules. This co-operative effect is, in fact, a manifestation of what Monod called the *allosteric modification* of enzyme behaviour, the addition of a molecule, or atom, to an enzyme molecule modifying the reactivity of its active site. The essential feature of this process is that the modifier does not react with the active site, which is reserved for the reactant, but, by combining with the enzyme molecule at a different point, is able to influence the reactivity of the active site. In order to assess how the uptake of an O_2 molecule may influence the reactivity of the haemoglobin molecule, so that it reacts with another O_2 molecule with a different affinity or equilibrium constant, we must examine the detailed structure of the molecule.

Structure of Haemoglobin

Haemoglobin is a combination of the pigment, *haem*, with the protein, *globin*. Haem contains a porphyrin ring made up on a basis of pyrrole units:

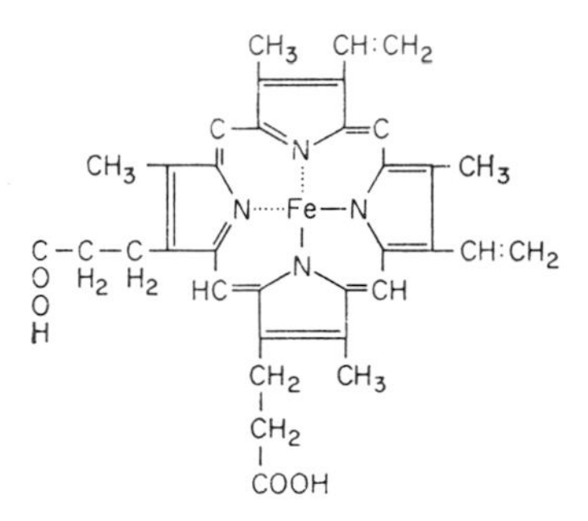

Chlorophyll

It thus shares a strong similarity with chlorophyll with an Fe-atom substituted for the Mg-atom of chlorophyll:

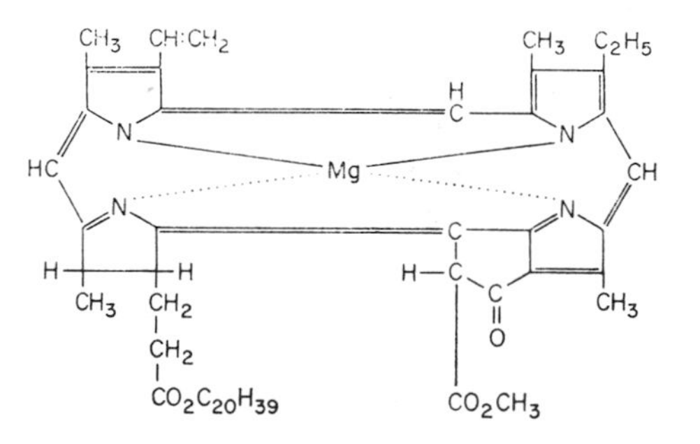

Ferrous Iron. In haemoglobin the Fe is in the ferrous state, and remains in this condition both when the molecule is oxygenated—*oxyhaemoglobin*—and deoxygenated—*deoxyhaemoglobin* (or reduced haemoglobin). If the ferrous iron is, in fact, oxidized, either experimentally or pathologically, to the ferric form, the colour changes to a brown, and the resulting *methaemoglobin* has no power of combining reversibly with O_2, and thus is useless as an oxygen-carrying molecule.

Globin. The protein moiety, globin, is built up of four separate polypeptide chains or subunits, held together in its quaternary structure by hydrogen bonding and salt linkages; each unit contains an atom of Fe.

Types of Haemoglobin. When we speak of haemoglobin it must be appreciated that there are, in fact, several haemoglobins in normal human blood, and this number may be increased if we include pathological bloods, such as in sickle cell anaemia. *Haemoglobin A* represents some 98 per cent of normal adult human haemoglobin, whilst *Haemoglobin* A_2 represents the remainder, which can vary in amount in pathological states, whilst *Haemoglobin F* forms more than half the haemoglobin in the newborn.

Protein Subunits. Examination of the various haemoglobins has revealed the presence of four different protein subunits, differing in the sequence and total number of amino acids; they have been designated α, β, γ, and δ. Haemoglobin A contains two α-units and two β-units, and is written conventionally as: $\alpha_2\beta_2$, Haemoglobin A_2 is $\alpha_2\delta_2$, and Haemoglobin F is $\alpha_2\gamma_2$. Thus haemoglobins have the common feature of two identical α-subunits.

Molecular Configuration. The amino acid sequences of these chains have been worked out (that of the α-chain, containing 141 residues, is illustrated in **Fig. 7.12**, where its relation to the haem unit is also shown), the connection being between two histidine residues, only one of which actually makes a chemical bond so that the two histidine residues are called the *proximal* (making the link) and the *distal* (free) residues. The β-chain contains 146 residues and possesses many regions of identical sequence to those in the α-chain, suggesting a common genetic origin. Myoglobin contains 153 residues. The amino acid residues are organized largely in the form of α-helices with intervening "pleated sheet" regions, and the chains so formed are bent into a tertiary structure in such a way as to produce a "globular" subunit (Fig. 7.13). Because of the similarities in amino acid sequence in certain critical regions, the shapes of the subunits, and also of myoglobin, are very similar. The subunits are linked together by hydrogen bonding between amino acid residues, and also by salt formation between the free carboxyl endings of α-chains and the terminal amino groups of β-chains, so that if the individual chains are regarded as snakes, the head of one bites the tail of another.

Location of Fe-Atoms. The important point brought out by

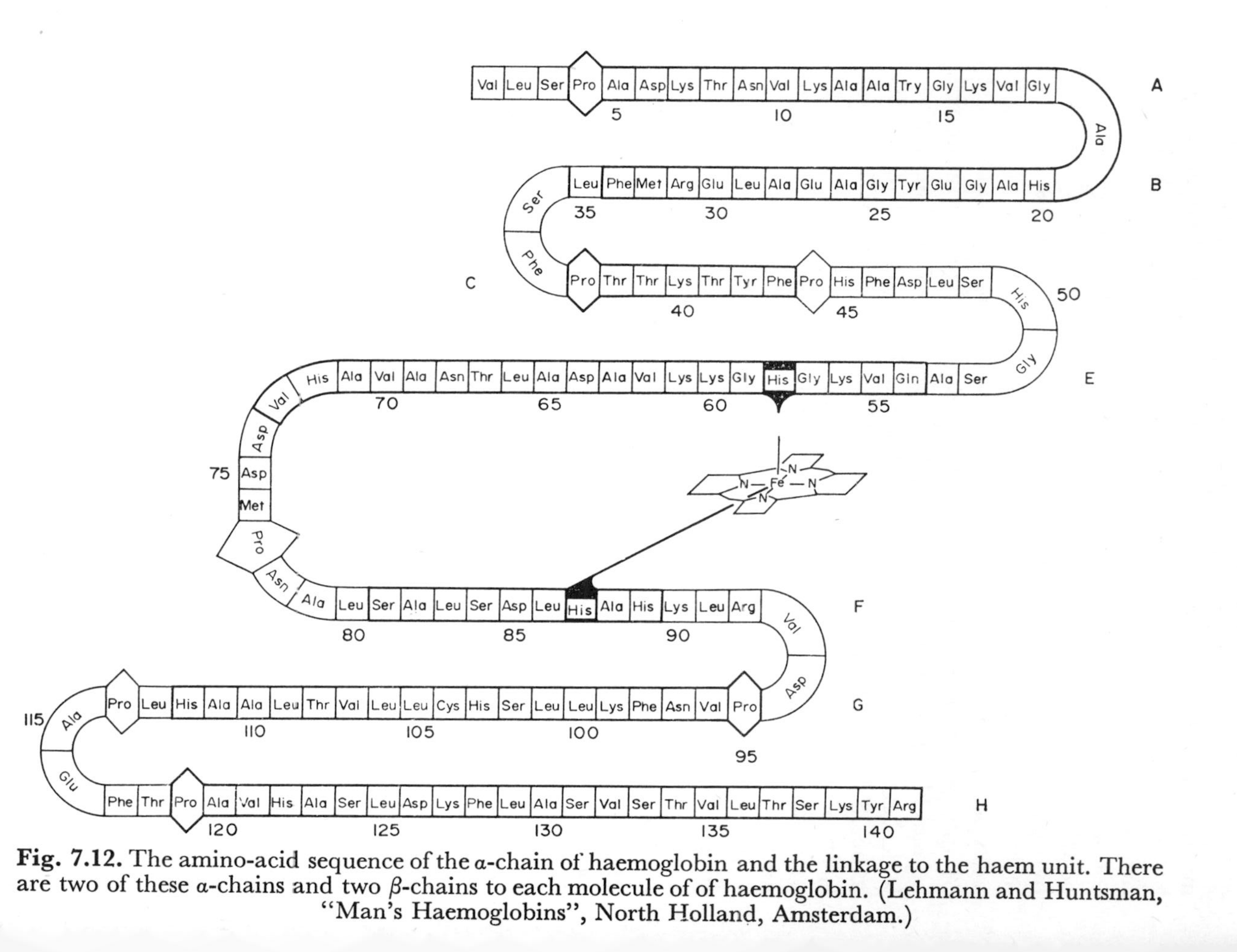

Fig. 7.12. The amino-acid sequence of the α-chain of haemoglobin and the linkage to the haem unit. There are two of these α-chains and two β-chains to each molecule of of haemoglobin. (Lehmann and Huntsman, "Man's Haemoglobins", North Holland, Amsterdam.)

Perutz' analysis of the tertiary structure of the polypeptide subunit is that the Fe-atom is contained within a pocket of non-polar portions of the amino acid sequence, i.e. it is in an essentially hydrophobic environment, and it is because of this that it can exercise its reversible O_2-carrying powers; remove the haem molecule from its attachment to the globin unit and this power is lost. In fact, much less drastic changes

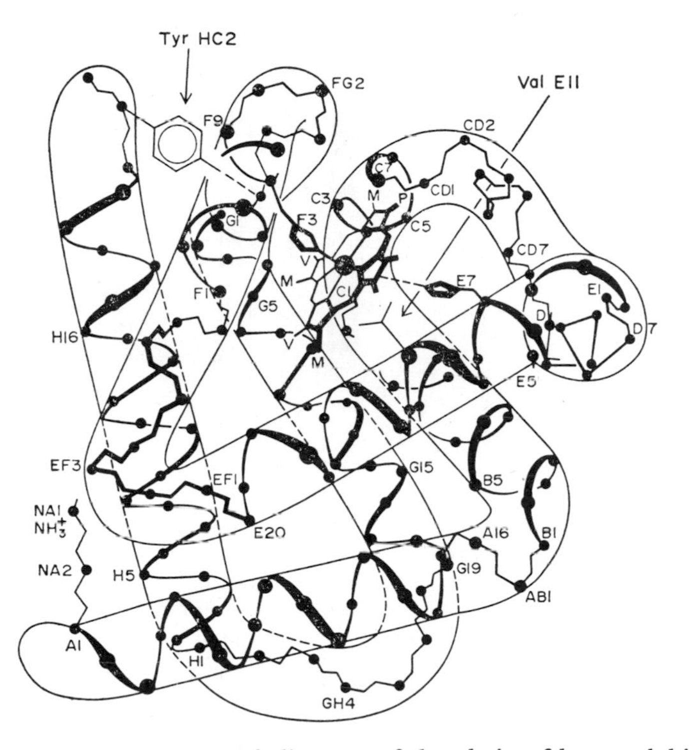

Fig. 7.13. A three-dimensional diagram of the chain of haemoglobin. The black molecules are amino acids; the helical residues are denoted A to H while non-helical stretches of chain are denoted AB, BC, etc. (Blake, *Prog. Biophys.* 1972, **25,** 2.)

in the structure, by altering the folding of the polypeptide chain, can prejudice this reversible O_2-reactivity.

Access of O_2 to Fe. As illustrated by Fig. 7.14 the O_2 must pass into the hole, squeezing past the histidine-haem link, and any change in the dimensions of this hole can be expected to influence the ability of the molecule to take up O_2.

We have already seen that the affinity of the haemoglobin molecule for O_2 is sensitive to two main influences, namely the pH of the medium

—the Bohr effect—and the state of oxygenation of the haemoglobin—the S-shaped dissociation curve. Neither of these effects is manifest with myoglobin which has a very similar chemical structure to that of a haemoglobin subunit, so that it is reasonable to seek the causes of these two effects in the polymeric structure of haemoglobin.

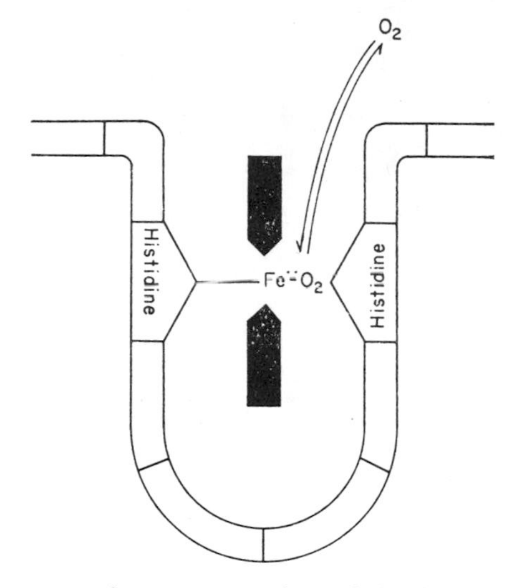

Fig. 7.14. A diagrammatic representation of the "pocket" into which the oxygen atom must pass in order to bind with the ferrous iron. The "pocket" is a fold in the amino acid backbone and is a mostly non-polar environment. The two histidine groups are the only ones that may become polar and oxygenation can only occur between the non-haem linked histidine and the iron atom. (Lehmann and Huntsman, "Man's Haemoglobins", North Holland, Amsterdam.)

Structural Changes in Haemoglobin with Oxygenation

Comparison of the X-ray diffraction patterns of haemoglobin crystals in the oxygenated and deoxygenated states showed that there were quite large changes in shape caused mainly by movements of the individual subunits in relation to each other, movements that must have been made possible by a rupture of many of the links that held the molecule in its more restricted deoxygenated state. In other words, there was a change in the *quarternary structure* of the haemoglobin molecule. These alterations were brought about by a primary alteration in the tertiary structure, i.e. in the relations of the individual amino acid residues to each other in the polypeptide chains, due to the binding of the O_2-molecules with the Fe-atoms.

The manner in which these tertiary structural changes are brought about is complex and cannot be entered into here in detail. Examination of the "pockets" into which the O_2-atoms must pass to link with Fe shows that in the α-subunits there is ample room, whilst in the β-subunits there is not, so that oxygenation of these must await structural changes resulting from the previous oxygenation of the α-subunits. In this way, then, we can explain the change in affinity of the haemoglobin molecule as it takes up oxygen.

Breakage of Inter-unit Links. The changes in the α-subunit, resulting from binding an O_2-molecule, consist, first, in a shift in the proximal histidine residue towards the porphyrin ring that may amount to as much as 0·95 Å. This, in turn, forces a tyrosine residue out of a pocket in which it had been shielding, thereby breaking a salt link between adjacent α-subunits with the liberation of a H^+-ion. Successive uptake of O_2-atoms leads to further rupture of linkages until all the subunits become free to adopt their new, oxygenated, configuration, and during the process H^+-ions are released, thereby accounting for the Bohr effect. The important steps are illustrated schematically in Fig. 7.15. It will be noted that one of the links between subunits is through diphosphoglycerate; this is present in human erythrocytes and it lowers the oxygen affinity of haemoglobin in a physiologically advantageous way, binding specifically to the reduced form and being expelled from the quaternary structure on oxygenation.

Co-operative Effect Explained. In general terms, then, the "co-operative effect" seen in the S-shaped O_2-dissociation curve is a manifestation of the fact that the ease of oxygenation of each of the four Fe-atoms in a given molecule depends on how many have already been oxygenated. The uptake of O_2 in the β-subunits must await a change in the size of the "pocket" that will allow the molecule to approach close enough to the Fe-atom, and this change in shape only becomes possible when the linkages between subunits are loosened or completely broken; thus the uptake of the last O_2-molecule is much easier than that of the one before because the whole molecule has acquired great flexibility due to rupture of many inter-unit linkages.

Bohr Effect Explained. The Bohr effect, manifest as release of H^+-ions on oxygenation, is a consequence of the breakage of salt linkages and of changes in the shielding of critical groupings, such as the imidazole of histidine. In general, if oxygenation involves release of H^+-ions, addition of H^+-ions to the medium will tend to reverse the process, on the principle of mass-action, so that the haemoglobin molecule will tend to revert to its deoxy-state in an acid medium, i.e.

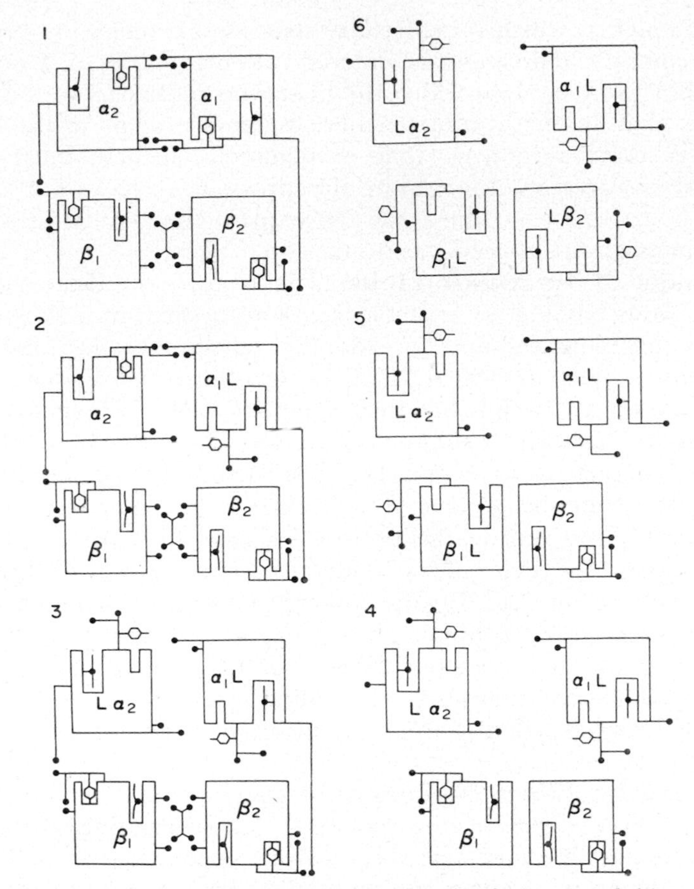

Fig. 7.15. A sketch of the steps in oxygenation of the haemoglobin molecule. The pockets for oxygen on the α-chains are exposed first as the intermolecular bonds are disrupted. Step 1 is the deoxygenated haemoglobin; step 2 exposes the pocket on the α_1-chain and step 3 on the α_2 chain. Step 4 splits the α-chains from the β-chains and steps 5 and 6 are the oxygenation of β_1 and β_2 respectively. Since the affinity for the binding sites is increased as each site on the 4 chains is occupied, the sigmoid shape of the dissociation curve can be readily explained. The rupture of bonds releases H^+ ions, and this accounts for the Bohr effect. (Perutz, *Nature*, 1970,. **228,** 726.)*

it will tend to give up O_2 to the medium. Thus the "reverse Bohr effect" is the necessary corollary of the Bohr effect.

Affinity for CO_2. We have indicated that a significant component in CO_2-transport is the reversible combination of CO_2 with haemo-

* Later evidence has shown that the salt bridges on the histidine 146 β and lysine 40 α and 94 β break when the quarternary structure changes, i.e. earlier than shown in the figure.

globin to form carbamino compounds. The important feature of this combination is that it depends on the degree of oxygenation of the haemoglobin, and very little on CO_2-tension; thus deoxy-, or reduced, haemoglobin has a higher affinity for CO_2 than oxyhaemoglobin, and thus loss of O_2 to the tissue is accompanied by increased uptake of CO_2 as carbamino-haemoglobin. There is little doubt that the changes in tertiary structure, leading to exposure or shielding of amino groups involved in the combination with CO_2, are the basis for the changes in CO_2 affinity with oxygenation.

THE SUPPLY OF OXYGEN TO THE TISSUE

The reversible reaction of O_2 with haemoglobin, and the special features of the oxygen-dissociation curve just described, namely the S-shape and the Bohr effect, represent adaptations that permit the blood to carry out its task of supplying O_2 to tissues in amounts that would be quite impossible were the system to rely simply on the solubility of O_2 in an aqueous medium. It is worth examining the general principles involved in this transport in more detail, so fundamental are these adaptations.

Diffusion Equation

Concentration Gradient

We have seen in our discussion of transport of dissolved material that the rate of diffusion of a substance, that is, the number of molecules that pass from one region to another in unit time (which we may indicate by $\Delta S/\Delta t$), is governed by the area through which this diffusion can occur, A, and the *concentration gradient*, i.e. the difference in concentration between the two points considered divided by the path-length, R. Thus, if we represent the concentration in the blood as C_{Cap} and that in the tissue cell as C_{Cell} then the concentration gradient is given by

$$\frac{C_{Cap} - C_{Cell}}{R}$$

Thus we have:

$$\frac{\Delta S}{\Delta t} = \frac{DA\,(C_{Cap} - C_{Cell})}{R}$$

The proportional constant in the equation, D, is the diffusion coefficient; and the classical studies of Krogh have shown that the cells of a tissue, including those of the capillary membranes, offer very little impediment to the diffusion of the respiratory gases through the tissue, so that the diffusion coefficients are not greatly different from those found in free aqueous solution.

Mean Path Length

Thus the diffusion equation shows us that the concentration gradient governs the rate of diffusion of material from the capillary to the tissue. As we can see, this is determined by three parameters: the concentration in the capillaries (C_{Cap}) and cell (C_{Cell}) and the mean path length, R. As Fig. 7.16 illustrates, R may be regarded as the average radius of action of a capillary which is imagined to be in the

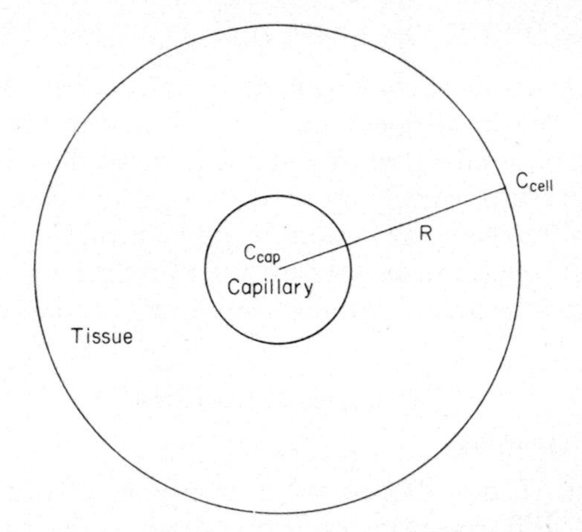

Fig. 7.16. A diagram of the mean path length for diffusion of respiratory gases from a capillary. C_{Cap} is the concentration in the capillary and C_{Cell} that in the cell. R is the average radius of tissue supplied by the capillary. If the capillary density is high and therefore R is low, the diffusion gradient will be steep.

centre of a cylinder of tissue. Clearly, the smaller R the greater will be the gradient; histologically this radius of action may be measured by estimating the "capillary density" of the tissue, i.e. the number of capillaries in a mm^2 of cross-section of tissue, from which R is obtained by the relation:

$$1/(2R)^2 = \text{No. of capillaries/mm}^2.$$

Variation in Capillary Density

One method, then, of increasing transport of gas is to increase the capillary density, and this is achieved by the opening up and closure of capillaries through the action of the pre-capillary sphincters discussed earlier. This shows a remarkable dependence on the P_{O_2} of the tissue, and thus represents a mechanism that is well adapted to meet the metabolic requirements of the tissue, the lower the P_{O_2} the greater the requirement for O_2 and the smaller the path-length, R. Thus the capillary density of the rat heart ventricular muscle, working under normal conditions, is about 2500/mm^2 giving a diffusion radius, R, of 10 μ; when the heart is working under stress maximally the number increases to about 3100 giving a value of 9 μ for R.

Concentration in Capillary and Cell

The two other parameters in the concentration gradient are C_{Cap} and C_{Cell}; and the ideal situation would be for C_{Cap} to remain at its highest level, corresponding to complete saturation of the blood with O_2, and C_{Cell} to be zero, corresponding to instantaneous utilization of the O_2 as it arrives. The partial pressure of O_2 in the mitochondrion, the site of intracellular utilization, is remarkably low, perhaps 1 mm Hg, and if the escape of O_2 from the capillary did not lower the concentration in the blood significantly we should, indeed, have an ideal arrangement for obtaining maximal diffusion in the tissue.

Significance of Haemoglobin

The development of a transporting molecule for O_2, namely haemoglobin, is the means whereby this extra requirement is partially met. It may be regarded as a means whereby the blood may give up considerable amounts of O_2 and yet maintain nearly the same concentration of O_2 as before. Thus the determining factor in diffusion is the concentration of O_2 dissolved in the plasma, and this would tend to fall rapidly as the O_2 diffused out of the capillary if it were not for the large reservoir of O_2 held by the haemoglobin. In order that this reserve supply may fulfil its proper function, however, it must be released when needed, i.e. as soon as the concentration in the plasma tends to fall. The S-shape of the O_2-dissociation curve can be regarded as the second adaptation that contributes to the maintenance of the steep concentration gradients that favour diffusion, the amount that can be released for a given requirement, i.e. a given decrease in tissue O_2-partial pressure, being large over the range of partial pressures normally occurring in a metabolizing tissue.

Oxygen Debt

On the basis of the known diffusion coefficient of O_2 and the measured capillary densities of tissues, it has been computed that tissues should be adequately supplied with O_2. It may be, however, that these calculations are too "generous" and that the supply of O_2

may become critical under conditions of severe work; this is certainly true of fast skeletal muscle which rapidly goes into oxygen-debt, producing lactate in large quantities.

Function of Myoglobin

Red and White Muscles

When discussing the action of muscle we shall see that mammalian muscles controlling the movements of the skeleton fall into two main categories, namely red and white, the red muscles appearing thus because of the presence of the protein *myoglobin* within the cytoplasm of the muscle cells. Red muscles are associated with sustained activity, and are therefore found in the wings of birds, such as the pigeon, whilst white muscles are associated with bursts of very intense activity that are not sustained for any length of time. Red muscles rely strongly on oxidative metabolism, involving the mitochondria, for the immediate supply of energy during work, by contrast with white muscle which rapidly goes into "oxygen debt", relying on anaerobic metabolism leading to the formation of lactate for the immediate supply of energy. In this respect heart muscle shares the characteristics of red skeletal muscle; it tolerates O_2-debt very badly and is characterized by a high concentration of myoglobin, whilst its high oxidative capacity is revealed by the dense packing of the muscle cells with mitochondria.

Myoglobin and Oxidative Activity. There is a strong correlation between the oxidative activity of muscles and their myoglobin content as Fig. 7.17 shows, so that it is reasonable to assume that myoglobin exercises a significant role in the supply of O_2 to the tissue. Two possibilities may be discussed: first that myoglobin acts as a reservoir for the tissue, preventing it from going into oxygen-debt when the blood-supply becomes critical; alternatively, or additionally, the myoglobin in some way facilitates transport of O_2 through the cell.

Myoglobin as O_2 Store. That myoglobin can act as a store is well proven by studies on diving animals; the muscles of the penguin, seal, whale, etc., are strongly red due to exceptionally high myoglobin contents which may provide an oxygen-binding capacity of as high as 11·8 volumes per cent, and are much higher than would be expected of the respiratory requirements of the tissue (Fig. 7.17). The studies of Irving and Scholander on the seal, for example, have shown that after a dive of 5–10 min the myoglobin is completely deoxygenated; since seals rarely dive for more than 4–5 min this means that the muscle does not become anaerobic, the O_2 bound to the myoglobin sufficing for this period. When a whale dives for long periods, which may extend for 120 min, the myoglobin-bound oxygen is first used up and only later do the muscles go into oxygen-debt, producing lactate. However, to extend these observations on diving animals to the situation in skeletal muscles of birds, and other red muscles, is hazardous, and the alternative explanation, namely that myoglobin facilitates diffusion of O_2 through the tissue, is better substantiated experimentally.

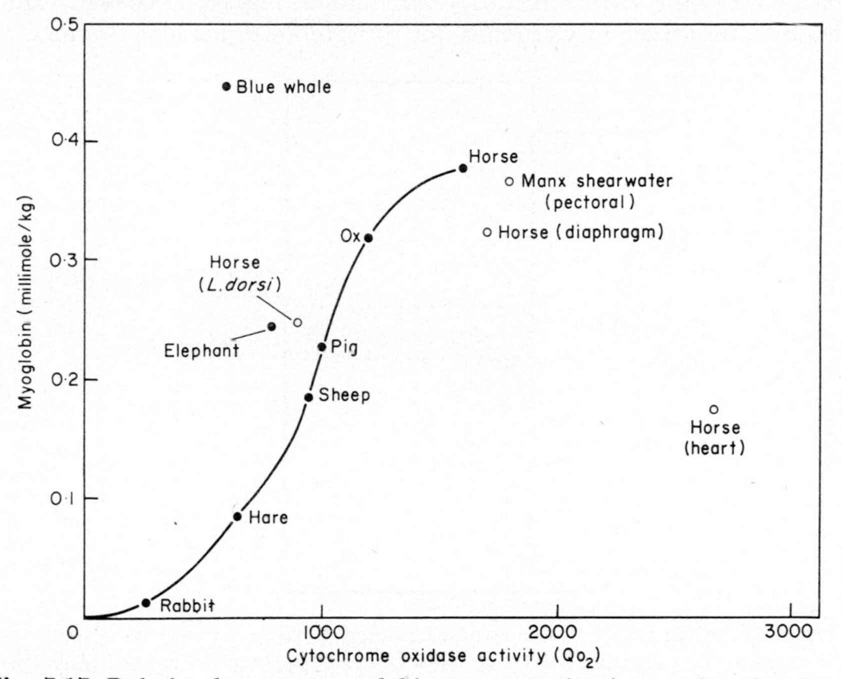

Fig. 7.17. Relation between myoglobin concentration in muscles of various animals and their oxidative activity as represented by the level of cytochrome oxidase activity. Solid circles psoas muscles, open circles other muscles. Manx shearwater is an oceanic bird that flies for great distances without rest. (Wittenberg, *Physiol. Rev.* 1970, **50,** 559.)

Facilitated Diffusion

Roughton was the first to suggest that the haemoglobin in the erythrocyte might act as a carrier, as well as a store, for movement within the cell, so that diffusion occurred by two processes—the movement of unbound O_2-molecules in solution and the movement of the O_2-haemoglobin complex. At first sight this second form of movement would be unlikely, since the diffusion coefficient of the protein is so much smaller; thus myoglobin has one of $11{\cdot}3.10^{-7}$ cm^2/sec compared with one of $2{\cdot}13.10^{-5}$ cm^2/sec for O_2. However, the efficiency of a carrier depends not only on the speed with which it moves—its diffusion coefficient—but also on its concentration in the medium compared with that of the molecule it has to carry. Because of the very low solubility of O_2 in water, the actual concentration in the cell water is remarkably small. Thus with a partial pressure of 20 mm Hg at 37°C the concentration of free O_2-molecules is $3{\cdot}5 . 10^{-5}$ M whilst that of myoglobin is 5.10^{-4} M, i.e. some 15-times higher, so that it is quite feasible, on theoretical grounds, for myoglobin to act as a carrier.

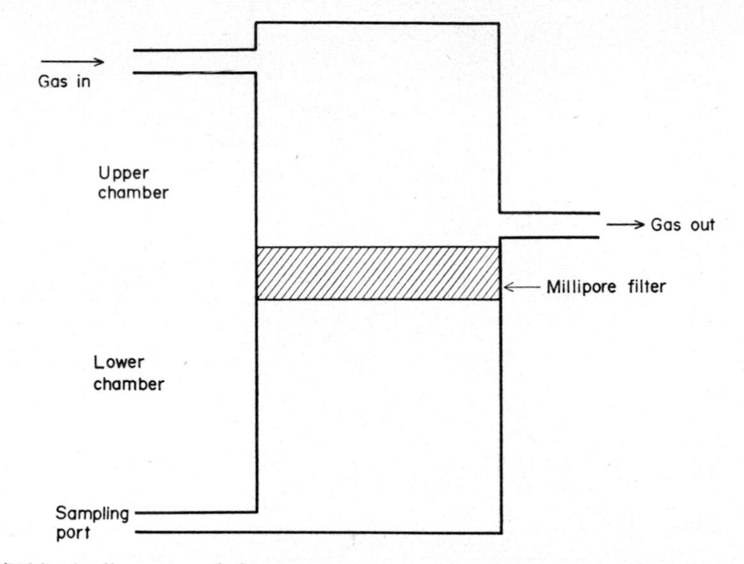

Fig. 7.18. A diagram of the apparatus used to demonstrate the carrier role of respiratory pigments. Two chambers are separated by a millipore filter. A stream of moist gas at a known P_{O_2} is passed through the upper chamber and the lower chamber contains only moist nitrogen. Samples of gas are taken at intervals from the lower chamber to determine the rate of entry of oxygen. The presence of haemoglobin or myoglobin in the millipore filter increased the rate of entry of oxygen into the lower chamber.

Experimental Proof. Practically, it can be shown that the presence of myoglobin (or haemoglobin) in a solution does, indeed, accelerate diffusion of O_2 through it; thus a solution of myoglobin or haemoglobin may be held in a millipore filter, and this may be made to separate two chambers as in Fig. 7.18; in the upper chamber a stream of moist O_2 of known partial pressure is maintained whilst in the lower chamber we have no O_2, e.g. moist N_2. At given times we measure the amount of O_2 appearing in the lower chamber after diffusion through the layer of myoglobin solution, and it is found that the flux of O_2, i.e. the volume passing in unit time, is very markedly accelerated by the presence of myoglobin (or haemoglobin) in the solution. Figure 7.19 illustrates the results obtained with haemoglobin (black circles) and methaemoglobin (open circles), methaemoglobin being a form of haemoglobin in which the iron is in the ferric state, when it

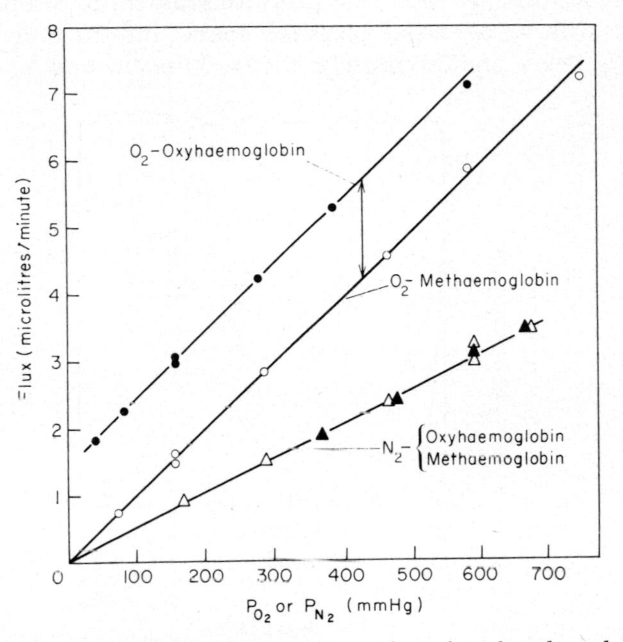

Fig. 7.19. The results of an experiment using the chamber described in Fig. 7.18. The volume of gas passing in unit time through the solution of haemoglobin is plotted against gas tension. The uppermost line indicates passage of O_2 through a solution of oxyhaemoglobin and this may be compared with the lower line indicating passage of O_2 through a solution of methaemoglobin, i.e. a solution of the pigment in which the iron is in the ferric state and thereby is unable to act as a reversible carrier of O_2. The lowest line indicates the passage of N_2 through solutions of either oxyhaemoglobin (▲) or methaemoglobin (△). The difference between the two upper lines, indicated by the thick arrow, represents the facilitated component of diffusion of O_2. (Wittenberg, *Physiol. Rev.*, 1970, **50,** 559.)

cannot operate as a binder of O_2. It will be seen that the flux is uniformly higher with haemoglobin in solution. The lowest line relates to diffusion of N_2, which is unaffected by changing haemoglobin to methaemoglobin. Similar results are obtainable with myoglobin.

Effects of O_2-saturation. If the acceleration is really due to the formation of a complex and its carriage from the side of high O_2-tension to that of low tension, we may expect the effect to disappear if we maintain the solution fully oxygenated. This can be achieved by maintaining a definite partial pressure of O_2 in the lower chamber. Thus we may compare the flux under the same gradient of partial pressure but with a different pressure in the lower chamber, e.g. we may compare the flux when the pressure gradient is 20 mm Hg in all cases but with zero partial pressure below and 20 mm Hg above; 10 mm Hg below and 30 mm Hg above, 20 below and 40 above, and

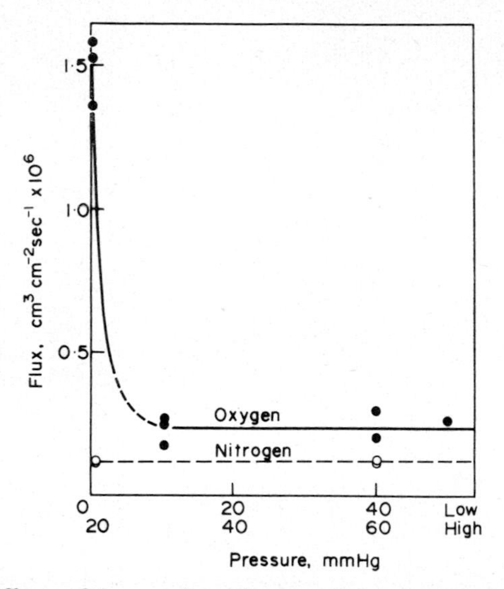

Fig. 7.20. The effect of increasing the partial pressure of oxygen in the lower chamber of Fig. 7.18 on the flux of oxygen through a millipore filter containing myoglobin. As can be seen, the flux of oxygen rapidly falls to a low level as the oxygen partial pressure in the lower chamber increases. The abscissae represent the pressures in the two chambers with constant pressure differentials across the membranes. The nitrogen flux is unaffected and remains constant. (Hemmingsen, *Acta Physiol. Scand.* 1965, **64,** Suppl. 246.

so on; if only simple uncomplicated diffusion operated, the fluxes should all be the same, but if the process relied to any extent on oxygenation of the myoglobin, we should expect the transport to be lower when O_2 is in the lower chamber. As Fig. 7.20 shows, the flux is highest when there is no O_2 in the lower chamber, but as soon as the tension reaches 10 mm Hg here the flux falls to a steady minimum valuc; the diffusion of N_2 is quite unaffected. The O_2-tension of 10 mm Hg in the lower solution corresponds with a tension necessary to saturate the myoglobin with O_2 completely.

Deoxygenation of Myoglobin *in vivo*. Experimentally, then, myoglobin and haemoglobin can accelerate the movement of O_2; moreover, Millikan showed a long time ago that, during work, the myoglobin of skeletal muscle is, in fact, deoxygenated, so that the necessary gradient of oxidized and reduced myoglobin is established for facilitated diffusion; a similar situation prevails in the beating heart, so that the physical basis for this facilitated diffusion is present.

Inactivation of Myoglobin. When the supply of oxygen to a muscle is reduced below a critical partial pressure, the amount it uses depends critically on how much it can obtain, so that the utilization of O_2 under these conditions is a measure of how quickly it reaches the mitochondria from the blood; Bindels showed that when the myoglobin was put out of action (by treatment with nitrite) the respiratory rate was approximately halved.

Diffusion Out of the Erythrocyte

The rate of release of O_2 from a sample of blood is considerably increased by destroying the erythrocyte membranes and causing the haemoglobin to pass into the plasma. This suggests that escape of O_2 from the cell is a rate-limiting factor. This could be due to a low permeability of the erythrocyte membrane to O_2 or to the fact that the gas must diffuse through the bulk of the cell contents; since these are not stirred in the same way as plasma during flow, this latter factor may well be highly significant. If it is, then any factor that causes convection of fluid within the cell should favour O_2 release.

Effects of Intracellular Stirring

We have seen that, when the blood flows at high shear-rates, there is every reason to believe that the shear is transmitted to the interior of the cell by virtue of the "looseness" of the cell membrane; hence causing blood to flow at high shear rates, e.g. in a rotating viscosimeter, should increase rate of release of O_2. In fact Zander and Schmid-Schönbein were able to increase the rate of release in this way so that it became comparable with that in haemolysed blood.

Cell Permeability

It would seem, on this basis, that restraint across the cell membrane is not very significant, and this is confirmed by more direct studies.

Thus Stein *et al.* measured the diffusion of O_2 in a packed suspension of erythrocytes consisting of over 90 per cent cells, and he found the diffusion coefficient for O_2 to be $6–8.10^{-6}$ cm^2/sec comparing with almost identical values for haemoglobin solutions containing the equivalent concentrations of haemoglobin.

Shape of the Red Cell

The biconcave shape of the red cell is important rheologically, i.e. in facilitating flow along very small channels (Chapter 5). It was argued by Hartridge that the biconcave shape facilitated diffusion through the erythrocyte, ensuring a more even uptake than that which would be obtained in a spherical cell. In fact, however, calculations have shown that the difference between rates of equilibration in spheres and biconcave discs of the same volume would be barely detectable.

CHAPTER 8

Exchanges in the Lungs

AERATION OF THE BLOOD

Perfusion and Aeration

During its passage through the lung the blood, after supplying the tissues with O_2 and removing some of their CO_2, is recharged with O_2 and discharged of its extra CO_2. Thus, the blood passing from the right ventricle along the pulmonary arteries is of the "venous" type with a P_{CO_2} of, say, 40 mm Hg and a P_{CO_2} of 47 mm Hg; the blood leaving the lungs in the pulmonary veins is of the "arterial" type

TABLE 1

Partial Pressures of Respiratory Gases in Blood

	Arterial blood mmHg	Mixed venous blood mmHg
O_2	100	40
CO_2	40	46
N_2	573	573
H_2O	47	47

having values of P_{O_2} and P_{CO_2} of about 100 and 40 mm Hg respectively (Table 1). The basic process taking place within the lungs is thus the *perfusion* of the tissue with venous-type blood and its *aeration* by exposure to atmospheric air, or rather the *alveolar air* which, because of the exchanges of CO_2 and O_2 taking place between it and the blood, and the saturation with water-vapour, has a lower P_{O_2} and higher P_{CO_2} than that of the atmospheric air drawn into the respiratory system.

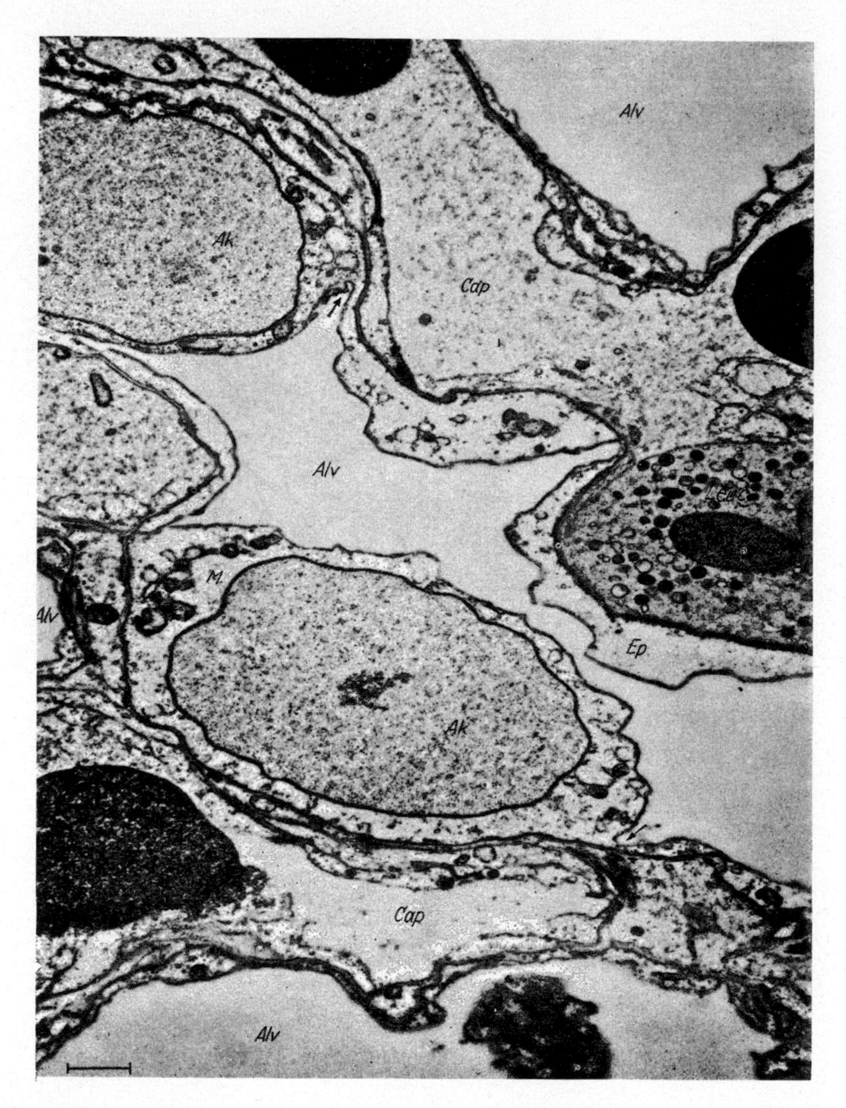

Fig. 8.1. An electron micrograph of the relations between the pulmonary capillary and alveolar wall. The capillary is exposed on both sides to air in adjacent alveoli. (Schultz "Pulmonary Structure and Function", Ciba Symposium, 1962, **203.**)

The Alveolus

The structure in which this aeration occurs is the *alveolus*, a minute epithelium-lined sac some 166 μ in diameter, the wall of which contains a capillary. As the electron micrograph of Fig. 8.1 shows, the blood in the capillary is exposed on both sides to air in adjacent alveoli; it will be clear from the schematic illustration of Fig. 8.2 that the "blood–air pathway" requires that the gases diffuse across the epi-

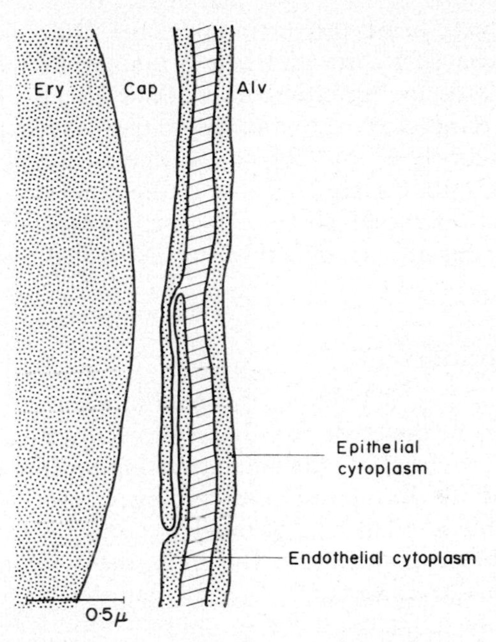

Fig. 8.2. A diagram, drawn to scale, of the pathway from air in the alveolus (Alv) throughout the lumen of the capillary (Cap) to the erythrocyte (Ery). The hatched area represents the basement membranes common to the capillary endothelial cell on the left and the alveolar epithelial cell on the right. (After Schultz, Ciba Symposium, 1962, p. 203, Churchill.)

thelium, across the capillary endothelium and thence through the plasma into the erythrocyte. As indicated earlier, exchanges occur between plasma and environment, whether this is the alveolar air or the tissue fluid, but because the O_2 is bound to the haemoglobin in the erythrocyte, and the CO_2–HCO_3^- interactions must also largely occur here, the passage into this cell is an important step and is, in fact, the rate-determining one so far as the carriage of CO_2 is concerned. As we have seen, this is not true of O_2.

Gaseous Exchanges

During the period in which the erythrocyte remains in this alveolar capillary (about 0·75 sec in a resting subject), it must take on O_2 and lose CO_2; the physical factors concerned in this process are just the reverse of those encountered in the metabolizing tissues and need not be repeated in detail here. To recapitulate, the blood with a P_{O_2} of about 40 mmHg is exposed to alveolar air which has a P_{O_2} of about 100 mmHg; because of the difference in P_{O_2}, oxygen will diffuse into the blood, the rate being proportional to the difference. The process of uptake is favoured by the shift in pH taking place by virtue of the loss of CO_2 that occurs simultaneously (Bohr effect). The loss of CO_2 occurs in accordance with the same principles; the P_{CO_2} in the venous blood is high, namely 47 mm Hg, whilst that in the alveolar air is less than this, namely 40 mm Hg. Because of the difference in gas tensions, CO_2 diffuses out of the blood into the air, a process favoured by the conversion of reduced haemoglobin into the stronger acid, oxygenated haemoglobin.

Rate of Equilibration

In general, as Fig. 8.3 shows, the equilibration of the blood with the alveolar air, in the sense that it acquires almost exactly the same gas tension, is remarkably efficient, occurring within about one-third of the time available during resting breathing; in exercise, too, although the time available is reduced by as much as a third because of the much greater rate of blood-flow through the lungs, the exchanges are virtually complete. Thus, during severe exercise in man the oxygen requirements may increase by a factor of 22 or more—from 250 ml O_2/min to 5500 ml/min; this means that the blood must carry a correspondingly larger amount to the tissues. This may be achieved partly by giving up a greater percentage of its total content (increased extraction) but, since this only permits an increase by a factor of about two to three, the rest of the increase must be provided by a more rapid blood-flow, and this is indicated by the increase in cardiac output from the average resting value of 5–6 litres/min to 25 litres/min in very severe exercise. Such an increase in cardiac output must mean a great increase in rate of flow through the pulmonary vascular bed and a consequent reduction in the circulation time, and hence the time available to a corpuscle to exchange its gases with the alveolar air. In spite of these oxygen requirements, the blood leaving the lungs from a normal healthy subject during severe exercise is virtually in perfect equilibrium with alveolar air.

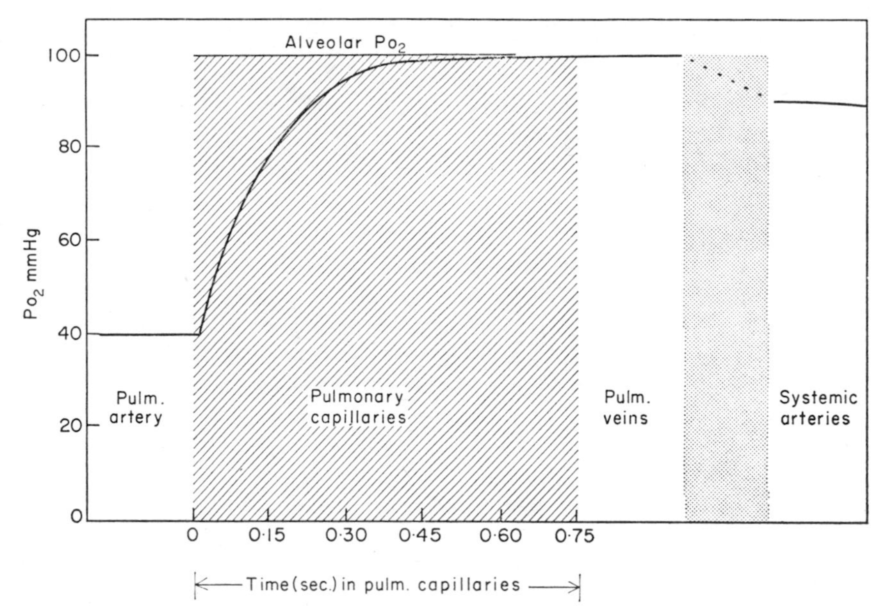

Fig. 8.3. The equilibration of blood with alveolar air during passage through pulmonary capillaries. Note that the blood is fully saturated in some 0·3 sec although the blood is in the capillaries for some 0·75 sec. The extra time allows the blood-flow to increase five-fold in exercise yet there is still sufficient time for the blood to always be fully oxygenated. (Comroe "The Lung", Year Book Med. Pub., 1962.)

The V_A/Q Ratio

The important feature in achieving this is what West has called the *V_A/Q or Ventilation/Perfusion Ratio*, defined as the ratio of the ventilation rate (V_A) over the rate of blood-flow through the lung (Q). The normal resting alveolar ventilation in man is about 5 litres/min, and the pulmonary blood-flow or cardiac output is 6 litres/min, giving a ratio of 0·85. As Fig. 8.4 shows, with severe exercise the two parameters increase in proportion to the oxygen used, but the rate of ventilation becomes steeper, so that the V_A/Q ratio tends to increase, achieving a value of $55/25 = 2{\cdot}2$ at the severest exercise.

Changes in V_A/Q

Under resting conditions the value of the ratio is a useful diagnostic tool since changes can indicate the nature of any impairment in aeration of the blood. Thus at the one extreme we may consider blood flowing through a completely unventilated part of the lung; the

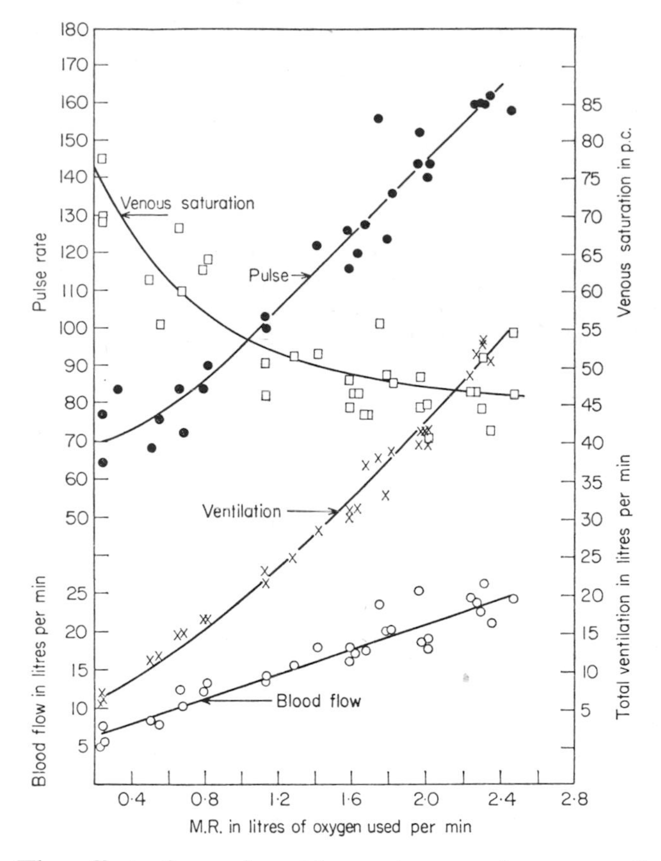

Fig. 8.4. The effect of exercise of increasing severity on ventilation and cardiac output. The cardiac output and hence pulmonary blood-flow increases from 5 L/min to 25 L/min, whereas ventilation increases from 10 L/min to 55 L/min, so that the ratio Va/Q increases markedly during exercise. V_A is alveolar ventilation L/min, Q is blood-flow L/min. Work periods 20 min on bicycle ergometer. Blood-flow indirect Fick. (Bock *et al.*, *J. Physiol.*, 1928, **66**, 136.)

ratio clearly becomes zero and the average V_A/Q ratio for the whole lung will be reduced. At the other extreme blood-flow may cease and the ratio becomes infinite. The effects of intermediate conditions may be easily inferred. Thus a restriction on blood-flow through a part of the lung will tend to increase the efficiency of gas exchange in this part so that the blood returning from it to the lung will be well aerated; however, unless more blood is shunted into the healthy part of the lung, the net efficiency of the lungs in aerating the blood will be decreased because the lung is handling less blood than normally. Shunting of blood to the healthy region may overtax its ability to aerate the larger volume of blood unless its V_A/Q is correspondingly increased.

Alveolar Air and V_A/Q

Corresponding with alterations in the degree of aeration of the blood, associated with altered V_A/Q ratio, there will be alterations in the composition of the alveolar air in those parts of the lung responsible

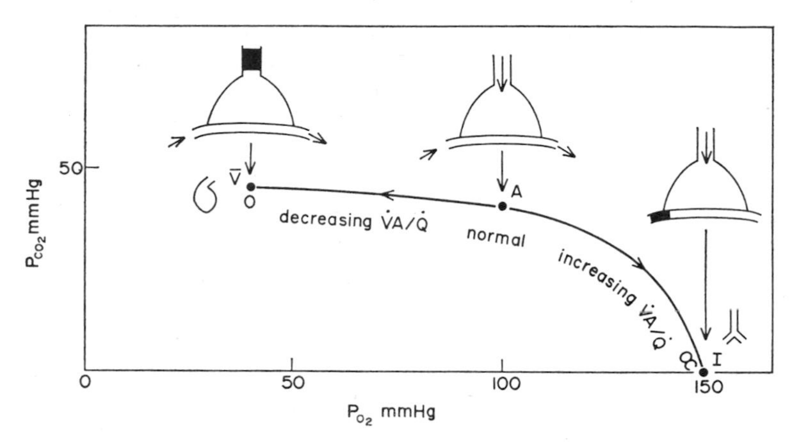

Fig. 8.5. The oxygen-carbon dioxide diagram. P_{CO_2} on the vertical axis and P_O on the horizontal axis. The compositions of three lung units are shown. In the uppermost one there is no air-flow so $V_A = O$. In the lowest there is no blood-flow so $Q = O$. In the middle one $V_A = Q$ and the line joining these points represents all the possible alveolar gas compositions from $V_A/Q = O$ to ∞. Thus by examining the composition of alveolar gas it is possible to investigate the efficacy of blood aeration. (West, "Ventilation/ Blood-Flow and Gas Exchange", Blackwell, 1967.)

for the changed ratio. With a normal ratio the P_{O_2} will be about 100 mm Hg and the P_{CO_2} 40 mm Hg in the alveolar air; with an increasing ratio, less O_2 will be removed from the air and less CO_2 given up, so that the P_{O_2} increases and the P_{CO_2} decreases; the opposite changes occur with decreasing V_A/Q as Fig. 8.5 shows. Thus analysis of the alveolar air provides information on the efficacy of blood aeration, as indicated by the V_A/Q ratio.

Topographical Variations in Blood-flow

It must be emphasized, however, that in estimating a ratio of this sort we are averaging the whole behaviour of the 750 million-odd alveoli of the human lung, and even in the normal condition there are some very large local variations in the blood-flow (perfusion) and air-flow (ventilation), variations due to the special anatomical features of the lung. Thus the relative rates of blood-flow through different parts of the lung may be measured very simply by causing the subject to inhale a single breath of a radioactive gas and measuring the radioactivity emitted from different levels of the chest by a suitable scanning device. If the breath is held, the decrease in radioactivity with time is a measure of the blood-flow in the region scanned. Fig. 8.6 shows the

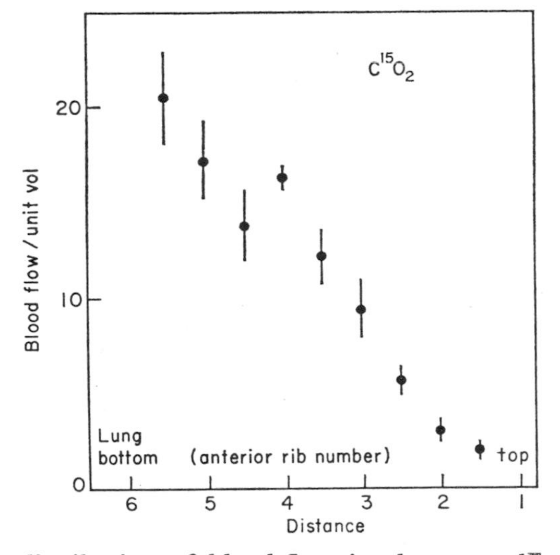

Fig. 8.6. The distribution of blood-flow in the normal upright lung as measured with radioactive carbon dioxide. Data from 16 normal subjects, means and standard errors of clearance rates. Note that the blood-flow decreases steadily from the bottom to the top of the lung. (West, "Ventilation/Blood-Flow and Gas Exchange", Blackwell, 1967.)

remarkable fall in blood-flow/unit volume of lung as we pass from the base of the lung to the apex, blood-flow at the apex being very small indeed. That gravitational forces are largely responsible for this gradient is shown by the tendency for flow to equalize when the subject adopts the supine position, and for the tendency to reverse when a man is held upside down.

Gravity Effect. The manner in which gravity influences blood-flow is through its effect on pulmonary arterial pressure, the pressure in the arteries at the base of the lung being higher than in the apex by virtue of the column of blood which may be some 30 cm in height.

Thus in the erect position the arterial pressure falls as we proceed up the lung, and a point may be reached where the pressure is actually lower than that in the alveoli, which is fairly uniform throughout the lung. At this point, then, the capillary and venous pressures must be

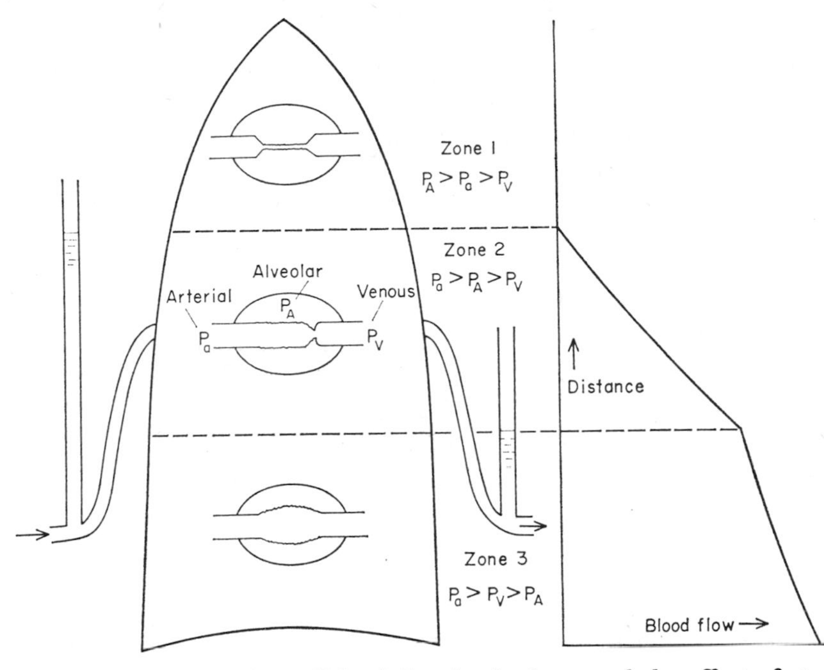

Fig. 8.7. The distribution of blood-flow in the lung and the effect of atmospheric and hydrostatic pressure. In zone 1 the alveolar pressure exceeds the arterial pressure and no flow occurs. In zone 2 the arterial pressure exceeds the alveolar pressure but the alveolar exceeds the venous pressure and the flow is intermittent. As soon as flow starts in zone 2 the pressure falls and the vessels collapse, gradually pressure builds up and then flow starts again but only to stop as before. The arterial-alveolar pressure differences increase down this region and flow increases until in zone 3 the venous pressure exceeds the alveolar pressure so that flow now is determined by the arterial-venous pressure difference. (West, "Ventilation/Blood-Flow and Gas Exchange", Blackwell, 1967.)

lower than the alveolar pressure, and these thin-walled blood vessels will collapse; if the arterial pressure remains below the alveolar pressure throughout the cardiac cycle this part of the lung will be unperfused.

Pulsatile Blood-flow. If, during a part of the cardiac cycle, the arterial pressure is higher than the alveolar pressure, then we may expect the capillaries and veins to be forced open to give some flow, but this will be intermittent, and, in general, we may distinguish three conditions as in Fig. 8.7 where we have, at the apex, the alveolar pressure greater than both arterial and venous pressures, in which case there is no perfusion. At the intermediate level arterial pressure is greater than alveolar pressure, but the venous pressure is less than the alveolar pressure, so that the pressure in the capillary therefore tends to fall below that surrounding it; it will tend to fall, moreover, as soon as flow begins and thus it will collapse intermittently. After collapse, the pressure from behind builds up, forcing it open; the flow so induced causes a drop in pressure in accordance with Poiseuille's law, so that the capillary collapses at a point downstream, as illustrated in the mid-portion of Fig. 8.7.

Topographical Variations in Ventilation

The changes in ventilation with position of the lung are not so marked but run in the same direction, so that the greater requirements

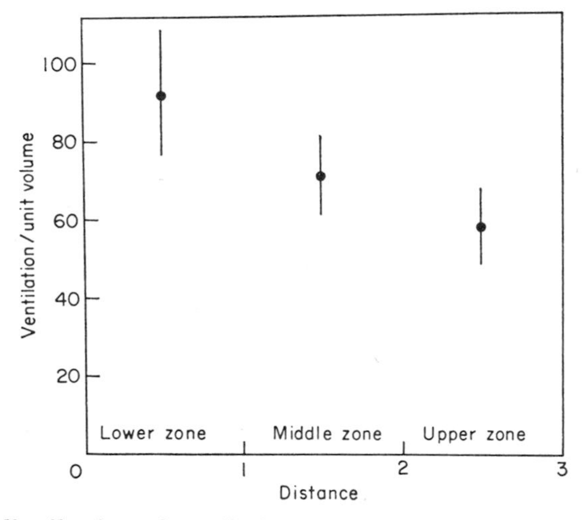

Fig. 8.8. The distribution of ventilation in the upright human lung. Ventilation decreases from bottom to top of the lung but the rate of change is much less than for blood-flow. (West, "Ventilation/Blood-Flow and Gas Exchange", Blackwell, 1967.)

of the base of the lung are met by an increased ventilation. Thus Fig. 8.8 may be compared with Fig. 8.6. The distribution is affected by changes in posture in the same way as distribution of blood-flow. A current explanation for the distribution of ventilation is the variation in intrapleural pressure; this increases down the upright lung, i.e. at the end of expiration the pressure within the lung is less negative at the base than at the apex. In consequence, the part of the lung at the apex is at a different position on the pressure–volume curve of the lung, being already half-way up before beginning inspiration so that

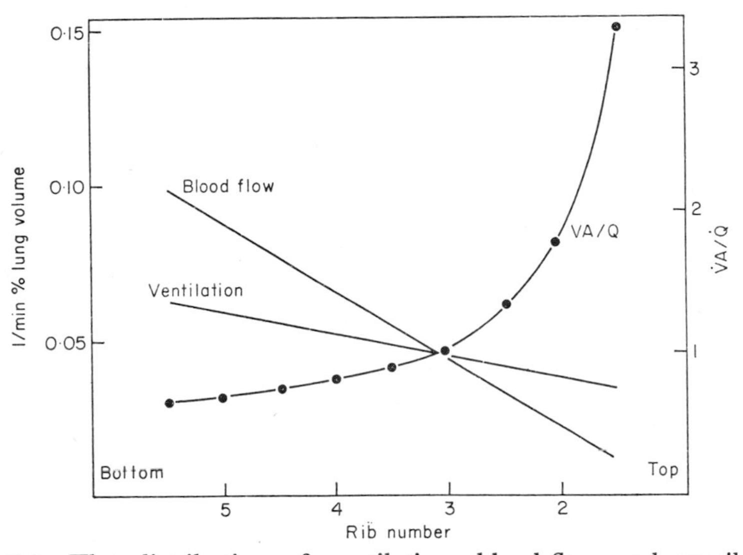

Fig. 8.9. The distribution of ventilation, blood-flow and ventilation-perfusion ratio from the base to the apex of the normal upright lung. Blood-flow decreases more rapidly than ventilation from bottom to top of the lung so that the ventilation perfusion ratio rises slowly at first and then rapidly. (West, "Ventilation/Blood-Flow and Gas Exchange", Blackwell, 1967.

it can only expand by about 25 per cent compared with the much greater expansion at the base which begins at a much lower initial volume.

Topographical Variation in V_A/Q

Finally, then, we may plot the changes in V_A/Q ratio as we progress from base to apex of the lung, as in Fig. 8.9. It is seen that the ratio increases as we pass from base to apex, indicating that the aeration of blood improves as we pass upwards in the lung. Thus although both blood-flow and ventilation decrease as we pass upwards, the ratio, V_A/Q increases because the *slope* of decrease of ventilation is not so steep as that of blood-flow. At the base the ratio is low, 0·6 compared with an overall value of 0·85 for the whole lung.

MECHANISM OF VENTILATION

In considering the general principles of aeration of the blood we have anticipated a great deal of the anatomy and physiology of breathing, i.e. the manner in which atmospheric air is brought into close relation with the blood in the alveoli; the chemical composition of this air in the alveoli during steady-state breathing; and the main features of the pulmonary vascular circulation. Let us consider, now, the ventilatory process.

Structure of the Lung

Ramifications of the Bronchial Tree

The lung consists of a series of branching conducting tubes, beginning with the single trachea which bifurcates to give the right and left *main-stem bronchi* carrying air to the right and left lungs. The

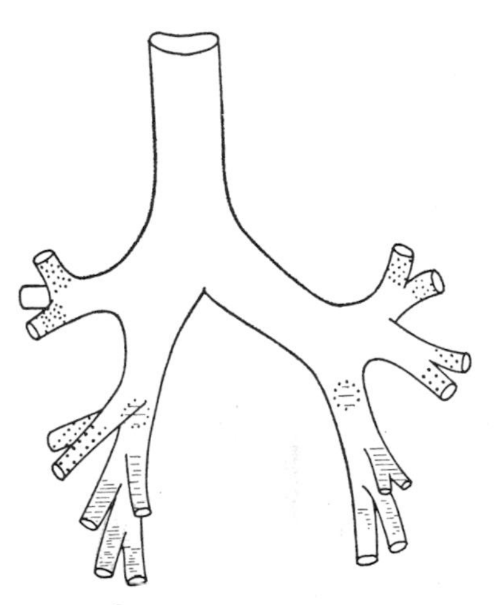

Fig. 8.10. Ramifications of the bronchial tree. The main airway, the trachea, divides into two main bronchi carrying air to left and right lungs. Further branching now occurs through twenty-three generations to arborize into the terminal bronchioles. (Slonin and Hamilton, "Respiratory Physiology", Mosby, St. Louis, 1971.)

main-stem bronchi divide into *lobar bronchi* (Fig. 8.10), and these branch successively through twenty-three generations to arborize into *terminal bronchioles* of 0·6 mm diameter; these open into *respiratory bronchioles* (Fig. 8.11) which branch again into alveolar ducts, which form a branching network of tubes (the *atria*) opening into the *alveolar sacs* in whose walls are the minute sacculations called the *alveoli* (see Fig. 8.1, p. 352). A few alveoli also open from the walls of the alveolar ducts and respiratory bronchioles so that these vessels participate in the exchanges of gases with the blood, contrasting with the remaining bronchioles and bronchi, whose function is purely one of conduction. In man the two lungs contain some 750 million alveoli with a total surface area of some 50–80 square metres.

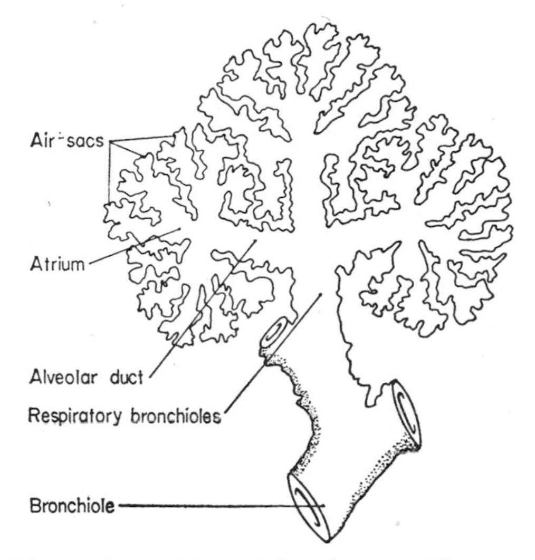

Fig. 8.11. The ultimate branching of the airways. The terminal bronchioles open into the respiratory bronchioles and these further divide into the alveolar ducts. Each alveolar duct in turn opens into several atria, the final passageway to the alveolar sacs on whose walls reside the minute sacculations, the alveoli. (Miller, "The Lung", Thomas, Springfield, Illinois, 1947.)

Cartilage

The walls of the conducting system are strengthened by the presence of cartilage, but this has disappeared by the eleventh order of branching, beginning with the trachea as zero, and the conducting tube becomes, by definition, a bronchiole. Collapse of the bronchiole is prevented by radial traction on it exerted by elastic connective tissues.

Mucus

The trachea and main conducting channels contain, in their epithelial lining, two types of mucus-secreting glands, namely surface goblet cells, reacting to surface irritation, and submucosal glands responding to nervous stimulation (vagus). About 100 ml of mucus are secreted daily by the healthy adult and this is slowly swept upwards along the "mucus escalator", driven by the beating of the cilia projecting from the epithelial cells, each ciliated cell having some 275 cilia which beat continuously in a well co-ordinated fashion in such a way as to propel a sheet of mucus towards the mouth at a rate of about 2 cm/min. Here the mucus with its entrapped particles, including cells such as leucocytes (Vol. 2), is removed by expectoration or swallowing. This protective ciliated epithelium extends as far as the terminal bronchioles.

Alveolus

The basic structure of the alveoli has already been briefly indicated (Figs. 8.1 and 8.2, pp. 352–3). They arise from the alveolar ducts and the alveolar sacs; they are thin-walled polyhedral formations lacking one side so that air diffuses freely from alveolar ducts into alveolar sacs and thus into the cavities of the alveoli. Within the alveolar walls is a dense capillary network anastomosing freely and arranged so that the greater portion of the capillary surface is exposed to alveolar air. The

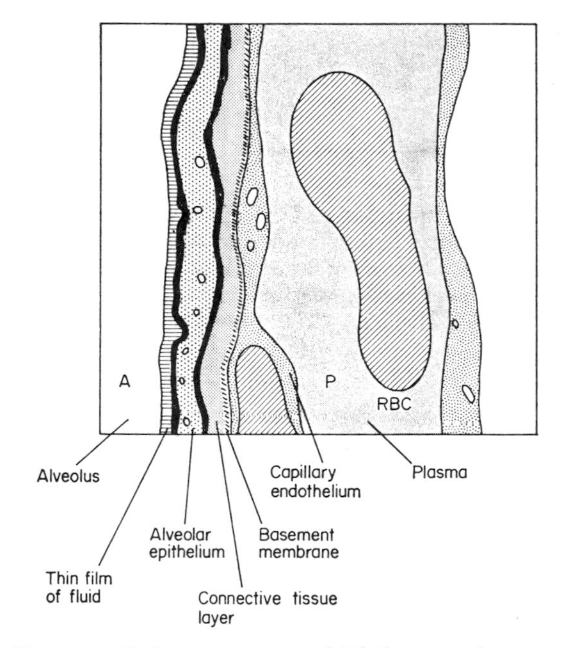

Fig. 8.12. A diagram of the structures which have to be traversed in going from alveolar air to the red cell. The respiratory gases first dissolve in the thin film of lung liquid, then diffuse through the alveolar epithelium and the capillary endothelium which are separated by a thin connective tissue layer. The final path is through the plasma and into the red cell.

walls contain a close meshed network of elastic and reticular fibres that provide structural rigidity to these thin-walled air vesicles. The surface of the alveolus facing the air is the squamous *pulmonary surface epithelium* only 0·1 to 0·5 μ thick; it is perforated in places by *alveolar pores of Kohn*, which may be caused by local desquamation of the epithelial cells; they provide openings from one alveolus to another adjacent one. The various layers separating air from the capillary blood are: (a) the thin film of material lining the alveolus, constituting what has been called the *surfactant system* of the lung (p. 377), (b) the alveolar epithelium, (c) the connective tissue layer, (d) the basement membrane of the blood capillary, and (e) the capillary endothelium (Fig. 8.12). The total thickness however is only about 0·7 μ. The pulmonary surface epithelium is continuous with the epithelium of the terminal bronchioles, whilst the superficial reticular layer of the alveolus is continuous with a similar reticular layer in the bronchiole.

Blood Supply. The alveolar capillaries are derived from small terminal arteries (not arterioles) which divide into capillaries at the level of respiratory bronchioles and alveoli. It is estimated that their total surface area in man is 45 m^2. The blood is collected in pulmonary venules from the alveolar walls, and also from the pleura and bronchial artery. As will be indicated later the supply of blood to the lungs falls into two categories: (i) that reaching the alveoli, and engaging in exchanges with the atmosphere—the *pulmonary circulation;* (ii) that supplying the nutritional requirements of the lung tissue—the *bronchial circulation*. The two systems anastomose by way of the capillaries at the level of the respiratory bronchioles.*

Cellular Types. The main cellular type in the alveolus is the attenuated alveolar epithelial cell (Type I); it is extremely thin, varying from 0·7 μ in the region of the nucleus down to 0·1–0·2 μ in the protoplasmic extensions. The cells rest on a basement membrane and, in areas where they come into close relation with capillaries, the two basement membranes become apposed. A second (Type II) cell occurs singly but is definitely a part of the epithelium; its characteristic feature is the intracytoplasmic *laminated body* which apparently contains the surfactant responsible for maintaining a low tension on the alveolar surface (p. 377). The remaining categories are concerned with protective functions, such as alveolar macrophages or scavenging cells that phagocytose particulate matter inhaled or derived from the blood; granulocytes, lymphocytes, and so on. Thus leucocytes may leave the

* The respiratory bronchioles, alveolar ducts and alveoli are supplied only by the pulmonary artery, whereas the rest of the pulmonary structures are supplied by way of the systemic circulation.

blood capillaries by diapedesis and penetrate the interstitial tissue of the alveolar septum; thence they may enter the alveolar spaces and be removed via the airway.

Expansion and Contraction of Thoracic Cage

The basic principle of aeration is the establishment of a negative pressure within the thorax by enlargement of its cavity. Because the only route for air to enter the thorax is through the trachea, it flows in through this channel and expands the elastic walled cavities,

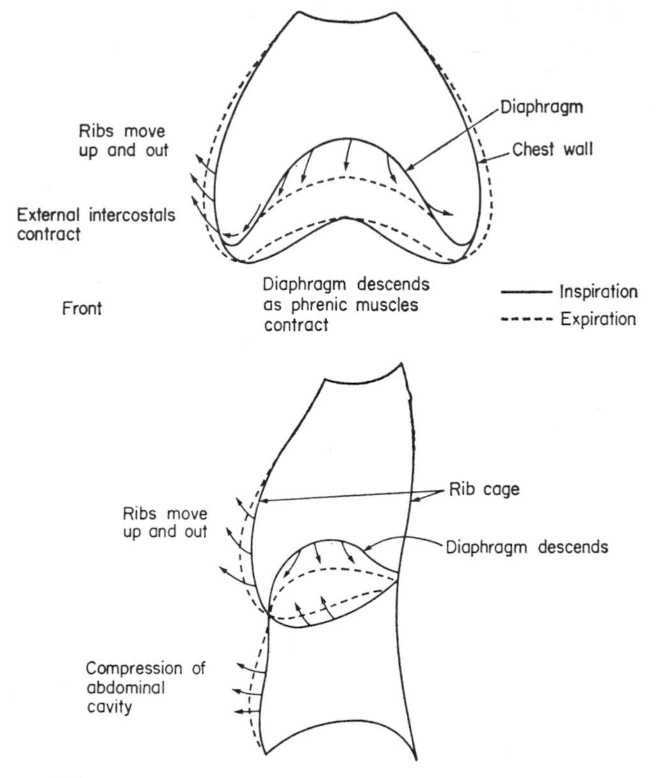

Fig. 8.13. The movements of the chest wall and diaphragm during respiration. The volume of the thorax is increased during inspiration by the upward and outward movement of the rib cage brought about by contraction of the external intercostal muscles. At the same time the diaphragm flattens from its dome-like shape. These processes establish a negative pressure within the thorax and atmospheric air is drawn in through the trachea. Expiration is usually a passive process, depending on elastic recoil forces to expel the air from the thorax.

thereby increasing the total volume of the lung. The thorax increases in volume by virtue of the upward movements of the ribs brought about by contraction of the external intercostal muscles and the downward movement of the *diaphragm*—the muscular membrane that separates the thoracic cage from the peritoneal cavity. At the end of the inspiratory movement the ribs fall to their previous position as a result of inhibition of the external intercostal muscles. The diaphragm moves up into the thoracic cage, and the lungs return to their pre-inspiratory volumes (Fig. 8.13). It must be emphasized that it is the thoracic cage that is forcibly enlarged in inspiration, and that the lungs enlarge passively by virtue of their distensibility, or *compliance* as it is called in this context. Normal expiration is brought about by the tendency for the ribs and diaphragm to return to their pre-inspiratory positions, aided by the elastic recoil of the lung tissue.

Pleurae

The lungs do not float freely in the thoracic cavity, but are attached by the *pleurae;* these are two membranes, continuous with each other—the *visceral pleura* closely investing the surface of the lung and reflected back on to the wall of the thorax to form the *parietal pleura* (see Fig. 8.14). Thus the two pleurae are normally in contact, and movements of the lungs are associated with a sliding of the visceral pleura over the parietal; the *intrapleural cavity*, therefore, is only a potential cavity, but may be enlarged pathologically.

Intrapleural Pressure

The pressure within this potential cavity may be measured by inserting a needle and connecting to a suitable manometric device; this is normally negative, indicating a tendency for the lung to collapse through its elastic properties. Such a collapse would tend to separate the pleurae and the negative pressure is a measure of this tendency. Hence if the needle were connected to the atmosphere instead of the manometer, air would flow into the intrapleural space until the pressure became atmospheric (pneumothorax).

Expansion of the thoracic cage obviously tends to separate the pleurae still further, and so the intrapleural pressure becomes more negative during inspiration. Thus in the expiratory position the pressure is some 5 mm Hg below atmospheric; in normal inspiration it is about 5–10 mm Hg below, and in maximal inspiration it may be 30 mm Hg below atmospheric, these being measures of the elastic

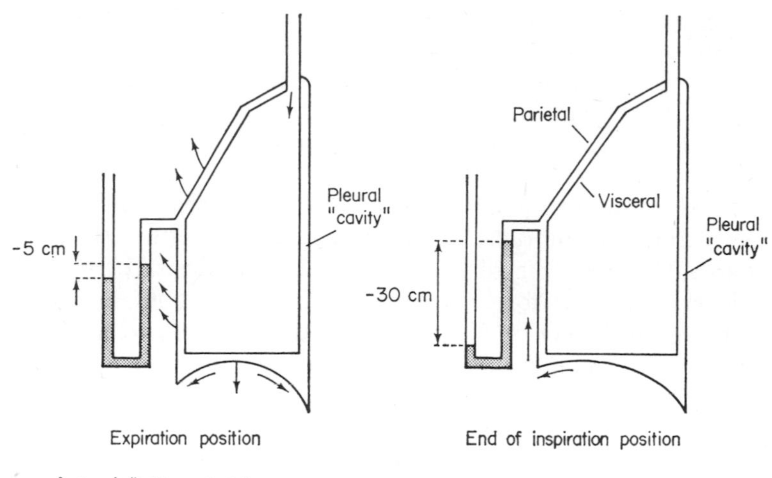

Fig. 8.14. A diagram of the pleural cavity of the chest and variation in intrapleural pressure during respiration. At rest the intrapleural pressure is negative due to the elastic forces in the lung and chest wall working in opposition. The lung is stretched at the position of rest while the chest wall is "compressed", as its resting volume is larger. During inspiration the intrapleural pressure becomes more negative as the volume of the thorax increases. The intrapleural cavity in reality is a "potential space" and in normal life there is only a thin film of fluid between the visceral and parietal pleurae.

recoil of the stretched lung (Fig. 8.14).* During forced expiration, when the volume of the chest contents is reduced by active contraction of the internal intercostal and abdominal muscles, the intrapleural pressure becomes positive.

THE FORCES INVOLVED IN RESPIRATION

The forces responsible for bringing air into the lungs are derived from the contractions of the muscles that increase the capacity of the bony cage, namely the external intercostal muscles and the diaphragm.

* This negative pressure is transmitted to all structures within the thorax. Thus, a cannula placed, say, between the lung and heart would register this negative intrathoracic pressure, and air from the atmosphere would enter here as well as into the intrapleural space. For the same reason, the pressures in the great veins of the chest are usually also negative. The oesophagus is likewise subjected to the intrathoracic pressure; its walls offer little resistance to deformation, so that the *intra-oesophageal pressure* is a good measure of intrathoracic, and thus of intrapleural pressure.

Compliances and Resistances

These forces are required to overcome three factors, or resistances, that oppose the inflow.

Airflow Resistance

First we have the viscous resistance to the flow of air through the system of tubes culminating in the respiratory bronchioles and alveolar ducts. This is similar to that which we have already discussed in relation to the flow of fluid through tubes and is given by Poiseuille's law if flow is laminar.

Elastic Resistance

Secondly we have the elastic resistance to expansion of the lung. Thus the lung contains a considerable amount of elastic tissue, and has a natural unstretched relaxation volume when the elastic elements are neither stretched nor compressed. For example the unstretched volume of the human lungs is about 1 litre, i.e. this is the volume they

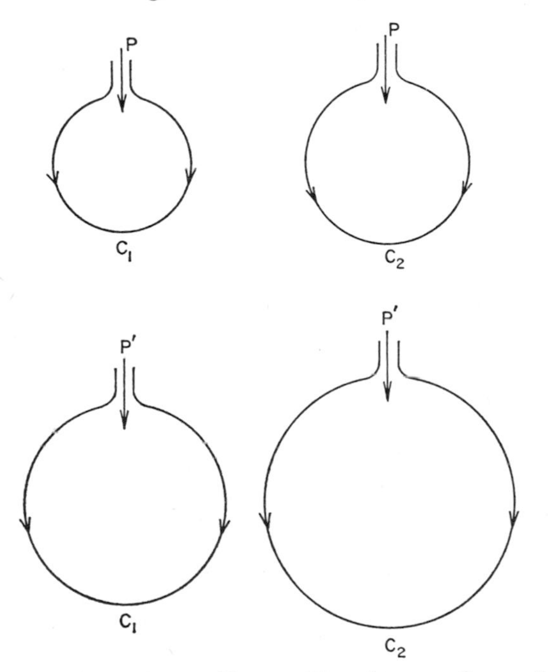

Fig. 8.15. The meaning of compliance. Two lungs, of compliance C_1 and C_2 are in a relatively deflated condition with a transmural pressure of P. The transmural pressure is raised to P′ and the lung with the larger compliance increases in volume to a greater extent.

would take up on removal from the thorax. At the end of expiration this volume is actually very much larger, namely 2·5 litres, so that the elastic tissue of the lungs *in situ* is always under tension, and tends to oppose any expansion, such as that involved in inspiration.

Compliance. Rather than speak of the resistance to expansion it is customary to speak of its reciprocal, namely its *compliance* which is given by $\Delta V/\Delta P$, i.e. the ratio of the expansion to the difference of pressure required to produce this (see Fig. 8.15, where two lungs of different compliance are represented as elastic balloons). The lung with the greater compliance expands to a greater extent when the same difference of pressure is applied to both.

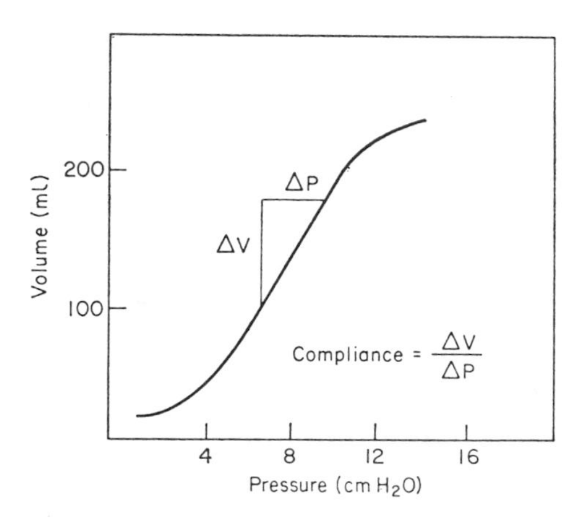

Fig. 8.16. The static compliance of the lung. The volume of the lung is plotted against intrathoracic pressure. The slope of the volume–pressure relationship of the lung is the compliance.

It will be clear that the compliance is a measure of the *distensibility* of the tissue, a term used to describe the ease of expansion of blood vessels (Chapter 5). In Fig. 8.16 the volume–pressure relationship of the lung is represented as a curve, the compliance over any given change of pressure being given by $\Delta V/\Delta P$.

Thoracic Resistance

Thirdly, the muscles and elastic tissues of the thoracic cavity likewise have a resistance to expansion, whose reciprocal is described as the *thoracic compliance.* It is interesting that the natural unstretched volume of the thorax is much larger than that of the lungs, i.e. 5 litres, so that at the end of expiration, when the volume of the contained lungs is

2·5 litres, the elastic tissues of the thorax are compressed. In this state the lungs are tending to *contract* under their elasticity and the thoracic cage is tending to *expand* under its elasticity (Fig. 8.17).

Pneumothorax. If, under these conditions, the muscles controlling the expansion or contraction of the thoracic cage are completely relaxed, the steady state is maintained by the mutual antagonism of these two elastic forces, the pull of the lungs being just balanced by the

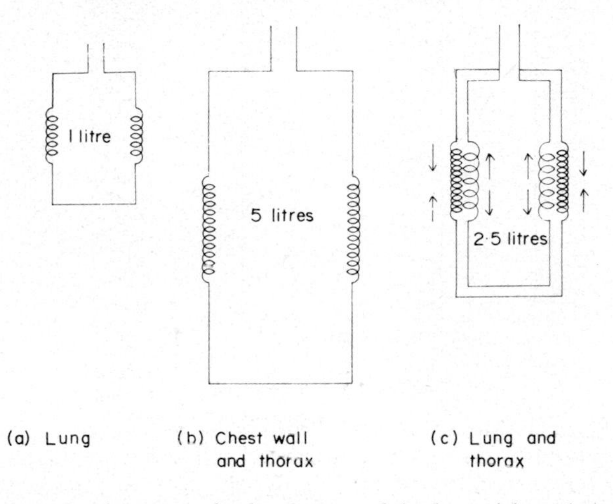

Fig. 8.17. The natural unstretched volumes of the lung (*a*) and the thorax (*b*) and the resting volume of the lung and thorax *in situ* (*c*). The unstretched volume of both lungs is approximately 1 litre and that of the thorax is 5 litres. At end-expiration the resting volume of the chest is some 2·5 L so that it can be seen that the lungs are stretched whereas the thorax is in effect compressed. The lung and thorax are held together by the integrity of the pleural cavity and pleural fluid which acts rather like water between two microscope slides, these can be moved easily over each other, but cannot easily be pulled apart.

expansive tendency of the thoracic cage. The "link" between these two forces is given by the pleurae, any expansion of the cage, or contraction of the lung, being opposed by the negative pressure, created by these tendencies. This balance is most clearly seen by the simple expedient of inserting a cannula connected to the atmosphere into the space between the pleurae (pneumothorax); air will, under these conditions, enter the intrapleural space until the two opposing tendencies are satisfied, the lungs contracting to their minimal volume and the thoracic cage expanding to its resting position.

Mechanical Analogy

The intrapleural pressures may be represented mechanically by two linked springs, as in Fig. 8.18. In (A) air has been admitted to allow equilibrium, and neither spring is under tension. In (B) we have end-expiration with muscles relaxed, and both springs stretched to produce equal and opposite forces. In (C) we have full inspiration,

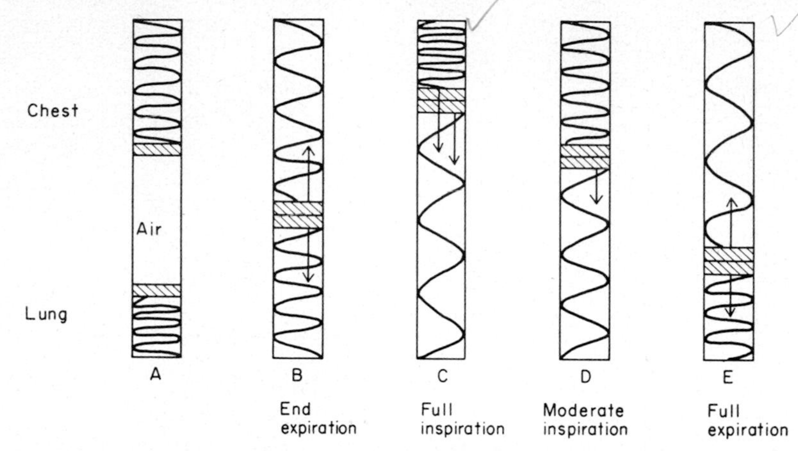

Fig. 8.18. A mechanical analogy of the forces acting in the chest wall and lung, represented as springs. In (A) chest wall and lung have assumed their unstretched states by allowing air to enter the pleural cavity. (B) Represents end-expiration; the lung spring is stretched a little and the chest wall spring is also stretched, both springs exerting equal and opposite forces. Both springs tend to separate the pleurae and contribute equally to the negative intrapleural pressure. (C) Represents full inspiration with the lung spring fully stretched and the chest wall spring compressed, so that the negative intrapleural pressure created by the lung spring is partly compensated by the positive pressure created by the tendency for the chest wall to collapse. In (D) the chest wall is at its resting volume so that its spring is relaxed; the negative intrapleural pressure is governed by the stretch of the lung spring. In full expiration, (E), the lung is still expanded above its resting state so that the lung spring is stretched a little; the chest spring is stretched considerably, so that both springs contribute to the negative intrapleural pressure.

with the lung spring greatly stretched and the thoracic spring compressed, so that both recoils are working in the same direction. The negative pressure developed by the recoil of the lung is partially balanced by the positive pressure that the recoil of the thorax tends to develop. In (D) a moderate inspiration is indicated, with the thorax now in its relaxed condition but the lung spring extended, so that the negative pressure is governed entirely by the lung spring. Finally, in

(E) we have full expiration; the lungs are still expanded so that the lung spring tends to lower the intrapleural pressure, while the thoracic spring is expanded, its recoil likewise tending to lower intrapleural pressure.

Measurement of Airway Resistance

We may apply Poiseuille's law to the flow of air through the airways and define the airway resistance by the equation:

$$\text{Airway resistance} = \Delta P_a / Q$$

where ΔP_a is the difference in pressure between the mouth and the alveoli, and Q is the minute–volume of air. To measure the airway

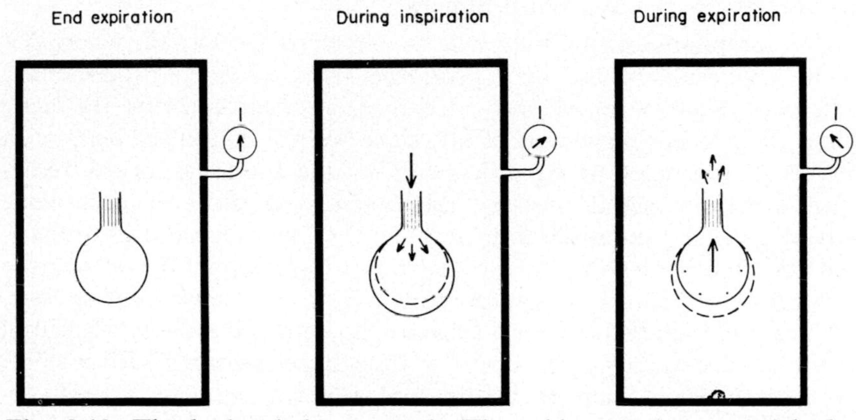

Fig. 8.19. The body plethysmograph. The subject sits in an airtight box and during inspiration the pressure rises in the box by a small amount. This small rise in pressure can be recorded by sensitive transducers, converted into a volume change by suitable calibration, and used to calculate alveolar pressure. By the measurement of the volume of air inspired, using a flow-meter, it is then possible to measure airways resistance. (Comroe *et al.*, "The Lung", Year Book Med. Pub., 1962.)

resistance, then, we must determine the rate of flow into the lungs at a given moment, together with the difference in pressure between the alveoli and the mouth, or atmosphere. During inspiration the pressure in the alveoli is greater than the intrapleural pressure, since some of the intrapleural negativity is required to expand the lungs; hence the alveolar pressure requires a more direct measurement. This is done by the use of an air-tight body-plethysmograph, as indicated in Fig. 8.19. The subject, whose lung only is pictured, is seated within an air-tight box and pressure is measured continuously in this box. During inspiration the pressure rises; this is in spite of the fact that the air inspired has been taken out of the box to fill the expanding lung, and this is because during its stay in the lung it has occupied a rather larger volume than it would have occupied in the box, due to the fact that it is at a slightly lower pressure while in the lungs. This change in pressure is minute but measurable accurately, and by suitable calibration can be converted into a volume representing the change due to the difference between the pressure in the box and the subject's alveolar pressure. From this it is easy to calculate the actual alveolar pressure; the volume of air inspired is measured by a suitable flow-meter and the ratio: Pressure difference/Flow gives the airway resistance.

Measurement of Lung Compliances

The compliance of the lung will be measured by $\Delta V/\Delta P$, where ΔV is the expansion of the lung produced by a given transpulmonary difference of pressure, ΔP, provided that the measurements are made when there is no movement of air, since we are concerned only with the elastic resistance to expansion and not the frictional airway resistance. The difference of pressure, ΔP, when there is no air movement, will be given by the difference between the intrapleural pressure, I_{Pl}, and the mouth-pressure, at the end of an inspiration. If we take the pressure in the mouth as atmospheric, then the *transpulmonary pressure*, ΔP, is equal to the intrapleural pressure, I_{Pl}, and this may be measured as the intraoesophageal pressure by inserting a catheter with a small partly inflated balloon at its tip and connecting it to a pressure-transducer.

The subject rebreathes into a spirometer (described later) so that the volume of air brought into the lung, ΔV, is measured, and at the end of an inspiration the pressure is recorded. By taking progressively larger inspirations, a curve of ΔV against ΔP, as in Fig. 8.16, is obtained, the slope over the linear region being a measure of the compliance.

Lung Plus Thorax Compliance

A simple way of measuring the total compliance of lungs plus thorax is to exert a positive pressure on a resting subject, as in an iron lung, and measure the change in volume. Alternatively the method, employing the spirometer used for lung compliance, may be modified. The subject inspires to produce a given degree of inflation, ΔV, as before; the tube connecting his mouth to the spirometer is clamped, and the subject is told to relax his muscles of inspiration. Under these conditions the intrapleural pressure represents the balance between elastic recoil tendencies of lung and thorax. Hence the values of ΔV and intrapleural pressure under these conditions give us the parameters for calculating the combined compliance:

$$C_{T+L} = \Delta V/(\text{Intrapleural pressure} - \text{Atmospheric pressure})$$

The combined compliance of two linked springs is related to the individual compliances by:

$$1/C = 1/C_1 + 1/C_2$$

Thus $$1/C_{T+L} = 1/C_L + 1/C_T$$

so that $1/C_T$ may now be calculated. The average value of lung compliance is 0·20 litres/cm H_2O whilst the combined compliance is 0·13 litres/cm H_2O, whence the compliance of the chest-wall is 0·37 litres/cm H_2O.

Surface Tension Forces

Van Neergard in 1929 pointed out that a part of the resistance to expansion of the lung would be the forces required to expand the surfaces of the alveoli, i.e. the forces required to overcome the surface tension at the alveolus–air interface. This interface would be essentially a water–air interface and, if so, would have a surface tension of some 72 dynes/cm. Thus the surface tension is a measure of the tendency for a surface to contract, or, put the other way, it is a measure of the force necessary to increase the surface area. Van Neergard concluded that some two-thirds to three-quarters of the resistance to expansion of the lung was derived from interfacial forces, and suggested that, in view of the high surface tension of an air–water interface, there might be a surface-active material on the surface of the alveolus that reduced this tension.

Saline-filled Lung

That interfacial tension is, indeed, an important factor in determining the resistance to expansion (or its reciprocal, the compliance) is shown by the simple experiment of measuring the pressure–volume relationship in lungs filled with saline and comparing the same relationship when they are filled with air. Figure 8.20 shows the result; with saline, a much smaller pressure is required to expand the lung than with air. Moreover, the relation between pressure and volume is simpler, with little or no hysteresis, in the sense that the same volume is reached whether the pressure has been brought from a low

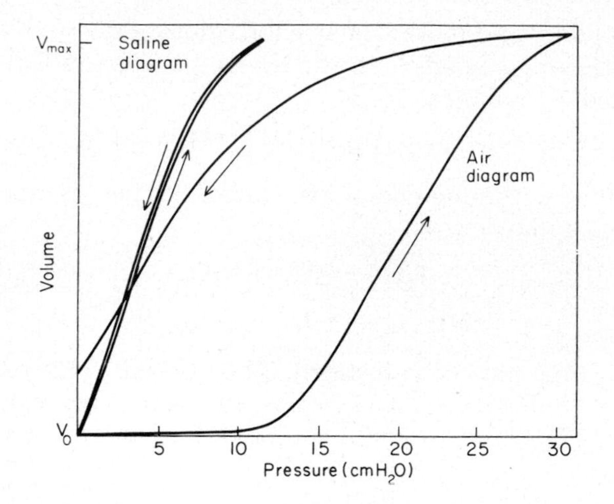

Fig. 8.20. The volume pressure diagrams for air-filled and saline-filled lungs. Inflation is from the degassed state V_0 to the maximum volume V_{max}. It is easy to see that much more pressure is required to inflate the air-filled lung than the same lung when it is filled with saline. The extra force expended is needed to overcome the surface tension in the alveoli. Note there is little hysteresis in the saline-filled lung when compared with the air-filled trace. (Scarpelli *et al.*, "The Surfactant System of the Lung", Lea and Febiger, Philadelphia.)

value to a higher one, or *vice versa*, whereas with air there is a pronounced difference in the curves according to whether the pressure is rising or falling.

Alveolar Surface Tension

From the difference between the pressures required to fill the lungs with saline and with air, an estimate of the surface tension at the alveolar air–water interface can be made. It is clear that this is lower than that of an air–water interface and, moreover, becomes smaller as the lungs are expanded.

Bubbles from Lung

Such a reduction in surface tension could be brought about by the presence of a thin film of a surface-active material, e.g. a phospholipid. Evidence for this was provided by Pattle's observation that air-bubbles squeezed out of a lung were remarkably stable. Thus an air-bubble in water is unstable for the simple reason that the pressure within it is higher than atmospheric—the difference in pressure, P being given by the Laplace formula:

$$P = 2T/r$$

being twice the surface tension, T, divided by the radius, *r*. The smaller the bubble the greater the pressure. Figure 8.21 illustrates a bubble in water which has equilibrated with the atmosphere so that the pressure of gases in the water is the same as that in the atmosphere, P_{ATM}. The pressure in the bubble, P_O, is greater than this, by the

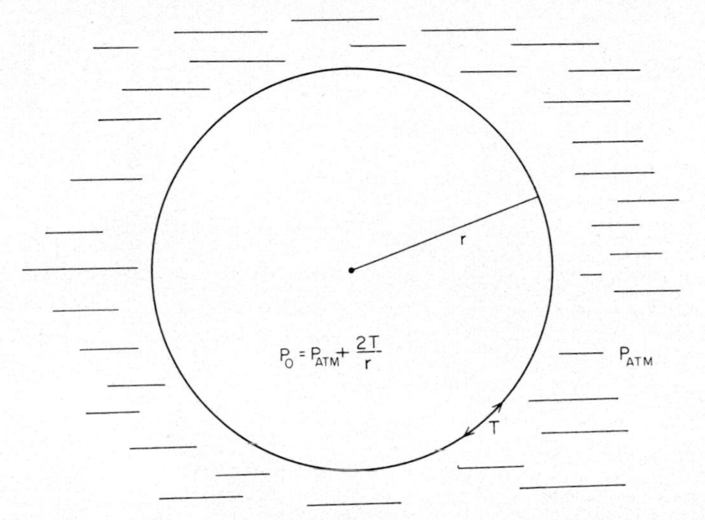

Fig. 8.21. The instability of a bubble of gas in water. The water surrounding the bubble has equilibrated with atmospheric pressure and is at P_{ATM}. The pressure in the bubble is greater than this by $2T/r$ due to the surface tension forces. Air diffuses out of the bubble which shrinks at a gradually increasing rate as r decreases and hence T/r becomes large.

factor $2T/r$, so that the gases diffuse out of the bubble into the surrounding water and thence into the atmosphere. As the gases escape from the bubble the pressure increases, since r decreases, so that the bubble shrinks at an accelerating pace. If T were very low the bubble would be much more stable, in the sense that it would diminish in size slowly; and this is found with the minute bubbles expressed from a lung.

Surfactant

Subsequent work in which the lungs have been extracted directly, or in which the foamy fluid obtained by washing out the lung system with saline has been analysed, has demonstrated the presence of surface-active material, called *surfactant*, which, when added to water, causes a remarkable lowering of its interfacial tension with air. The major constituent of the material, which is predominantly phospholipid, is dipalmityl phosphatidyl choline (dipalmityl lecithin). The Type II cell of the alveolar epithelium is almost certainly responsible

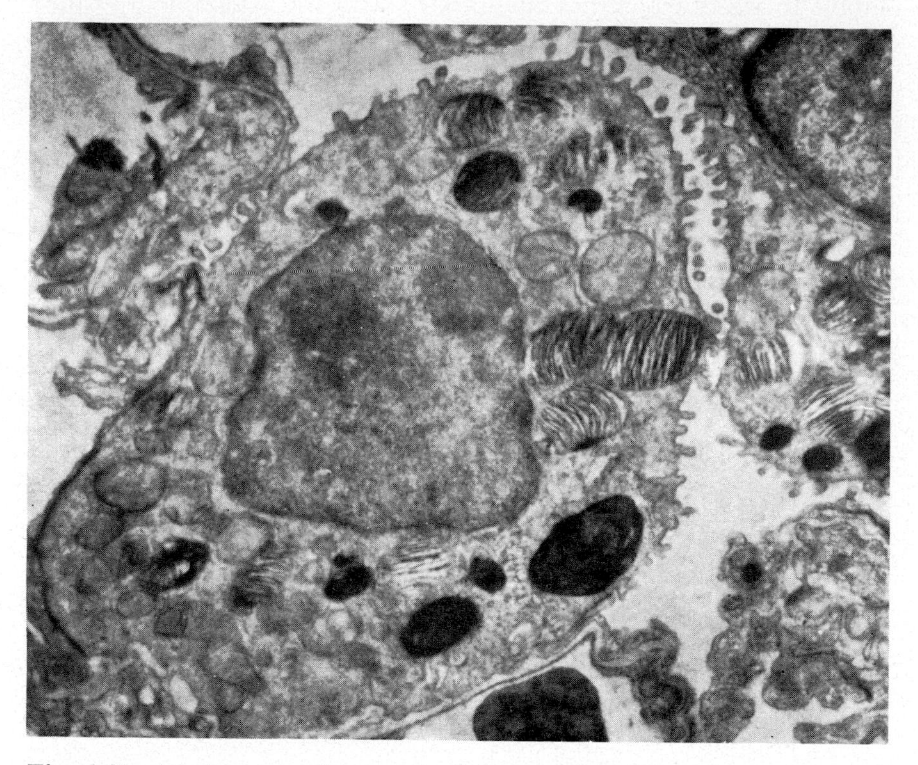

Fig. 8.22. An electron micrograph of the type II cell of the alveolar epithelium, which is thought to be responsible for the secretion of lung surfactant. The cells contain characteristic oval lamellar bodies limited by a unit membrane lining the alveoli. (Hatasa and Nakamura, *Z. Zellforsch.* 1965, **68,** 266.)

for secreting the lipid material. As illustrated by Fig. 8.22, the cell contains characteristic oval lamellar bodies ranging in size from 0·3–0·5 μ to 1·0–1·5 μ, the lamellae consisting of osmiophilic material. They are limited by a typical unit membrane, and in the electron microscope can be seen emptying their contents into the air-space of the alveolus by a process of exocytosis (Fig. 8.23). When the alveolar tissue is fixed by perfusion through the pulmonary circulation, a layer

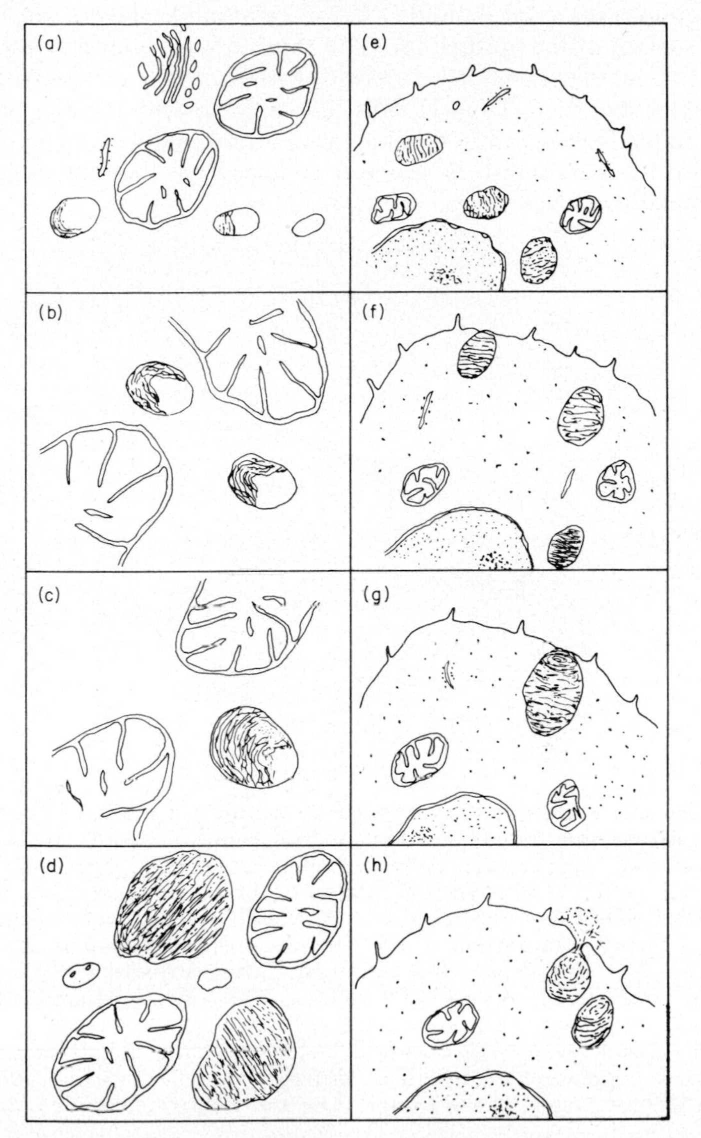

Fig. 8.23. The stages of development of the lamellar bodies of the type II cells of the alveolar epithelium. (*a*) to (*d*): stages of development of the lamellar bodies; (*e*) to (*g*): fusion of the lamellar body with the cell wall; (*h*): the discharge of the contents of the lamellar body into the fluid lining the alveolar cavity—the process of exocytosis. (Hatasa and Nakamura, *Z. Zellforsch*, 1965, **68,** 266.)

of amorphous material, bounded by an osmiophilic layer, can be seen on the surface of the epithelium. The thyroid secretions are important in lipid metabolism and it is interesting that, when rats were treated with thyroxine, their Type II cells increased in size and the numbers of lamellated bodies within them increased. Parallel with these changes there was an increase in the amount of surfactant material that could be extracted from the lungs.

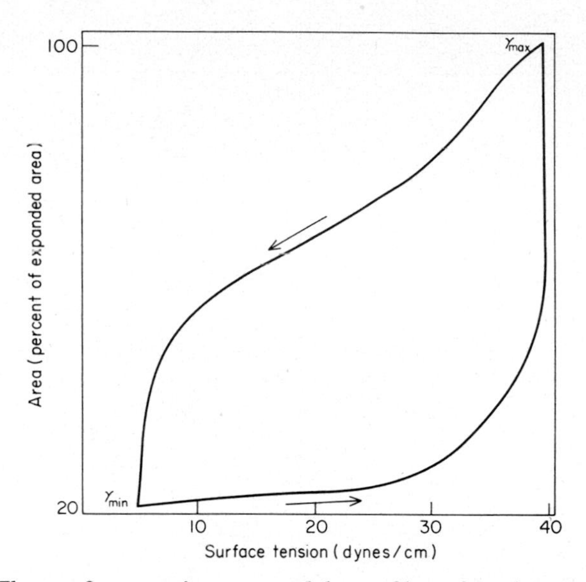

Fig. 8.24. The surface tension exerted by a film of surfactant on water is plotted against the area of the film as it is allowed to expand. It will be seen that during expansion from a highly compressed state the surface tension rises steeply, but when a tension of about 30/dynes/cm is reached a large degree of expansion takes place with only a small increase in surface tension. During compression of the film the surface tension increases but the curve does not follow that obtained during expansion, i.e. there is hysteresis reminiscent of the hysteresis in the lung expansion curve of **Fig. 8.20.** The character of the surfactant that permits the surface tension to decrease with decreasing area is a factor favouring the stable existence of a mixed population of alveoli of different sizes. (Scarpelli *et el.*, "The Surfactant System of the Lung", Lea and Febiger, Philadelphia.)

By virtue of this layer of surfactant on the alveolar epithelium, the force required to expand the surfaces of the alveoli is reduced, and the compliance of the lung has a practicable value. The absence of this film, as in certain conditions of the newborn, leads to respiratory distress, the forces required to expand the lung being abnormally large.

Replacement of Surfactant. The "wear and tear" of this surfactant material means that there must be continual replacement, and it can be shown by labelling the phospholipids with ^{14}C that they are indeed rapidly synthesized within the lung. In general, the replacement-time for dipalmityl lecithin in rat lung was computed to be some 17·5 h.

Surface Area and Tension. An important feature of the surfactant is that the interfacial tension between water and air is not only lowered by its presence but also this surface tension decreases as the surface becomes smaller, and increases when it is expanded. This is illustrated by Fig. 8.24 which shows how, when the surface film of surfactant is

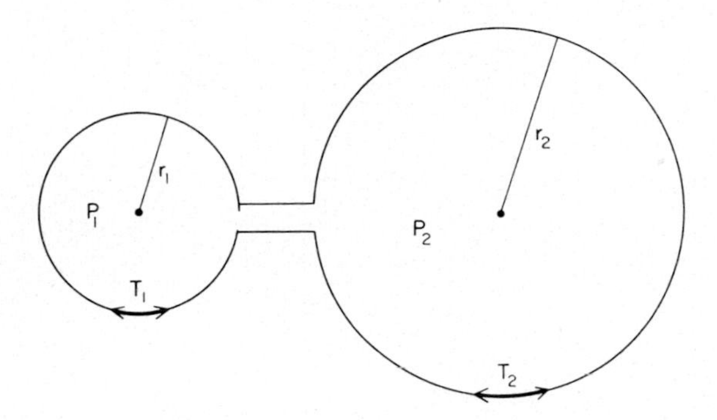

Fig. 8.25. The diagram shows two bubbles in communication with each other; in order that the system be stable the pressures within them must be equal, i.e. $P_1 = P_2$. If the surface tensions are equal this cannot be true because the pressure is given by $2T/r$, so that the smaller bubble has the higher pressure and it will be absorbed by the larger one. For stability, then, T_1 must be less than T_2.

allowed to expand, the surface tension at first rises rapidly, and only slowly increases further during the rest of the expansion. On compressing the film the surface tension falls, but more gradually, so that the cycle exhibits hysteresis, the surface tension reached by an expanding film being less than that reached on compression.

The importance to the lung of the variation of surface tension with area of the exposed film concerns the ability of adjacent alveoli of different size to remain open.

Thus we have seen that the smaller a bubble, other things being equal, the greater the pressure within it; if the alveolus can be considered as a bubble, then a small one will tend to collapse, giving up its air to a neighbouring larger one. However, if the larger one had the higher surface tension, as in Fig. 8.25, we could envisage a condition where both were stable. Thus we have:

$$P_1 = T_1/r_1 \text{ and } P_2 = T_2/r_2.$$

If P_1 and P_2 are to remain the same, the tensions must be related as the reciprocals of the radii; and it is apparently this characteristic of the surfactant, namely its lowering the surface tension more at low surface-areas than at high areas, that permits this adjustment of surface tension to lung volume.

Hysteresis. The hysteresis in the response of surface tension to expansion and compression of the surface has a counterpart in the

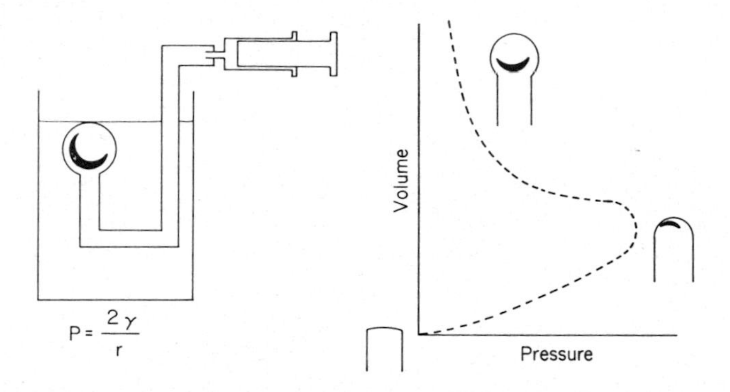

Fig. 8.26. A plot of the force required to inflate a bubble from a flat film. As can be seen from the volume-pressure plot, more force must be exerted to make the bubble spherical and then further increases in volume are made with decreasing amounts of pressure, showing "negative" compliance. (Clements, *Physiologist*, 1962, **5,** 11.)

pressure–volume relation of the lung. Thus it has been recognized for a long time that, when the pressure required to expand the lung is plotted against the volume, for a given volume the pressure is much higher when the lung is being compressed than when it is being expanded. This is shown in Fig. 8.20 (p. 376) and the analogy with the surface-tension area curves is striking. That the hysteresis depends on the presence of surfactant is clear from the curves obtained when the lungs are filled with saline. It will take us beyond the scope of this book to analyse in detail the special shape of the pressure–volume curve of the lung (as Fig. 8.20) and its hysteresis. It is likely that, as expansion occurs, there is *recruitment* of more and more alveoli, rather than the gradual expansion of all the alveoli, together; if it were not for this recruitment we might expect a simple linear relation between expansion and pressure.

An additional factor is the change in force required to expand a bubble, for example a soap-bubble blown from a pipe as in Fig. 8.26. The pressure–volume relation for such a process is illustrated at the right, and this would help to explain how such large increases in volume are attained with small increases in pressure at the high-pressure end of the curve.

The hysteresis reflects the fact that it is harder to open an alveolus than to close one over the low-pressure range; thus over the range zero to 5 cm H_2O there is only a small increase in lung volume, but over the same range during collapse of the lung there is a very large change in volume. In the high range of pressures the reverse holds, so that over the range of, say, 20–25 cm H_2O the change in volume during expansion is considerably larger than during contraction.

Elastic Tension in Alveoli. As we have seen, such a hysteresis does not occur in the fluid-filled lung, so that it is reasonable to seek the cause in the hysteresis in the surface-tension area curves of Fig. 8.24. It must be emphasized, however, that the alveoli are not mere soap-bubbles but have an inherent elastic resistance to expansion, so that this must enter into the final picture. The really important feature of surfactant is the reduction in the force required to expand the lung through lowering the surface-tension that resists expansion of a liquid surface.

THE RENEWAL OF THE GASES

Respiratory Capacities and Volumes

The total capacity of the lungs of a normal young adult is about 6000 ml, and may be indicated by the height of the block (TLC) on the extreme left of Fig. 8.27. Under resting conditions, even at the end of expiration, the lungs contain a considerable volume of air which may be called the *functional residual capacity* (*FRC*). It is found, moreover, that even after a maximal expiratory effort not all this air can be expelled, so that there is a fundamental *residual volume* that cannot participate directly in the respiratory movements. Thus in the right-hand column we may indicate the air normally breathed in and out (the *tidal air*), by the small block labelled TV. The air expelled maximally after a normal tidal expiration is marked ERV, standing for *expiratory reserve volume*. Just as we can increase expiration above the tidal position, so we can increase inspiration, the extra volume of air brought in during maximal inspiration being represented by the block marked

IRV, the *inspiratory reserve volume.* To the right of the blocks there is a graph indicating the resting tidal air, which is increased during maximal inspiration and expiration; the total tidal air under these maximal conditions is called the *vital capacity*, being the total volume of air that can be expelled from the lungs by forceful effort after maximal inspiration.

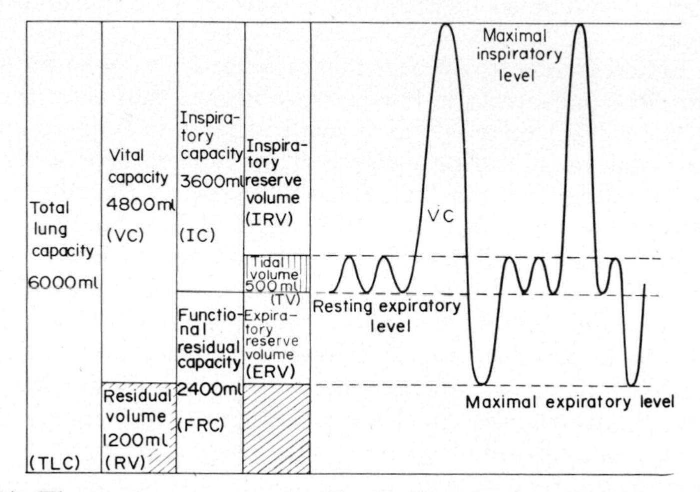

Fig. 8.27. The respiratory volumes. The total lung volume is some 6 L which can be divided into the *vital capacity* of 4·8 L, which can be expired, and the *residual volume* of 1·2 L which cannot be expired. The *inspiratory capacity* is about 3.6 L and this leaves 2·9 L of *inspiratory reserve volume* with quiet respiration at a *tidal volume* of 500 ml. The *functional residual capacity* is divided into the *expiratory reserve volume* of 2.4 L and the *residual volume.*

Dead Space

The functional reserve capacity, which amounts to some 2400 ml, may be looked on as the normal reservoir of air in the lungs, which is diluted by successive inspirations of tidal air; since this new air, about 500 ml in volume, will mix with some 2400 ml of the functional reserve, this means that a single inspiration and expiration will produce only a relatively small change in composition; moreover, of the 500 ml of tidal air, some 150 ml are represented by the volume of the conducting tubes, the *dead space*, and this dead-space air does not contribute to dilution of the air already in the alveoli.

Buffer Action of FRC

Thus the effective ventilation of a single inspiration and expiration only amounts to that caused by the introduction of 350 ml of atmospheric air into a volume of 2400 ml. This single breath would only reduce the tension of CO_2 by a factor of about 15 per cent; as a result of this mechanism of only partial dilution with atmospheric air, the air in the alveoli maintains a composition considerably different from that of atmospheric air (Table II). We may consider the FRC, then, as a buffer mechanism that prevents violent changes in composition of the air to which the alveolar blood is exposed at any moment.

TABLE II

Compositions of Atmospheric, Expired and Alveolar Airs (partial pressures in mmHg)

	I Atmospheric		II Expired	III Alveolar
	Dry	Satd. H_2O at 37°C		
Oxygen	159·2	149	117	99
Carbon dioxide	0·3	0·285	29	40
Nitrogen	600·5	563·7	567	574
Water	0·0	47	47	47
Total	**760**	**760**	**760**	**760**

This buffer mechanism must be regarded as an important homeostatic adaptation, since alveolar air comes rapidly into equilibrium with the arterial blood, and if the composition of this gas varied widely in composition the arterial pH and gas tensions would likewise fluctuate. It is interesting that fish do not have this buffer mechanism. so that their blood gases reflect the composition of the water passing over their gills.

Definitions

It will be seen from Fig. 8.27 that certain volumes of air are described as "volumes" and others as "capacities". Conventionally the term capacity is reserved for a compartment that is measured by tests of pulmonary function. Thus *inspiratory capacity* is the maximum volume of air that can be inhaled from midposition, and comprises tidal volume plus inspiratory reserve volume. *Functional residual capacity* is the volume of gas remaining in the lungs at the end of a spontaneous expiration when lungs and chest are in midposition. *Vital capacity* is the maximum volume of gas that can be exhaled after a maximal inspiration.

Measurement

Spirometer

Volumes of air inspired or expired may be conveniently measured with the spirometer illustrated in Fig. 8.28. It consists of a cylinder sealed at one end, floating on a water seal. The subject breathes in and out of the tube which passes up inside the apparatus; the gas cylinder is counterbalanced, so that respiratory movements are recorded by an ink-pen on a variable speed kymograph. The system is usually so calibrated that graduations ruled on the recording paper directly correspond to fractions of a litre, and volume changes can easily be ascertained. Both static and

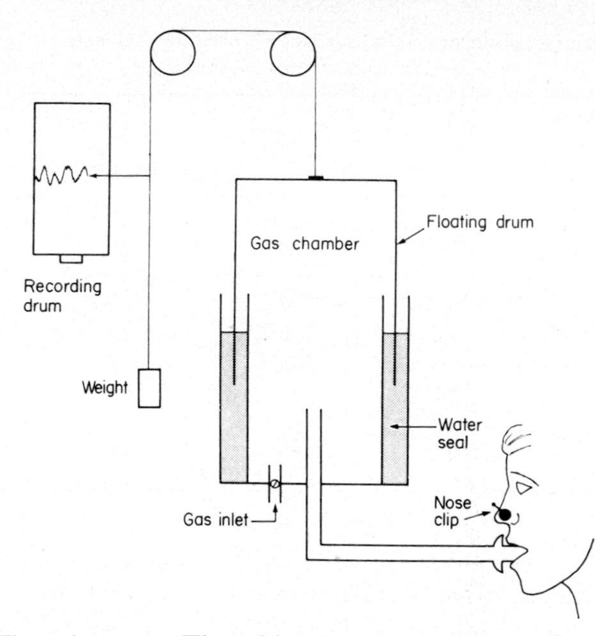

Fig. 8.28. The spirometer. The subject breathes into a gas chamber floating on a water seal as shown. The gas chamber is counterbalanced by a weight and has a recording pen to measure the excursions of the chamber.

dynamic volume changes can be measured with this apparatus and by filling the bell with oxygen, and including a CO_2 absorber in the circuit, measurements of the respiratory quotient can be made.

Gas Dilution

To estimate volumes that contain the fraction of the air that cannot be expelled from the lungs, such as the functional residual capacity, a less direct method is necessary. In the closed-circuit method, the dilution of a foreign gas inhaled into the lungs is measured; the subject breathes into a sealed system containing an exactly known volume, V, of air plus a gas such as helium at concentration $[He]_1$; after a time a steady concentration of helium $[He]_2$ remains in the system, and from the initial and final concentrations $[He]_1$ and $[He]_2$ and the volume of the system, the volume of air in the lungs, V_L, can be calculated using the empirical formula:

$$V_L = \left(V \times \frac{[He]_1}{[He]_2}\right) - V + C$$

C being a correction factor to allow for changes in volume due to various respiratory alterations in the composition of the air, such as N_2 excretion and loss of helium to the blood.

THE GAS TENSIONS

Expired Air

By breathing into a bag with an appropriate valve so that only the air expelled from the lung passes into the bag, the *expired air* may be collected and analysed. The typical composition for the resting state is shown in Table II (Column II) where that of the atmospheric air (Column I) is included. The rise in tension of CO_2 and fall in that of O_2 are obvious.

Alveolar Air

Because of the dead-space air, however, the composition of the expired air is different from that prevailing in the alveoli. Thus the conducting passages at the beginning of expiration contain atmospheric air that has not been exposed to the blood, and this will be the first air to enter the collecting bag with each expiration; consequently the tension of O_2 will be higher than that of the true alveolar air, and the tension of CO_2 will be less. If a normal expiration is made into a long wide-bore tube (Haldane tube) and a sample of gas is withdrawn from a side-arm close to the mouth, a good representative sample of alveolar air may be obtained, and its composition is shown in Table II (Column III).

Equilibration of Blood

At first thought it might be considered that oxygenation of blood would not be complete even under resting conditions; consultation of the O_2-dissociation curve of blood, however, shows that a tension of 100 mm Hg is more than adequate for complete oxygenation. Similarly, the extra CO_2 taken up from the tissues can be removed from the blood as a result of the difference of tension in the two phases (46 mm Hg in venous blood compared with 40 mm Hg in alveolar air), and the final result is that the blood returning from the lung has almost the same tensions of O_2 and CO_2 as in alveolar air.

Respiratory Alkalosis and Acidosis. Quite clearly, by forcibly inspiring and expiring a larger tidal volume than physiologically necessary, the composition of the alveolar air can be altered, the hyperventilation causing a lowered tension of CO_2 and a raised tension of O_2. The tension of CO_2 in the returning blood will be decreased leading to a *respiratory alkalosis* with the blood becoming more alkaline than normal; conversely a reduced ventilation would cause a *respiratory acidosis* and, if the O_2 tension fell sufficiently, a respiratory hypoxaemia. (Although hyperventilation increases alveolar P_{O_2} this is not reflected in a changed oxygenation of the blood since it is already fully oxygenated during normal ventilation.)

Ventilation and Respiratory Rate

The degree of ventilation of the lungs quite obviously depends on the volume of the tidal air and the respiratory rate. Thus with an average rate of 12/min and a tidal air of 500 ml, the total amount of air brought into and out of the lung per minute is 6000 ml. The *effective* ventilation must take account of the dead-space air, however, so that the true ventilation is given by $12 \times 350 = 4200$ ml/min. If the rate of breathing were increased, but the total minute–volume remained the same, the efficiency of ventilation would be decreased precisely because of the dead space. Thus if the minute–volume of 6000 ml were now made up of 24 inspirations of 250 ml, the effective ventilation would only be $24 \times (250 - 150) = 2400$ ml/min. Hence shallow breathing is inefficient from the point of view of ventilation. However, where breathing is employed in thermoregulation, as in the panting of the dog, then shallow breathing is necessary if the lungs are not to be hyperventilated.

Exercise

In exercise, the ventilation increases almost linearly with the increased oxygen consumption (Fig. 8.29); and this is achieved by both increase in tidal volume and respiratory rate. Of most interest is the alveolar P_{CO_2} which, after increasing slightly as the rate of work increased, tended to decline indicating an extremely efficient ventilation process that not only keeps pace with the larger volumes of gases to be removed but actually exceeds the requirements, so that P_{CO_2} is less than at rest. The P_{O_2} of the blood returning from the lungs, even in severe exercise, is not significantly reduced, so once again we have adequate ventilation of the lungs permitting the oxygenation of the greatly increased minute–volume of blood flowing through the pulmonary vessels.

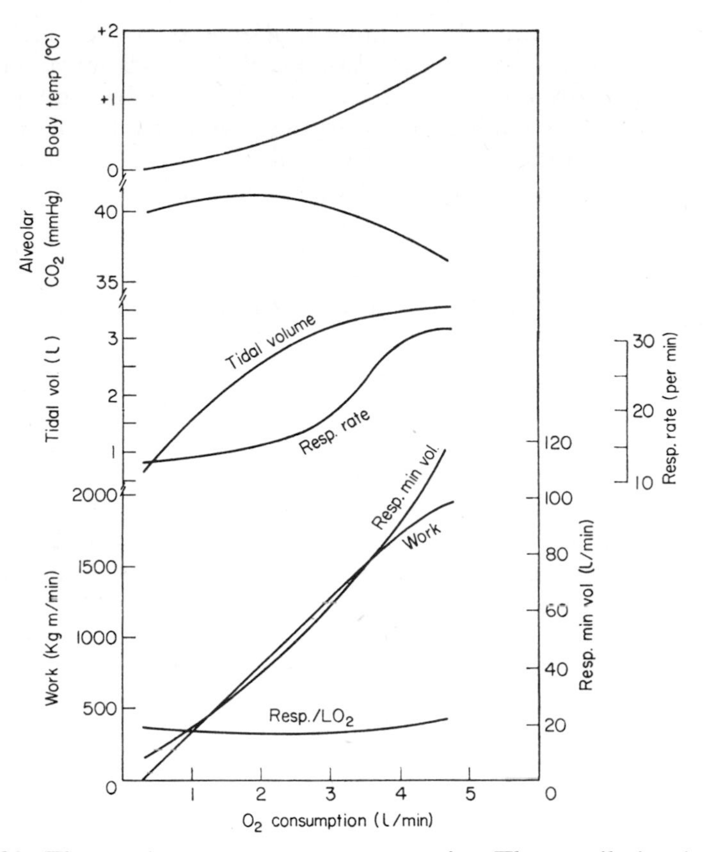

Fig. 8.29. The respiratory responses to exercise. The ventilation increases almost linearly with the work rate; the alveolar P_{CO} although increasing at first tends to fall at the higher rates of respiration. The respiratory rate and tidal volume both increase to enable the large increase in ventilation to be made. (Christensen, *Arbeits Physiologie*, 1932, **5**, 463.)

Re-breathing Air

The control mechanisms concerned in adaptation of the respiration and circulation to the increased demands of severe exercise will be discussed in Chapter 3, Volume 3. For the moment we may note that the primary influence on the rate of respiration is exerted through the tension of CO_2 in the blood supplying the brain; thus exercise increases the production of CO_2 by the muscles of the body, and the increased P_{CO_2} acts as a powerful stimulus to ventilation of the lungs. This is illustrated by the simple experiment of re-breathing air; as the P_{CO_2} in the inspired air increases, ventilation increases through increased depth and increased rate. If the same experiment is performed but the CO_2 produced is absorbed with soda-lime, then the stimulus to breathing can only be through oxygen-lack. This stimulates breathing but the effects are more complex because the hyperventilation tends to remove CO_2 from the blood and so produce a "hypocapnic stimulus" that competes with the "hypoxic stimulus" (Vol. 3).

THE PULMONARY VASCULAR CIRCULATION

Pressures

The circulation of blood through the lungs is brought about at an altogether lower range of pressures than those in the systemic circulation. This is illustrated by the pulses recorded from different regions (Fig. 8.30), and it will be seen that the pressure in the right ventricle reaches a maximum of only 25 mm Hg in systole, whilst the pressures in the pulmonary veins are of the order of 5 mm Hg. Although the capillary pressures were not measured, it will be clear that they will not be greater than arterial pressure and only a little greater than venous pressure. This means that normally the pressure available for filtration of blood plasma is not high enough, so we may expect, in general, a tendency for the capillaries to absorb fluid rather than to give it up to the tissue. When the pulmonary venous pressure rises, as in failure of the left ventricle, a condition of pulmonary oedema can develop, due to the raised capillary pressure.

Bronchial System

The blood involved in the ventilation process passes to the capillaries in the alveoli. A separate *bronchial system* of vessels is required for the nutrition of the lung tissue, and this is provided by bronchial arteries branching from the thoracic aorta. These arteries follow the ramifications of the bronchial tree, anastomosing to form plexuses from which ultimately capillaries are derived to supply the bronchial tissue as far as the respiratory bronchioles. Some of this blood is drained by way of broncho-pulmonary veins into the pulmonary vein, so that some systemic blood actually mixes with the pulmonary venous blood. In addition, some capillaries in the respiratory bronchioles anastomose with alveolar capillaries, giving a further occasion for mixing of the two types of blood. This mixing is indicated in Fig. 8.30 by the shunt. The admixture of systemic blood with the pulmonary venous blood

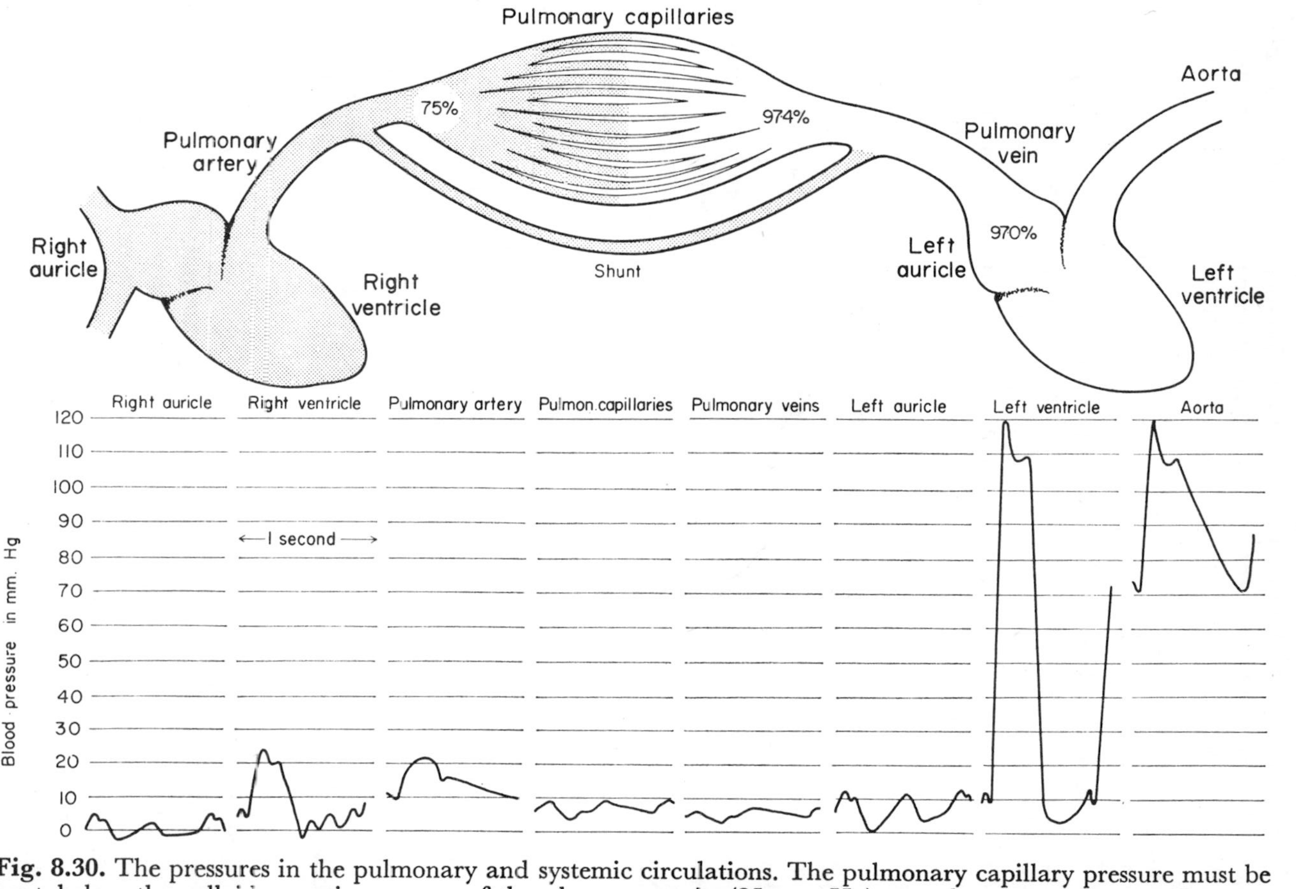

Fig. 8.30. The pressures in the pulmonary and systemic circulations. The pulmonary capillary pressure must be kept below the colloid osmotic pressure of the plasma protein (25 mm Hg) or pulmonary oedema will occur. (Comroe *et el.* "The Lung," Year Book Med. Pub., 1962.)

probably does not amount to more than 1·5 per cent of the total cardiac output, but this amount may account in part for the fact that arterial blood is not usually more than 97 per cent saturated with O_2.

Pulmonary Oedema

Lymphatic vessels accompany the bronchial tree, the pulmonary arteries and veins; the vessels do not reach the alveoli, however, so these have no lymphatic drainage. This is probably not important normally because of the low capillary pressure militating against formation, and favouring absorption, of fluid in the alveolus. In pulmonary oedema caused by raised venous and capillary pressures, however, the fluid must be carried away, and it is found that lymphatic flow is, indeed, significant. This process is slow; thus Courtice found that saline solutions inhaled as aerosols are quite rapidly absorbed,* presumably across the alveolar membrane into the blood capillaries; protein-containing solutions, on the other hand, were absorbed very slowly and this occurred by way of the lymphatics. This may account for the severity of the pulmonary oedema that occurs when pulmonary capillary permeability is increased, as in phosgene poisoning.

Changes in Pulmonary Flow

In general, the same factors operative in the systemic circulation governing flow of blood are operative in the lungs. Two aspects may, however, be emphasized. Thus, changes in cardiac output, and therefore pulmonary flow, require remarkably small changes in arterial pressure, and this is because of the great distensibility of the blood vessels, so that as pressure rises the resistance to flow tends to decrease. For example, Daly and Wright found a pulmonary resistance, measured as the ratio of ΔP (in mmHg) over ΔV (in ml/min), of about 0·03 when the blood-flow through the perfused dog lung was 400 ml/min, and about twice this, 0·06, when the flow was reduced to 180 ml/min.

Pulsatile Flow

A second factor is the pulsatile, and even intermittent, flow that is the consequence of the relatively high pressure surrounding the veins and capillaries of the pulmonary system compared with the pressure within the vessels, i.e. it is because the *transmural pressure* is low. This has been indicated earlier, and is reflected in a pulsatile absorption of gases from the lungs. Thus Lee and DuBois kept a human subject in a

* Drinker found that saline, trickled into the lungs of a dog by a bronchial catheter, was absorbed at the rate of 1·5 litres per hour.

body plethysmograph, a chamber which recorded any changes in volume of the subject plus surrounding air. On breathing from a bag within the chamber little or no pressure-change occurs because the O_2 used up is about equal to the CO_2 evolved; if a mixture of nitrous oxide and oxygen is breathed, however, the rapid absorption of nitrous oxide without corresponding liberation of CO_2 causes a reduction in pressure which is recorded continuously. With the subject holding his breath the absorption is reflected in a series of waves synchronous with the arterial pulse (Fig. 8.31).

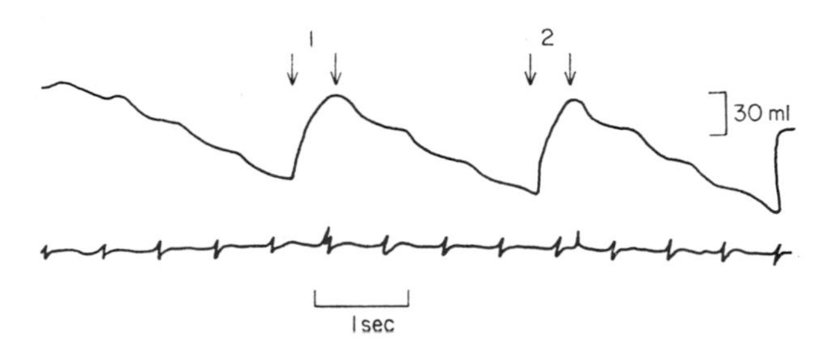

Fig. 8.31. Records from a body plethysmograph indicating pulsatile absorption of inspired nitrous oxide with the subject holding his breath. The lower record is that of the electrocardiogram. The volume-changes take place in a pulsatile manner. Note that between arrows 1 and 2 the plethysmograph was opened to the atmosphere. (Lee and Dubois, *J. Clin. Invest.* 1955, **34,** 138.)

Pulmonary Blood Volume

The *intrathoracic blood volume*, or blood in the heart and lungs, is 1·8 litres on average, representing about 45 per cent of the total blood volume; whilst the *intrapulmonary volume* is about 1·2 litres or 29 per cent of the total. The amount in the pulmonary capillaries can be estimated from the rate of absorption of a gas, such as carbon monoxide, from the lungs, and the cardiac output; this comes out at only about 100 ml, so the great bulk is in the large conducting vessels, mainly the large veins.

Furthermore, the volume of blood in the lungs can vary quite considerably, so that it is customary to regard the lungs as a reservoir of blood; thus on lying down blood passes from the limbs into the thorax, and the reverse occurs on standing. Again, positive pressure breathing is associated with a considerable diminution in lung blood volume.

Pulmonary Resistance

Finally, it must be appreciated that the pulmonary vascular system contains no "resistance vessels", in the sense used for the systemic circulation, the arterioles responsible for this being absent. We shall see that during severe exercise there is some change in the distensibility of the vascular system that favours increased blood-flow, but this is brought about not by a dilatation of small arteries so much as by a tendency for the walls of the large vessels to *constrict*, or rather to *resist expansion*, so that they tend to act as better conducting tubes. Thus a tube made of hard rubber would respond to a pulsatile driving of fluid better than one made of a soft rubber, the latter tending to balloon out with each systole.

CHAPTER 9

The Digestion and Absorption of the Foodstuffs

BREAKDOWN OF FOODSTUFFS

We have so far considered the distribution of materials throughout the body largely from the point of view of the blood gases, O_2 being the gascous substrate for the energy-giving reactions, and CO_2 the gaseous product of these reactions; we must now consider the other components in these reactions, their mode of ingestion, their conversion into absorbable products, and the mechanism of this absorption into the blood-stream.

Basic Principles

During their passage along the alimentary tract, the foodstuffs are broken down to classes of relatively simple substances, principally hexoses from starch and other polysaccharides, amino acids from proteins, fatty acids and glycerol from fats, and pentoses, purines and pyrimidines from nucleic acids. In these relatively simple forms they may be removed into the blood whence they are carried primarily to the liver to be converted back to complex substances of the type required by the animal. Thus the starch of the diet is converted to glucose which is converted to the animal's storage polysaccharide, *glycogen*. The many animal and plant proteins in the diet are broken down to some 26 amino acids, which act as building blocks for the animal's own specific proteins; and so on. Additional elements of the diet, salts, vitamins, etc., may require no chemical change in the alimentary tract, and to a greater or lesser extent the degree of their absorption into the blood will be governed by the animal's requirements, e.g. those of Ca^{2+} and phosphate. Where no control is, or can be, exerted on absorption, highly efficient excretory mechanisms are available in the kidney and liver for removing the unnecessary materials from the blood, whilst the residues from the diet, remaining unabsorbed, are removed as faeces.

Carbohydrates

The structures of the polysaccharides, proteins and nucleic acids have been briefly indicated. They are essentially polymers, made by the elimination of water from a given pair of unit molecules, e.g. two

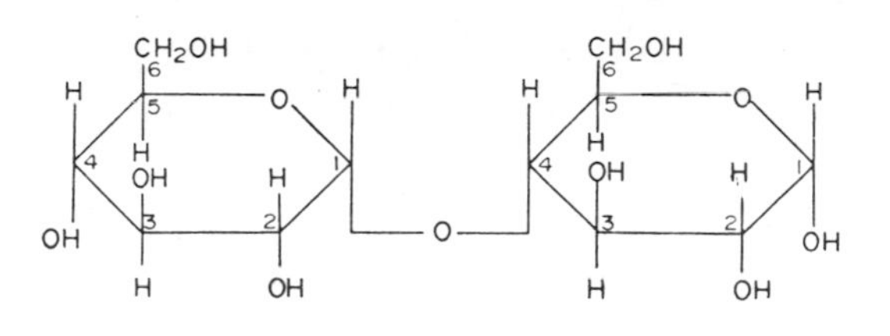

(a) Maltose α 1 : 4

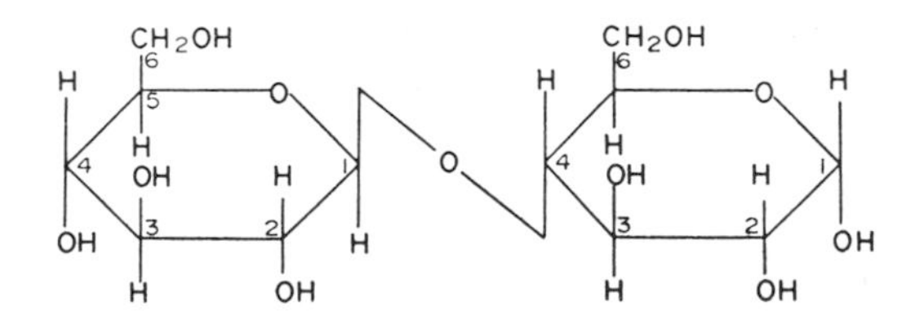

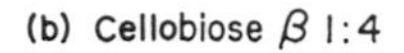
(b) Cellobiose β 1 : 4

Fig. 9.1. The two forms of the 1:4 linkage by which glucose molecules are joined together to form polysaccharides. (*a*) The α-1-4 linkage between two glucose molecules forming the disaccharide maltose. This linkage is split by digestive enzymes present in the small intestine. (*b*) The β-1-4 linkage between two glucose molecules forming cellobiose. This linkage, found in cellulose of plant walls, can only be split by enzymes released from bacteria in the caecum of herbivorous animals.

glucose molecules eliminate water to form maltose; maltose plus glucose eliminate water to form a trisaccharide, *maltotriose*, and so on (Fig. 9.1). The linkages of glucose molecules can occur in two principal ways according to the positions of the H- and OH-groups on the 1-C-atom, to give an α-disaccharide (*maltose*), or a β-disaccharide (*cellobiose*) as in Fig. 9.1. Furthermore, the 1 : 4 type of linkage shown in Fig. 9.1 is not the only type, a 1 : 6 linkage being possible, so that branching of chains can occur when two of the OH-groups of a single

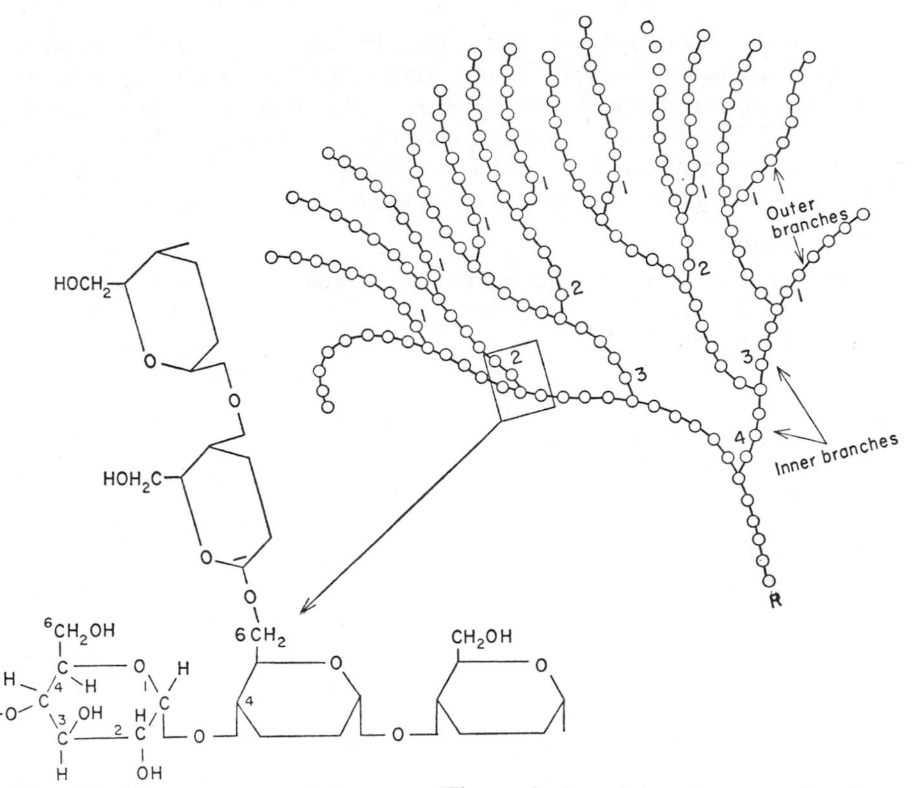

Fig. 9.2. The structure of glycogen. The majority of the glucose molecules are joined by the α-1-4 linkage (open circles). Branching of the chain is achieved by joining the glucose molecules through the 1:6 position (filled circles) and detailed by the inset. (Harper, "Review of Physiological Chemistry", 14th Ed., 1973).

hexose molecule are involved in the polymeric process (Fig. 9.2). Thus starch exists in a branched form, called *amylopectin*, and an unbranched form called *amylose*; glycogen is similar to amylopectin.

Hydrolytic Enzymes

The breakdown of these polymers consists in hydrolysis, and enzymes are required for the process. These enzymes show some specificity in so far as they will attack only certain forms of linkage, so that if a sugar polymer is held together by, say, α- and β-linkages, then a given enzyme will not be able to break the polysaccharide down completely to its glucose components. Again, some enzymes will only attack a polymer from its ends, or if, as in protein chains, one end is different from another, then it will attack only one type of end. Others will only attack linkages if they are within the chain; so far as the polypeptide chains are concerned, for example, the enzyme that attacks the ends is called an *exopeptidase* and one that attacks the inner links an *endopeptidase*.* The chains of a polypeptide end as either a carboxyl group or an amino group, and two different exopeptidases are concerned with breaking off terminal groups—*carboxypeptidase* and *aminopeptidase*.

Final Products

In general, then, as the enzymatic digestion proceeds, we may expect to find mixtures of differing sized polymers of the particular types of molecule being considered, until finally the mixture consists only of the absorbable forms of the particular substance. At one time this was considered to be the unit, i.e. an amino acid or hexose from a protein or carbohydrate, but more recent work suggests that it is the dimeric form that is taken out of the lumen of the intestine, the final stage of breakdown occurring in the lining epithelial cells.

Absorption

The digestive process begins in the mouth and continues in the stomach and small intestine; since the absorption of the digestive products occurs in the small intestine only, we may expect to find the conversion to absorbable products to be completed by the time the food has reached this region. In general, this is true, but since absorption and hydrolysis go on at the same time, the contents of the small intestine after a meal consist of a mixture of fission products, not all of which are in a suitable condition for absorption. This is especially true of the fats, and it seems that an important factor here in causing digestion to proceed is the removal of the absorbable products as they are formed, this favouring further hydrolysis in accordance with the Le Chatelier principle.

Outline of Digestive Process

The processes of digestion may be briefly summarized as follows.

Mouth

In the mouth there is only a limited breakdown of starch into maltose, which continues for some time in the stomach until the gastric juice, with its acid pH, makes the salivary amylase inactive.

* The endo-amylase that attacks only the links within the carbohydrate chain is called an *alpha-amylase* whilst the exo-amylase, attacking end-links, is called a *beta-amylase*. This is an unfortunate terminology, since α- and β- are used to characterize the linkages of glucose in carbohydrate; thus an α-glucosidase will hydrolyse an α-glucosidic link but a beta-amylase also will attack an α-glucosidic link, the beta this time referring to its tendency to attack end-links.

Stomach

In the stomach, pepsin, in conjunction with hydrochloric acid, attacks proteins at specific sites within the molecules and produces a range of polypeptides. At the end of the gastric phase of digestion only a limited breakdown of carbohydrates and protein has occurred, and the mixture of food products undergoes another pH change on entry to the alkaline medium of the duodenum.

Duodenum

In the duodenum the pancreatic enzymes trypsin and chymotrypsin reduce the polypeptides to smaller peptides, the amylases reduce larger sugar molecules to smaller units, and the lipases attack the fats, which are solubilized by the bile salts, and begin the degradation of these fats to fatty acids and glycerol. By the end of the duodenum, proteins have been reduced to di- and tripeptides, sugars to mono- and disaccharides and some fats to mono- and diglyceride.

Jejunum

In the jejunum the final stages of digestion are completed by the enzymes of the succus entericus prior to the absorption processes. Specific enzymes reduce peptides to amino acids, saccharide to single units and the di- and triglycerides to fatty acids and glycerol to provide the substrate for the various absorptive processes, which are carrier modulated with respect to amino acids and sugars.

In this chapter the main physiological processes that will occupy us are the secretion of the appropriate digestive juices and the absorption of the products of digestion into the blood. The more mechanical aspects involved in chewing the food, and swallowing, are largely problems for the anatomist and will be treated rather more summarily, whilst the details of the chemical changes taking place in the food, and the general nutritional aspects of the diet, are problems that fall within the sphere of the biochemist and will be largely ignored. Finally, the control mechanisms governing the act of deglutition and the transport of the food from one end of the alimentary tract to the other involve a knowledge of the basic features of the nervous system, so that their description will be deferred to Vol. 3.

Breakdown of Carbohydrate

Salivary Amylase

Digestion begins in the mouth by virtue of the action of the saliva which contains an amylase, *ptyalin*, which splits 1-4-glucosidic linkages.

Thus, during chewing, the saliva is secreted from three salivary glands —the *parotid*, *sublingual* and *submaxillary*—into the cavity of the mouth where it becomes mixed with the bolus of food. Because the food is in the mouth for only a short time, and because of the limited potentialities of the amylase, digestion of carbohydrates is only partial here. Nevertheless there is no doubt that the processes initiated in the mouth are continued for some time in the stomach. The strongly acid nature of the gastric juice, however, finally inhibits the amylase, which has a pH optimum of 6 to 7, which is the approximate pH range of the saliva. The limited potentialities of amylase are due to its failure to hydrolyse α-1,6-linkages (Fig. 9.2) at the branching points of the starch (amylopectin) molecule, and has little affinity for the adjacent α-1,4-linkages, so that the salivary digestion results in a mixture of *dextrins*, i.e. relatively large polymers, and some maltose.

Salivary Lubrication

Probably far more important than its digestive function is the lubricating action of the saliva that promotes attrition by the teeth and permits the formation of a smooth bolus that will pass easily down the oesophagus. This lubricant action is promoted by the presence of "mucins", i.e. combinations of mucopolysaccharide with protein synthesized within the salivary gland.

Intestinal Enzymes

The further breakdown of carbohydrate is largely reserved for the small intestine, although the strongly acid medium of the gastric juice favours a little hydrolysis of polysaccharides; the intestinal breakdown is achieved by an amylase, secreted by the pancreas into the small intestine together with *maltase* which breaks down the maltose molecule to two molecules of glucose. The breakdown of the 1-6-linkage which, we have seen, the amylases cannot carry out, is brought about by the enzyme *oligo*-1,6-*glucosidase*, secreted by the intestine. Another important enzyme is the *invertase* that breaks up sucrose into glucose and fructose.

Breakdown of Protein

Gastric Enzymes

The breakdown of protein begins in the stomach, a process favoured by the strong acidity of the gastric juice, and the presence of *pepsin*. The enzymes responsible are given the general name of *peptidases*, having the common feature of hydrolysing the peptide bond:

$$R_1-\overset{\overset{\displaystyle O}{\|}}{C}-\underset{\underset{\displaystyle H}{|}}{N}-R_2 + H_2O \longrightarrow R_1-C\begin{matrix}\nearrow O \\ \searrow OH\end{matrix} + H_2N-R_2$$

but, as indicated, they are individually quite specific in the particular peptide linkage they can attack, as manifest, for example, in the *endopeptidases* attacking linkages within the chain and *exopeptidases* attacking terminal linkages. Pepsin is an endopeptidase, so that the products of peptic digestion alone would be relatively large polypeptides. Moreover, pepsin has a remarkable preference for bonds in which an aromatic amino acid, such as tyrosine, provides the amino group, and the result is that the products of peptic digestion alone consist of a mixture of polypeptides containing about seven amino acid residues. The pH optimum for pepsin is about 2, a degree of acidity that is easily reached by the secretion of certain cells of the stomach the *oxyntic* or *parietal* cells (p. 429). Another gastric enzyme is called *gastricsin*, with a pH optimum of about 3; in addition, *rennin* is present in the juice of some ruminants, converting the casein of milk to a curd that leaves it amenable to attack by pepsin and other proteolytic enzymes. *Gelatinase* is another enzyme, specially effective in hydrolysing gelatine.

Intestinal Enzymes

After leaving the stomach, the *chyme* is subjected to the digestive juices of the small intestine; these contain additional peptidases, which operate at the much less acid pH range of the intestinal contents; some of these are, like pepsin, endopeptidases, namely trypsin and chymotrypsins A and B. These are specific not only in being endopeptidases but also in the type of bond that they break, e.g. trypsin will cleave a bond where the CO-group is provided by the basic amino acids arginine or lysine. The final breakdown to amino acids is brought about by exopeptidases; the *carboxypeptidases*, secreted by the pancreas, attack terminal links where the free end is a carboxyl group; *the aminopeptidases* are present in the *intestinal juice* which, we shall see, is a mixture of the secretions of glands of Lieberkühn in the mucosa of the intestine and of epithelial cells shed from the epithelium and disintegrating within the intestinal lumen; these attack a residue with a free amino group. The exopeptidases show a further specificity in the size of the polypeptide that they will attack, so that we have tripeptidases and dipeptidases. Figure 9.3 illustrates the specificities of the different enzymes attacking proteins.

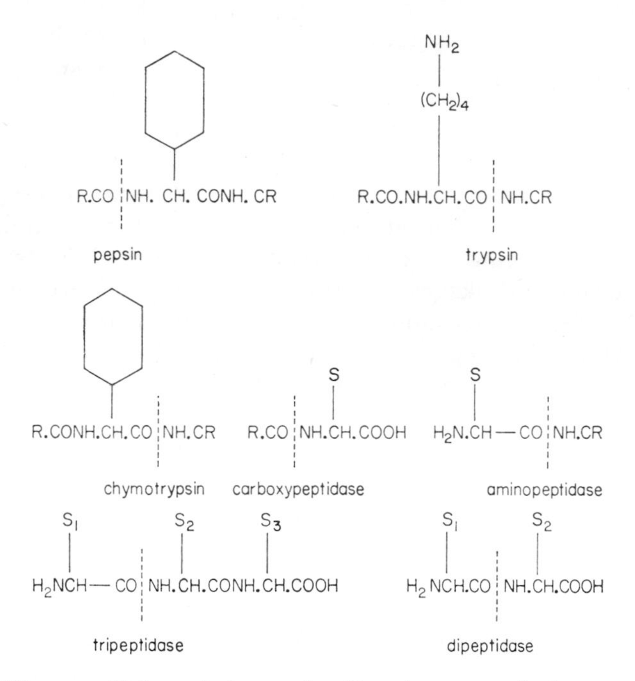

Fig. 9.3. The specificity of the main digestive proteolytic enzymes. Sites of action are indicated by the dotted lines. R— means the rest of the protein molecule consisting of a number of amino acids. S_1, etc., indicate the rest of the amino acids.

Breakdown of Fats

The hydrolysis of fats is brought about by lipases, mainly those of the pancreatic juice, although there is some activity in gastric and intestinal juices. The breakdown of fat is so intimately concerned with its absorption that we may defer consideration of the hydrolytic process until then.

THE GLANDULAR SECRETION

Exocrine and Endocrine Glands

By a gland we mean a group of cells organized in such a way that they are able to form a *specific secretion*, which is liberated either into a cavity of the body, as with the salivary gland, or directly into the blood, as with the suprarenal gland. The distinction between the two types is reflected in the different names—*exocrine* and *endocrine*.

Thus the exocrine salivary gland organizes a fluid secretion (the saliva) which is conducted along a salivary duct into the mouth; the endocrine adrenal gland, on the other hand, does not form a characteristic fluid, liberated through ducts into the blood, so that its "secretion" is not something that can be collected in bulk and analysed. Its cells synthesize certain specific compounds, e.g. adrenaline and noradrenaline, and these appear in the blood flowing away from the gland, so that it is essentially the specific compounds liberated into the

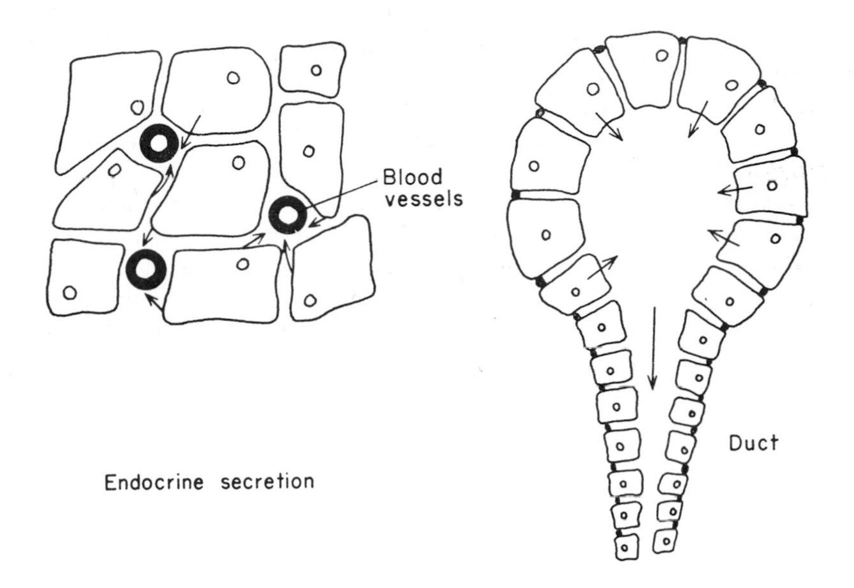

Fig. 9.4. Endocrine and exocrine secretion. (*a*) Endocrine secretion occurs from the secreting cells directly into the capillary network of the tissue. (*b*) Exocrine secretion occurs from the secreting cells into the lumen of a duct system and thence into the cavity for which the enzyme is required.

blood that are called the secretions of the endocrine or ductless glands. These substances have been given the general name of *hormones*.

To analyse this secretory activity experimentally we must collect the blood flowing from the gland, or, since secretory activity is accompanied by rapid synthesis of the endocrine secretion, we may remove the gland and extract the secretion by appropriate techniques. The basic difference between the exocrine and endocrine glands is illustrated schematically in Fig. 9.4.

The Secretory Process

Active Transport

The anatomy and physiology of the endocrine glands will be discussed in Vol. 2. It must be appreciated that the two types of secretion have little more in common than their name; the word secretion has come to mean, today, not only the exocrine type of *fluid* formed by glands such as the saliva, the tears, or gastric juices, but also the fundamental ionic and fluid transport mechanisms that bring about the transfer of fluid secretions from blood into the cavities. In fact, very often the secretory process has come to mean essentially *active transport*, namely the transport of ions and molecules against gradients of concentration or electrochemical potential. Where this transport is accompanied by the movement of water, the secretory process is usually obvious, as with the formation of saliva, but where the transport merely leads to the accumulation of ions or molecules against concentration gradients, the secretion, in its common sense, is not obvious although the active process that leads to this transport is often called secretory. Certainly we may say that, where all the obvious secretions are concerned, be it the tears, the saliva or the gastric juice, the fundamental mechanism involving their elaboration is one of active transport of solutes, but in addition the process involves the synthesis of special substances, notably enzymes, that are carried in the fluid.

Glandular Structure

Acinus

In general, the exocrine glands may be regarded as arrangements of epithelial cells that have invaginated into a primary epithelial surface, as illustrated in Fig. 9.5. It is clear that many degrees of complexity can be developed from this basic pattern of closely packed epithelial cells, coming into relation, on the one hand, with the blood capillaries and the interstitial fluid, and on the other with a cavity, the *acinus* or *alveolus*, whose walls are formed by the cells. Thus Fig. 9.6 illustrates the structure of the salivary gland based on the acinus. These cells produce a fluid which we may call the primary secretion; this passes along a series of confluent epithelium-lined tubules or ducts, the largest of which opens into the mouth.

During its course along the system of ducts the fluid's composition may well change, either as a result of passive exchanges with the blood vessels lying on the opposite side of the duct, or through active transport processes operating across the epithelial lining.

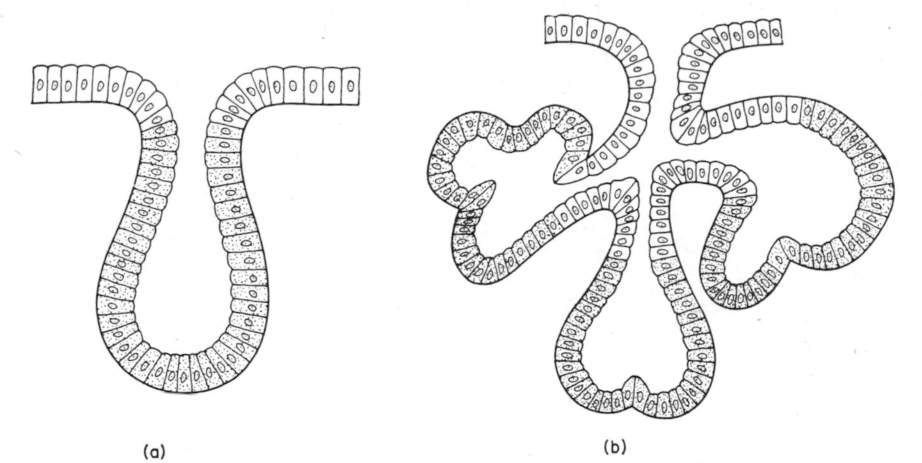

Fig. 9.5. Illustrating invagination of an epithelium to form (*a*) a simple cavity and (*b*) a more complex arrangement of cavities lined with specialized secretory cells (Davson, "Encyclopaedia of Biological Sciences", Van Nostrand Reinhold.)

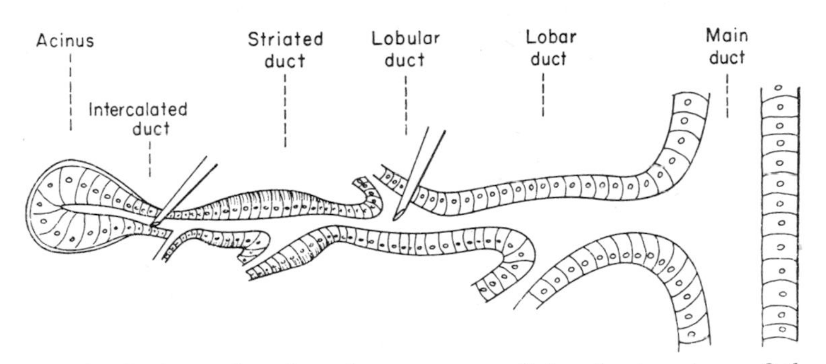

Fig. 9.6. A diagram showing the segments of the duct system of the rat parotid salivary gland. Micropipettes shown can be used to obtain samples of fluid from various parts of the system. (Mangos *et al.*, *Pflug. Arch.*, 1966, **291,** 99.)

Elimination of Secretory Products

The manner in which the materials synthesized in the glandular cells are ejected into the spaces outside the cells (be it the cavity of the acinus or duct, or the extracellular space adjacent to the blood capillaries) varies according to the type of secretion. Where the secretion is accumulated into granules within the cytoplasm of the cell the contents of the granule, which really consists in a membrane-bound vesicle stuffed with the secretion, are ejected by exocytosis, as indicated earlier

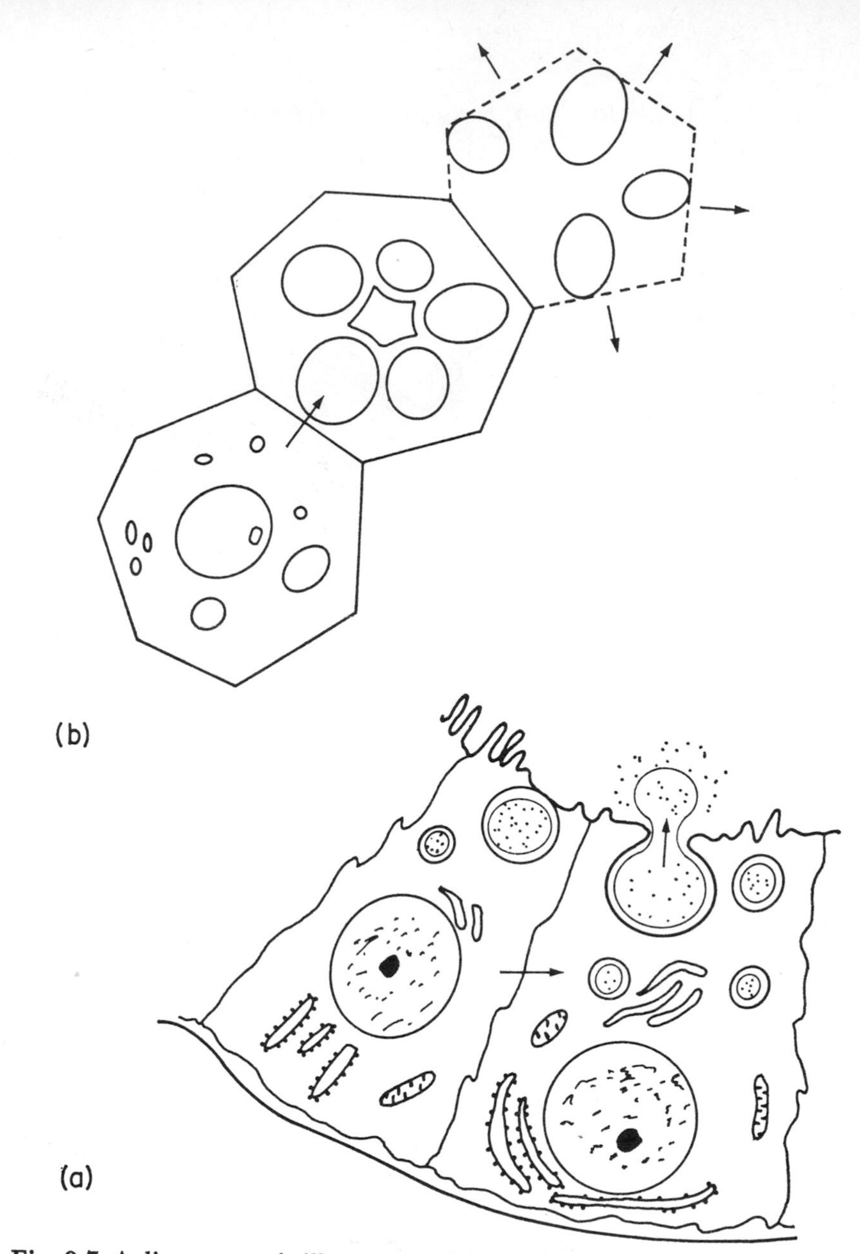

Fig. 9.7. A diagrammatic illustration of the two types of extrusion mechanism of secretion observed with the electron microscope. (*a*) Exocytosis of a granular vesicle. The vesicles formed by the Golgi apparatus drift towards the luminal cell membrane. Fusion of the vesicle membrane with that of the cell allows the contents of the vesicles to pass out of the cell. (*b*) Holocrine secretion. In this process, vesicle formation occurs until the cytoplasm is so filled with vesicles that disruption of the cell wall occurs and the vesicles are released. The cell itself is destroyed in this form of secretion. (Kurosumi, *Int. Rev. Cytol.*, 1961 **11,** 1–124.)

(Chapter 1, p. 48), the membrane of the vesicle fusing with that of the cell (Fig. 9.7a). The enzymes contained in the exocrine cells of the pancreas are secreted in this fashion, and the same is true of the secretion of salivary enzymes by, for example, the parotid gland, the enzymes being synthesized on the endoplasmic reticulum and stored in cisterns of the Golgi apparatus, which ultimately develop into mature granules. The granular secretions of endocrine hormones, e.g. insulin by the pancreas, are also eliminated in the same manner.

Apocrine, Eccrine and Holocrine Secretions

When discussing the secretion of milk in the mammary gland (which is to be regarded as a highly differentiated sweat gland) we shall find that the secretion of the sugar and protein into the acinus takes place by exocytosis, but the elimination of the fat globules from the cell takes place by a pinching off, the resulting globule containing a part of the plasma membrane of the cell. In this latter process the cell may be regarded as having sacrificed some of itself, and the secretion might be described as *apocrine*. This contrasts with the *eccrine* type of secretion, in which no such sacrifice has taken place, and the *holocrine* type of secretion in which the cell apparently sacrifices itself completely when delivering its secretion, the cell membrane disintegrating and allowing the cytoplasm and disintegrated nucleus to enter the acinus (Fig. 9.7b). An example of this holocrine secretion is that of the sebaceous gland whose duct leads into the hair follicle.

The Composition of the Secreted Fluid

Active Transport of Na^+

The eccrine type of secretion is essentially a transport of fluid brought about by a primary active transport of Na^+ from the extracellular fluid, bathing the basal and lateral parts of the cell, into the cavity in which the secretion is ejected, e.g. the acinus of the salivary gland. As discussed earlier, this active transport of Na^+ into the intercellular clefts gives rise to an osmotic flow of fluid and, according to the various parameters of the situation, the fluid may be iso-osmotic, hyperosmotic or hypo-osmotic with the blood as it flows into the acinus or other cavity. The osmolality of a body fluid is, with rare exceptions, governed by the concentration of Na^+ in it, so that the *hypo-osmolal* sweat or saliva has a low concentration of Na^+, the *iso-osmolal* cerebrospinal fluid has a concentration of Na^+ approximately equal to that in plasma, and the *hyperosmolal* secretion of the salt-gland has a very high concentration of Na^+. In general, the concentrations of other ions—K^+, Mg^{2+}, Ca^{2+}, Cl^-, HCO^-_3—in a fluid secretion, such as saliva, are also different from those in plasma so that it is usually necessary to invoke active transport processes directed towards accumulating or excluding some or all of these ions, in addition to the primary active transport of Na^+.

Plasma Filtrate

It will be recalled, however, that when an ultrafiltrate of plasma is formed, the removal of the pasma proteins from the filtered fluid, as in, say, the formation of the interstitial fluid of the tissues, gives rise to the Gibbs–Donnan distribution of ions so that the concentrations of cations are less in the filtrate than in the parent plasma, and the concentrations of anions are greater. Thus the Gibbs–Donnan ratio for univalent ions is 0·96 approximately:

$$[Na^+]_{Fluid}/[Na^+]_{Plasma} = [K^+]_{Fluid}/[K^+]_{Plasma} = [Cl^-]_{Plasma}/[Cl^-]_{Fluid} \ldots = 0{\cdot}96$$

Hence, if in a secreted fluid the concentration of K^+ were some 4 per cent less than in plasma it might be unnecessary to invoke active transport processes to account for this, so that when considering the concentrations of ions in a secretion it is desirable to compare these with those in a filtrate of plasma rather than in plasma itself. If the concentration is the same as in a filtrate, then there is a presumption that there has been no active transport directed towards this ion, but even now we cannot be certain since the active transport of one ion may influence the distribution of another (e.g. by creating a difference of potential) and to be more certain we must take into account any difference of potential between the fluids we are comparing.

Electrochemical Potential

In other words, as indicated in Chapter 2, we must establish whether a given ion is at equal *electrochemical potential* in plasma and the secreted fluid. With interstitial fluid, formed as a plasma filtrate, the ion concentrations in plasma and the fluid are different, being governed by the Gibbs–Donnan distribution, but these differences are sustained by the electrical potential of about 1 mV, and the ions are, in effect, at equal electrochemical potential.

Summary

To summarize these general principles of secretion, we may say that the secretory process implies, basically, a separation of constituents of the plasma to produce a fluid of different composition. Implicit in the definition is that the process involves active transport, so that the fluid is different from a filtrate of plasma, which has its ions and non-electrolytes distributed in accordance with the Gibbs–Donnan equilibrium. Secretions, according to this viewpoint, may or may not contain specially synthesized components. Thus, the salt-gland secretion is concerned with eliminating as much salt as possible from the animal with a minimum of water, and there is no reason to expect the gland to synthesize materials that will be simply eliminated. The salivary secretions, whilst acting primarily as systems for the pumping of water into the mouth to moisten the food, do also contain a synthesized product, e.g. the enzyme *amylase* concerned with the preliminary digestion of starch, so that here we have the purely eccrine function of fluid transport combined with an apocrine emptying of amylase-laden granules (or vesicles) into the acinar cavity. The specific hormones synthesized by the endocrine glands, and ejected into the blood, are also described as secretions, but it must be emphasized that their inclusion in this category today is rather a matter of history than logic. Here the granules, or vesicles, containing the synthesized hormone are emptied out of the cell with presumably only the fluid in which the hormone is dissolved, so that fluid transport is negligible.

The Salivary Secretions

The Salivary Glands

In man, the bulk of the saliva is secreted by three pairs of glands—the *parotid*, the *submaxillary* and the *sublingual*. In addition, numerous small glands are scattered over the inside of the mouth. Each principal gland drains by a main duct in the mouth—the parotid duct opening opposite the upper second molar; the submaxillary at the side of the frenulum of the tongue; and the sublingual opens by way of some 10–20 openings on the floor of the mouth. The glands are of the compound tubular type lying in a stroma of connective tissue through which run the blood vessels, lymphatics and branches of sympathetic and parasympathetic nerves. The basic unit, as indicated above, is the acinus (Fig. 9.6), a cavity composed of a layer of epithelial cells and communicating with a ductular system that finally, through confluence of smaller channels, becomes the main duct. The acini are of two types, according as their cells secrete a watery fluid rich in enzymes—the *serous type*—or a viscous mucus—the *mucoid type*. The type of secretion of the whole gland depends on the proportion of serous to mucoid acini;

thus the parotid gland, having mainly serous cells, secretes a watery fluid whilst the submaxillary, containing mostly mucoid cells, secretes a highly viscid material.

Composition of Secretion

The composition of human parotid gland secretion is shown in Fig. 9.8; since the concentrations of the ions are dependent on the rate of flow these have been plotted against flow-rate. It is clear that the composition is very different from that of plasma or its ultrafiltrate, the most striking feature being the low concentration of Na^+ by comparison with that in plasma. Since this is only to a small

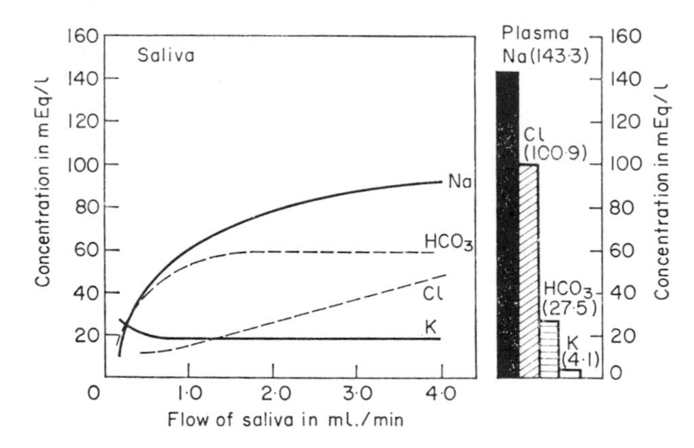

Fig. 9.8. The composition of human parotid saliva. Since the concentrations of ions in saliva are dependent on the flow rate, these have been plotted against the rate of flow. The columns on the right indicate the concentrations of these ions in plasma. (Burgen and Emmelin, "Physiology of the Salivary Gland", Arnold.)

extent compensated for by a raised K^+, the fluid is strongly hypotonic to plasma, so that there must be a strong tendency for water to pass by osmosis from the saliva to the blood through the interstitial fluid of the gland. By producing this hypotonic fluid, then, the gland is performing osmotic work. The extent of this work could be calculated on thermodynamic principles, as indicated above, but this would require a knowledge of the extent to which each ion had been moved away from its equilibrium distribution, and this depends on the electrical potential across the acinus and ducts as well as on the concentrations in the fluids.

Potential. It would seem that the potential across the epithelial linings is not high, of the order of 10 mV, inside negative, although it is modified by the gland's activity. In general, we would not be far wrong in equating differences from equilibrium distributions of ions in terms of active transport. In other words, the low concentrations of Na^+ and Cl^- and high concentrations of K^+ and HCO_3^-, found in parotid saliva are, indeed, indications of active processes.

Dependence on Secretory Rate. Furthermore, the composition varies very considerably in accordance with the rate of secretion. Thus the degree of hypotonicity, which is the striking feature of this human parotid saliva, decreases as the rate of secretion increases; it is as though the gland has not had time to establish the maximal degree of hypotonicity before the fluid is ejected into the mouth. With very slow rates of secretion we find the same phenomenon, the fluid becoming approximately isotonic with plasma; this time it looks as though the fluid spends too much time in the ducts.

Secretion and Reabsorption

On the basis of the histological differences in the epithelial cells of the gland, on passing from acinus to intercalated and striated ducts (Fig. 9.6), and on the basis of the effects of rate of secretion on composition, such as those illustrated in Fig. 9.8, Thaysen postulated that the secretory cells, constituting the acinus, produced a fluid very similar in composition to that of a simple ultrafiltrate of plasma, in the sense that the total osmolality was in the region of 300 mos/litre, the concentration of Na^+ about 150^+, that of K^+ about 6 meq/litre, and so on. This process would be an active secretory one, and not a simple ultrafiltration, because the formation of the fluid continues even if pressures as high as 80 mm Hg are exerted in the duct, whereas we know that a pressure of the order of 30 mm Hg would be enough to prevent filtration of plasma.

Reabsorption of Na^+. On its passage through the intercalated acinar ducts, the striated ducts, and so on, to the main duct it is postulated that there is an active reabsorption of Na^+ accompanied by Cl^- and HCO_3^-, whilst some K^+ passes from the duct-cells into the fluid. In these regions of reabsorption we may expect a low permeability to water, otherwise the difference of osmolality created by the removal of salt would be levelled out by osmosis. A similar problem is encountered in the diluting and concentrating mechanisms of the kidney (Chapter 10). The basic mechanism is illustrated by Fig. 9.9. Increasing the rate of flow would be expected to reduce the efficiency of this reabsorption, and this would account, in general terms, for the changes with flow-rate.

Very Slow Secretion. One difficulty is the very high osmolality and Na^+ concentration found with very slow rates of secretion indeed; in the more classical studies, the examination of fluid collected at extremely slow rates was not practicable, since fairly large quantities

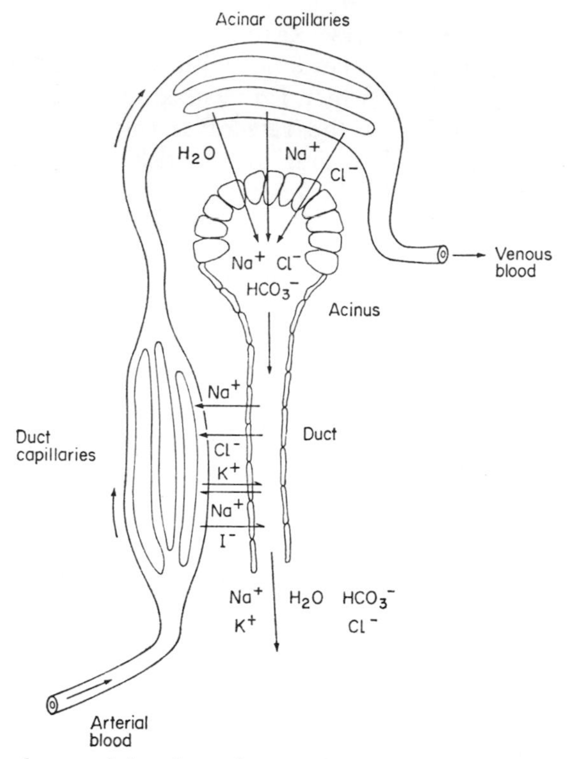

Fig. 9.9. The scheme of the electrolyte exchanges that occur during the secretion of parotid saliva. Active secretion of sodium draws water, bicarbonate and chloride into the acinus from the capillaries. In the duct system sodium is reabsorbed in exchange for potassium and the fluid then passes into the mouth. (Modified from Davenport, "Physiology of the Digestive Tract", Year Book Med. Pub., 1971.)

were required for analysis. With the development of modern methods of micropuncture, applied originally to the kidney, it was possible to examine fluid secreted at very slow rates, and in this event the curve of Fig. 9.10 was obtained. To explain this it is postulated that, at a later stage in its passage, the low-osmolality fluid comes into contact with a part of the tubular system relatively permeable to salts and water, permitting an equalization of salt concentrations when flow was very slow indeed.

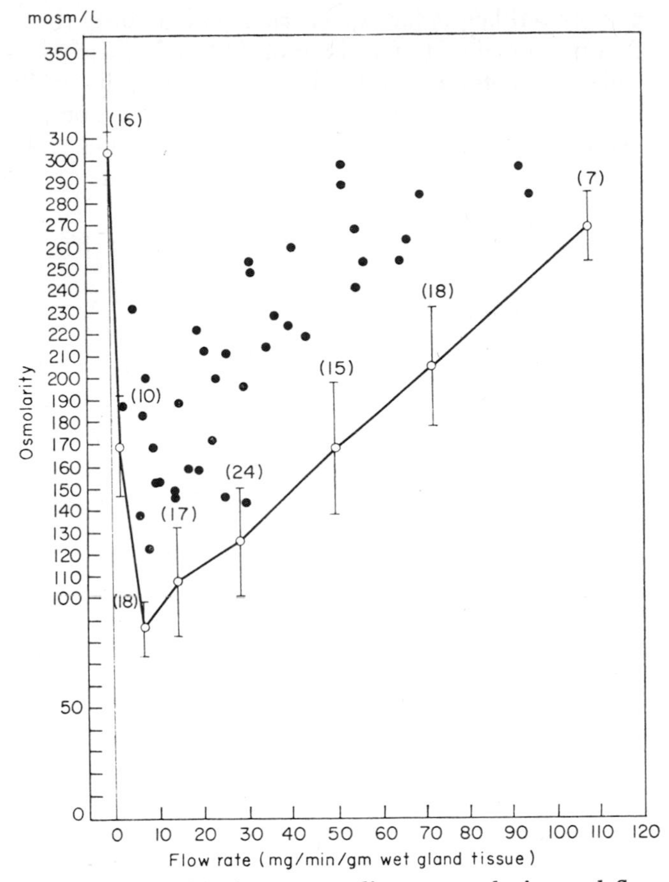

Fig. 9.10. The relationship between salivary osmolarity and flow rate. The filled circles indicate the effect of retrograde injection of ouabain into the duct system of the rat parotid. (Mangos and Braun, *Pflug. Arch.* 1966, **290,** 184.)

Micropuncture Studies

The micropuncture techniques not only allowed the collection of very slowly formed fluid from the main duct, but it allowed the collection from selected points in the gland. Thus the intercalated acinar ducts, carrying freshly formed secretion, are close to the surface and may be penetrated by a fine glass pipette of about 6 μ in diameter; quantities as small as 1 nl (10^{-6} ml) could be collected and analysed. Deeper in the gland, a striated duct (and larger ones) could be penetrated and finally fluid could be collected from the two ends of the main duct by pushing a cannula along this (Fig. 9.6). Experiments on rat sublingual, parotid and submaxillary glands, carried out mainly in German laboratories, have confirmed the basic hypothesis of a primary secretion with nearly the same composition as that of plasma. This is illustrated by Fig. 9.11 which shows the changes in Na^+ and K^+ composition as the fluid is collected at successively distal points. The broken lines indicate samples from unstimulated glands, producing secretion

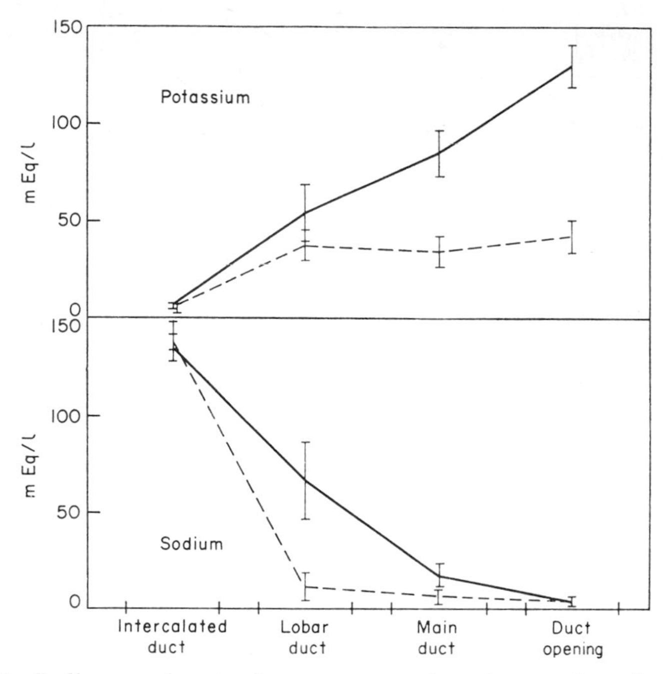

Fig. 9.11. Sodium and potassium concentrations in samples of saliva collected from four regions of the rat submaxillary duct system. Solid lines indicate mean values of samples from unstimulated glands, flow rate 3 μl/g/min. Broken lines indicate samples collected from glands 30 min. after stimulation with pilocarpine, flow rate 30–40 μl/g/min. (Young and Schlögel, *Pflug. Arch.* 1966, **291,** 85.)

slowly, whilst the solid lines represent secretions under the influence of the parasympathomimetic drug pilocarpine. Figure 9.12 shows the corresponding changes in osmolality.

Species Variations

It must be emphasized that the salivary secretions are not the same in composition, whether we compare different glands of the same species, or the same gland, e.g. the parotid, in different species. Moreover, the effects of flow-rate are not always the same; thus the dog's sublingual saliva is iso-osmotic with blood plasma at all measured rates of flow, and the same is apparently true of the cat, but not of man and the rat. In almost all cases studied, however, the primary secretion collected from the intercalated acinar ducts is approximately the same, i.e. like a filtrate of plasma, so that it may well be that the variations observed are due chiefly to the processes taking place later in the ductular system. An interesting variation is the sheep's parotid; the fluid has a greater osmolality than that of plasma, whilst the rate of secretion had no influence on the composition. When the animal

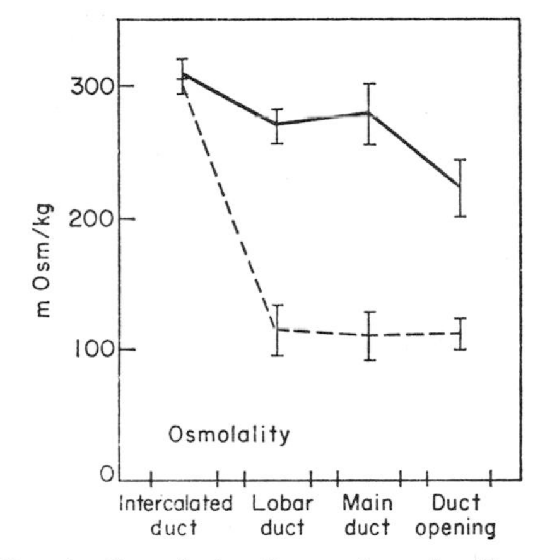

Fig. 9.12. Osmolality (mOsm/kg) of samples of saliva collected from four regions of the rat submaxillary duct system. Solid line unstimulated gland, broken line samples collected 30 min after stimulation with pilocarpine. (Young and Schlögel *Pflug. Arch.* 1966, **291,** 85.)

became depleted of Na^+, however, there was a remarkable fall in the Na^+-concentration with a reciprocal rise in the K^+ concentration, indicating that, with this gland, the secretory process could "manage" on either Na^+ or K^+; this is a rare situation in active transport processes.

The Active Transport Processes

The primary process of secretion by the acinar cells has been attributed to active transport, largely because flow continues against very high pressures in the lumen of the duct. When active transport inhibitors, such as ouabain or even cyanide, have been introduced into the gland there has been little influence on this aspect, the rate of flow being unchanged, whilst the secondary changes were inhibited, so that the osmolality rose to that of the primary secretion, or plasma. Again, treating the animal with Diamox, the carbonic anhydrase inhibitor which inhibits secretion of many fluids (e.g. the cerebrospinal fluid, aqueous humour of the eye, and so on), has no effect on rate of salivary secretion. Thus the mechanism of the primary secretion is uncertain, in the sense that we do not know what primary ion is actively transported, e.g. Na^+ or Cl^-.

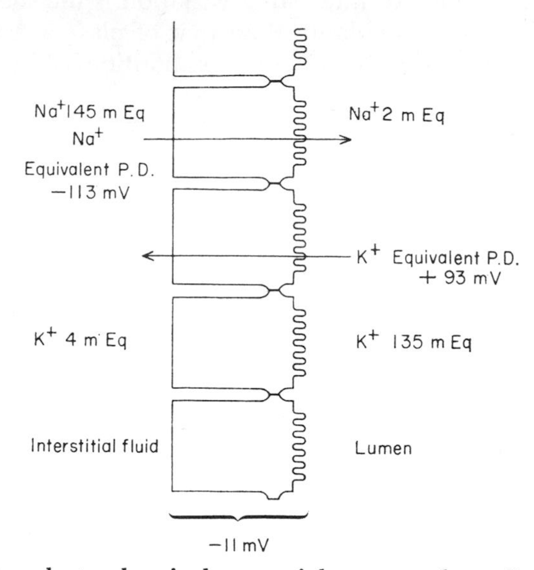

Fig. 9.13. The electrochemical potentials across the salivary duct. The potential difference is 11 mV, inside negative. The concentration gradient of Na^+ is equivalent to an electrical potential of 113 mV driving Na^+ into the lumen, so that its total electrochemical potential is 124 mV. The concentration gradient of K^+ is equivalent to a potential of 93 mV driving the ion outwards, whilst the measured electrical potential of 11 mV works in the opposite direction, giving an electrochemical potential of 82 mV. Active transport processes are thus necessary to explain the observed ionic distribution and potential.

Perfused Submaxillary Duct. The subsequent steps, namely removal of Na^+ and Cl^- and addition of K^+ and HCO_3^-, are largely active. Thus Young *et al.* perfused the main duct of the rat's submaxillary gland with a Ringer's solution similar in composition to a filtrate of plasma, and therefore comparable with the primary secretion of the acinar cells. In course of time the concentration of Na^+ decreased whilst that of K^+ rose; during this phase, when there was no movement of the two ions, the potential was some 70 mV, the inside negative.

Such a potential would tend to oppose the removal of Na^+ so that the net outward flux of Na^+, leading to a concentration gradient from inside to the outside, must have been active. Eventually a steady state was reached, with an internal concentration of only 2 meq/litre of Na^+ compared with 145 meq/litre in the interstitial fluid outside the lumen. The potential in the steady state was 11 mV, inside negative, and thus Na^+ was being held in the lumen of the duct at about one-seventieth of the concentration of the medium outside it and with a potential tending to attract Na^+ in. Thus both types of measurement indicate the intervention of active transport in determining not only the net movements but the steady state distribution. So far as K^+ is concerned, calculations indicated once again the operation of active transport; thus, at the steady state, the concentration in the lumen was 135 meq/litre compared with 4 meq/litre in the interstitial fluid outside the duct; the potential was 11 mV. The 11 mV tended to drive K^+-ions into the lumen, but this could not result in such a steep concentration gradient, so that, in fact, there is a difference of electrochemical potential (p. 120) equivalent to 82 mV driving K^+ out of the lumen; in other words, only if the inside potential were 93 mV would the distribution of K^+ be an equilibrium distribution. The situation in the steady state regarding Na^+ and K^+ is illustrated schematically in Fig. 9.13.

Analogy with the Nephron

We shall see that urine is formed by a primary elaboration of fluid in the glomerulus followed by active removal or addition of constituents as it passes along the tubular system of the nephron; it has been argued that the acinus is analogous with the glomerulus and that the striated duct (Fig. 9.6) is analogous with the tubular system. The cells of the duct are striated by virtue of the deep invaginations of the plasma membranes into the base of the cell and thus show analogy with kidney tubule cells; the cells also have a brush-border, consisting of minute microvilli projecting from the surface (p. 488), so that there is some real analogy, particularly with the distal tubule of the nephron.

Stimulation of Secretion

The control mechanisms over salivary secretion are discussed briefly in Vol. 3; the subject is of interest here primarily because the character of the secretion may be significantly modified by nervous and drug activity. An obvious way in which nervous activity can modify salivary composition is through its effect on flow-rate; thus powerful stimulation through the chorda tympani nerve,* or by administration of a parasympathomimetic drug like pilocarpine or carbachol, increases rate of secretion, and thus the osmolality, of the parotid or submaxillary secretions. The salivary gland is interesting in that the sympathetic division also excites secretion and is therefore not inhibitory to the parasympathetic in this respect. The effect of sympathetic stimulation, as mimicked with the β-stimulator isoproterenol, is to produce a fluid with a vastly different composition from that provoked by the parasympathetic, the concentrations of K^+ and HCO_3^- rising to very high levels, thereby making the fluid iso-osmotic with plasma, by contrast with the hypo-osmolality of carbachol-stimulated fluid at comparable flow-rates (Table I).

TABLE I

Concentrations of Ions (meq/litre) in Rat's Submaxillary Saliva provoked by Different Stimuli (Young, Martin and Weber, *Pflug. Arch.*, 1971)

Stimulus	Na^+	K^+	HCO_3^-
Parasympathetic nerve	10	46	28
Carbachol	26	29	43
Sympathetic nerve	21	9	91
Isoproterenol	6	150	129

Influence on Duct-exchanges. When the primary fluid from the acini was examined it was found that this was barely affected by the type of stimulus, so that it must be postulated that the drug-simulated nervous activity was affecting the re-absorptive and other changes in the duct system. Thus it appears that the β-adrenergic mechanism was promoting active transport of K^+ and HCO_3^- into the duct and inhibiting absorption of Na^+. Studies of the combined effects of carbachol and isoproterenol indicated that the two drugs actually exerted similar actions at the duct level. Moreover, it is because the sympathetically provoked primary secretion is so much less in volume than that provoked by the parasympathetic that the concentrations of K^+ and HCO_3^- rise to such high values in the final fluid.

Potential Changes with Secretion

It is well established that there are pronounced changes in potential, both across the secreting epithelium and across individual acinar cells, with the onset of secretion. This is understandable in general terms, since active transport involves accelerated movements of individual

* Containing the parasympathetic supply to the salivary gland from N VII.

ions, as for example in the frog skin. Moreover, it is likely that the stimulus provokes alterations in permeability of the secreting cells to ions and, since the potential across its membrane is a complex function of the permeabilities to the individual ions (p. 124), a change in these relative permeabilities will doubtless be reflected in a change in potential. Thus the change in potential across the gland, or its individual cells, may be taken as the consequence of provoked secretory activity, rather than the primary cause of it; this is the view of Yoshimura and Imai, who measured potentials across individual acinar cells of the dog's submaxillary gland before, and during, provoked secretion. It is also concordant with the striking escape of K^+ from the glandular cells both into the secretion and into the blood when the perfused gland is stimulated with acetylcholine, the parasympathetic transmitter. However, Lundberg considers that the altered potential, which corresponded with a hyperpolarization of the acinar cells, was the *cause* of increased active transport of Cl^- ion.

The Myoepithelial Cells

Injection of the peptide bradykinin (Vol. 2) causes a large rise in pressure in the cannulated duct of the dog's submaxillary gland, and since this is not accompanied by an increased flow of saliva, we may conclude that it is due to contraction of the myoepithelial cells, squeezing the existing fluid out of the acini. In the dog, secretion of the submaxillary gland may be provoked by stimulation of the sympathetic supply, and this also causes contraction of the myoepithelial cells; the two effects can be dissociated, since secretion depends on activation of a β-receptor (Vol. 2) whilst myoepithelial contraction depends on α-adrenergic activity. Thus, when myoepithelial contraction was abolished by an α-blocker, stimulation of the sympathetic caused secretion of saliva, but the emergence of the secretion from the duct was considerably delayed by comparison with that provoked in the absence of the α-blocker. Thus the rapid appearance of saliva in the mouth depends on activating both the *secretory* and *contractile* mechanisms. It was shown by Ludwig in 1851 that the secretion of saliva can take place against very large opposing pressures—150–200 mmHg—but whether this is because the active transport processes can, indeed, work against these pressures, or whether it is the myoepithelial contraction that permits the flow, has not been proved.

Secretory Granules

As indicated earlier, the saliva is not a mere solution of salts in water; it contains mucins and enzymes, the main one being the amylase that converts starch into sugar. These specific secretions are synthesized within the acinar cells where they are recognized as granules of microscopical size. When secretion was stimulated by pilocarpine the number of granules was reduced dramatically.

The mode of elimination of the granules is by exocytosis, namely fusion of the granule membrane with the cell membrane followed by emptying of the granule contents into the acinar lumen.

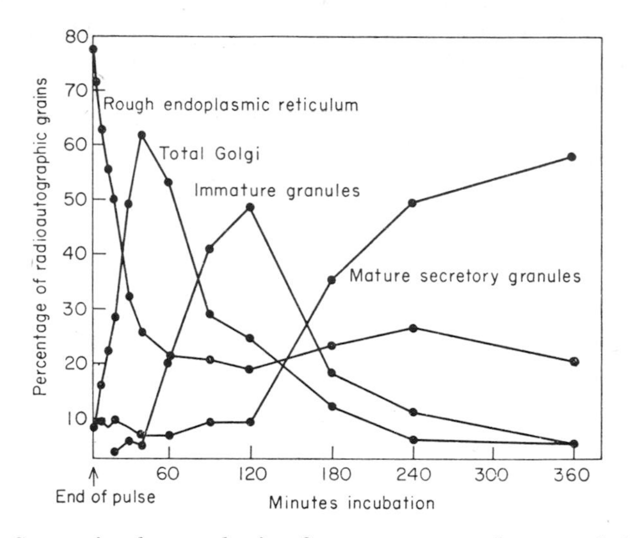

Fig. 9.14. Stages in the synthesis of secretory granules revealed by radioactive pulse labelling. The first radioactive peak is confined to the endoplasmic reticulum, indicating that these particles are the site of protein synthesis. The next peak of radioactivity is in the fraction containing the Golgi apparatus, and then in the swollen vesicles, which were immature granules. Finally, after some four hours, the radioactivity appears in the mature secretory granules. (Castle *et al.*, *J. Cell Biol.*, 1972, **53,** 290.)

The steps in the synthesis of the secretory granules, within the acinar cells of the parotid gland, have been studied by Palade using his technique of pulse-labelling the excised gland with ^{3}H-leucine and following the emitted radioactivity autoradiographically. Initially the radioactivity was confined to the rough-surfaced endoplasmic reticulum, indicating this as the site of synthesis of the protein (amylase); later this passed to the Golgi apparatus becoming concentrated in swollen vesicles that were clearly immature granules. Finally it passed to the mature granules. Figure 9.14 illustrates the changes in radioactivity in the various regions.

GASTRIC DIGESTION

Deglutition

Swallowing is brought about by a complex series of voluntary muscular contractions and relaxations which, involving as they do some 25 muscles, would be tedious to describe in detail. The muscular activities result in the elevation of the root of the tongue and pushing out of the fauces that force the food into the pharynx; the larynx is closed by its own muscles—the true and false vocal cords—together with the downward movement of the epiglottis. Entry into the nasopharynx is prevented by elevation of the soft palate, and the pharynx is lifted up to receive the bolus like the mouth of a sack. The bolus is driven forwards by contraction of the pharyngeal constrictors, which also close the pharyngeal cavity behind the bolus, which thus passes into the oesophagus through the relaxed cricopharyngeus, a motion that has been described as analogous with peristalsis where forward propulsion is preceded by a wave of relaxation. Successive stages in the act of deglutition are illustrated in Fig. 9.15 derived from X-ray cinematograph frames taken during the swallowing of a bolus of X-ray opaque material.

The Oesophagus

This portion of the alimentary tract, connecting the pharynx with the stomach, is a muscular tube, some 25 cm long in man; the muscle layer is distinctive since it consists at the pharyngeal end of striated fibres which are continuous with the striated muscle of the pharynx. Lower down the striated muscle is gradually replaced by smooth muscle so that the final third consists only of this type. The muscle forms two layers, an external longitudinal and an internal circular layer. The innermost—mucosal—layer is formed of stratified squamous epithelium with a few small mucous and serous glands whose secretions contribute to lubrication. The oesophagus is normally occluded at the oral and gastric ends by sphincters, the upper being the cricopharyngeal muscle and the lower an ill-defined muscular region some 2–3 cm above the entrance to the stomach. The upper sphincter effectively seals the oesophagus from the mouth, so that the intraoesophageal pressure is about equal to that in the chest cavity rather than atmospheric (p. 368).

Peristalsis

The presence of the bolus in the oesophagus is followed by a well co-ordinated propulsive wave, called *peristalsis*, that drives the bolus

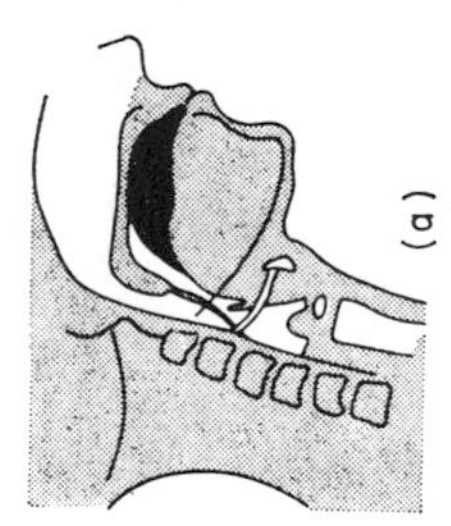

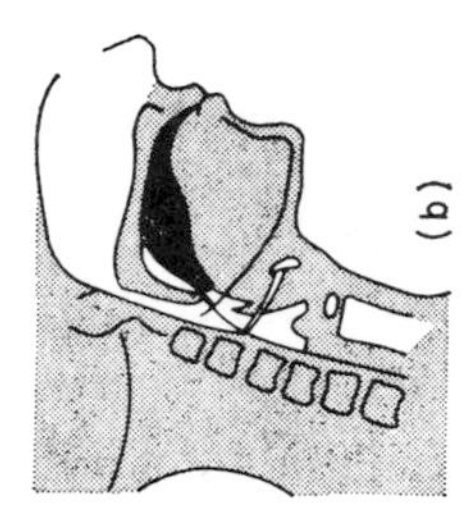

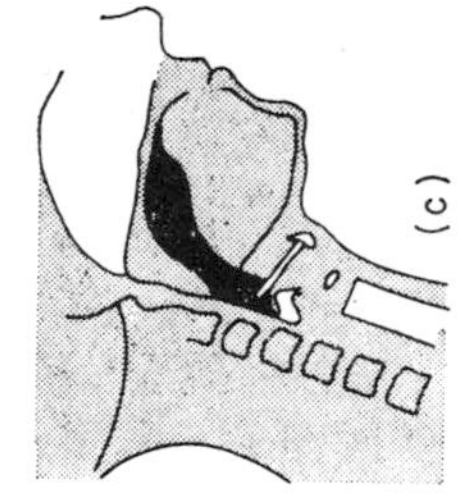

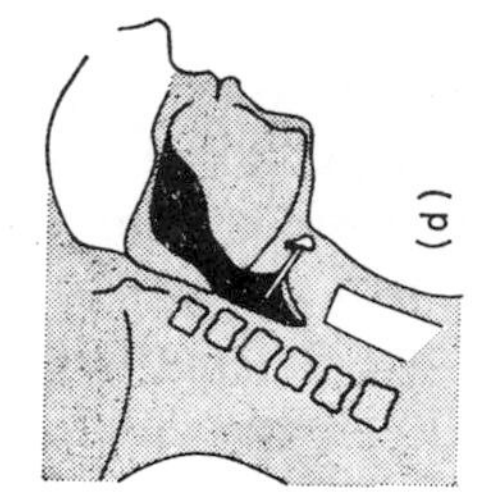

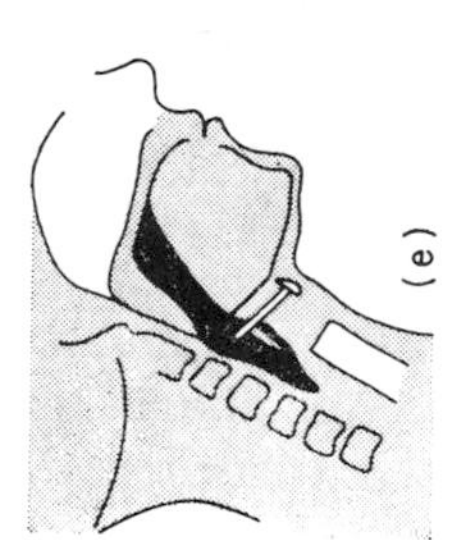

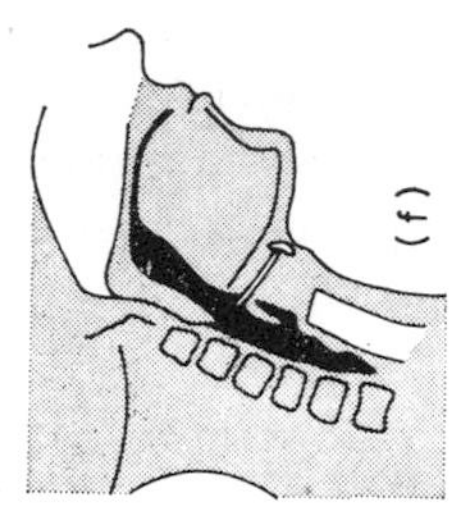

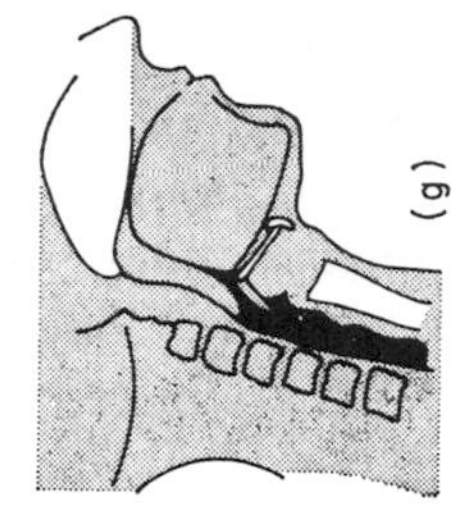

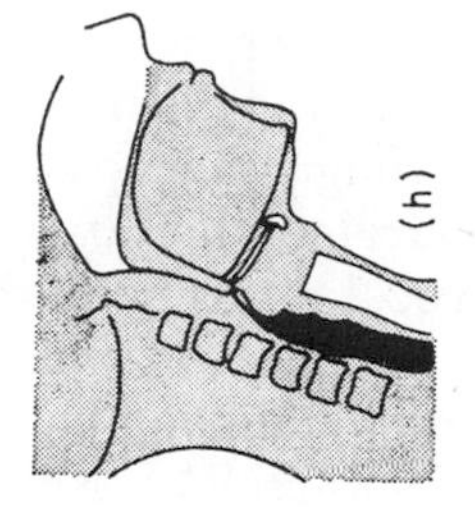

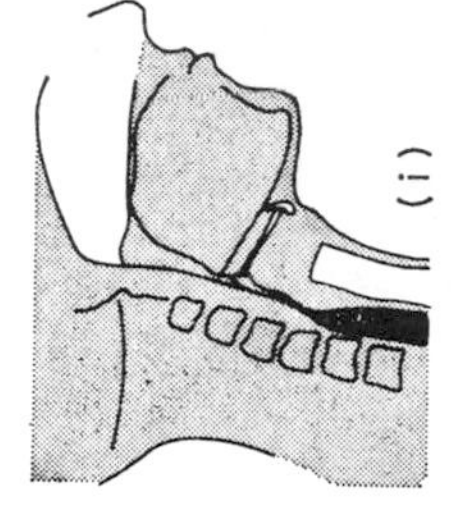

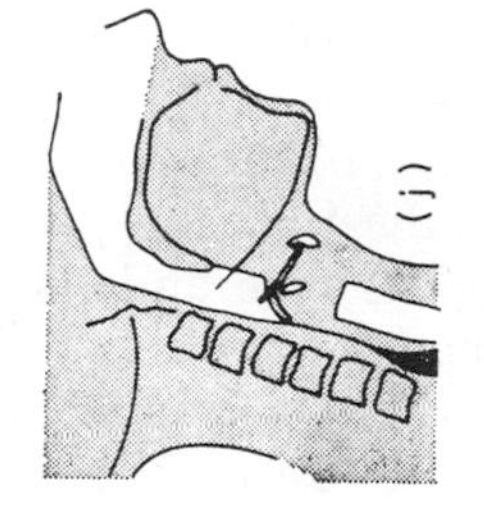

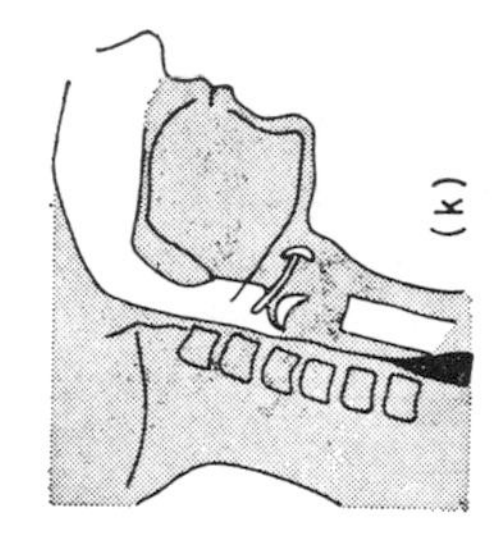

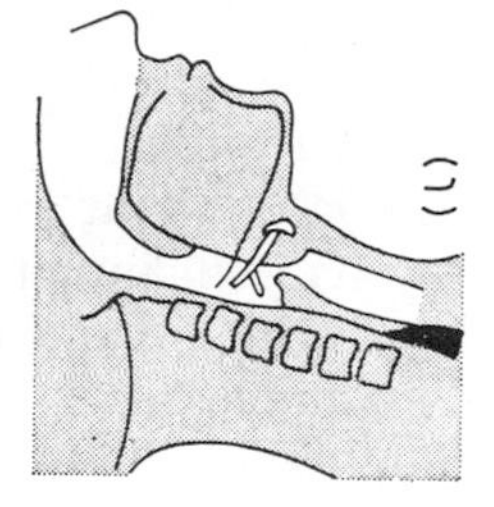

Fig. 9.15.

towards the stomach causing it to reach this, in man, in about 5–6 sec. This peristaltic wave is capable of generating pressures as high as 100 cm H_2O so that it is, in fact, possible for a man to drink while standing on his head.

The contents of a tube may be driven along it by simply constricting it locally at successive points along its length, and this is the basis of what has been called the peristaltic pump (Fig. 9.16a), the direction of flow being governed by the direction in which the successive constrictions take place. The tubes of the gastrointestinal tract are provided with a powerful circular muscle which, on contraction, tends to obliterate the lumen thus providing the mechanical basis for peristalsis; the *aboral*—away from the mouth—direction of movement of the contents is facilitated by the active relaxation of the muscle immediately ahead, so that peristalsis is defined biologically as a wave of contraction of the circular muscle preceded by a wave of relaxation of the muscle immediately in front (Fig. 9.16b). As we shall see, this is a well co-ordinated process, relying on nervous activation and inhibition of the circular muscle; moreover the longitudinal muscle, contraction of which tends to dilate the tube, is also activated and inhibited (Vol. 3).

Fig. 9.15. Drawings of single frames from cinefluorographic study of swallowing. Bolus, containing barium, is shown in black. Passages containing air, as well as hyoid bone, are unshaded. All other structures are stippled. (*a*) Bolus on dorsum of tongue with beginning elevation of soft palate. (*b*) Forepart of tongue elevated, with further elevation of soft palate. (*c*) Complete closure of nasopharynx by elevated soft palate. Further elevation of tongue, with some grooving posteriorly. Epiglottis turning down ahead of bolus. Beginning closure of larynx. (*d*) More grooving of posterior tongue, with further turning down of epiglottis. Complete closure and elevation of larynx. (*e*) Further elevation of anterior tongue with flattening and grooving posteriorly. Hyoid bone and larynx are displaced anteriorly to accommodate bolus. (*f*) Posterior pharyngeal wall is contracting and soft palate is being pulled down by contraction of posterior faucial pillars. Bolus has passed cricopharyngeus muscle. (*g*) Strong contraction of posterior pharyngeal soft tissue has occurred to assist tongue in stripping oropharynx of barium. Further pulling down of soft palate by posterior faucial pillars. Indentation of anterior margin of barium column by cricoid cartilage may be seen in(*f*), (*g*) and (*h*). (*h*) Further progression of posterior pharyngeal thickening moves bolus through pharynx into oesophagus. (*i*) Further progress of contraction wave of posterior pharyngeal soft tissues. (*j*) Pharyngeal contraction waves continues into oesophagus with relaxation above. Opening of nasopharyngeal region and entry of air from above. Beginning flip-back of epiglottis with entry of air into larynx. (*k*) Further progress of oesophageal peristaltic wave and flip-back of epiglottis. (*l*) Resting stage. Flip-back of epiglottis and opening of larynx still incomplete. (Ramsey *et al.*, *Radiology*, 1955, **64,** 498.)

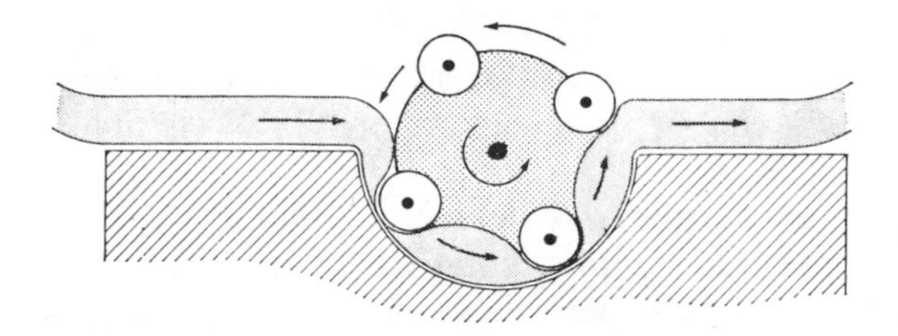

(a) A peristaltic pump

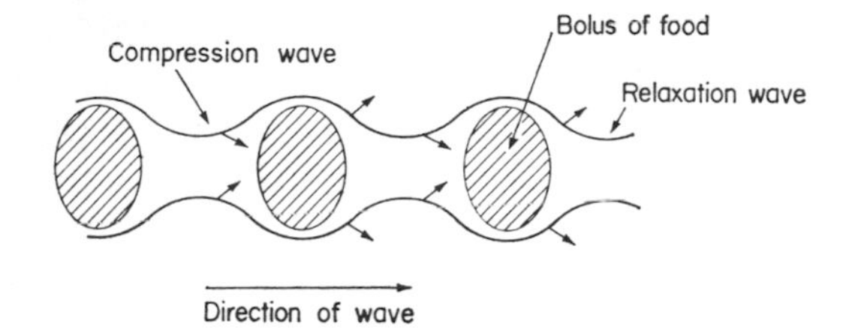

(b) Peristalsis in the gut

Fig. 9.16. (*a*) The peristaltic pump. The tube is compressed locally by the wheels as the drive shaft rotates. The direction of flow is governed by the direction in which successive constrictions take place. (*b*) Peristalsis. The bolus is propelled forward by ring-like constriction of the circular muscle behind the bolus. The muscle ahead of the bolus relaxes actively, and this facilitates the movement of food in the aboral direction along the digestive tract.

Relaxation of Sphincters

When pressure-measuring devices are present in the upper and lower oesophageal sphincters, it is found that both relax—the lower usually some 2 sec after the upper—when the buccopharyngeal phase of swallowing is initiated, so that by the time the wave of peristalsis has reached the lower—oesophageogastric—sphincter, the "door is open" for passage into the cardia of the stomach. That the upper sphincter is normally closed is shown by the fact that when a subject swallows fluid in the upside-down position this does not regurgitate; as we shall see, when discussing the control of the alimentary tract, this closure is probably more an elastic recoil of the surrounding tissues than a maintained tonic contraction of the cricopharyngeus muscle. When a rapid series of swallows is made, the peristaltic contraction of the oesophagus is inhibited until the last swallow, whilst the lower sphincter remains open; the final wave of contraction results in closure

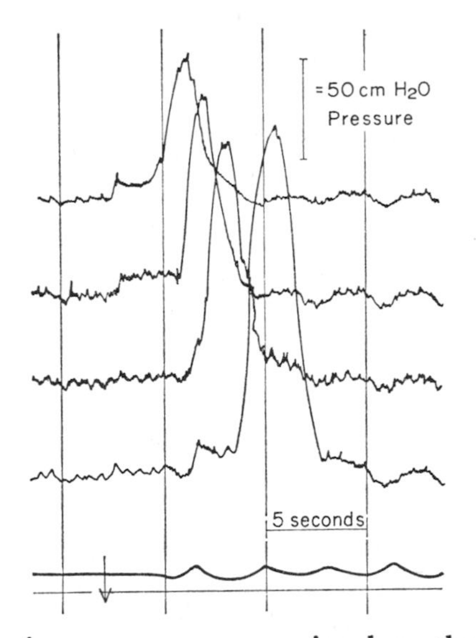

Fig. 9.17. The successive pressure waves passing down the human oesophagus Pressure detectors at 5 cm intervals. The arrow on the lower pneumograph trace indicates the start of swallowing, each excursion on this trace being a swallow. (Code and Schlegel, "Handbook of Physiology", Section 6, Vol. 4, 1821.)

of the lower sphincter. In Fig. 9.17 the successive pressure-waves proceeding down the oesophagus are shown.

The Stomach

The shape of the stomach is illustrated by Fig. 9.18, its parts being divided into the *fundus*, *body*, *pyloric antrum* and *canal*, the pylorus being the region of junction with the beginning of the small intestine. The wall of the stomach, along with that of the intestine, is built on the general plan illustrated by Fig. 9.19, where the innermost layer is called the *mucosa* and the outermost layer the *serosa;* between are the *muscular layer* and the *submucosa.* The innermost cellular layer of the mucosa is the *epithelium*, which may become differentiated to form secretory glands which may be in the mucosa, as with the glands of the stomach, and some intestinal glands; they may occur in the submucosa (as in the duodenum) and the third group include glands lying outside the wall of the alimentary tract, opening into the tract through long ducts; these are the pancreas, salivary glands, and the liver, which empties the contents of its gall-bladder into the small intestine.

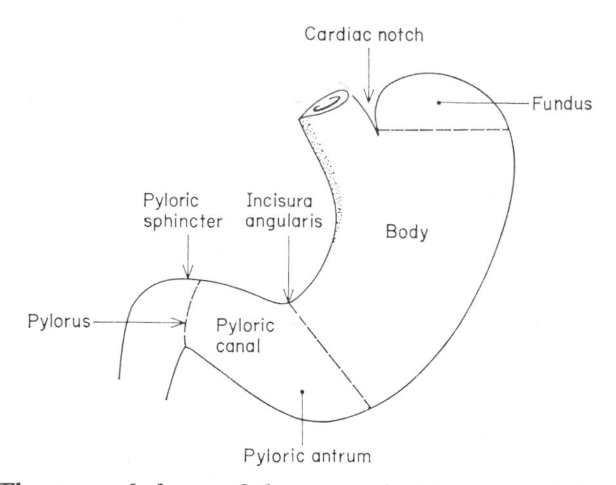

Fig. 9.18. The general shape of the stomach. The stomach is divided into three sections—the fundus, body and pyloric region. The pyloric region is subdivided into the pyloric antrum, the canal, and finally into the pylorus which connects the stomach to the small intestine.

Glands

The secretory structures of the stomach are contained within the mucosa and are essentially folds of its epithelium. Thus Fig. 9.20 illustrates a longitudinal section through a gastric gland; its length is some 1·3–3·5 mm and it consists of a number of tubules made up of cells of different types; these tubules open into surface folds or foveolae or crypts. Along its length the gland is differentiated histologically into four regions from the surface downwards as *crypt, isthmus, neck* and *body* regions. The lining cellular layers of the tubules are made up of four types of cell, doubtless subserving different functions. These are: the *columnar* or *surface cells,* on the surface of the mucosa, which secrete mucus; the *chief cells of the neck* of the gland, probably also secreting mucus; *chief cells of the body* of the gland, secreting the proteolytic enzyme, pepsin; and the *parietal* or *oxyntic* cells secreting hydrochloric acid. In certain regions of the mucosa the glands are different; thus in the pyloric area the surfaces of the glands are made up of cells resembling neck chief cells with almost or complete absence of oxyntic (acid-secreting) cells. Chief and oxyntic cells are absent from the glands of the cardiac area, a ring of mucus-secreting glands at the junction of the oesophagus and stomach.

Mucus. The mucus secreted by the surface cells is a highly viscous gel; it normally covers the surface of the resting stomach in the absence of acid secretion, but the amount may be increased by mechanical

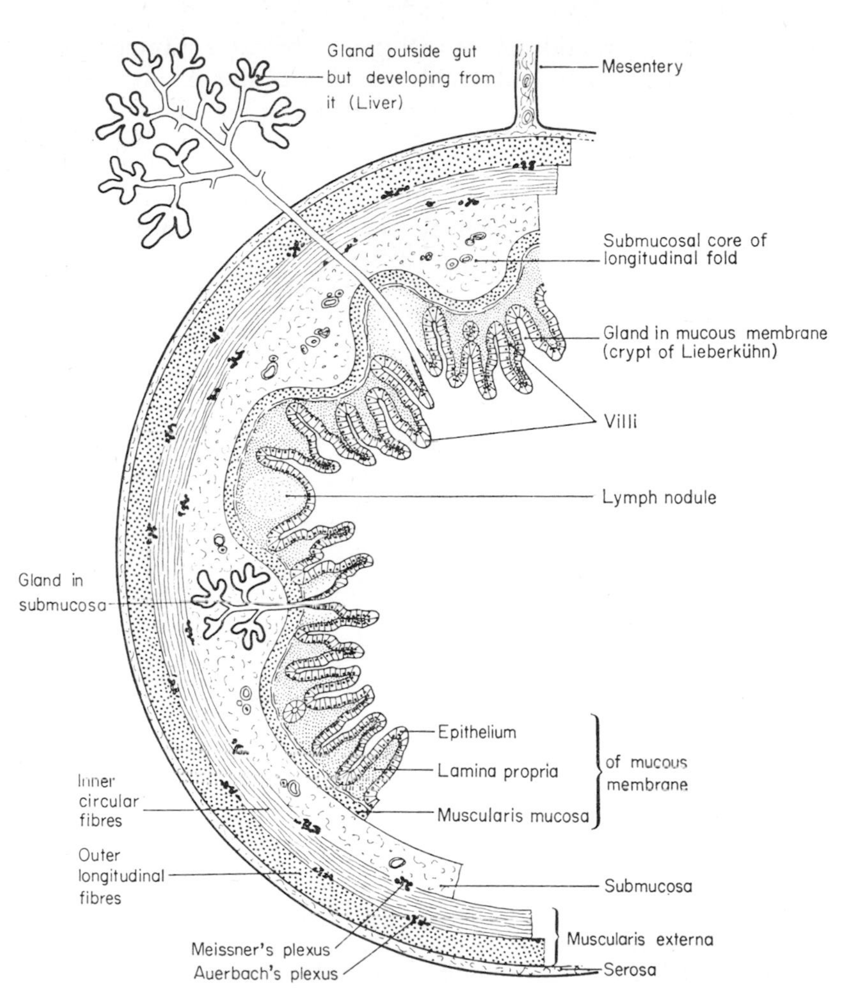

Fig. 9.19. A general plan of the gastro-intestinal tract. The wall is composed of four layers, the innermost layer being the mucosa with an epithelial lining in contact with the food. Beneath the mucosa is a submucosal layer containing glands and connective tissue. The muscle layer contains fibres running in a circular fashion on the innermost side and an outer longitudinal layer. The outermost coat of the digestive tract is the serosa which consists mainly of connective tissue. (Ham, "Histology", Lippencote and Co., 1957.)

stimuli, and chemical and nervous influences that cause acid secretion. The mucus secreted by the neck chief cells is chemically different from that secreted by the surface cells; unlike the latter it is only secreted in response to a stimulus. It is reasonable to believe that the mucous secretions, in addition to acting as lubricants, protect the cells of the mucosa from the highly acid secretion of the oxyntic cells and, perhaps, also, from the digestive action of the proteolytic enzymes.

Chief Cells

These last are secreted by the chief cells, where they may be seen as granules which disappear from the cell during secretion; if secretion

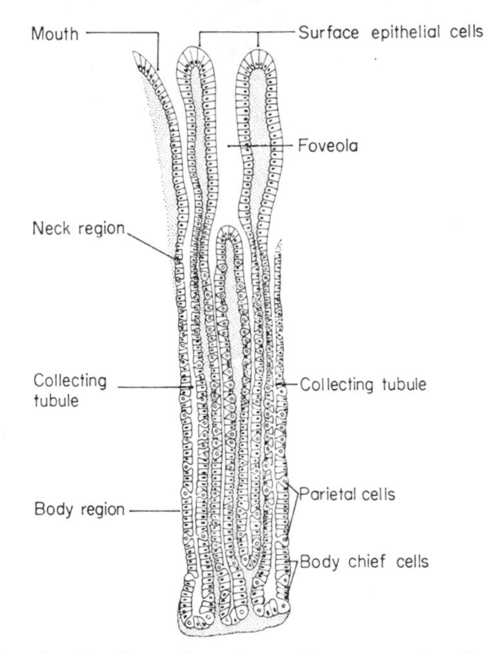

Fig. 9.20. A longitudinal section through a gastric gland. The gland is built up of a number of different types of epithelial cells and is subdivided into four regions: the crypt, isthmus, neck and body regions.

continues for some time the cells become free of granules, although secretion of enzymes continues, indicating that the granule is essentially a storage form of the enzyme, ready for immediate use. When demands are great the enzyme apparently leaves the cells in the form of submicroscopic units.

Pepsins. The enzymes in the secreted juice are pepsins, but in the cells they exist as inactive precursors; and it is presumably for this reason that they do not digest the cytoplasm of the cells. As a result of contact with the acid secretion of the oxyntic cells, the *pepsinogens* are converted to *pepsins*. Thus the best known of the pepsins, from pig mucosa (pepsin II), is derived from pepsinogen II, with a molecular weight of 42,000, by the splitting off of some nine amino acid residues to leave a protein of molecular weight 34,500. This enzyme is synthesized in the chief cells in the oxyntic gland area, whilst the location of the other enzymes is uncertain.

Secretion of Acid

To the physiologist, the striking feature of the gastric secretion is its acidity which may reach a value equivalent to decinormal hydrochloric acid. We may therefore devote some space to a description of the basic research that has been carried out in an attempt to elucidate the mechanism of secretion of acid in physico-chemical terms.

The Oxyntic or Parietal Cell

That this cell secretes the acid was demonstrated by Linderstrom-

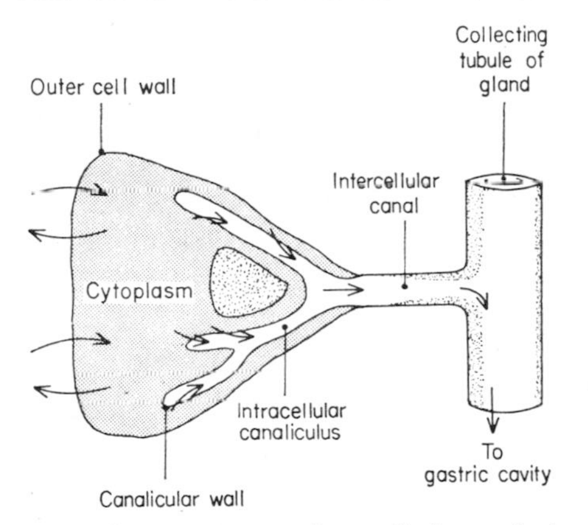

Fig. 9.21. Diagrammatic representation of the relationship between a parietal cell and the gastric tubule. (After Hollander, *Gastroenterology*, 1961, **40,** 477.)

Lang, who correlated the amount of acid secreted by a portion of mucosa with the number of oxyntic cells in it. The oxyntic cell has a unique structure, in so far as it contains within it a microscopically observable *intracellular canaliculus*, into which the acid is secreted (see Fig. 9.21). This canaliculus either opens directly on to the surface of the gland or else into an intercellular canal, which opens into a collecting tubule opening into the gastric cavity. The important point is that the elaboration of the secretion in this particular instance is not into an intercellular cleft, as with, say, the salivary gland, but into a canal that begins within the cell, a device that may well protect the cell cytoplasm from the destructive effects of its own secretory activity, although the absence of intracellular canaliculi in the bull-frog's oxyntic cells must give us pause.

Parietal Secretion

The fluid collected from a gastric pouch in response to an appropriate stimulus, e.g. the injection of histamine, is clearly a mixture of the secretions of the various types of cell constituting the secretory apparatus of the gastric mucosa, hence to identify the products of the oxyntic cells is a matter of deduction rather than of direct experiment. The main experimental finding is that the composition of the juice alters as the rate of secretion changes.

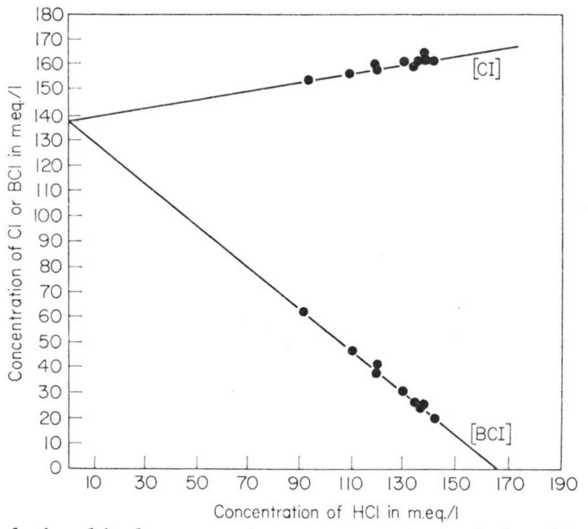

Fig. 9.22. Relationship between the concentrations of the chloride fraction of gastric juice. As the acidity of the juice increases, the concentration of neutral chloride (BCl) falls. (Gray, *Gastroenterology*. 1943, **1**, 390.)

Variations in Composition

Thus, if the amount of alkali required to neutralize the secretion is measured, this increases as the rate of secretion increases; since the concentration of chloride remains fairly constant this means that the proportions of neutral chloride (NaCl, KCl, etc.) to acid chloride (HCl) change (see Fig. 9.22, in which is plotted the concentration of HCl in the secretion as abscissa, and as ordinate the concentration of neutral chloride.) The reciprocal relation between the two is clear; if we extrapolate to zero concentration of neutral chloride we reach a figure of 0·168 M HCl, i.e. a solution of about isotonic hydrochloric acid.

"Diluting Fluid". Davenport concluded from this finding that the secretion, as originally formed, was in fact pure HCl, and that the increasing proportions of neutral chloride found with slower rates of secretion represented the effects of neutralizating this acid by the secretions of the other cells of the mucosa. The "diluting fluid" was considered to be mainly a mixture of $NaHCO_3$ and $KHCO_3$, which would react to give NaCl and KCl, the CO_2 liberated passing into the blood.

Teorell Hypothesis. The alternative hypothesis, suggested by Teorell, is that the primary secretion of the oxyntic cell is, indeed, largely pure HCl, but that during its passage over the mucosa exchanges between blood occur, resulting in loss of H^+-ions and gain of Na^+ and K^+. Thus, at rapid rates of secretion, the time for exchanges would be small and the collected fluid would have nearly the same composition as that of the primarily secreted fluid. As the rate of secretion slowed there would be more time for exchanges, and the proportion of neutral chloride would increase. Teorell's hypothesis seems eminently reasonable, but cannot be accepted without further work because studies on the exchanges of H^+ and Na^+ across the gastric mucosa suggest that, when the mucosa is secreting acid, these are remarkably slow, as if some mechanism came into play to prevent the loss of acidity that would otherwise occur; in the resting, i.e. non-secreting stomach, the exchanges were relatively rapid.

Mechanism of Acid Secretion

There is no doubt from many histological studies that the oxyntic cells do not accumulate acid which is subsequently ejected into the canaliculus. Instead, we must assume that the oxyntic cell actively transports acid, or rather H^+-ions, from its cytoplasm into the canaliculus; and this is confirmed by observation of the colours of dyes, presented to the isolated frog's mucosa, that acted as pH indicators. These dyes appeared in the gastric secretion, and the red colour of, for example, toluene-azo-amino-toluene-2:1:1:4:3 indicated a pH of less than 1·4 in the tubules; the site of secretion seemed to be at or near the walls of the canaliculus.

Splitting of Water

If, then, the oxyntic cell behaves as a transporter of H^+-ions into the canaliculus, we must assume that these are ultimately derived from H_2O:

$$H_2O \longrightarrow H^+ + OH^-.$$

Consequently, if the secretion is not to be neutralized as soon as formed, the OH^--ions must either be retained by the cytoplasm, making it impossibly alkaline, or these must be passed on to the blood. That the latter occurs, is suggested by the observation that the gastric venous blood is some 0·07 to 0·09 pH units more alkaline than arterial blood, when the stomach is secreting, a difference that is absent at rest. Again, it was observed that the amount of CO_2 that could be collected from a stomach, by blowing a current of N_2 through it, was considerably less when it was secreting acid than at rest; since it had been found that the amount of CO_2 carried out was proportional to the P_{CO_2} of the blood, the finding of a diminished outflow indicated a lower P_{CO_2}, i.e. the blood had been taking up OH^--ions.

Isolated Mucosa

The best demonstration, however, was made by Davies, who exploited the fact that the frog's gastric mucosa may be isolated in the form of a sac and suspended in a nutrient medium; under these conditions the amount of acid formed in the sac was equal to the amount of alkali liberated into the medium. The OH^--ions liberated from the splitting of water by the oxyntic cell presumably diffused back into the medium in exchange for Cl^-, which accompanied the H^+-ions into the secretion.

Carbonic Anhydrase

More probably, the OH^- reacted with metabolic CO_2 to form the bicarbonate ion, HCO_3^- (Fig. 9.23). Thus, the secretion of acid is strongly dependent on the activity of the enzyme, carbonic anhydrase, which, as we have seen, catalyses the reaction:

$$CO_2 + H_2O \rightleftharpoons H_2CO_3$$

so that inhibiting this catalyst with acetazoleamide (Diamox) inhibits secretion. We may presume, therefore, that the OH^--ions produced by the oxyntic cells combine with carbonic acid, so that we have the chain:

$$\begin{array}{ccccc} CO_2 + H_2O & \longrightarrow & H_2CO_3 & \longrightarrow & H^+ + HCO^-_3 \\ & & & & + \\ & & & & OH^- \\ & & & & \downarrow \\ & & & & H_2O \end{array}$$

Thus the effect is to remove CO_2 from blood, converting it into bicarbonate, thereby reducing the P_{CO_2} of the blood.

Gastric Potentials

By analogy with the frog skin, which actively transports the positive Na^+-ion from outside to inside to produce a difference of potential with the inside positive, we might expect the gastric mucosa to produce a similar potential by virtue of transporting H^+-ions from the outside to the inside. However, this is by no means a necessary hypothesis, as has been argued earlier. Thus simultaneous transport of Cl^- ions might result in a "neutral pump", or alternatively the secretion of

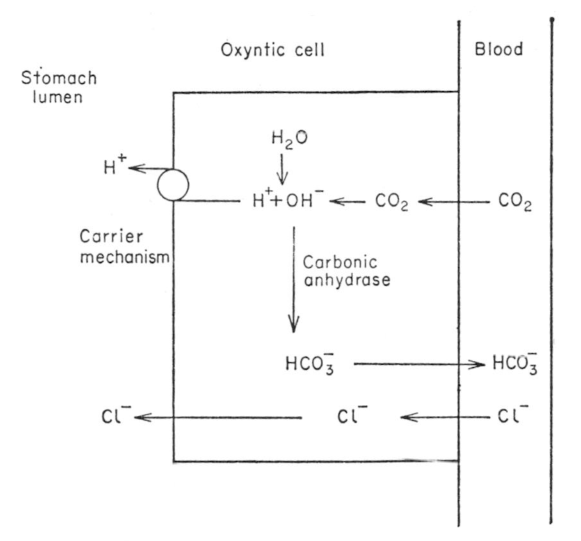

Fig. 9.23. The mechanism of H^+ ion secretion by the stomach. Carbon dioxide diffuses into the oxyntic cell and is converted into carbonic acid, a reaction catalysed by carbonic anhydrase. H^+ ions, derived from carbonic acid, pass by a "H^+ pump" into the lumen of the stomach and HCO_3^--ions diffuse into the blood stream.

HCl might result, not from an active transport of H^+-ions, but of Cl^--ions with the H^+-ions following passively. In the intact animal a potential of some 40–60 mV may be measured across the mucosa, the mucosa being *negative* in relation to the outside (serosa); this potential is measured in the resting (non-secreting) stomach; when secretion is provoked by administering pilocarpine or histamine the potential falls. The potential is thus related to the secretory process in so far as it falls during secretion, but the fact that it is high and stable when there is no acid secretion is not easy to interpret.

Chloride and Hydrogen Ion Pumps. According to a study by Hogben on the isolated mucosa, the short-circuit current (p. 126)

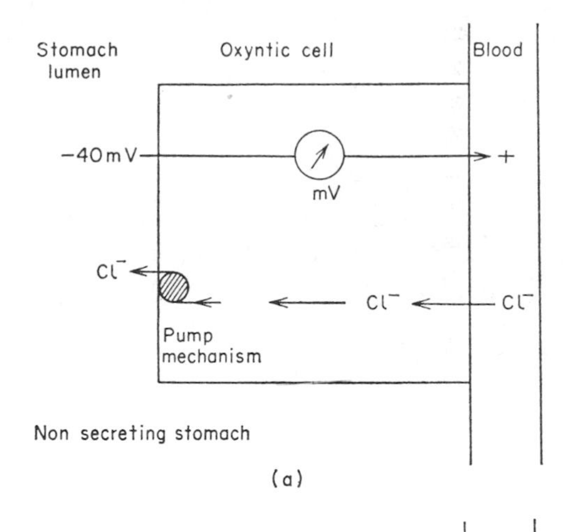

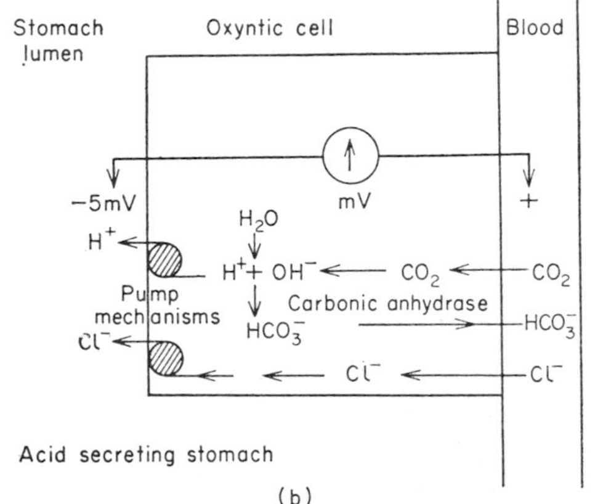

Fig. 9.24. The secretion of acid by the stomach and the potentials associated with secretion. (*a*) The active transport of chloride ions in the non-secreting stomach is thought to produce a potential difference of 40 mV negative, stomach lumen to blood. (*b*) In the secreting stomach the potential falls as H^+ ions are moved into the lumen of the stomach to accompany the Cl^--ions by a pump mechanism as illustrated.

indicates a continuous active transport of Cl^--ions by the resting mucosa; and this would account for the internal negativity of the stomach (Fig. 9.24a). Thus it is postulated that the non-acid-secreting stomach continuously secretes Cl^- accompanied by metallic cations such as Na^+ and K^+. When secretion is provoked, the H^+-ion pump, actively transporting H^+ into the stomach, comes into play. This could produce an oppositely directed potential, the acceleration of the H^+-ions by the active pump causing an internal positivity (Fig. 9.24b).

This would account, then, for the fall in gastric potential at the onset of secretion. Confirmation of this view is provided by replacing the Cl^- available to a secreting mucosa by an ion such as SO_4^- that would, on theoretical grounds, be unlikely to be able to replace Cl^- in the active transport of negative ions. The secretion of acid continued at about two-thirds its original rate, but a potential of opposite sign could be measured, presumably that associated with the active transport of H^+-ions.

Metabolic Inhibitors

We have seen how acetazoleamide inhibits gastric secretion by virtue of its attack on the carbonic acid synthesis stage in neutralizing the OH^--ions formed as by-products of the acid. As with other active

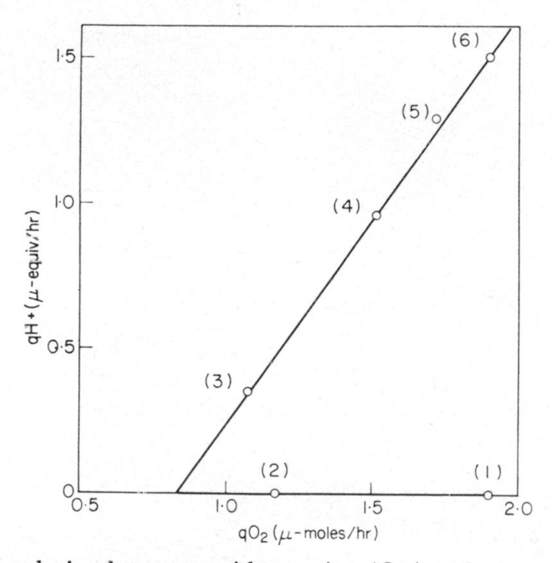

Fig. 9.25. The relation between acid secretion (Q_H) and oxygen consumption (Q_{O_2}) in an everted sac of frog gastric mucosa. Numbers refer to period of incubation. The slope of the line is 2–3 and the intercept on the Q_{O_2} axis represents the basal oxygen consumption of the non-secreting stomach. (Bannister, *J. Physiol.* 1965, **177,** 429.)

transport processes, the secretion is inhibited by more general metabolic poisons, such as DNP, cyanide, etc. The energy utilization in secretion can be estimated, as indicated earlier, by the extra oxygen consumption provoked by secretion. In Fig. 9.25 we see that the rate of acid secretion (Q_H) of a frog gastric mucosa is linearly related to the rate of oxygen uptake, Q_{O_2}, the slope of $\Delta Q_{O_2}/\Delta Q_H$ being a theoretical

2·3, and the intercept on the Q_{O_2} axis representing presumably the basal oxygen consumption of the non-secreting stomach. There is some evidence that some of this basal metabolic energy can be diverted into the secretory process; if this is true, then the extra oxygen consumption provoked by secretion gives an underestimate of the energy actually employed.

Gastric Digestion

Muscles

As illustrated by Fig. 9.26, the stomach is provided with three layers of muscle—longitudinal, circular and oblique—at successive depths. The circular muscle is especially pronounced at the pyloric end, forming the pyloric sphincter which on contraction blocks off con-

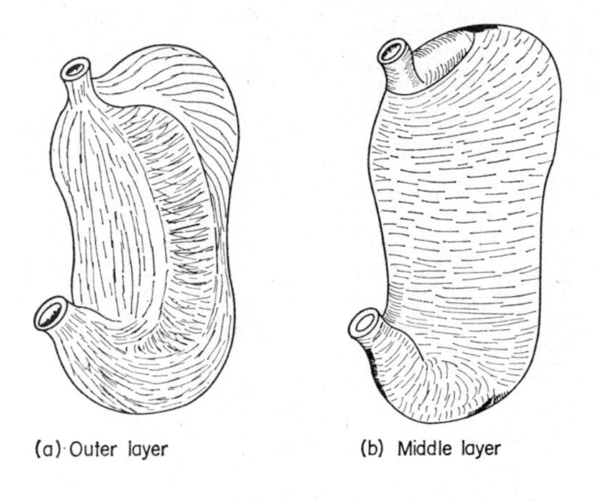

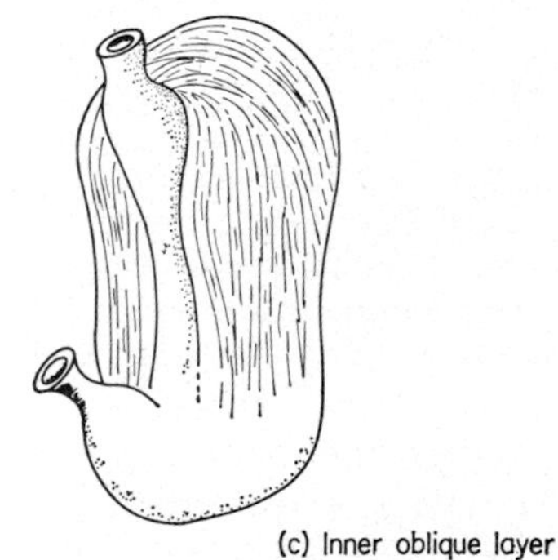

Fig. 9.26. Illustrating the muscle layers of the human stomach. (*a*): outer longitudinal layer; (*b*): middle circular layer; (*c*): inner oblique layer. (From Lockhart *et al.*, "Anatomy of the Human Body".)

nection with the duodenum. The longitudinal and circular muscles are continuous with the corresponding muscles of the oesophagus, whilst some fibres of the longitudinal coat are continued into the duodenum. An additional muscular layer is the muscularis mucosae immediately beneath the glands of the mucosa, contraction of the muscle fibres presumably assists in expulsion of the secretions.

Cardiac Sphincter

The bolus of food enters the stomach easily at the cardiac end. Regurgitation, during the powerful contractions of the stomach that take place during digestion, does not normally take place, and this is due to the sphincter-like action of the few centimetres of oesophagus leading into the stomach, the so-called *cardiac sphincter.* Physiologically this valve-like action can be demonstrated by measuring the pressures required to force fluid from oesophagus to stomach and in the reverse direction; these are 5 and 80 mmHg respectively. The anatomical basis for this sphincter-action is difficult to determine. There is certainly no ring of muscle that actively contracts to close off the entry, as with other sphincters of the alimentary tract; and it may simply be the oblique entry of the oesophagus into the stomach that produces a valvular flap which tends to close as the pressure in the stomach rises, and open when that in the oesophagus rises. The rise in pressure due to the wave of contraction of the oesophagus is presumably adequate to cause opening, but, as we have seen, the act of deglutition is accompanied by a relaxation of the sphincter region that occurs well before the arrival of the peristaltic wave.

Digesting Vat

The history of the food in the stomach is best understood if we return to the basic structure of this organ; essentially it is a hollow viscus that acts as a digesting vat, a process that requires time. The stomach must be able to retain its contents for some time, it must be able to stir these, and finally it must expel them into the more distal or caudad portion of the gastrointestinal tract. The storage functions of both fundus and body are illustrated by Fig. 9.27 which shows the degrees of expansion resulting from ingestion of larger and larger volumes.

Receptive Relaxation

In general, we may say that the fundus acts as a storage depot only, adapting its walls to increases in volume without appreciable increase in pressure. It is enabled to do this by a reflex relaxation of its musculature as a result of the act of swallowing; this was shown by making a pouch of a portion of fundus and observing the decrease in pressure in a balloon following deglutition movements. The phenomenon has been called *receptive relaxation;* it is part of the complex series of events comprising swallowing and is presumably controlled by the "swallowing centre" in the brain. The body, or corpus, also acts as a storage space, adapting to changes in volume without development of large tensions, but it mixes its contents and propels them towards the antrum.

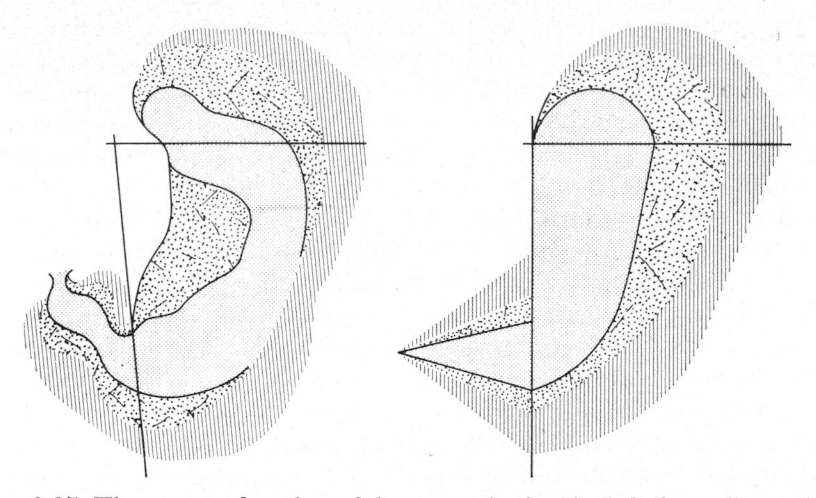

Fig. 9.27. The storage function of the stomach. On the left the radiographic silhouette of the stomach of a human after the ingestion of 2 oz (stippled area), 10 oz (intermediate zone) and 20 oz (dark outer area) of barium meal. On the right the volumes ingested reproduce the silhouette of the stomach reasonably well when assigned to the fundus as a hemisphere, the corpus as a cylinder and the antrum as a core. The fundus and the body are the principal storage regions of the stomach and reflexly relax to accommodate ingested food. (Code and Carlson, "Handbook of Physiology", Section 6, Vol. 4, 1903.)

Antral Systole. The antrum also changes in volume, but the changes are not concerned with storage but with propulsion and what Code has called *retropulsion.* Thus peristaltic waves of propulsion pass down the antrum; if the pyloric channel remained open during the whole duration of this wave the result would be simply the propulsion of gastric contents into the duodenum. However, the pyloric muscle also participates in this wave so that very soon the communication between antrum and duodenum is cut off and the energy of the wave is expended in retropulsion of the contents back towards the body, at the same time the antral contents tend to be triturated. Thus it is customary to speak of *antral systole*, by analogy with the contraction of a cardiac chamber following its expansion. However, the contraction of the antrum, following its expansion, is not followed by expulsion of all its contents in the same direction, by contrast with the situation in a cardiac chamber.

Types of Muscle Contraction

In general we may divide the contractions of the stomach into two functional types, the one being static and the other migratory. The migratory contraction has been called peristalsis and consists in the contraction of a ring of muscle; at any given moment only a few millimetres in width of this ring contract, but the process is propagated behind a wave of relaxation giving rise to the propulsive movement. The non-propagated types of contraction are characteristically seen as a rhythmic increase in tone of the body of the stomach, leading to a small, if any, rise of pressure in a balloon incorporated into the lumen.* They may be regarded as base-line changes in tension of the stomach's wall on which are imposed the more vigorous types of contraction concerned with peristalsis.

Terminal Antral Contractions. The contractions occurring in the antrum are especially marked and are described as *terminal antral contractions*, mainly because they represent the terminal phase of a wave of peristalsis. They consist in a simultaneous vigorous contraction of the terminal segment of the antrum and pyloric canal. which is clearly related to a wave of peristalsis from the more cephalad portion of the stomach. As a consequence of this contraction, some contents are expelled into the duodenum, but as Fig. 9.28 shows, the pyloric canal is shut off early in the process so that retropulsion of the gastric contents towards the fundus is the main result.

Gastric Emptying

Examination of the fate of radio-opaque meals shows that gastric emptying and digestion go hand in hand, so that within a few minutes some of the gastric contents have passed into the duodenum. In general, the pattern of emptying is remarkably constant, the rate at any moment being proportional to the amount present, so that a straight line is obtained if the logarithm of the volume of gastric contents is plotted against time (Fig. 9.29).

* Code has named the types of contraction in the stomach as: Type I, the increase in tone; and Types II and III, the peristaltic type which differ in the vigour of their contraction —Type II being the more vigorous and typically seen in the antrum.

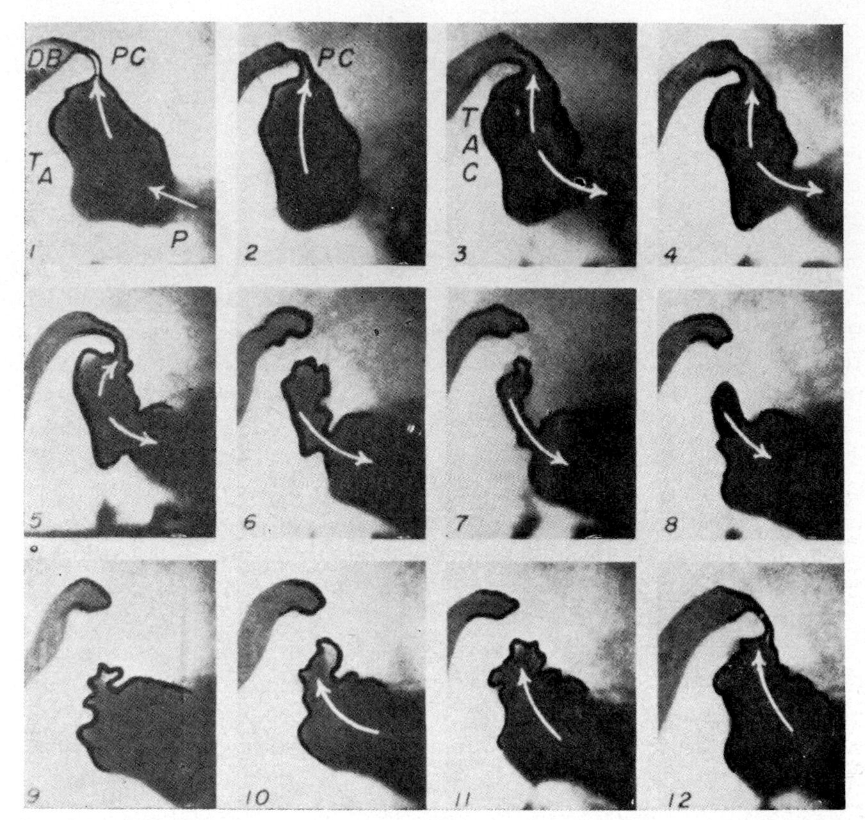

Fig. 9.28. Cineradiographic sequence of an antral cycle. In frame (1) a peristaltic type II contraction (P) is approaching the terminal antrum (TA). The pyloric canal (PC) contains barium. Some barium is entering the duodenal bulb (DB). Arrows define direction of movement of barium. With the terminal antral contraction (TAC, frame 3), barium is propelled into the duodenum and retropelled into the antrum and corpus of the stomach. The pyloric canal closes (frame 6), and the remainder of the barium in the terminal antrum is retropelled. The antrum and terminal antrum are then filled by an approaching peristaltic contraction (frames 10–12). (Code and Carlson, "Handbook of Physiology", Section 6, Vol. 4.)

Antral Pump. The transport from stomach to duodenum is one that has not been completely elucidated in spite of many observations; Fig. 9.30 shows the anatomy of the gastroduodenal junction, and it will be seen that the initial region of the duodenum has the form of a cap— the *duodenal cap* (D.C.). A true sphincter, in the sense of a specialized ring of muscle, probably does not exist, and an *antral pump* that forces the chyme both forwards and backwards when it contracts,

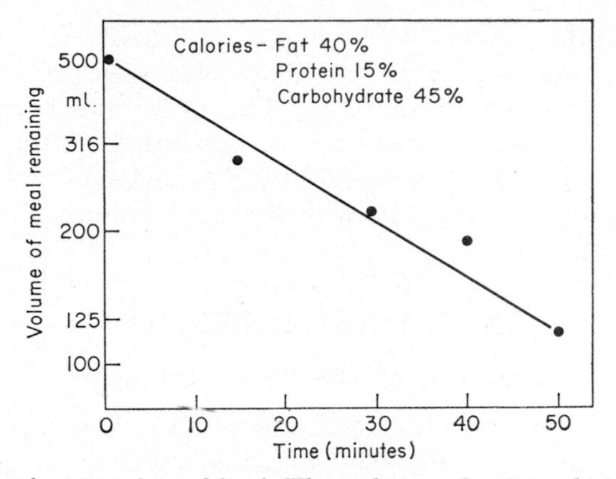

Fig. 9.29. Gastric emptying of food. The volume of a 500 ml meal, recovered at given times, has been plotted against time. Note the logarithmic scale of the ordinates. The linear graph indicates that rate of emptying is proportional to the amount remaining in the stomach. (Hunt and Knox, "Handbook of Physiology", Section 6, Vol. 4.)

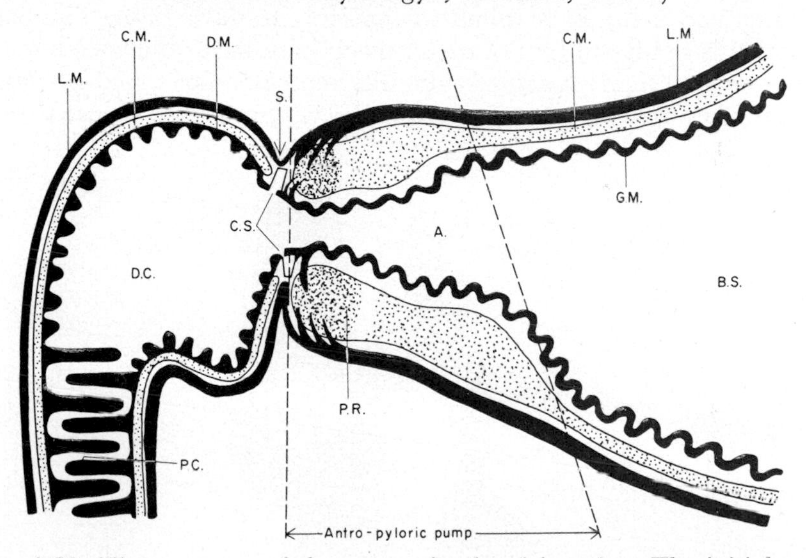

Fig. 9.30. The anatomy of the gastro-duodenal junction. The initial part of the duodenum (DC) has the form of a cap. A true sphincter probably does not exist, the antral pump (A) forces chyme both forwards and backwards into the duodenum and back into the stomach for further mixing. CM, LM: circular and longitudinal muscle; GM: gastric mucosa; A: antrum; PR: pyloric ring; S: shallow groove; CS: connective tissue septum; PR: pyloric ring; PC: plicae circulares; DM; duodenal mucosa; BS: body of stomach. (Edwards and Rowland, "Handbook of Physiology", Section 6, Vol. 4.)

seems the best description; in this way only small amounts of chyme are forced at any time into the duodenum whilst the retropulsive movement ensures further mixing. The very acid contents of the stomach, and the relatively neutral or alkaline medium required for duodenal digestion, make a transfer by repeated short squirts the best manner of transport consistent with the maintenance of a reasonably steady pH level.

INTESTINAL DIGESTION

Structure of the Small Intestine

In the small intestine the digestion of food is completed and its hydrolytic products are absorbed into the blood stream, a residue of unabsorbed material passing into the large intestine or colon. The small intestine is conventionally divided by anatomists into three sections from the stomach onwards, namely the *duodenum*, *jejunum* and *ileum* but the differences are not large, and the transition between the jejunum and ileum in mammals is not easy to distinguish. The basic structure is similar to that of the stomach in so far as the wall is made up of an innermost mucosa, a middle muscular layer and an outer serosal layer, but the absorptive function of the tissue is revealed by a

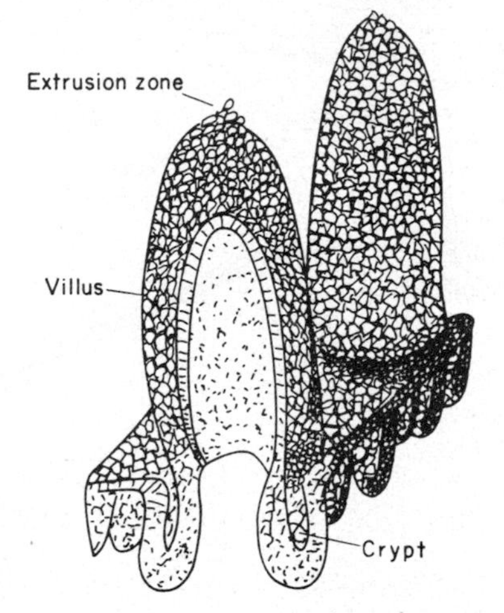

Fig. 9.31. A three-dimensional diagram of an intestinal villus. The whole of the internal surface is covered with these finger-like villi. At the base of the villus are the crypts of Lieberkühn. The epithelial cells of the crypts are continuous with those covering the villi, and the cells migrate from the base of the crypt to the apex of the villus where they are shed at the extrusion zone. (Greep, "Histology", McGraw, Hill, 1966.)

characteristic alteration in the mucosa, which is now thrown into circularly arranged folds that include the submucosa, the *plicae circulares* or *valves of Kerckring*. The whole surface of the mucosa is studded with innumerable *villi;* at their bases are tubular depressions or pits extending to the mucosal muscle; these are the *intestinal glands* or *crypts of Lieberkühn* (Fig. 9.31).

The Villus

The villi are the absorptive units of the intestine, whilst the crypts provide some of the digestive enzymes. In addition, the cells lining the crypt are the source of renewal of the epithelial lining to the villus. Thus the simple columnar epithelium that lines the crypts, and covers the villi, is a continuous sheet, most of which is constantly renewed. It contains several distinct types of cell; in the crypt the principal one is the undifferentiated *columnar cell*, which divides frequently; the daughter cells move on to the villus as *absorptive cells* and are finally cast off at the apex at a site called the *extrusion zone*.

Turnover of Epithelial Cells. This turnover of epithelium is best demonstrated by tagging the dividing cells with radioactive DNA. This may be done by feeding a radioactive precursor to DNA to the animal, e.g. tritiated thymidine. When a cell undergoes mitotic division its DNA is replicated by synthesis of new DNA, so that the daughter of a dividing cell contains newly synthesized DNA; if tritiated thymidine is present in the blood at the time of synthesis, the daughter cells become labelled with tritium, when they can be identified by autoradiography. It is found that the life-span of an epithelial cell is remarkably short, some 2–3 days; in fact the whole epithelial lining of the gastro-intestinal tract, from stomach to rectum, is completely renewed every 2–4 days in man. During migration of the absorptive cells they differentiate progressively, and it would seem that the absorptive capacity of the epithelial cells increases as they approach the apex of the villus.

Intestinal Glands

In addition to these two main types of cell we may mention the *goblet* and *argentaffin*, or *enterochromaffin* cells interspersed between the main types both in the crypt and villus; these are mucus-secreting cells. The bottom of the crypt is lined with a cluster of *Paneth cells;* in the electron microscope these can be seen to contain characteristic zymogen granules, and there is little doubt that these cells are the source of the digestive enzymes secreted by the small intestine, in addition to those provided by the pancreas, so that the crypts of

Lieberkühn are described as the *intestinal glands*. The epithelial cells, shed into the lumen of the gut, will, however be mixed with this secretion, and there is no doubt that these cells contain within them large concentrations of digestive enzymes; thus the tiny microvilli covering the surface of the absorptive cell (Fig. 9.31, p. 442) may be separated mechanically from the rest of the epithelium, and it is found that they contain nearly all the invertase [which converts sucrose to glucose and fructose (p. 400] and maltase, and 75 per cent of the aminopeptidases and alkaline phosphatases; these last are enzymes that split phosphate from sugar, as in the breakdown of nucleic acids.

Brünner's Glands. The goblet cells participate in the renewal process described above, but the Paneth cells or the epithelium of Brünner's glands apparently do not. These are composed of ramifying tubules located either in the mucosa or the submucosa, emptying the numerous ducts into the base or sides of the crypts of Lieberkühn in the overlying mucosa. The cells of the acini and tubules probably secrete a mucus and no digestive enzymes; their high concentration at the pyloric region of the duodenum would enable protection of the small intestine from acid secretions; certainly the upper end of the duodenum is more resistant to acid insults.

The Pancreas

The main supply of digestive enzymes, together with an alkaline juice, is provided by the pancreas, a gland lying outside the intestine, emptying its secretion into the *common bile duct* before entering the duodenum. The pancreas is a gland of both internal and external secretion; the hormones *insulin* and *glucagon* are elaborated by cells in the *islets of Langerhans*, and carried directly into the blood in capillaries related to the islets. The *exocrine cells*, constituting the acini, contain characteristic secretory granules that are discharged into the acinus, as will be described later.

Liver and the Bile

The Liver

The bile is the secretory product of the liver, an organ with a large variety of functions. As indicated in the general scheme of the circulation on p. 212, it lies on the route taken by the intestinal blood to the heart, the blood being carried from the intestine in the *portal vein* to the liver, where it mixes with the arterial blood of the hepatic system, as illustrated in Fig. 9.32.

Sinusoids and Canaliculi. The two streams mix in the *sinusoids*, which are endothelium-lined channels that come into relation with bile canaliculi, minute cell-lined tubes that, by confluence, become larger and which finally open into the bile duct. The relation of the bile canaliculi to the sinusoid is illustrated by Fig. 9.33 which shows that the canaliculi are essentially grooves in the contact surfaces of the

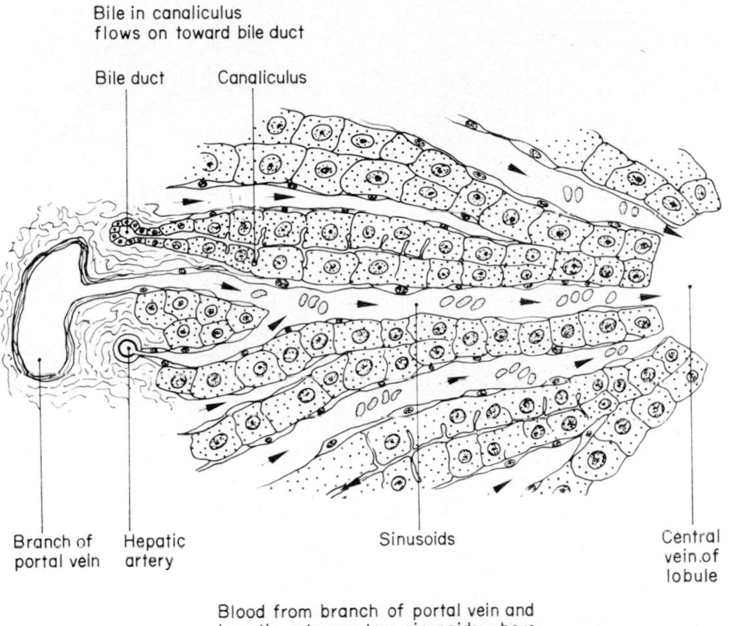

Fig. 9.32. A drawing of the blood flow in the sinusoids of the liver. The blood from the portal vein and the hepatic artery flows into sinusoids, lined by reticulo-endothelium, that lie between liver cords and empties into the central vein on the right of the diagram. Bile travels in the opposite direction in the canaliculi to empty into the bile ducts. (Ham, 1957, "Histology", Lippencott.)

hepatic cells; microvillous portions of the surface of the hepatic cell project into the lumen of the canaliculus.

The blood from the liver sinus flows into the hepatic vein which carries it towards the heart. Thus, from an absorptive point of view, the liver plays a primary role, having first access to the materials taken into the portal blood from the intestine.

Synthesis of Bile. From the digestive point of view it is the synthesis of bile salts, required to stabilize the fats in a fine emulsion in the intestine, that is of interest; and this is carried out by the *hepatic cells* of the liver lining the canaliculi.

Küpffer Cells. The other type of cell making up liver tissue is the Küpffer cell; these line the sinusoids and are phagocytic, being part

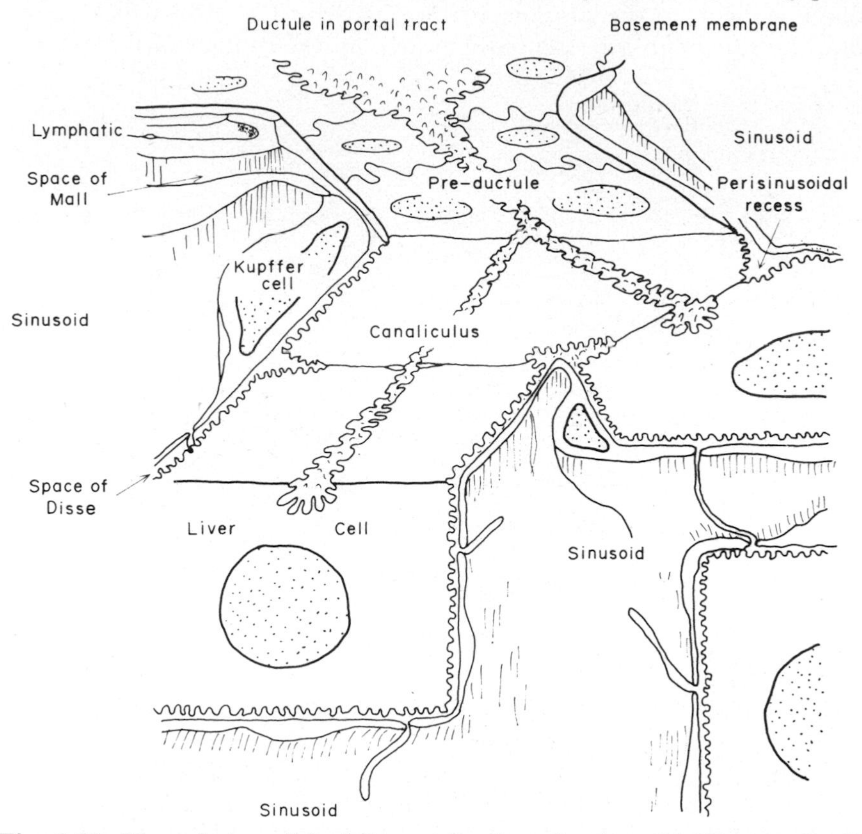

Fig. 9.33. The relation of the bile canaliculi to the sinusoid. The canaliculi are essentially grooves in the contact surfaces of the hepatic cells, lined with microvilli which project into the canaliculi. Sinusoids run in separate planes as shown in the diagram. (Hargreaves, "The Liver and Bile Metabolism", North Holland, Amsterdam, 1968.)

of the reticulo-endothelial system (Vol. 2) and playing a part in the breakdown of red blood cells.

Bile and Cystic Ducts

The bile flows down the *bile-duct* which opens into the duodenum; on its way it is joined by the *cystic duct*, a channel leading from the *gall-bladder*, which is the body that stores the freshly secreted bile and concentrates it by absorption of water and salts (p. 449). Entry of bile into the intestine is controlled by a ring of smooth muscle, the *sphincter of Boyden* (Fig. 9.34); contraction of the muscular wall of the gall-bladder raises the pressure sufficiently to open the sphincter. With the muscle of the gall-bladder relaxed, the bile secreted by the liver finds its way along the cystic duct into the gall-bladder.

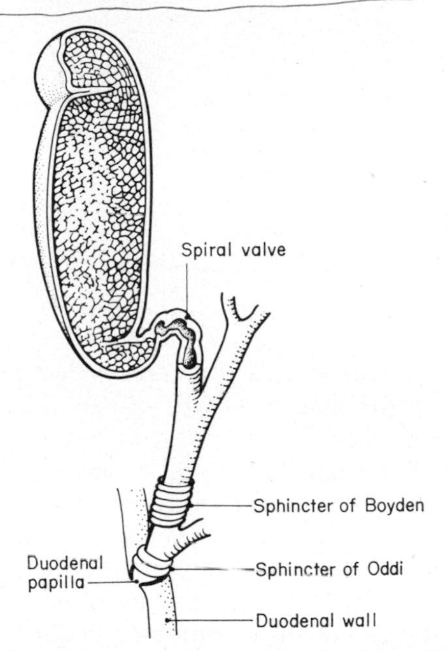

Fig. 9.34. The gall bladder and duct system. The gall-bladder stores bile from the liver and this is passed down the cystic duct to be released into the duodenum. The entry of bile is controlled by the sphincter of Boyden, which is a ring of smooth muscle in the duodenal papilla through which both bile and pancreatic secretion must flow to reach the duodenum. (Grants' "Method of Anatomy", 1958, Baillière Tindall and Cox.)

The Bile

The digestive secretion of the liver is the bile containing, in addition to the non-colloidal constituents of plasma such as Na^+ and K^+, etc., the *bile salts*, which are formed from the end-products of cholesterol metabolism, *cholic* and *deoxycholic acids*, by conjugation with taurine and glycine respectively to give *taurocholic* and *glycocholic* acids (p. 448.) Injection of bile salts into the blood stimulates the formation of bile, and the salts are thus called *choleretics*. After the bile salts have served their digestive function, namely that of emulsifying the fats in the intestine prior to absorption, they are absorbed from the intestine and

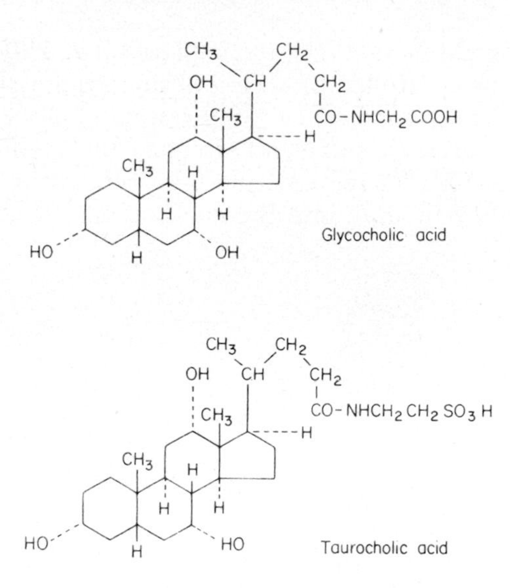

taken up by the liver where they promote further production of bile; they thus serve in a positive feed-back system that replaces the bile secreted in the digestive process.

Concentration Process. Bilirubin, a product of breakdown of haemoglobin, is produced continuously by the liver and must be removed from the blood circulation; hence a continuous secretion of bile is necessary to avoid accumulation of bilirubin in the blood. However, the injection of bile into the intestine is only required at intervals—after feeding; and these opposing requirements are met by the storage of the bile in the gall-bladder. During its stay here it becomes concentrated, as indicated by Table II, which compares the composition of bile collected directly from the bile duct and that collected from the gall-bladder. The total amount of bile salts in a gall-bladder does not change during this concentration process, which means that the volume of the fluid is reduced to about 10 per cent of its original value, giving rise to the large increase in concentration of the bile salts. The two biles have the same osmolality, so that the process of concentrating bile is essentially one of removal of an iso-osmotic solution consisting largely of NaCl and $NaHCO_3$.*

Active Transport. The gall-bladder, in which the bile is concentrated by removal of salts and water, is a remarkably simple organ,

* It will be noted that the concentration of Na^+ is considerably higher than in plasma after this concentrating process; this does not mean that gall-bladder bile is hyperosmotic, however, since the bile-acids are behaving essentially as polyvalent anions, so that the total number of anions in unit volume is less than in plasma.

consisting of a single layer of epithelial cells—the mucosa—separating the bile from the muscular and connective tissue layer (Fig. 9.35). The epithelial cells are connected together by tight-junctions, and we may envisage an active transport of Na^+ across the apical surfaces associated with a passage out of the cell into the intercellular clefts (as described in Chapter 2), in fact the process is presumably basically similar to transport of salt and fluid across the frog-skin. The interesting feature of this transport process, which differentiates it from that taking place across the frog-skin and urinary bladder, is the negligible

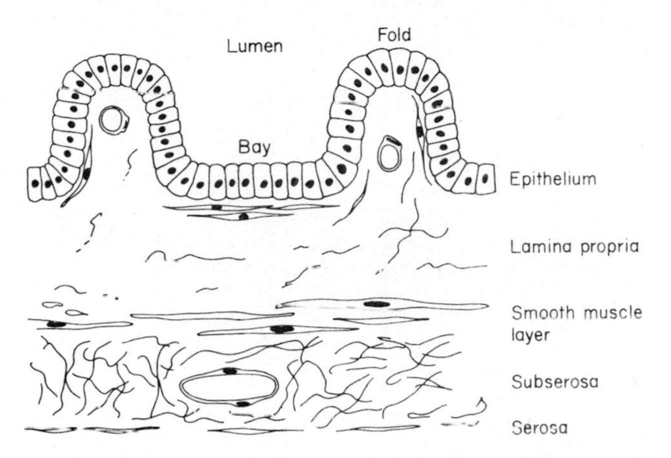

Fig. 9.35. A diagram of the structure of the gall-bladder wall. Lining the lumen is a single layer of epithelial cells; beneath this is a vascular serosal layer which is composed of connective tissue and smooth muscle cells. Active transport processes are thought to occur across the epithelial layer. (Tormay and Diamond, *J. Gen. Physiol.* 1967, **50,** 2031.)

difference of potential across the gall-bladder, and it was originally postulated by Diamond that the active transport of Na^+ was linked with that of Cl^-, so that no significant separation of charges could take place. However, subsequent studies have shown that this linkage is unlikely, and that, in fact, the failure to develop a significant separation of charge at the boundary between mucosa and serosa is the high permeability to anions, such as Cl^-. Although the Na^+-ions are "helped across" by active transport, this help is insufficient to create a significant positivity on the serosal side, as occurs in frog-skin, any separation being rapidly neutralized by passage of Cl^-- or HCO_3^--ions. The evidence suggested, too, that the site of penetration of the negative ions was the intercellular junction, as illustrated by Fig. 9.36; thus the actively transported Na^+ passed through the cell into the cleft but the passively transported Cl^- passed through the fused cell membrane.

Effect of Flow-rate. An increased flow of bile induced by intravenous infusion of bile salts may be expected to modify the electrolyte

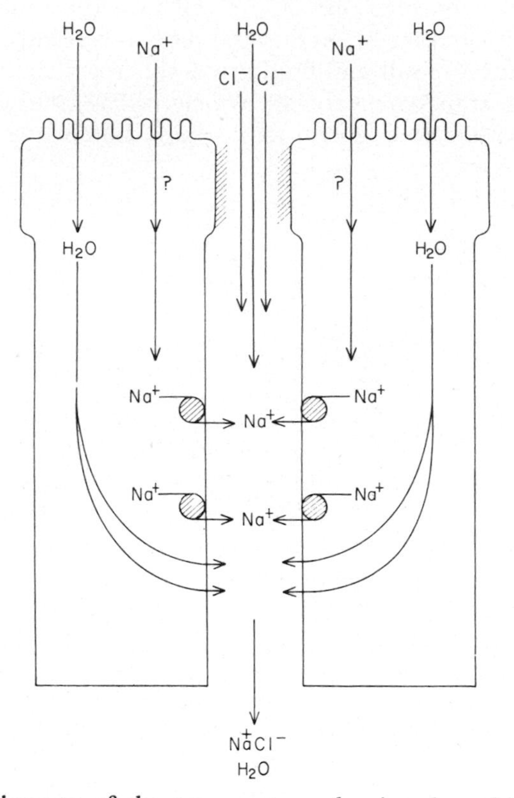

Fig. 9.36. A diagram of the transport mechanism by which bile is concentrated by the gall-bladder epithelium. Sodium is actively transported into the intercellular cleft and water is drawn in by the osmotic gradient. The water influx increases the hydrostatic pressure and a flow is set up towards the serosal end of the intercellular cleft. Chloride passes in via the tight junction complex with little restraint so that a negligible electrical potential is set up across the epithelium by the transport process.

composition of the fluid, if exchanges across the bile ductules with the vascular system take place. In fact, an increased flow in the rabbit causes the concentration of Cl^- to fall and that of K^+ to rise. In general, however, the changes in ionic composition, including that of bicarbonate, are variable from species to species and with the choleretic agent employed, the hormone *secretin*, for example, which in most species provokes a large increase in rate of flow, tending to decrease the rise in bicarbonate concentration caused by bile-salt infusion.

Mechanism of Secretion. When the bile is collected directly through a cannula in the duct, avoiding storage in the gall-bladder, it is found that the rate of flow is related in a linear fashion to the rate of secretion of bile salts; if the line relating these quantities is produced back to zero bile salt excretion, it extrapolates to zero flow. This close relation between bile salt secretion and flow of bile has suggested that the primary event in the secretion is the active transport of the bile salt out of the hepatic cell into the canaliculus, analogous with the primary active transport of Na^+ in the transport of fluid across the frog-skin, or of H^+ and Cl^- in the gastric secretion, and so on.

TABLE II

Composition of Canine Bile taken from Liver directly (Hepatic Bile) and after being stored in the Gall-bladder (Gall-bladder Bile). (Diamond. *In* "The Biliary System.")

Ion	Concentration (millimoles/litre) Hepatic bile	Gall-bladder bile
Na^+	174	220–340
K^+	6·6	6–10
Cl^-	55–107	1–10
HCO_3^-	34–65	0–17
Ca^{2+}	6	25–32
Mg^{2+}	3·6	—
Bile acids	28–42	290–340

Analogy with Kidney. This active transport shows strong analogies with the secretion of closely related acidic compounds by the proximal tubule of the kidney, substances such as p-aminohippurate, phenol red and so on (Chapter 10). Renal excretion of these substances shows mutual competition, the presence of one in the blood tending to suppress the excretion of another, and the same applies to the elimination of bile salts. Thus dehydrocholic acid depresses the excretion of cholate; taurocholic and glycocholic acids depress the secretion of phenol red in the bile, and so on.

Bile as Excretory Pathway. In this connexion we must note that the bile, besides its digestive function, may be regarded as the excretory pathway for many substances that are chemically similar to the bile salts. Thus bile contains blood pigments derived from the breakdown of blood cells; principal among these is bilirubin, which is converted into a water-soluble derivative by conjugation with glucuronic acid and is eliminated in this form. Again, the excretion of thyroxine, the hormone of the thyroid gland, occurs to a large extent (30–40 per cent.) by way of the bile after conjugation with glucuronic acid to form a water-soluble product. The process involves an active transport mechanism since Hillier has shown that the isolated perfused liver can accumulate thyroxine in the bile to a concentration some 1800 times its concentration in the perfusion fluid.

Many dyestuffs are preferentially secreted in the bile, such as sulphobromphthalein, fluorescein, phenol red, and so on. These dyestuffs apparently utilize the same "mechanism" for accumulation in the bile as that employed in secretion of bile salts, since there is competition amongst these solutes for secretion. The active secretion is not confined to organic acids, since Schanker has shown that there is a group of organic cations that are also actively transported into the bile; these use a different "mechanism" since there is no competition between anionic and cationic compounds.

A striking feature of the bile is the ready penetration of large water-soluble molecules from blood into it; e.g. inulin injected in the blood appears in equal concentration in the bile, and the same is true of erythritol and mannitol, substances that would be excluded from other secretions.

The Digestive Process

Movements of Intestine

The hydrolytic processes concerned with digestion of protein and carbohydrate were indicated earlier; these are catalysed by the pancreatic and intestinal enzymes, whilst the necessary mixing is achieved by vigorous contractions of the smooth muscle of the intestinal wall. These movements are of two fundamental types, i.e. *segmenting* movements which, by localized constriction, tend to cut the intestinal tube into segments, and *peristaltic* movements, the waves of contraction preceded by relaxation that force the chyme onwards towards the colon. Although the special feature of smooth muscle, as we shall see, is its capacity to contract and relax spontaneously, the highly co-ordinated nature of these movements, at any rate of peristalsis, indicates an overriding nervous control.

Segmentation

The most prominent activity in the small intestine is the segmenting contraction, consisting of a localized circumferential contraction predominantly of the circular muscle, of the bowel wall. It tends to divide the contents of the lumen, and has an obvious mixing function. Sometimes the contractions occur in a characteristically rhythmic fashion, and this is correlated with a basic rhythm of changes in potential of the surface of the gut. At other times the movements may be quite irregular. The rhythmic segmentation was described by W. B. Cannon as follows: "A string-like mass of food is seen lying quietly in one of the intestinal loops. Suddenly an undefined activity appears in the mass, and a moment later, constrictions at regular intervals along its length cut it into small ovoid pieces (Fig. 9.37).

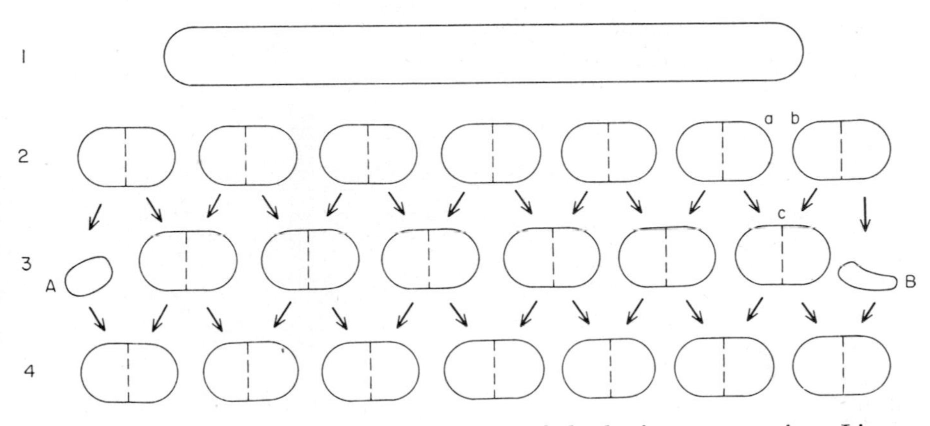

Fig. 9.37. Diagram representing process of rhythmic segmentation. Lines 1, 2, 3, 4 indicate sequence of appearances in the loop of intestine. Dashed lines mark regions of division. Arrows show relation of particles to the segments they subsequently form. (Hightower, "Handbook of Physiology", Section 6, Vol. 4.)

A moment later each of the segments is divided into two particles, and immediately after the division, neighbouring particles rush together, often with the rapidity of flying shuttles, and merge to form new segments. The next moment these new segments are divided and neighbouring particles unite to form a third series, and so on." According to Hightower, it seems likely that this type of movement was interpreted by some investigators as a pendular movement or swaying of the whole loop of intestine, so that there is frequent reference in the literature to this third type of motion. It is difficult to see what function a mere swaying of the gut could serve, and it seems that this type of motion is the result of an optical illusion. The rate of rhythmic segmentation diminishes from the duodenum to the caecum, and also from species to species; in the dog it is about 18/min in the duodenum and 12/min in the terminal ileum.

The Pancreatic Secretion

Effect of Flow-rate

The ionic composition of the pancreatic secretion, like that of the saliva, varies with the rate of flow. In Fig. 9.38 we can see that, whereas the Na^+ and K^+ concentrations remain constant, those of HCO^-_3 and Cl^- show a reciprocal relationship, so that at high rates of secretion the bicarbonate concentration is high, and very much higher than that in

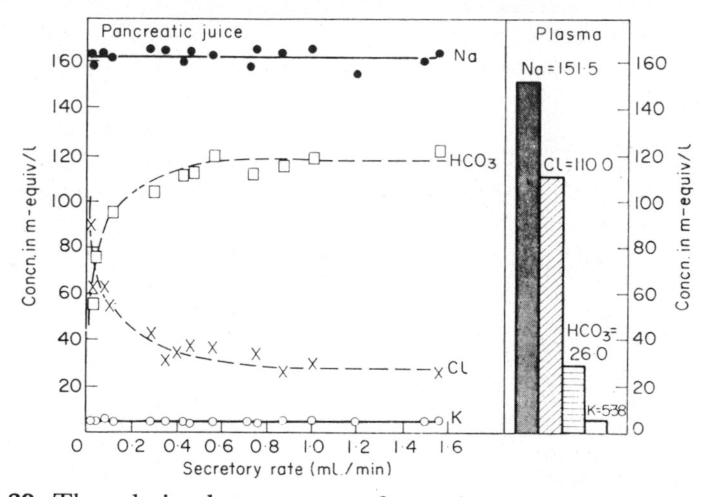

Fig. 9.38. The relation between rate of secretion and concentration of ions in the pancreatic juice collected from an anaesthetized dog injected with secretin. The concentrations of Na^+ and K^+ remain constant, whereas those of HCO_3^- and Cl^- vary, in a reciprocal manner, with rate of secretion. Plasma concentrations are indicated at the right. (Gregory, "Secretory Mechanisms of the Gastro-intestinal Tract", Arnold, 1962.)

the plasma, whilst the Cl^--concentration is low. Thus at very low rates of secretion the composition is little different from that of a plasma filtrate; as secretion is provoked, active transport processes come into play giving rise to highly discrepant concentrations of anions; these could not be explained on the basis of any difference of potential, since this should affect both negative ions to the same extent.

Since the P_{CO_2} remains the same, this means that the pH of the juice varies with the rate of secretion; in fact it is about 8·2 at high rates and 7·6 at low rates. This ensures an adequately alkaline reaction for the enzymatic processes.

Mechanism of Secretion

The mechanism of secretion of the pancreatic juice may well involve the active transport of Na^+, as with so many other processes. An interesting feature is the strong dependence on the presence of bicarbonate in the blood; thus when the gland is perfused with a saline

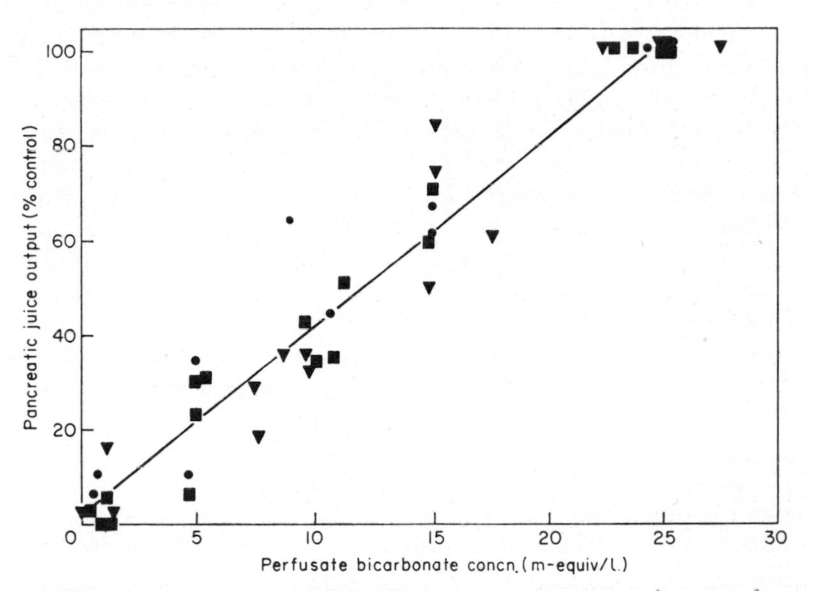

Fig. 9.39. The effect of perfusate bicarbonate concentration on the rate of secretion from an isolated pancreas. The rate of secretion is expressed as the percentage of the control value, with a bicarbonate concentration of 25 mEquiv/l. The line is the calculated regression line. Triangles and squares represent Ringer perfusate solutions with added colloid. (Case, Scratcherd and Wynne, *J. Physiol.* 1970, **210,** 1.)

medium instead of blood, and the concentration of bicarbonate in the fluid is varied, the rate of secretion varies from about zero, at zero fluid concentration, to 100 per cent of its control value when the level is approximately that in normal plasma, namely 25 meq/litre (Fig. 9.39). When the carbonic anhydrase inhibitor, acetazoleamide, was added to the perfusion fluid, secretion was inhibited by about 65 per cent, the fall in secretory rate being paralleled by a fall in bicarbonate concentration. The fact that secretion was inhibited when extraneous bicarbonate was absent means that the endogenous production of bicarbonate is a negligible quantity so far as secretion is concerned, and this was confirmed by placing ^{14}C-labelled bicarbonate into the perfusion fluid, when it appeared rapidly in the secretion.

Fluid and Enzyme Secretion

Secretion may be provoked by both nervous and hormonal agencies. Stimulation of the vagus nerve causes the production of a very scanty secretion rich in enzymes, whilst the injection of the hormone, *secretin*, into the blood causes the formation of a copious secretion containing very little enzyme, in fact this could be simply that flushed through by the secreted fluid. This dissociation of enzyme and salt and water secretion suggests different sites for secretion of these components, and the evidence indicates that it is the acinar cells that secrete the enzymes whilst the ductular cells, lining the ducts leading away from the acini, produce the salts and water. The discovery of a second hormone, *pancreozymin*, brought the dissociation of activities into greater prominence; this, which like secretin may be extracted from the intestinal wall, causes a secretion of enzymes, which may be observed by the loss of zymogen granules from the acinar cells.

Zymogen Granules. By breaking up pancreatic tissue and submitting it to fractional sedimentation (p. 14), the zymogen granules may be isolated and analysed; and it is found that their enzymatic composition is the same as that of pancreatic juice as secreted.

Synthesis and Secretion of Enzymes. The formation of the zymogen granule, and its emptying into the acinus, have been followed by Palade in the electron microscope; the granules quite clearly originate in the Golgi zone as "condensing vesicles" which grow into granules of increasing size and electron density; finally, the fully developed granule may be seen emptying into the acinar lumen by the process of exocytosis. By the use of radioactive precursors to the synthesized enzymes, Palade was able to show that the synthesis took place on the rough-surfaced endoplasmic reticulum, and that the products were carried to the Golgi zone, appearing first on the smooth surfaced vesicles of the Golgi apparatus, next in the condensing vacuoles and finally in the zymogen granules.

Intestinal Secretion

Intestinal Loops

The intestinal secretions may be collected by preparing some form of isolated intestinal pouch, or loop, choosing the portion of intestine in which either Brünner's glands or intestinal glands predominate. Thus

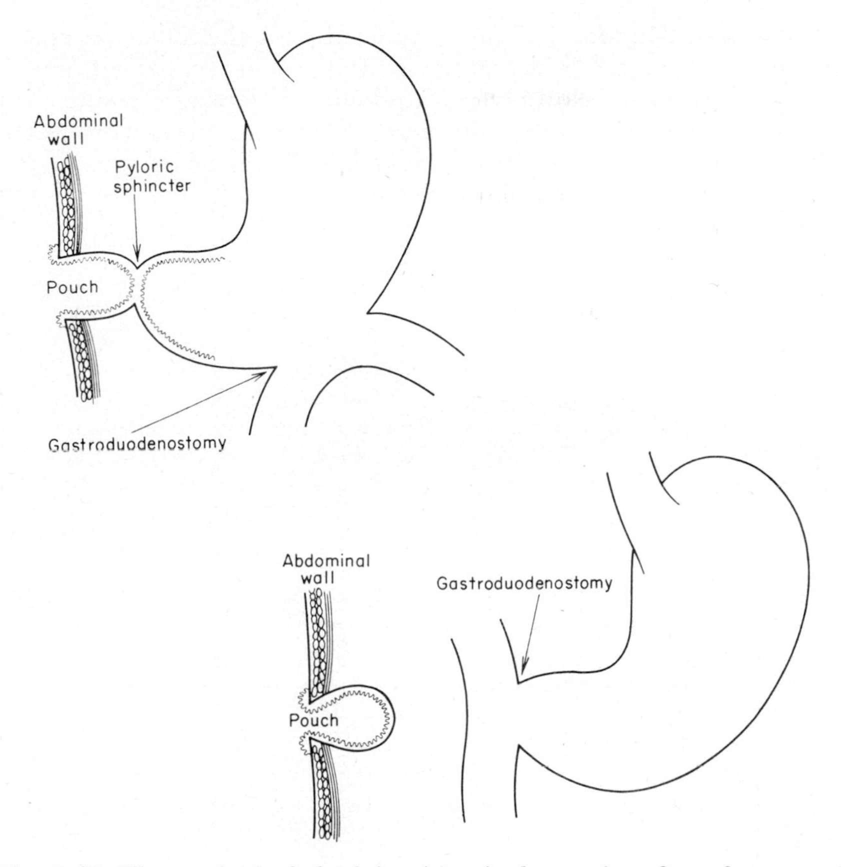

Fig. 9.40. The method of obtaining intestinal secretions free of pancreatic secretion. In the upper figure a gastro-duodenostomy has connected the stomach to the lower duodenum and a pouch has been formed of the upper duodenum which is still innervated by the vagus. In the lower figure the pouch is isolated from the influence of the vagus. (Gregory, "Secretory Mechanisms of the Gastro-intestinal Tract", Arnold, 1962.)

Fig. 9.40 illustrates a pouch of duodenum from which secretion of Brünner's glands can be collected free of pancreatic secretion, the stomach being connected by a gastroduodenostomy to a lower part of the duodenum, so that the animal can take food into its stomach and pass it on to the rest of the intestine. Another preparation is the *Thiry-Vella loop*, suitable for the study of the jejunal and ileal glands; thus Fig. 9.41 illustrates how both ends of a transected piece of intestine are brought to the surface of the abdominal wall, whilst alimentary continuity is maintained by end-to-end anastomosis.

Brunner's Glands. The fluid obtained from the duodenal pouch, rich in Brünner's glands, is highly viscid, its major component being a mucoprotein; there is little reason to doubt, in view of the absence of digestive enzymes in it, that the mucous secretion plays a protective role, and it is interesting that the duodenal region is much more resistant to acid insult than lower regions.

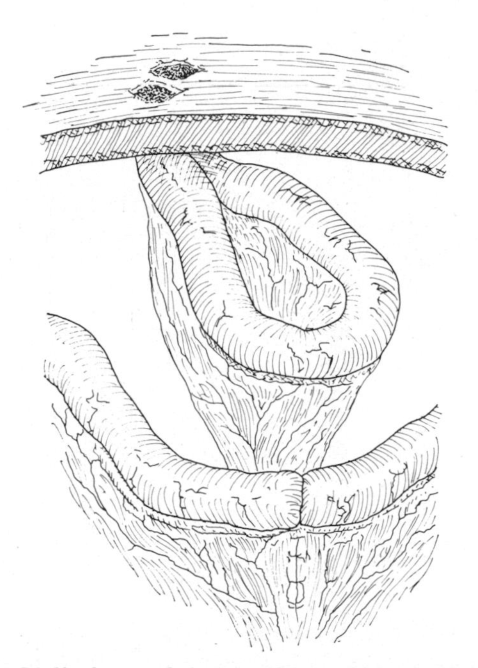

Fig. 9.41. A Thiry-Vella loop of the small intestine suitable for the study of jejunal and ilial glands. Both ends of a transected piece of the intestine are brought to the surface of the abdominal wall. Alimentary continuity is maintained by an end-to-end anastomosis. (Gregory, "Secretory Mechanisms of the Gastro-intestinal Tract", Arnold, 1962.)

Intestinal Gland. The intestinal gland secretion has an ionic composition very similar to that of a filtrate of plasma; it was thought to contain most of the digestive enzymes found in the intestinal tract, but in fact the major part of the enzymes found in the fluid collected from a Thiry-Vella loop are actually derived from the desquamated cells of the epithelium, so that the actual enzyme constituents are probably only enterokinase and amylase.

ABSORPTION FROM THE INTESTINE

The Intestinal Villus

The absorptive unit of the small intestine is the villus, a minute prolongation of the mucous membrane some 0·5–1·5 mm long. This villus is a composite organ, as illustrated by Fig. 9.42 which shows

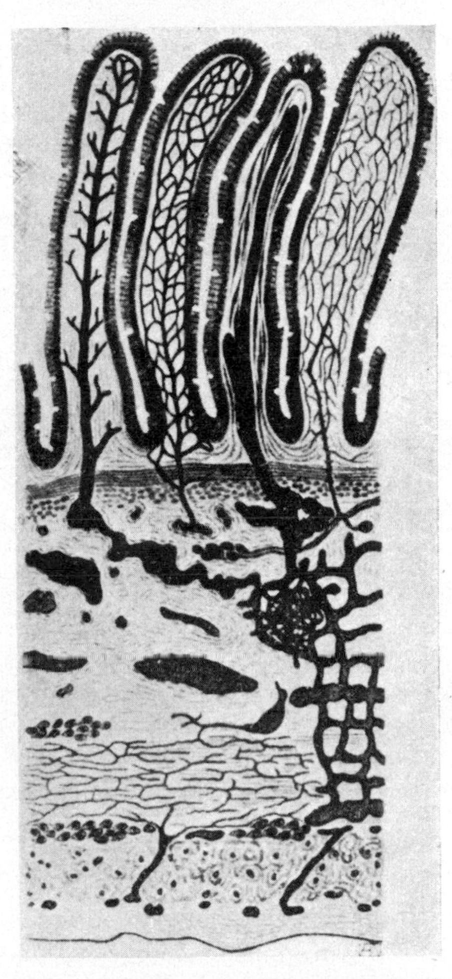

Fig. 9.42. A diagrammatic section through the wall of the small intestine. Four villi are shown illustrating, from left to right, the venous drainage, the arterial supply, the lymphatic drainage and the neural network. The lymph capillary, or lacteal, drains into a diffuse lymphatic plexus gathering the lymph from many villi. (Verzar and McDougall, "Absorption from the Intestine", Longmans, London.)

four villi schematically, in one of each of which the venous, arterial, lymphatic and nervous supplies are illustrated. The lymph capillary, or lacteal, begins under the epithelium covering the tip of the villus and leads into a lymphatic plexus gathering lymph from many villi; thence the lymph flows into larger channels that run with the mesenteric blood vessels.

Brush Border or Microvilli

The epithelial covering of the villus is made up of columnar cells resting on a basement membrane; the apical regions of cell contact are sealed by tight junctions, as we should expect of an epithelium that has

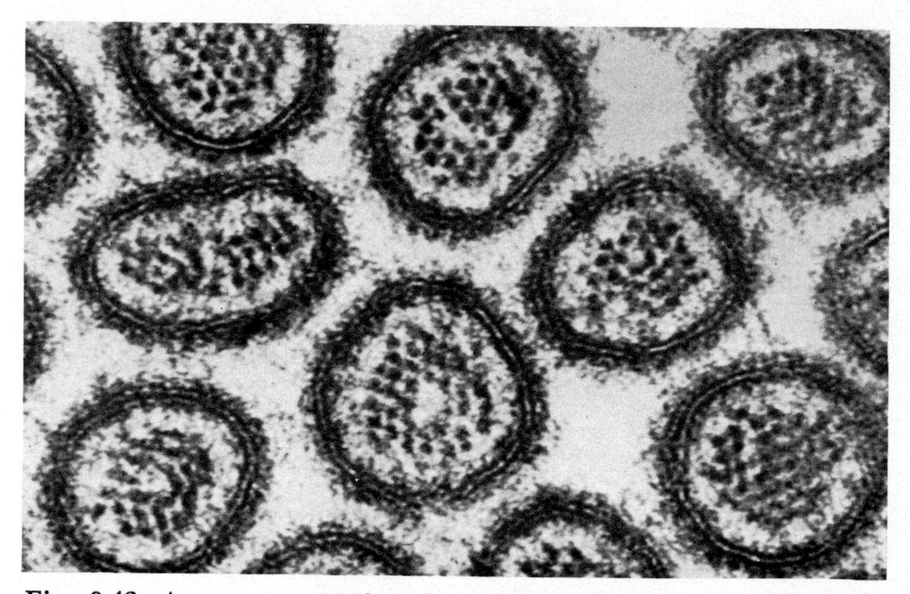

Fig. 9.43. A transverse section through the microvillous border of the jejunal epithelium. The trilaminar plasma membrane surrounds a filament-free zone with a central core of microfilaments. ×275,000. (Courtesy: T. M. Mukterjee.)

to exert some selective permeability. The area of the apical surfaces is increased greatly by what the classical histologists called the brush border, which, in the electron microscope, appears as a large number of microvilli or evaginations of the cell surface; Fig. 9.43 illustrates a transverse section through the apex of an epithelial cell, cut across the microvilli.

Blood and Lymphatic Pathways

The capillaries come into very close relation with the epithelial cells, being separated by little more than their respective basement membranes; these capillaries are of the fenestrated type (p. 278). Absorption from the lumen of the intestine into the blood take splace across the luminal surface of the epithelial cell, through its cytoplasm and thence through its basal surface into the subjacent tissue; from here it may pass into the blood capillary lumen and be carried into the portal circulation. Alternatively material may pass from the epithelial cells directly into the lacteals and find its way less directly into the main blood-stream through the lymphatic pathway; this latter pathway is taken by most of the fats of the diet.

Absorption of Water and Salts

When an isotonic solution of NaCl is placed in the intestine, this is absorbed quite rapidly; it could be argued that this process was equivalent to the absorption of plasma ultrafiltrate by plasma when the two are separated by a membrane permeable to the salts and water but not to the plasma proteins (Chapter 2). The epithelial layer covering the villus is certainly impermeable to the plasma proteins, so that the physical conditions are suitable for such a colloid osmotic absorption.

Active Transport of Ions

In fact, however, there is no doubt that active transport mechanisms come into play. This was early shown by Visscher when he placed a mixture of isotonic NaCl and Na_2SO_4 in the intestine and found that under these conditions, because the sulphate-ion could not be easily absorbed, the concentration of Cl^- in the intestinal fluid fell to very low values, far below that in the plasma, so that the Cl^--ion was being absorbed against a strong gradient of concentration (Fig. 9.44). The phenomenon can be interpreted in terms of an active removal of Na^+ from the intestinal contents; such a removal would demand the removal of accompanying anions, so that, in effect, NaCl would be absorbed (Fig. 9.44). Removal of salt would cause the contents to become hypotonic and this would cause the movement of water; in fact many studies indicate that absorption is *iso-osmotic*, in the sense that when the amount of solute removed in a given time is divided by the amount of water removed, the concentration so computed is equivalent to the osmolar concentration of plasma.

Analogy with Fluid Secretion. We may presume that the mechanism of absorption is similar to the mechanism of formation of a secretion, such as that of the pancreas or salivary acini, but in this case the secretion occurs backwards, in so far as the fluid passes from the body cavity into the apex of the epithelial cell and out at the basal side into the blood capillary. Diamond and Bossert's analysis of the dynamics of the secretory process, discussed earlier (p. 136), has been shown to be applicable to this type of activity as well as to the production of a secretion in the reverse direction.

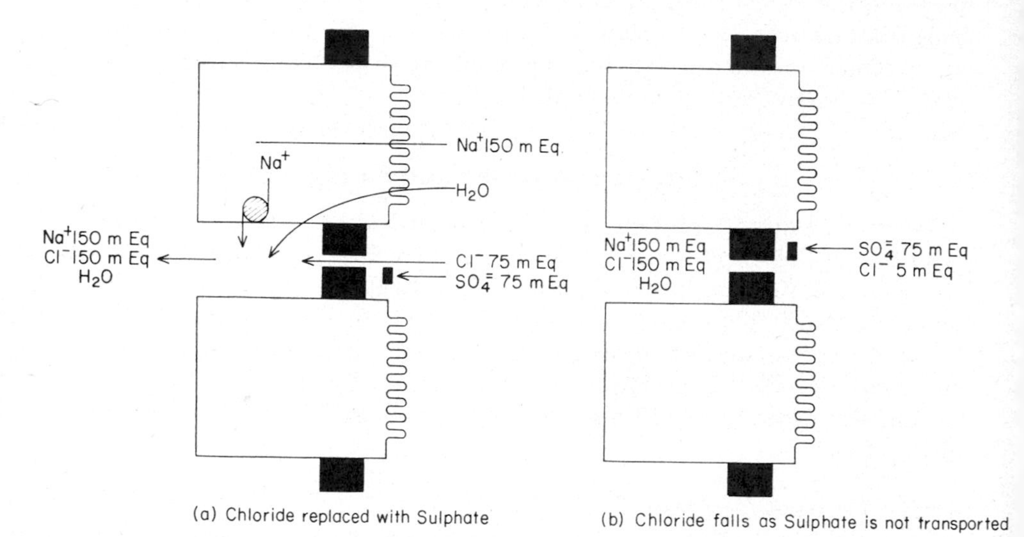

Fig. 9.44. The active transport of ions across the small intestine. Sodium ions are actively absorbed as an iso-osmotic solution of sodium chloride. If a part of the Cl^- ions in the solution in the lumen are replaced by $SO_4^=$ ions, the Cl^- concentration falls to very low values since the $SO_4^=$ ion cannot be transported and the iso-osmotic absorption of NaCl continues.

Potential. It will be recalled that, to define the active transport of an ion exactly, we must know the electrical potential across the transporting surface. The potential between electrodes in the lumen of the intestine and the outside serosal surface is of the order of 8mV in the ileum, the inside being negative; this might well be due to the active transport of Na^+ outwards, the negative ions following passively under the influence of the potential created by the movement of Na^+-ions (Fig. 9.45).

Short-circuit Current. By measuring short-circuit currents across the isolated rabbit ileum, Schultz and Zalusky showed that this could be largely accounted for by the active transport of Na^+, in the same way that the frog-skin short-circuit current could be accounted for by transport of Na^+ from the outside to the inside. If it is appreciated that the epithelial lining of the gut is really equivalent to the surface of the body, the gut being formed by invagination of the surface ectoderm, the transport processes for Na^+ are in the same functional direction, namely from outside to inside. Ouabain, the specific inhibitor of Na^+-K^+-activated ATPase, inhibited the short-circuit current when applied to the serosal—i.e. to the blood—side, and it is interesting that a preparation of isolated brush-border microvilli contained this ATPase.

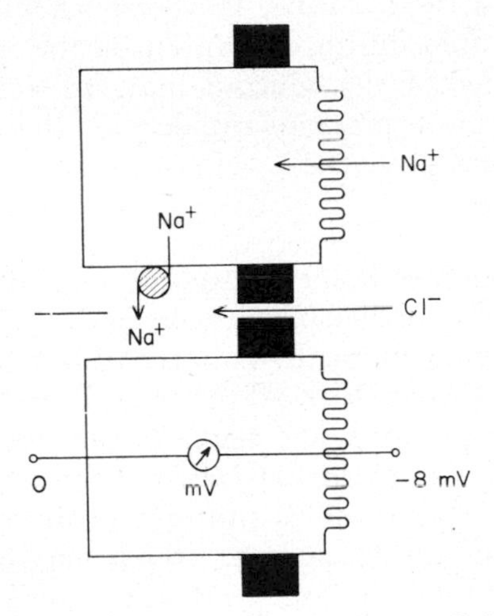

Fig. 9.45. The potential across the wall of the small intestine. A potential difference of some 8 mV, lumen negative to the interstitial fluid, exists across the wall of the small intestine. The potential has been related to the active transport of sodium ions by short-circuit current measurements.

Absorption of Calcium and Phosphate

The absorption of Na^+ and its accompanying anions has occupied most of the experimenter's attention because of its relation to the concurrent absorption of water. The evidence suggests that K^+ is absorbed passively as a result of gradients of electrochemical potential. By contrast, the absorption of Ca^{2+} is active and seems to be related to the physiological requirements of the animal; thus isolated sac preparations from pregnant and young growing rats were able to accumulate Ca^{2+} to a greater extent; again, experimental vitamin D deficiency decreased transport. Phosphate is likewise actively absorbed and requires the presence of Ca^{2+}; vitamin D increases absorption, perhaps by making Ca^{2+} available. Iron is an important element of the diet by virtue of the requirements of haemoglobin synthesis; its absorption is active and, like that of Ca^{2+}, is related to the requirements of the animal.

Absorption of Sugars and Amino Acids

Because the concentrations of sugars and amino acids in the blood plasma are held at relatively low values, a passive absorption down concentration gradients is feasible, the more so since the absorption of water and salts must necessarily increase the concentrations of the remaining solutes in the intestine. However, chemical analysis of the composition of the fluid during absorption in the intact animal shows that the concentration of, say, glucose, may fall well below that in the plasma whilst absorption proceeds rapidly, i.e. the transport is against a gradient of chemical potential.

Everted Sac

By employing pieces of isolated intestine tied into sacs, the process of absorption may be studied more conveniently, the sac being suspended in a nutrient medium. In the preparation shown in Fig. 9.46, the fluid actually absorbed drips off the serosal surface and is collected. In the *everted-sac* preparation, the sac is turned inside out so that the fluid transported by the epithelial cells appears inside the sac which at the beginning contains only a nutrient medium. Thus absorption takes place from the outside medium, which may be large in volume, into the inside medium, which is necessarily small in volume. In this way absorption results in quite large changes in concentration in the sac, and these are estimated by appropriate chemical or isotopic techniques.

Carrier Mediated Absorption

When the absorption of sugars and amino acids is examined by these techniques, remarkable specificities in the choice of substances for absorption appear, this choice being manifest by rapid absorption against concentration gradients. Thus, with the sugars the rates were as shown in Table III, and the absorptive process seemed to involve a carrier-type of transport, in so far as one sugar could interfere with the absorption of another, so that a series of "affinities" for the carrier could be drawn up, that of galactose, for example, being very high and that for mannose and the pentoses much lower. With the amino acids a similar competition was found, but here it seemed that at least three carrier mechanisms were employed, so that basic amino acids,

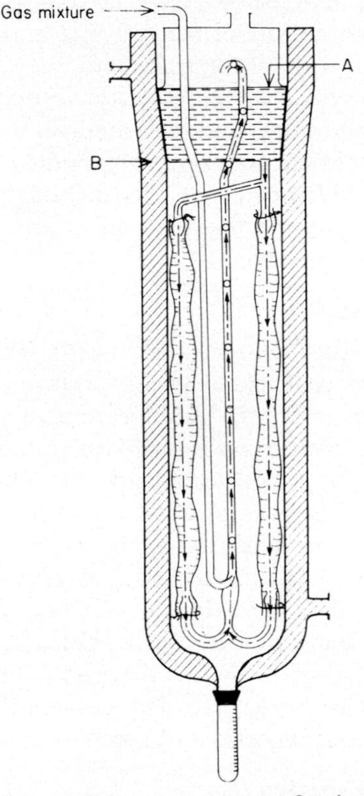

Fig. 9.46. Apparatus for studying water transfer in the intestine *in vitro*. The part of the apparatus carrying the intestine is provided with a grooved glass joint which fits into the outer water jacket at A. The loop of intestine is suspended in air and the fluid transferred is collected in the graduated tube. Water at 38°C is circulated through the water jacket B. (Smyth and Taylor, *J. Physiol.* 1957, **136,** 632.)

TABLE III

Relative Rates of Absorption of Sugars from Small Intestine

	Cori and Cori	Wilbrandt and Laszt Normal	Wilbrandt and Laszt Poisoned
Galactose (C_6)	110	115	53
Glucose (C_6)	100	100	33
Fructose (C_6)	43	44	37
Mannose (C_6)	19	33	25
Xylose (C_5)	15	30	31
Arabinose (C_5)	9	29	29

such as L-lysine, and acid amino acids such as glutamic acid, failed to interfere with the absorption of the neutral amino acid glycine; and many studies confirm that the amino acids may be arranged in three groups according to the acidity, basicity or neutrality of their side-chains. Converting glutamic acid to a neutral amino acid by neutralization of its side-chain carboxy-group permits it to compete with the neutral amino acid glycine. L-cystine, although not a basic amino acid, competes with L-lysine for absorption and so probably uses the basic amino acid mechanism.

D- and L-Enantiomorphs

The specificity of the carrier mechanism, whatever it may be, is revealed by the different rates of absorption of *dextro-* and *laevo-* enantiomorphs; thus Gibson and Wiseman introduced racemic mixtures of thirteen amino acids into the intestines of rats, and in every case observed that the *laevo*-acid was absorbed more rapidly than the *dextro*. That this was not due to preferential metabolism of the *laevo*-acids was shown by analysis of the venous blood flowing from the portion of intestine into which the mixture of amino acids had been introduced; the concentrations of the *laevo*-acids was some three times greater than that of the *dextro*-acids. When the portion of intestine was cooled, thus reducing the metabolic activity and hence the capacity for active transport, the concentrations of *dextro-* and *laevo*-acids in the venous blood tended to become equal.

Michaelis–Menten Kinetics

If a substance, in order to pass across a membrane, must combine with some sort of carrier, the kinetics of passage across the membrane will be analogous with the kinetics of an enzyme-catalysed chemical reaction, since in both situations, the fundamental step is the combination with some reactive grouping. In the case of the catalysed reaction this means combination with a grouping on the enzyme in such a way that the reacting molecule becomes more susceptible to chemical change; and in the case of the carrier-mediated transport it is presumably the combination with some reactive groupings in the cell membrane that permits it to pass through.

It will be recalled (Chapter 4) that Michaelis and Menten developed an equation describing the rate of an enzyme-catalysed chemical reaction, v, i.e. the number of molecules reacting in unit time, as a function of the concentration of the reactant, [S]:

$$\mathrm{v} = \frac{\mathrm{V_{max}} \times [\mathrm{S}]}{\mathrm{K_m} + [\mathrm{S}]}$$

V being the maximum velocity achieved at saturation of the enzyme complex, and K_m, the Michaelis–Menten constant, being the equilibrium constant for the reaction between the substrate and enzyme. This Michaelis–Menten constant is the reciprocal of the affinity of the substrate for the enzyme, the larger the value of K_m, the smaller the

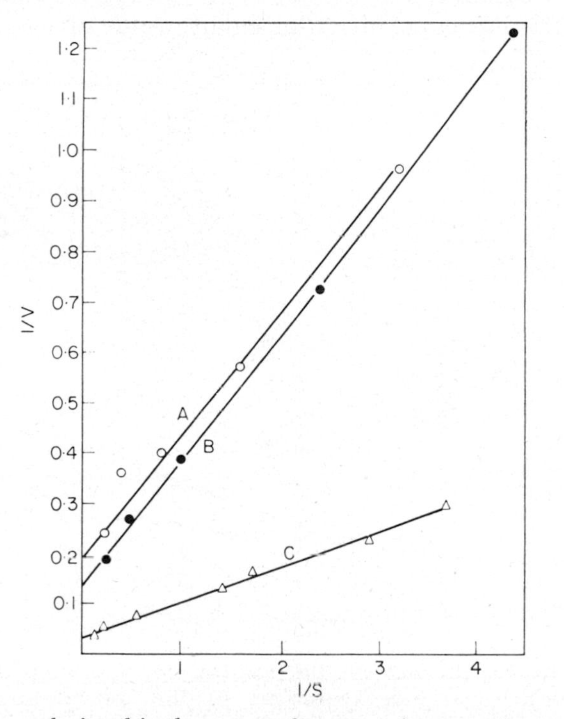

Fig. 9.47. The relationship between the rate of sugar absorption and the external concentration of the sugar isolated strips of small intestine. (*a*) Glucose KM = 1·25 mM. (*b*) Galactose, KM = 2·2 mM. Since the plots are straight lines, the sugar transport mechanism is following Michaelis-Menten kinetics and the slope of the line indicates the degree of affinity M = KM/V of the sugar for the carrier molecule. Hence it can be seen that glucose has a much greater affinity for the carrier than galactose and a much greater rate of absorption. (Crane, *Biochem. Biophys. Acta*, 1960, **45,** 477.)

affinity, and *vice versa*. K_m may be deduced from the Lineweaver–Burk plot, where $1/v$ is plotted against $1/[S]$.

Lineweaver–Burk Plots. If we accept the analogy between the absorptive process and a catalysed chemical reaction, we may plot the reciprocal of the velocity of absorption against the reciprocal of the concentration of the solute being absorbed. Some typical Lineweaver–

Burk plots for absorption of glucose (lines A and B), and galactose (line C), are shown in Fig. 9.47. The value of such plots is that the intercepts on the axes allow estimates of the maximum velocity, V, and the affinity constant, $1/K_m$. Thus when $1/v$ is zero, the value of $1/[S]$ is equal to $-1/K_m$ and when $1/[S]$ is zero, $1/v$ becomes equal to $1/V$. It is clear from Fig. 9.47 that the line with the shallower slope, namely that for galactose, gives the smaller value of $1/K_m$, and hence

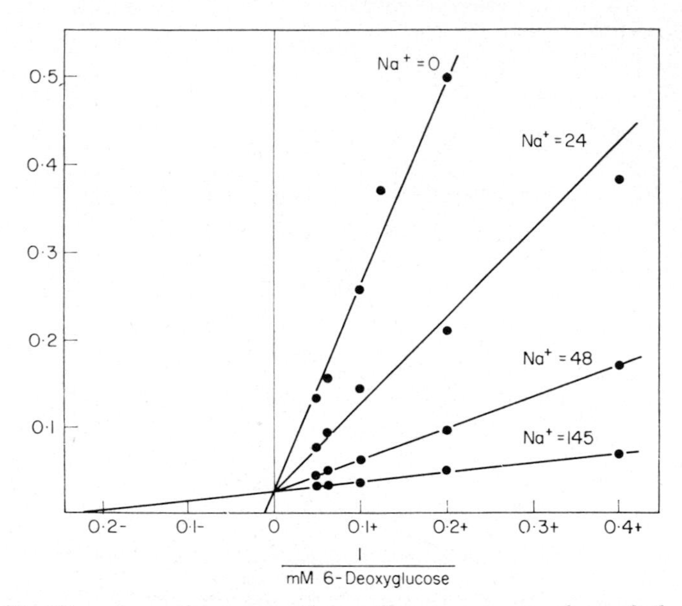

Fig. 9.48. The dependence on the sodium concentration of the rate of transport of the sugar 6-deoxyglucose. As the concentration of sodium increases the affinity for the carrier is decreased, although the maximum velocity is unchanged since all lines intersect at the same point on the y axis when $1/v = I/V$. (Crane *et al.*, *Biochem. Biophys. Acta*, 1965, **109,** 467.)

galactose has the smaller affinity for the transport mechanism (the actual values of K_m for glucose and galactose were 1·25 mM and 2·2 mM respectively).

Affinity and Transport. In general, a good correlation was established between the degree to which an isolated piece of intestine could create concentration gradients of a sugar and the reciprocal of K_m; thus Crane showed that the estimated values of K_m for glucose, galactose and 1·5-anhydro-D-glucitol were 1·5, 2·2 and 5·8 mM respectively, and the maximum concentration ratios developed were 203, 22 and 5·8 respectively.

Relation to Sodium. A very interesting finding was that the rate of transport of a sugar or amino acid was strongly affected by the concentration of sodium in the medium. Moreover, the transport of the sugar or amino acid was inhibited by the typical inhibitor of active sodium transport, ouabain, so that it seemed that in some way the transport of sugar or amino acid was linked to the active transport of sodium. This dependence is illustrated by Fig. 9.48 where we have Lineweaver–Burk plots for absorption of 6-deoxyglucose; the lines indicate that the value of K_m is increased at low concentrations of Na^+, i.e. that the affinity for the hypothetical carrier is decreased although the maximum velocity, whose reciprocal is indicated by the intercept on the vertical axis, is unchanged.

Allosteric Modification of Carrier. The full meaning of the effects of Na^+ is by no means clear; if the affinity of the sugar or amino acid is, indeed, increased by the presence of Na^+, then the efficiency of the carrier process will presumably be increased. It is known from enzyme-catalysed reactions that the presence of one substrate on an enzyme may modify quite considerably its reactivity in relation to another substrate, the so-called *allosteric modification.* It must be appreciated that the transports of both Na^+ and the organic solute are active, i.e. carrier-mediation is only one aspect; the carrier helps the solute through the membrane but of itself it cannot force it up a concentration gradient. At some point in the process, energy-giving chemical reactions must occur; it is very likely that the energy-giving reactions are primarily concerned with the transport of Na^+, and that the transport of the organic solute is so linked with this that the energy for the one process is adequate for the other.

Disaccharides and Dipeptides

Experimental studies on absorption of sugars and amino acids by the intestine have largely assumed that, in the natural state (i.e. during digestion of a meal), these molecules are absorbed as monosaccharides and amino acids rather than as disaccharides or peptides. However, it is well established that the rate of absorption of sugars is far greater than can be accounted for by hydrolysis to monosaccharides; and the finding that the enzymes responsible for breaking down disaccharides to monosaccharides are actually an integral part of the structure of the microvilli emphasizes the role of *intracellular hydrolysis*, and consequently the importance of absorption of the disaccharides, maltose and sucrose. Similarly, the presence of peptidases in the microvilli, and the demonstration that dipeptides may be actively absorbed, indicate the importance of intracellular hydrolysis.

DIGESTION AND ABSORPTION OF FATS

Hydrolysis

Neutral fats of the diet are mainly glycerol esters of long-chain fatty acids; in the intestine they are hydrolysed by the pancreatic enzyme, lipase. Complete hydrolysis would result in the formation of glycerol and the constituent fatty acids, but at any time in the intestine there will be a mixture of partial and complete hydrolytic products (Fig. 9.49) and modern work has emphasized the dynamic condition of the constituents, being determined not only by breakdown but by resynthesis. For example, if a labelled fatty acid is fed at the same time

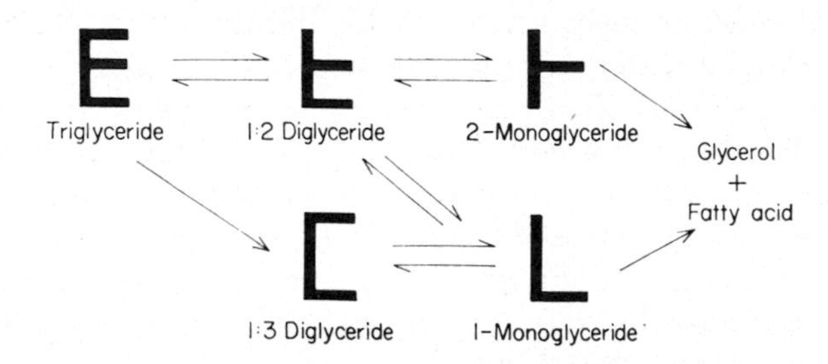

Fig. 9.49. A scheme of the changes that may take place in a triglyceride during its stay in the intestine. The E shape represents a triglyceride, with the glycerol molecule with three fatty acids attached. The fatty acid groups can be broken off and replaced in the intestine so there is a dynamic interchange of groups, and the composition of the mixture will depend on the relative rates of the various processes.

as triglyceride, it is found, on withdrawing the intestinal contents after a period, that all the material (triglycerides, diglycerides, etc.) is labelled with the new fatty acid. Only the hydrolysis of monoglyceride to give glycerol and fatty acid is irreversible in the intestine, and this is because the glycerol set free is much more rapidly absorbed into the blood-stream than the fatty acid. Ultimately, of course, we may expect the final product of digestion to be fatty acids and glycerol, but in fact the main processes are the splitting off of fatty acids from the 1- and 3-positions of glycerol, whilst the remaining 2-monoglyceride is relatively resistant to the action of lipase. Thus, after a period of digestion, the fat will contain predominantly 2-monoglycerides and fatty acids, but the actual composition of the mixture will be determined both by the digestive process and the absorptive process.

Solubilization of Fat

When the intestinal contents, after a meal of fat, are removed, and the lipase is inhibited to prevent further hydrolysis, high-speed centrifugation causes the material to separate into an aqueous phase which is quite transparent, and an oily phase; the aqueous phase is some 20–50 times the larger and thus, although its concentration of fat is lower, the amounts of fat in the two phases are about the same. In some way, then, a part of the fat has been "solubilized", and it becomes of great interest to examine how this occurs since, on physical grounds,

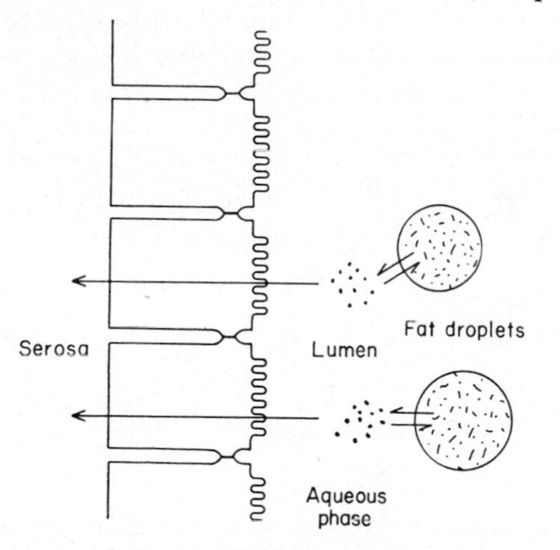

Fig. 9.50. A diagram of the equilibrium that exists for fats between the oily and aqueous phases in the intestine. Fat in the aqueous phase has been solubilized and contains most of the monoglycerides, a large amount of free fatty acids, a little diglyceride and no triglyceride. Absorption of fat takes place from the aqueous phase so that more fat can move out of the oily phase.

we should expect it to be the solubilized fat that was absorbed. Thus we may presume that there is an equilibrium between the oily phase and the watery phase; absorption of fat from the watery phase will favour passage of fat from the oily phase into the watery phase, so that, as absorption proceeds fat will be solubilized to replace the absorbed material (Fig. 9.50). Chemical examination of the watery phase shows that it contains most of the monoglycerides, a large amount of free fatty acids, a little diglyceride and no triglyceride.

Emulsions and Micellar Solutions

Bile Salts as Detergents

It has been known for a long time that the bile salts favour absorption of fat, and this is entirely by virtue of their "solubilizing" activity rather than in promoting hydrolysis, since they are effective in favouring absorption of free fatty acids. The conjugated bile salts, glycocholate and taurocholate, are *detergents*, having *amphipathic molecules*, i.e. molecules with a water-attracting polar group at one end of a long hydrophobic chain. Thus a typical detergent is a *soap* with an ionized-COO^- polar group at the end of a long hydrocarbon chain. When these substances are shaken with water they will form *micelles*, consisting of packed groups of molecules with their polar groups facing the watery phase, as illustrated in Fig. 9.51A; their diameters will be about twice the length of the molecule. This mixture of detergent and water is described as a *micellar solution*, its micelles being too small to scatter

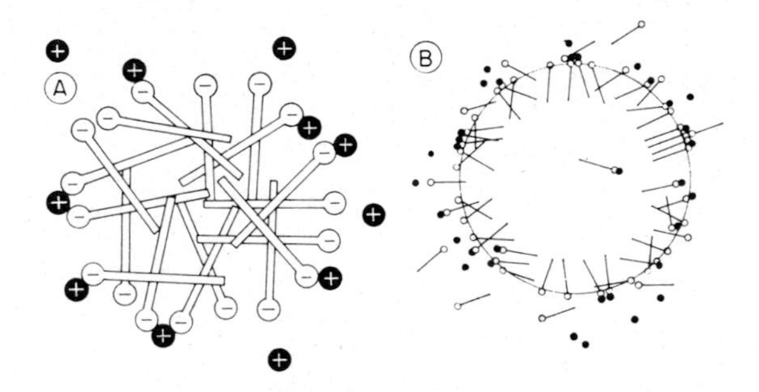

Fig. 9.51. (A) Diagram of a lipid micelle; the diameter is about twice the length of an individual molecule. (B) Particle of an emulsion at much lower magnification than (A). The amphipathic molecules accumulate on the surface of the fat particle with their polar groups facing the aqueous phase. (Hofmann and Borgström, *Fed. Proc.* 1962, **21,** 43.)

light appreciably, and hence the solution is transparent. Such a micellar solution is said to be stable because the individual micelles have little tendency to coalesce to form larger droplets. This stability is the result of two main factors; first the charges on the particles, so that two must approach each other with high velocity to get close enough to coalesce; and second the low surface tension at the oil-water interface. Thus the surface tension of a droplet measures, in essence, the free energy of its surface and this will tend to decrease spontaneously in accordance with the Second Law of Thermodynamics. The sum of the free energies of the surfaces of two droplets, when they have coalesced, will be less than the sum of their energies before coalescing because the total area of the fused droplets decreases. If the surface tension is high, coalescence will result in liberation of a large amount of energy and the tendency for coalescence will be high, and *vice versa*.

Emulsions. Shaking a neutral oil with water does not produce a micellar solution; an emulsion, consisting of relatively large droplets of oil surrounded by the watery phase may, however, be obtained provided that there is a sufficient quantity of an amphipathic molecule present. In this case a single layer of the surface-active material will stabilize each droplet, its polar end-groups facing the watery phase and its non-polar hydrocarbon chains being buried in the oil phase (Fig. 9.51B).

Amphiphiles. In the intestine there will be a rather complicated mixture of detergents—including the bile salts—and neutral fats; the detergents will stabilize an emulsion of the neutral fats in the manner indicated above. In addition to these two main classes of substance, we have another group called *amphiphiles;* these include phospholipids such as lecithin, and the monoglycerides resulting from partial fat hydrolysis. These amphiphiles are similar to the detergents in having polar end-groups and long hydrocarbon tails, but they are so insoluble that they will not form micellar solutions; they will, however, co-operate with detergents in forming micellar solutions, their polar groups being held in the watery phase beside those of the detergent, and their non-polar groups lying between the non-polar groups of the detergent.

Polar Solubilization. Thus the detergent, instead of making an emulsion of the amphiphile, as it would with a neutral fat, solubilizes it, a process described as *polar solubilization.* In this way the mono-glycerides and other amphiphiles are brought into the transparent watery phase of the intestinal contents. An additional factor favouring solubilization of the fat is the circumstance that the mixed micellar solution now becomes receptive to completely non-polar lipids like the neutral fat, and thus brings some of this into the watery phase.

Formation of Chylomicra

There is every reason to believe that it would be the fat in the watery phase of the intestinal mixture that would have better access to the epithelial cells, and it would be from this phase that they passed into these cells. We may now ask: In what way is the absorbed fat carried into the blood? After a fat meal, the lacteals and lymphatic ducts are filled with small fatty droplets called *chylomicra,* with diameters ranging from 1 to 0·33 μ; these contain some 95 per cent. of *triglyceride,* i.e. of *neutral fat,* the remainder being phospholipid and protein which probably form a film over the surface, reducing the tension sufficiently to stabilize the droplets.

Pflüger Hypothesis

Thus the fat leaving the mucosa has been *reconverted to neutral fat*, as suggested by Pflüger, obviously by reaction of fatty acids with mono- and di-glycerides and with free glycerol. The alternative hypothesis, namely that the fat in the lacteals and lymphatic pathway represents neutral fat that was absorbed directly as minute particles, cannot now be sustained, either on biochemical or morphological grounds.

Morphology of Fat Absorption

In the electron microscope there is very little evidence of the engulfment of lipid droplets by a process of endocytosis; instead, after a fat meal the most conspicuous change in the appearance of a jejunal epithelial cell was the appearance of spherical bodies in the apical cytoplasm. At first these were small but later they were larger and more numerous and they seemed to be being formed within the tubules of the smooth endoplasmic reticulum, which became much more prominent. In general, the appearance seemed to be that of synthesis of fat droplets by the endoplasmic reticulum under the apical web, their passage deeper into the cell, and finally their escape across the lateral cell membranes into the intercellular clefts, which became engorged with fat micelles. There was a gradient of sizes in the cell, those in the apical region being largest as though the size depended on access to the newly penetrating monoglycerides and fatty acids at the apical region. Convincing proof that the absorptive and synthetic processes were separate was provided by allowing fat absorption to proceed in the isolated intestine kept at 0°C; in this case the cells were no different from those of the fasting animal. When the isolated intestine was subsequently warmed to 37°C, thereby permitting metabolic processes to occur, fat droplets appeared in the cytoplasm. Thus Pflüger's hypothesis, put forward as long ago as 1901, has been amply substantiated by modern work.

Direct Absorption into Blood

When the products of hydrolysis in the intestine are sufficiently water-soluble, e.g. as with the triglyceride of butyric acid, the glycerol and fatty acid escape the process of resynthesis, and they pass directly into the blood stream. As the length of the carbon-chain increases, a point is reached, namely at decanoic acid, where some fatty acid appears in the lymph.

Fate of Bile Salts

The total amount of bile salts appearing in the intestine during the digestion of a meal is some two times the total amount in the body; this means that the bile salts are used again and again, thanks to a highly efficient absorptive mechanism. Physiological studies confirm that absorption is active and is confined largely to the ileum, i.e. beyond the region for maximum absorption of fat, namely the jejunum. Thus the bile salts are carried in the portal circulation to the hepatic cells where they are secreted into the bile and returned to the intestine.

ABSORPTION OF PROTEINS

In many species, including man, the transfer of immune bodies—globulins—from mother to progeny occurs *in utero*. In others this is delayed till after birth when the immune bodies are absorbed in the *colostrum*, as the milk is called in the early post-partum days. The absorption occurs through the small intestine, the epithelial lining of the distal portion being, at this early stage, characteristically different, exhibiting vacuoles through which the absorption probably occurs. After some 18 days (in the rat) the power to absorb proteins falls off abruptly, and it is interesting that it is only then that the epithelium is replaced by new cells from the crypts, as demonstrated histologically.

THE LARGE INTESTINE OR COLON

The small intestine leads into the large intestine, or colon, through the ileo-caecal valve or sphincter; this part of the gastrointestinal tract is divided into the *caecum*, to which the appendix is attached, the *ascending colon*, *transverse colon*, *descending colon*, *pelvic* or *sigmoid colon* and *rectum* (Fig. 9.52). The ileo-caecal valve is normally closed, and the residue of a meal, delivered to the terminal ileum, may remain there for an hour or so, undergoing segmentation and absorption until a wave of peristalsis drives it through the valve into the caecum.

By the time the chyme has reached the ileo-caecal valve, absorption of the foodstuffs is virtually complete, and the colon is the site for absorption of water and salts; such absorption is favoured by churning movements that knead the contents.

Haustra

The basic feature of the morphology of the colon is the appearance of semi-permanent partial segmentations or *haustra*. This is the result of three factors: an innate sacculation that persists after death; the concentration of muscular tissue at the constrictions forming the haustra; and, finally, mucosal folding.

Motility

Motility is accompanied by an increase in the haustration, the function of which seems to be rather to delay the progress of the

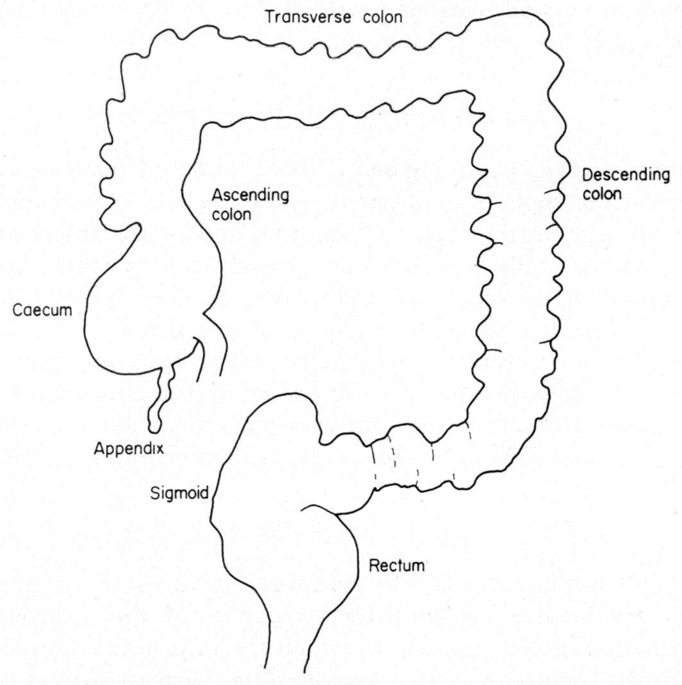

Fig. 9.52. A diagram of the colon. The colon is divided into three main sections—the ascending, the transverse and the descending colon. Continuous with the ascending colon is the caecum with the appendix; at the other end are the pelvic, or sigmoid, colon and the rectum.

colonic contents, or faeces, towards the rectum. It is thus the analogue of the segmentation movements of the small intestine, but confined to fixed points—the folds of the haustra. When propulsion occurs, this may take the form of successive emptying of adjacent haustra into each other, as illustrated in Fig. 9.53, which shows the movements of a barium sulphate suspension ingested some 12–24 h earlier. The propulsion may be aboral, directed towards the rectum, or it may be retropulsive. In other forms of movement several haustra are involved together, forcing their contents into distal haustra where the inter-haustral folds are opened up to form a cylindrical tube.

Receptive Relaxation. As the contents are more solid than in the small intestine the receptive relaxation of the more distal regions becomes more necessary and it may well be that the "mass movements", whereby large amounts of faeces are carried rapidly quite large distances, are due to the coincidental contraction of several haustral portions and the relaxation of a similar number distally. Thus it could be that the cessation of segmentation below provided the sufficient condition for large propulsive movement by removing the resistance that the segmented, or haustrated, state would otherwise have presented.

Transit Time

The total transit-time of food from mouth to the distal colon may be estimated by feeding appropriately labelled compounds that are unlikely to be absorbed, a label typically employed being ^{51}Cr; the appearance of radioactivity in the colon from this label occurs in about 48 h, compared with some 12 h for the small intestine.

Defaecation

When the contents of the colon are driven into the rectum, which is normally kept empty, the desire for defaecation is aroused. In man the process of evacuation is under voluntary control and is initiated by a relaxation of the anal sphincters accompanied by an increase in intra-abdominal pressure due to descent of the diaphragm into its lowest inspiratory position; with the glottis closed the chest muscles contract on to the full lungs, raising both intrathoracic and intra-abdominal pressures. There is no true valve between the sigmoid colon and the rectum, and the reason why the rectum is kept empty in the absence of a sphincter is probably the greater tonic activity of its smooth muscle than that of the sigmoid colon, maintaining a consistently higher pressure in the rectum. This may be assisted by actual retro-pulsive movements of faeces.

Anal Sphincters

The anal orifice is well protected by rings of muscle, the *internal anal sphincter* proximally and, at the terminal end, the *external anal sphincter;* the internal anal sphincter is equivalent to the circular muscle of the rectum, and therefore made up of unstriated fibres. The external anal sphincter is made up of bundles of striated fibres; the two deeper layers

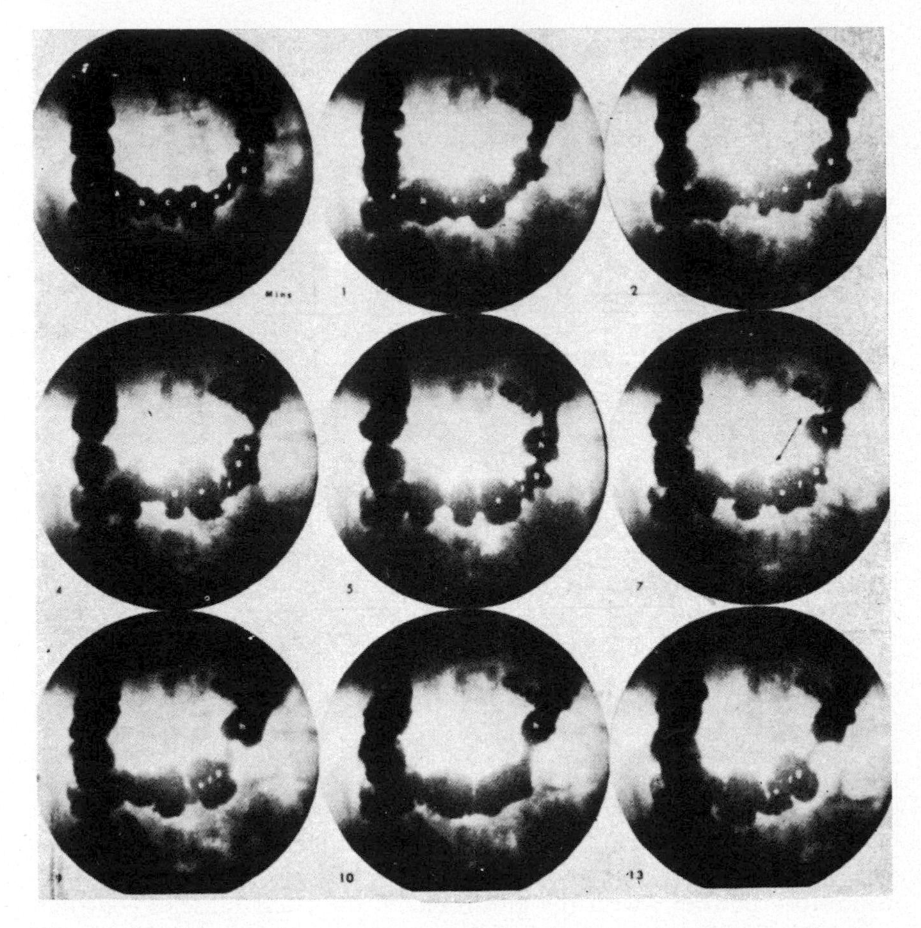

Fig. 9.53. Segmental propulsion and the development of co-ordination. X-ray pictures and line drawings of large bowel of a subject injected with a barium meal some 12–24 h previously. Successive haustra in the transverse colon of a subject with normal bowel function have been lettered for identification. The time intervals between cinefluorograms are indicated in minutes. After 1 min, contraction of part of the transverse colon, including

cover the internal sphincter whilst the outermost lies caudad to the internal sphincter and encircles the terminal portion of the anal canal.

Tone

Both sphincters, but especially the internal one, exhibit considerable tone, as revealed by the pressures developed when balloons are drawn

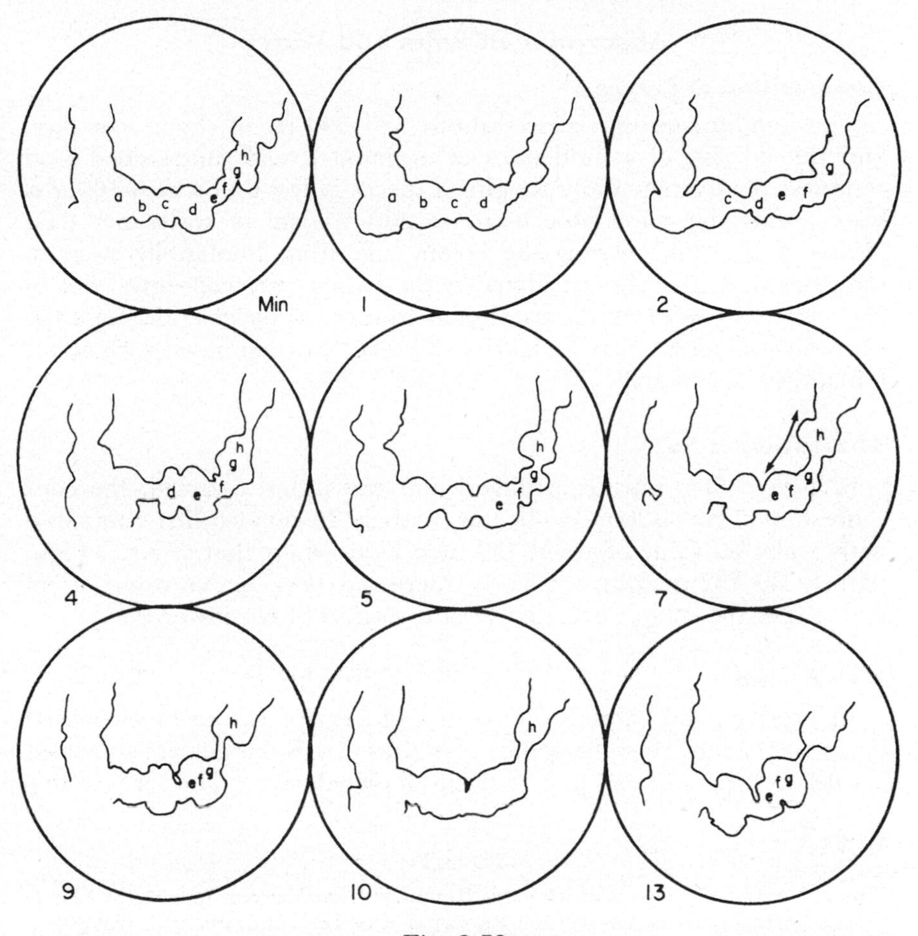

Fig. **9.53**

haustrum *a*, transferred additional barium to haustra *b*, *c* and *d*. When *c* and *d* contracted a minute later, a small proportion of their barium moved orally to *b* and the rest aborally to *e*, *f* and *g*. In the same way, when *f* contracted two more minutes later, its contents were shared between *e*, *g* and *h*. After another minute *g* also contracted and most of its contents went on to distend *h*. (Ritchie, *Gut*, 1968, **9,** 442).

through. This tone is an important element in the continence shown by the normal human; it is assisted, however, by reflexes that increase the activity of the sphincter muscles when, for example, intra-abdominal pressure is high, or when small amounts of faeces enter the rectum.

Absorption of Salts and Water

Composition of Faeces

The human caecum receives about 300–500 ml of chyme per day, and this consists of a fluid mass of undigested and unabsorbed food residues; the average daily weight of faeces is 150 g of which 100 g is water; thus the colon absorbs some 200–400 ml of water per day. Study of the fluid *entering* the colon, e.g. from human ileostomies, indicates that it is about isotonic with plasma; the concentrations of Na^+, K^+, Cl^- and HCO^-_3 are similar to those of plasma, although the concentration of K^+ may be high, 6–30 meq/litre compared with about 6 meq/litre for plasma.

Absorption of Na^+

If human faeces are centrifuged and the water analysed, the concentration of Na^+ is found to be low, namely 25–50 meq/litre compared with a plasma value of about 150 meq/litre, whilst that of K^+ is high, namely 80–132 meq/litre. Clearly there has been active transport of Na^+ out of the colon associated with transport of K^+ inwards.

Na^+–K^+-link

In experimental animals the process of absorption can be examined in greater detail; thus the results of a study by Edmonds are shown in Table IV for normal and sodium-depleted rats. The progressive

TABLE IV

Water, Na and K Content of Material Removed from Various Sites in the Gut of Six Normal and Six Na-depleted Rats on a Rice Diet. (Edmonds, *J. Physiol.*)

	Control			Na-depleted		
Site	Water (% of wet weight)	Na (m-mole/kg water)	K (m-mole/kg water)	Water (% of wet weight)	Na (m-mole/kg water)	K (m-mole/kg water)
Terminal ileum	81±3·1	141±8.5	24±2·6	83±1·2	137±3·5	29±2·2
Caecum	85±2·1	111±12·5	25±2·1	87±2·0	71±3·6	53±7·3
Cm along colon						
0–2	85±2·2	78±9·5	26±3·6	84±1·8	36±2·9	66±8·2
2–4	77±3·0	64±11·1	36±6·1	82±2.3	20±4·7	64±8·8
4–6	74±3·8	52±10·6	56±5·9	72±2·5	17±1·8	82±7·9
6–8	72±4·1	48±11·8	58±4·8	69±1·5	20±2·1	84±7·6

decrease in Na^+-concentration, and increase in K^+-concentration, along the colon are obvious; in sodium-depletion the absorption of Na^+ is more effective and is associated with an increased excretion of K^+, and it would seem that, as in the kidney, the absorption of Na^+ is linked to movement of K^+ in the reverse direction, i.e. to an actual secretion into the lumen. As with the kidney, the adrenal cortical hormones exert an influence on absorption of Na^+ and K^+, so that the absorption of these ions is an element in the salt-conservation mechanisms of the whole organism.

Potential

The potential across the colon is about 8mV, the inside being negative; in sodium-depletion this increases, perhaps reflecting the increased active transport of Na^+ out of the lumen.

CHAPTER 10

The Mechanism of the Kidney

General Mechanism of Renal Action

It is through the kidney that the major control over the composition of the blood is exerted. The basic mechanism employed by the vertebrate kidney is the selective elimination of specific solutes and water from the blood flowing through it. The elimination is selective in the sense that each of the major constituents is treated separately, and the elimination may vary from zero, i.e. complete retention within the blood flowing through the kidney, to a maximum rate, employed against substances that are usually not wanted in the blood at all. Under these conditions the blood may be "cleared" of the substance at a high rate, the maximum being governed, of course, by the rate at which the blood flows through the kidneys; so efficient is the clearance of some substances that the amount appearing in the urine can be used as a measure of blood-flow through the kidneys.

Filtration and Reabsorbtion

The selective mechanism is based on the indiscriminate removal of fluid from the circulating blood into a tubular system. After this "sequestration", the various constituents of the fluid—which is a filtrate of the blood plasma similar to tissue fluid—are returned to the blood to varying degrees. Certain unwanted substances like creatinine may not be returned at all, in which case we may say that the blood clearance is relatively high; other substances may be completely returned, and this is true of such important metabolites as glucose and amino acids unless the concentrations in the sequestered fluid are "too high", in which case only a portion of the solute is returned and the remainder is held in the tubular system to be finally passed out of the body as urine.

Tubular Secretion

With still other substances, the process of elimination is augmented by the transfer of the substance directly from the blood into the sequestered fluid, so that not only is there retention in this sequestered fluid but a further removal, the augmenting process being called *tubular secretion.*

General Scheme

The basic principles are illustrated in Fig. 10.1. The basic sequestration or filtration occurs in Bowman's capsule, and the fluid passes

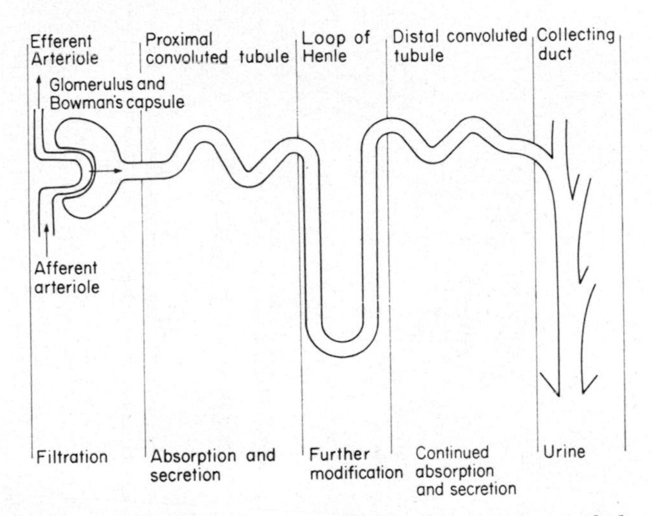

Fig. 10.1. Processes taking place in the various sections of the nephron. Filtration is from the glomerulus into Bowman's capsule. The filtered fluid is "processed" during its passage to emerge in the ureter as urine.

into the tubular system to be eliminated finally as urine; reabsorption occurs from the tubular system. The blood, after passing through the glomerulus and losing some of its fluid as a filtrate, passes on in vessels that ramify over the external surface of the tubular system, and it is thus able to take back from the filtrate such constituents as are due for reabsorption. Moreover, active transfer of some solutes may take place directly from this blood supply into the tubular lumen, thus augmenting elimination (tubular secretion).

STRUCTURE OF THE KIDNEY

The Nephron

The kidney consists of a large number of regularly arranged units—*nephrons*—one of which is illustrated schematically in Fig. 10.2 which shows the *renal corpuscle*, or *Malpighian body*, communicating with a tubular system made up of two segments—the *proximal* and *distal convoluted tubules*—separated by a hair-pin portion called the *loop of Henle*, with its descending and ascending limbs. The distal tubule

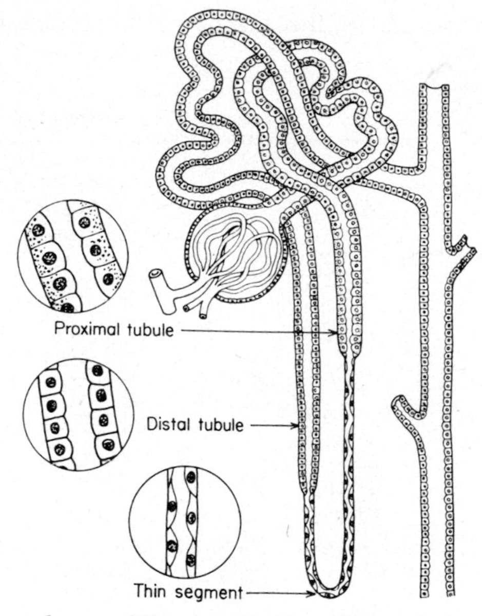

Fig. 10.2. The nephron. (Homer Smith, "Physiology of the Kidney", OUP, New York, 1951.)

connects with a *junctional tubule* which joins with a number of others to form a *straight collecting tubule*, which leads into a large *papillary duct*, which finally empties into the pelvis of the kidney which continues as the ureter. Each kidney has one ureter which empties into the bladder.

Renal Corpuscle

The renal corpuscle is the region where the filtration of blood occurs; the blood is brought in by the *afferent arteriole* and, after breaking up into a skein of capillaries—the *glomerulus*—it leaves in the *efferent arteriole*. The capillaries of the glomerulus come into close relation with the epithelial lining of the tubular system, which in this region has become dilated to form Bowman's capsule (Fig. 10.3). Thus, let us follow the proximal tubule back to its dilated end; it, like the rest of the tubular system, consists of a single layer of epithelium resting on a basement membrane, the apices of the cells pointing into the lumen and the basement membrane forming the outer wall of the tubule. In Bowman's capsule, the epithelium reaches up to the afferent and efferent arterioles, and is reflected over the surface of the capillaries to form the *visceral layer* of epithelium, the *parietal layer* being the wall

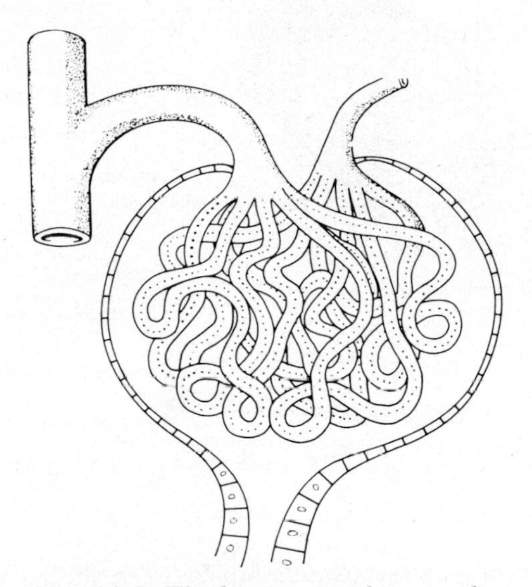

Fig. 10.3. The human renal corpuscle.
(Richards after Vimtrup, *Proc. Roy. Soc. B*, 1936, **126,** 398–432.)

of the capsule. The filtrate, exuding from the capillaries of the glomerular tuft, is carried across the visceral layer into thc space between it and the parietal layer, and passes out of the capsule into the proximal tubule.

Capillary–Epithelial Relationship

The relations between the capillary and visceral epithelium are illustrated schematically in Fig. 10.4; in the electron microscope it is easy to resolve five layers (Fig. 10.5). Firstly the endothelial lining of the capillary (Layer 5) which is of the fenestrated type, the endothelial cells becoming highly attenuated in localized regions; above this is a complex of basement membrane material (Layers 4, 3, 2) representing

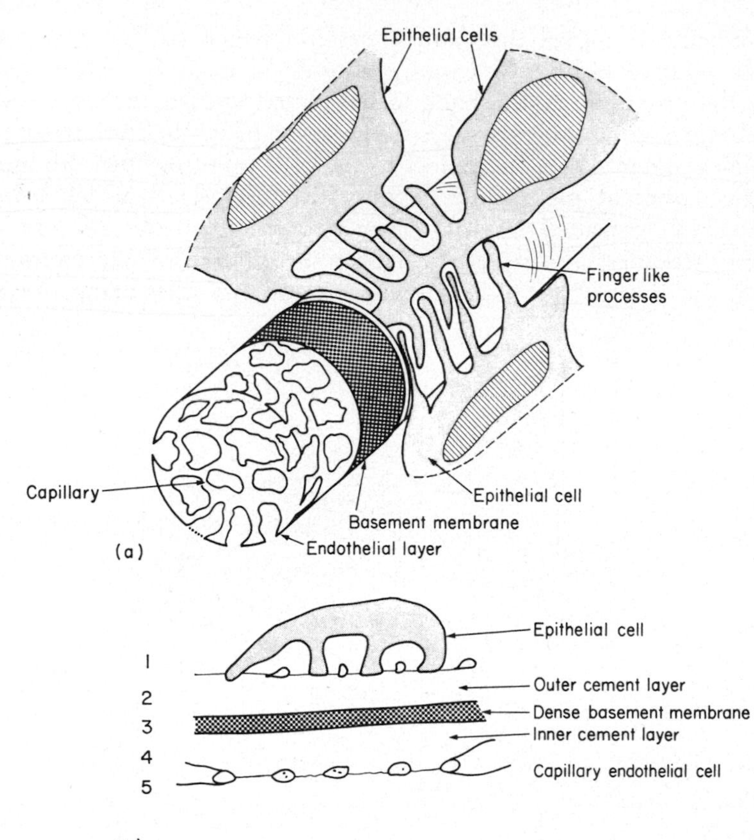

Fig. 10.4. A diagram of the relationship between the renal capillary and the epithelial cell of the nephron. (*a*) The capillary is composed of a fenestrated endothelial layer covered by a basement membrane. The epithelial cells wrap around the capillary with interdigitating finger-like processes. (*b*) A section through the capillary and epithelial wall. (1) The finger-like projections of the epithelial cell rest upon the outer cement layer (2) and are separated by the basement membrane (3). The endothelial cells make up layer (5) with the cell walls attenuated into thin regions or fenestrations. (Drawn from Pease, *J. Histochem. Cytochem.* 1955, **3,** 295.)

the apposing basement membranes of the capillary endothelium and the visceral epithelium of Bowman's capsule; finally, Layer 1 is made up of the epithelial cells, which form an elaborate system of podia resting on Layer 2, the spaces between the podia being called *pore-slits* which are probably covered by a thin membranous layer that helps to restrain passage of material from the basement membrane complex into the cavity of Bowman's capsule.

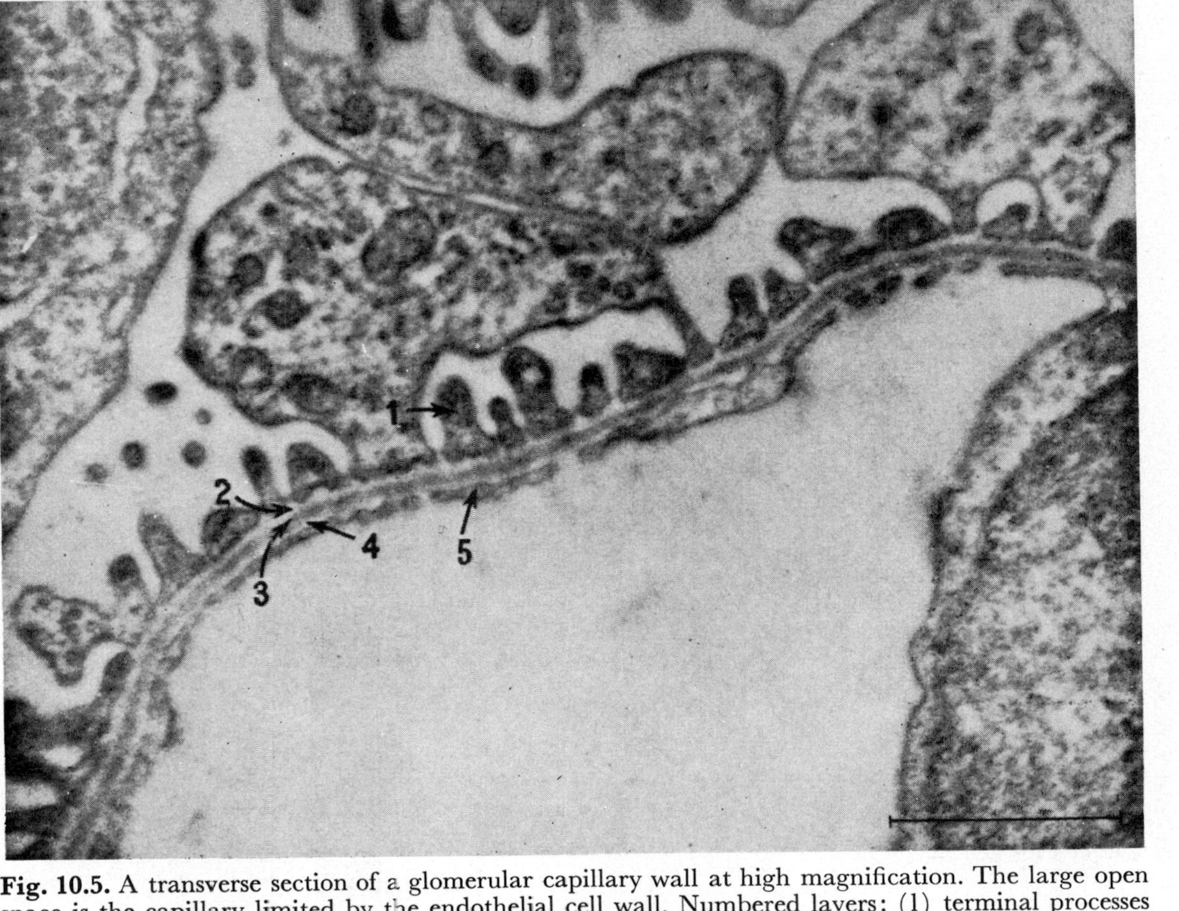

Fig. 10.5. A transverse section of a glomerular capillary wall at high magnification. The large open space is the capillary limited by the endothelial cell wall. Numbered layers: (1) terminal processes of epithelial cells; (2) outer cement layer; (3) basement membrane; (4) inner cement layer; (5) the fenestrated endothelial cell wall of the capillary. $\times 30{,}000$. (Pease, *J. Histochem. Cytochem.* 1955, **3,** 295.)

The Tubular System

Proximal Tubule

The proximal tubule is continuous with Bowman's capsule, and consists of a highly convoluted proximal portion which continues as a straight portion (the *pars recta*) into the descending limb of Henle's loop. The cells of the epithelium are large and have a characteristic "brush-border" which, when viewed with the high resolution of the

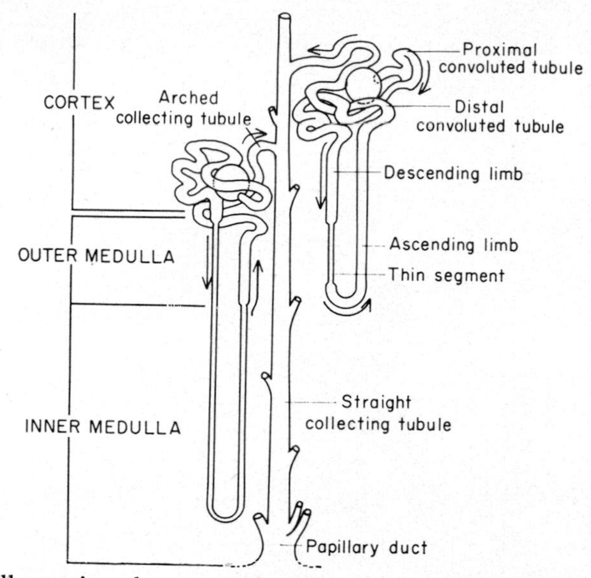

Fig. 10.6. Illustrating the two types of nephron, the one with a long loop of Henle and the other with a short loop. The loop of Henle is not a uniform structure, and has a thin segment of flattened epithelial cells. In the short loops the thin segment is confined to the descending limb, whilst in the long loops it extends into the ascending limb. (Greep, "Histology". McGraw Hill, 1966.)

electron microscope, is seen to consist of minute protrusions of the surface membrane, or microvilli, similar to those of the small intestine. The plasma membranes show characteristic infoldings, which represent interdigitations of the adjacent cells, a feature that is seen in epithelia that transport large amounts of fluid. In the pars recta the cells are flatter.

Loop of Henle

It must be noted that the loop of Henle is not a uniform tube, but contains a *thin segment*, which may be confined to the descending limb, or it may begin on the descending limb and extend part of the way up the ascending limb (Fig. 10.6); the nephrons located in the outer region of the kidney have the thin segment on the descending limb and those located deeper have the thin segment on both limbs. The transition from the pars recta of the proximal tubule to the thin segment is sharp, the epithelium consisting of the flattened cells rather similar to capillary endothelial cells with which they can be confused. The thin segment continues into the thick segment of the loop of Henle which has the same structure as that of the distal tubule with its pars recta; the lumen is larger than that of the proximal tubule and the cells have prominent interdigitations but lack a prominent brush-border.

Collecting Tubules

As we shall see, the collecting tubules are not merely a conducting system, their epithelial lining being concerned in the concentration of the urine; they, too, are composed of an epithelial layer of cuboidal cells, which become taller as the collecting tubule merges into papillary ducts.

Cortex and Medulla

Pyramids

The arrangement of the nephrons in the kidney is such that Bowman's capsule, with its related proximal and distal convoluted tubules, is in the outer layer, or *cortex*, whilst the pars recta and loop of Henle converge towards the papilla to form *medullary rays*. The nephrons are of two sorts according as they have short or long loops of Henle, and the arrangement is such that the short-looped nephrons are closer to the surface of the cortex, the long-looped ones being called juxta-medullary (Fig. 10.6). In the human kidney the nephrons are grouped into some 10 to 15 *Malpighian pyramids* and, corresponding with this arrangement, the large *papillary ducts* collecting their urine open into minor *calyces*, at the apices of the pyramids (Fig. 10.7), which are called *papillae*. In some animals, such as the rabbit and rat the nephrons are grouped into a single pyramid to give a single papilla at its apex.

Blood Vessels

The basic arrangement of the blood vessels is indicated in Fig. 10.8 the afferent arterioles of the glomeruli being derived from interlobular arteries. For most of the nephrons the arrangement is such that the efferent arteriole from the glomerulus breaks up into capillaries which ramify over the surfaces of the convoluted tubules and finally empty into the venous system.

Vasa Recta. The deeper-situated nephrons, however, have a different type of circulation, the efferent arteriole dividing to form a leash of long hair-pin loops parallel with the long loops of Henle, and

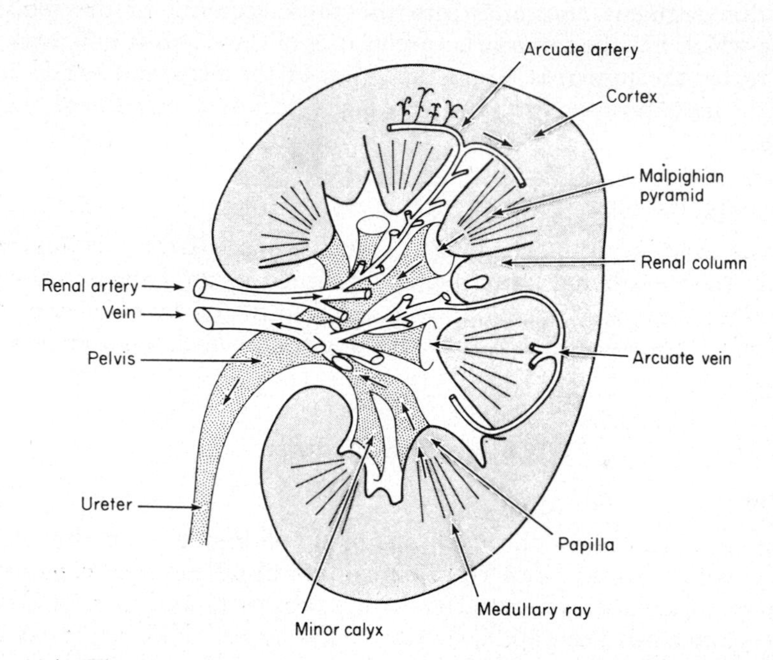

Fig. 10.7. The gross features of the kidney. In the human, the nephrons are grouped into some 10–14 Malpighian pyramids and these drain into the minor calyces which in turn empty into the renal pelvis and the ureter. The renal artery breaks up into a number of major branches which supply the arcuate vessels running in the cortex. Blood drains from the arcuate veins into the main renal veins.

called the *vasa recta*. These vessels are concerned with the specific functions of the loop of Henle in a countercurrent system of urine concentration (p. 516); the proximal and distal convoluted tubules belonging to this *juxtamedullary nephron* receive capillary networks from the efferent arteriole. It is interesting that blood may be shunted from the mainly medullary to the mainly cortical routes under certain conditions.

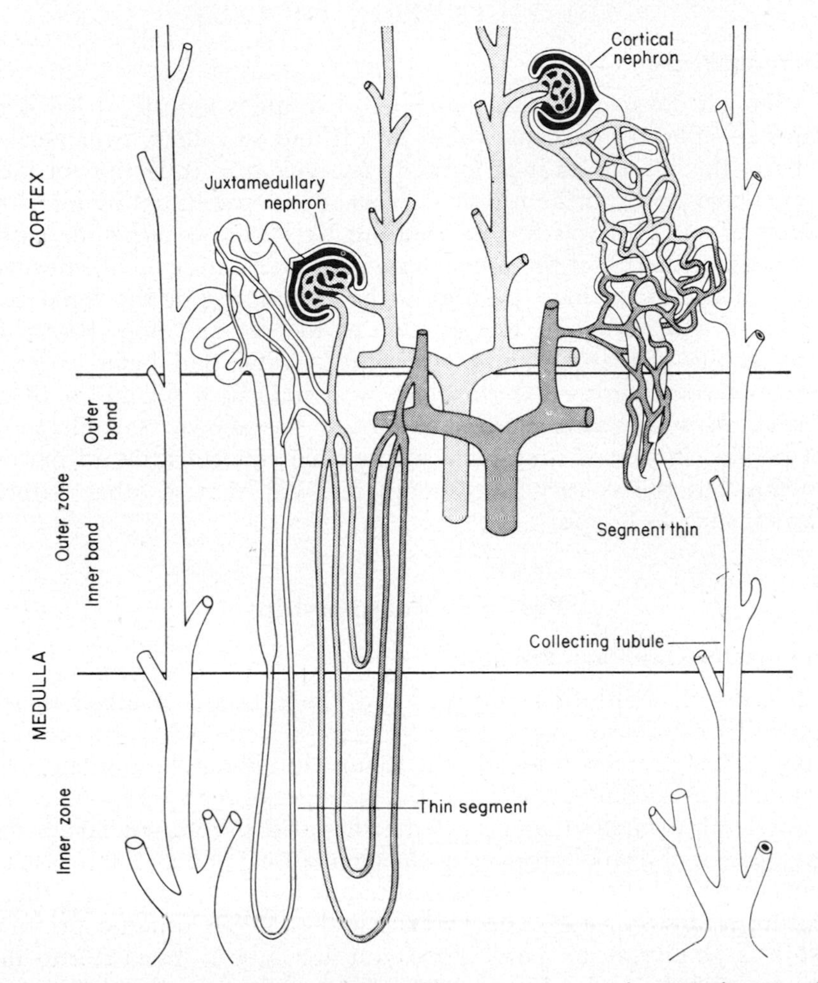

Fig. 10.8. The renal blood supply. The arcuate arteries running on the margin of the cortex and medulla arborize, and branches then supply individual glomeruli. The efferent network is wrapped around the proximal and distal convoluted tubules. Blood to the inner cortical nephrons passes through the glomeruli and then on to the vasa recta which accompany the loops of Henle into the medulla. The drainage of blood from the convoluted tubule capillaries and the vasa recta is by the arcuate veins and thence to the renal vein. (Homer Smith, "The Kidney", O.U.P., New York, 1951.)

THE FILTRATION PROCESS

Micropuncture

Although the space within Bowman's capsule is minute, it has been found possible to insert fine tubes into it and to collect, over periods of time, the filtrate as it is formed; to avoid the collection of fluid from farther down in the tubule, a blockade is established by injecting a drop of oil. This was first carried out by Richards using the easily accessible glomeruli of the mud-puppy, *Necturus;* subsequent improvements in technique have permitted the puncturing of the renal corpuscles of a mammalian species such as the rat and dog. Richards' study of the glomerular fluid established—what had before been matter of conjecture—that the fluid was, indeed, a filtrate of blood plasma. As subsequent mammalian work has shown, the filtrate is not perfect, in that it contains a measurable concentration of plasma proteins, but this is very small compared with that in other natural filtrates, such as lymph.

Pressure Relationship

Glomerular Capillary Pressure

In order that filtration may proceed, the pressure of blood in the glomerular capillaries must be greater than the colloid osmotic pressure of the plasma proteins, which in the mammal is about 25 mm Hg. The anatomical arrangement, whereby the diameter of the efferent arteriole is considerably smaller than that of the afferent arteriole, will favour the development of a large pressure in the glomerulus; and actual measurements of the pressure in the proximal tubule indicate that the capillary pressure is, indeed, adequate to exceed the colloid osmotic pressure with a large margin to spare. Thus, if a fine tube is inserted into the proximal tubule, and a little oil is forced into it, flow along this may be blocked, and a pressure develops in Bowman's capsule which is approximately equal to the capillary pressure minus the colloid osmotic pressure (Fig. 10.9). A second needle is placed proximally to the capsule and the pressure measured; as Fig. 10.10 shows, this depends on the arterial blood pressure, but has a plateau of around 60 mm Hg when the arterial pressure is above about 90 mm Hg. The capillary blood pressure is higher than these measured "stop-flow pressures" because it is opposed by the colloid osmotic pressure; the broken line shows the estimated capillary pressures when 25 mm Hg has been added. Thus the capillary pressure, over the plateau region, is about

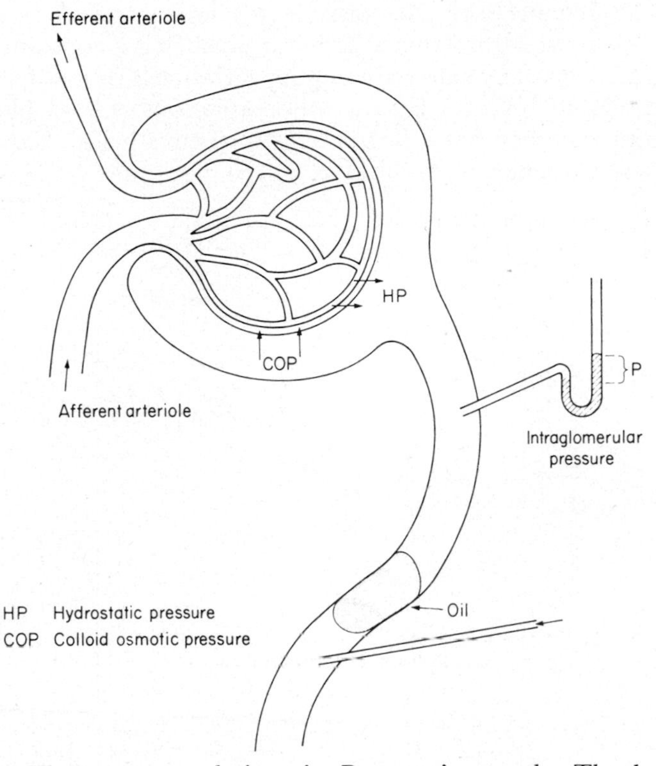

Fig. 10.9. The pressure relations in Bowman's capsule. The hydrostatic pressure (HP) in the glomerular capillaries is only partly balanced by the colloid osmotic pressure of the plasma (COP). The pressure is measured by closing off the capsule from the proximal tubule by insertion of a drop of oil and inserting a tube connected to a manometer. Under these conditions of stopped flow, the measured pressure is equal to the glomerular capillary pressure (HP) minus the colloid osmotic pressure (COP).

88 mm Hg, which compares with some 30–40 mm Hg at the arterial end of muscle or skin capillaries (p. 282).

Autoregulation

Figure 10.10 emphasizes a very interesting feature of the kidney, namely its capacity to *autoregulate*. Thus the glomerular capillary pressure remains independent of arterial pressure over quite a large range, and when the blood-flow through the kidney is measured it is found that this, too, remains constant over this range of arterial pressures—a constancy that is independent of the innervation of the kidney and is presumably a consequence of an adaptation of the blood vessels to altered pressure, the resistance tending to rise as pressure increases. We must regard this as a factor favouring a constancy of the filtration rate, or at any rate reducing its variations. In fact for arterial pressures down to 100 mm Hg the autoregulations of both blood-flow and filtration rate run hand in hand; at pressures below 100 mm Hg the filtration rate tends to decline.

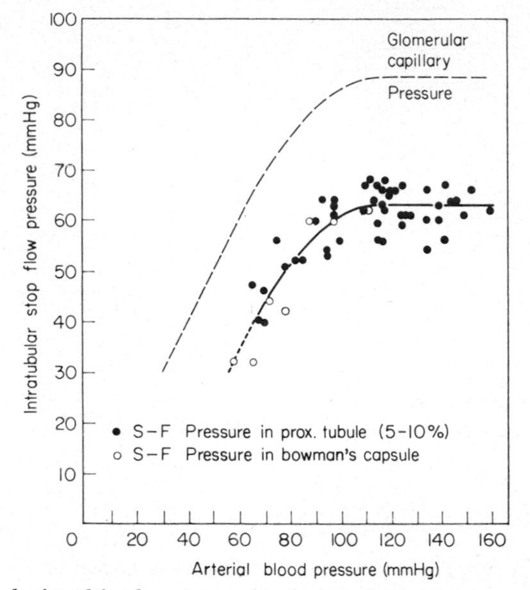

Fig. 10.10. Relationship between the intratubular pressure, measured by a stop-flow, and the arterial blood pressure. The interrupted line represents the glomerular capillary pressure, obtained by adding the plasma colloid osmotic pressure (25 mm Hg) to intratubular stop-flow pressure. Note that the intratubular pressure increases in step with the arterial pressure up to about 90 mm Hg, but above this remains independent of arterial pressure—*autoregulation.* (Gertz, Mangos, Braun and Pagel, *Pflüg. Arch.* 1965, **285,** 360).

Afferent Arteriolar Resistance. Clearly the dominant factor in filtration rate, for a given arterial pressure, is the resistance of the afferent arteriole, and it is likely that the autoregulation of blood-flow is largely brought about by parallel changes in afferent arteriolar resistance and arterial pressure, the higher the arteriolar pressure the higher must be the afferent arteriolar resistance to keep the blood-flow the same. As the resistance is raised, the pressure in the glomerulus tends to remain the same, and thus the filtration rate.

Efferent Arteriolar Resistance. The fact that, at the lower range of arterial pressures over which autoregulation pertains (namely 100 to 75 mm Hg), the glomerular filtration rate tends to decline, suggests that the resistance of the efferent arteriolar system tends to change as well as that of the afferent arteriole. Thus at 75 mm Hg the afferent arteriolar resistance has probably been reduced as much as it can be, and to maintain the constant blood-flow the efferent arteriolar resistance must also be decreased. Such a decrease would not help to maintain glomerular pressure, and so filtration rate might fall.

Filtration Pressure

The actual pressure determining filtration is the difference between the capillary pressure, on the one hand, and the colloid osmotic pressure plus tubular pressure on the other. The tubular pressure will depend on the frictional resistance to flow along the tubular system and on the amount of fluid that is reabsorbed on the way. The latter will vary considerably in accordance with the state of hydration of the body; if the body is overhydrated and water must be eliminated, the amount of fluid remaining in the tubular system tends to be large and this slows filtration, so we encounter the paradox that a high rate of urine formation may be associated with a diminished rate of filtration.

Filtration Fraction

If we measure both the flow of blood through the kidney, and the rate of filtration, we may estimate the fraction of the blood flowing through that is filtered into the nephrons. Thus if we accept an "average" filtration rate in man of 125 ml/min, and a minimum plasma flow of 700 ml/min, the filtration fraction is 0·18, or less than one-fifth. Quite large spontaneous variations in filtration fraction may be observed; for example Lewy and Windhager, in their study of the effects of filtration fraction on tubular reabsorption of fluid, found spontaneous variations from 0·28 to 0·44

The Nature of the Filter

As with the capillary systems elsewhere in the body, we may consider the filtration as occurring through intercellular clefts or pores. In the kidney we may expect the filter to have a rather smaller porosity since experiments show that proteins of the size of haemoglobin, or serum albumin, are far more efficiently excluded from the filtrate than from muscular or cutaneous lymph. In the glomerulus the capillaries are of the fenestrated type, in the sense that the endothelial cells show regions of great attenuation so that the thickness of the wall is actually less than that of the two plasma membranes in contact, i.e. it is as though all the cytoplasm had been squeezed away and that the two plasma membranes, on coming together, had somehow lost some of their material. These fenestrations may be expected to favour filtration but not necessarily to favour retention of the protein molecules, and it may well be that the additional layers (1–4 of Fig. 10.4) interposed between the endothelial cells and the lumen of Bowman's capsule, increase the selectivity of the system as a protein filter.

Pore-slits

By employing his horseradish peroxidase technique Karnovsky did, in fact, demonstrate that the main resistance to escape of this protein (which, because of its small size (M.W. 40,000), does pass into the glomerular fluid and appear in the urine) was apparently at the pore-slits of the epithelial layer, large accumulations being demonstrated on the capillary side of the slit. When a much larger protein—myeloperoxidase (M.W. 160,000–180,000)—was given into the blood, it failed to cross the filter into Bowman's capsule; it did, however, pass out of the capillaries, apparently through the fenestrae and basement membrane, but was held up at the pore-slits, which are therefore covered by a thin diaphragm since the diameters of the slits are much greater than that of the myeloperoxidase molecule.

THE REABSORPTIVE PROCESS

Micropuncture Technique

By micropuncture techniques we may examine the composition of the filtrate as it passes along the tubular system. Thus, by placing droplets of oil at a given point in, say, the proximal tubule, and collecting through a fine pipette placed closer to the glomerulus, the composition of the fluid as it reaches this region may be examined. Subsequent histological examination permits the exact identification of the site of puncture. In general, the proximal and distal convoluted tubules are relatively accessible by virtue of their superficial location.

Identification of Nephron Segments

They may be identified in the dissecting microscope by the expedient of injecting a dye, lissamine green, into the blood. Immediately after injection into the renal artery, the surface of the kidney becomes green due to the presence of the dye in the renal blood vessels; after filtration has occurred it is identified in the proximal tubules first, and is seen moving towards the loop of Henle, and ultimately it disappears from the surface. Later it appears in the distal tubule. Thus tubules that appear green later are distal tubules. This dye not only permits the identification of the proximal and distal parts of the nephron, but it enables the investigator to estimate the rate of flow of fluid along, say, the proximal tubule. Thus the first appearance of a flush indicates the arrival at the glomerulus; the time taken after this for the colour to appear at a fixed point in the proximal tubule is the amount of time it takes for the filtrate to reach this point, and this will be determined by the rate of reabsorption—the slower the rate the sooner will the column reach the point of observation. The papilla of the hamster is relatively accessible, so that fluid from the straight collecting duct can be collected; the ascending and descending loops of Henle are harder of access, but they also have been successfully punctured.

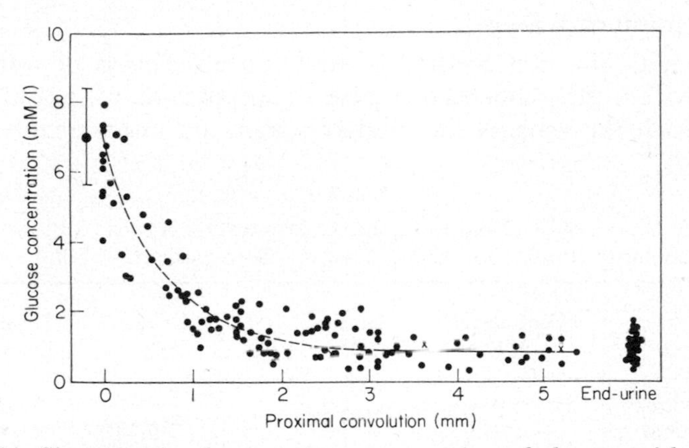

Fig. 10.11. Showing the decrease in concentration of glucose with distance along the proximal convoluted tubule. On the abscissa distances from Bowman's capsule are indicated. (Rohde *et al.*, *Pflug. Arch.* 1968, **302,** 219.)

Reabsorption of Glucose

As Fig. 10.11 shows, the concentration of glucose in the tubular fluid falls to about zero by the time it has reached the half-way line; the same is true of amino acids, so that under normal conditions neither glucose nor amino acids appear in the urine. The transport occurs across the tubular wall into the capillaries of the tubular system of blood vessels; clearly this must be against a concentration gradient, requiring the performance of work–active transport.

Iso-osmotic Reabsorption

When we examine the concentration of Na^+ we find that it remains the same as that in the filtrate at all points along the proximal tubule. This could be due to failure of absorption, or to the circumstance that absorption of Na^+ is accompanied by that of sufficient water to keep the concentration the same, i.e. what we have called *iso-osmotic reabsorption.* That the latter supposition is correct is shown by a variety of studies. Thus we may poison the reabsorption of glucose with phlorizin; as a result, it is found that the concentration in the fluid rises as we pass more distally, indicating the absorption of water, and since the concentration of Na^+ remains the same this means iso-osmotic reabsorption. Again, by placing a column of saline solution in a proximal tubule and imprisoning it between two drops of oil—the split-drop technique—the column can be seen to disappear gradually as the fluid is absorbed.

Reabsorption of Water

In general, the best method of studying absorption of water is to include within the glomerular filtrate a substance that is not absorbed, i.e. to which the walls of the tubular system are impermeable. Inulin

TABLE I

Average Values of the Fluid-to-Plasma Concentration Ratios (F/P) for Total Osmolarity, Inulin, and Urea. (Lassiter, Gottschalk and Mylle)

Source	Osmolarity F/P	Inulin F/P	Urea F/P	Urea Ratio/ Inulin Ratio
Early proximal	1·0	1·0	1·0	1·0
Late proximal	1·0	3·0	1·5	0·5
Early distal	0·7	6·9	7·7	1·1
Late distal	1·0	14·9	10·5	0·7
Ureteral urine	6·4	690	90	0·13

Note: The value of the Ratio, Urea Ratio/Inulin Ratio, is a measure of the fraction of the filtered urea absorbed; the fact that this becomes 1·1 in the early distal tubule indicates that there has been a gain of urea following the loss in the proximal tubule.

is a substance that fills this criterion and, since its molecule is small enough to permit it to pass into the glomerular filtrate from the blood, it is sufficient to establish a known concentration in the blood plasma, in which event the concentration in the glomerular filtrate will be the same. When fluid is taken from any locus, the amount of water lost by the filtrate when it has reached this locus is given by the rise in concentration of inulin. Table I shows the increase in concentration at various stages; by the end of the convoluted portion of the proximal tubule, the concentration of inulin has risen some three-fold, corresponding with a reabsorption of some 66 per cent of the filtrate. By the beginning of the distal tubule, i.e. after passing through the pars recta of the proximal tubule and the loop of Henle, the concentration is some seven times that in the filtrate, indicating a reabsorption of some 85 per cent of the filtrate; in the urine itself the concentration has risen nearly 700 times, corresponding with the reabsorption of over 99 per cent of the filtrate.

Obligatory and Facultative Reabsorption. In general, it is found that the degree of reabsorption in the proximal tubule is fairly constant, and of the order of 85 per cent, and this clearly places a limit to the amount of fluid that can be excreted in unit time since the rate of filtration is fairly constant. Under conditions where the body is removing water as fast as possible—*diuresis*—it is, indeed, found that the maximum does correspond to about 15 per cent of the filtration rate. Under conditions of water retention—*antidiuresis*—such as those illustrated by Table I, the reabsorption by the more distal parts of the nephron is considerable, and we may thus divide the reabsorption into an *obligatory fraction*, representing some 85 per cent of the filtrate, and a *facultative fraction* that may vary from zero to nearly 15 per cent of the filtrate according to the state of water-balance of the animal. As Table I shows, the osmolality of the urine rises to some six times that in the plasma, so that the urine is hyper-osmolal under these conditions of antidiuresis. Under conditions of diuresis, the urine is usually hypo-osmolal, the salts and some other solutes being preferentially reabsorbed in the distal nephron whilst the water is allowed to escape (p. 522).

Clearance

Calculation

We have already introduced the concept of clearance, if somewhat vaguely. More precisely it is the volume of blood plasma "cleared" of any given substance during unit time, usually expressed in ml/min. It is a fictitious volume, since usually none of the blood passing through the kidney is completely cleared; so it means, in effect, the volume of blood that contained the amount of the substance found in the urine in unit time. For example, if 50 mg of urea were found in the urine collected in one minute, and if the concentration of urea in the blood plasma was 50 mg/100 ml, then the clearance would be 100 ml/min, 100 ml of blood having been relieved of 50 mg of urea, all that it had. In general, the amount eliminated in the urine is given by $U \times V$, where U is the urine concentration and V is the flow of urine in ml/min. The volume of plasma that contains this amount is UV/P, where P is the blood plasma concentration. UV/P is thus the clearance in ml/min.

Clearance and Filtration

If a substance passes into the tubular system only by way of the glomerular filter, and if it is not reabsorbed during its passage through the nephron, then its clearance, UV/P, is by definition equal to the rate of filtration. Inulin fulfils this criterion, and so the inulin clearance is an accurate measure of filtration rate. In man this is reasonably constant, and amounts to some 120 ml/min of which, normally, only about 1 ml reaches the ureter as urine.

Clearance Ratio. If the substance we are considering is reabsorbed to a greater or lesser extent, then the concentration in the urine, U, will be reduced below that of inulin, and its U/P ratio will be less than that of inulin; hence its clearance is less than that of inulin, and we may use the magnitude of the *clearance ratio:*

$$\left(\frac{UV}{P}\right)_x \div \left(\frac{UV}{P}\right)_{In}$$

as a measure of reabsorption. Since the rate of formation of urine, V, is common to both fractions, we may use the ratio: $(U/P)_x \div (U/P)_{In}$ instead. As Table I shows, the ratio for urea varies with the position in the nephron, but in general is less than unity, indicating net re-absorption. When a substance passes into the filtrate but also passes directly from the blood across the tubular wall by the process called tubular secretion, the U/P will be greater than for inulin and the clearance ratio will be greater than unity. Thus, in the intact animal, by measurement of clearances, we may assess whether there is net reabsorption or secretion by the kidney. Sometimes the situation is complicated by both processes occurring together, as with K^+ where reabsorption occurs in the proximal tubule and secretion in the distal tubule; and to demonstrate this special methods must be employed.

Proximal Tubular Reabsorption of Na^+

The iso-osmotic reabsorption of water and salt in the proximal tubule has been examined in some detail by Solomon, employing the technique illustrated in Fig. 10.12 where the absorption of fluid is measured by establishing a blockade with oil above and below. After the fluid has been held in the tubule in this way for a given period it is removed and analysed. Solomon concluded that the primary process is, indeed, an active transport of Na^+ with its accompanying anions into the adjacent extracellular space surrounding the tubule; the consequent increase in osmolality causes the passage of water.

Electrochemical Potential. The active transport gives rise to a difference of potential,* as with the frog skin, but it is not so high, the outside being some 20 mV positive in relation to the lumen. Calculations indicated that the absorption of Cl^- might well be the passive consequence of the absorption of the positive Na^+-ion. That the absorption of Na^+ occurs actively, against gradients of electrochemical gradient, may be easily demonstrated by studying absorption when the filtrate contains a relatively high concentration of a solute to which the tubule is impermeable, e.g. mannitol; under these conditions Na^+ is reabsorbed, as before, but now the concentration of NaCl in the tubule is reduced so that the ion passes up a considerable gradient of concentration.

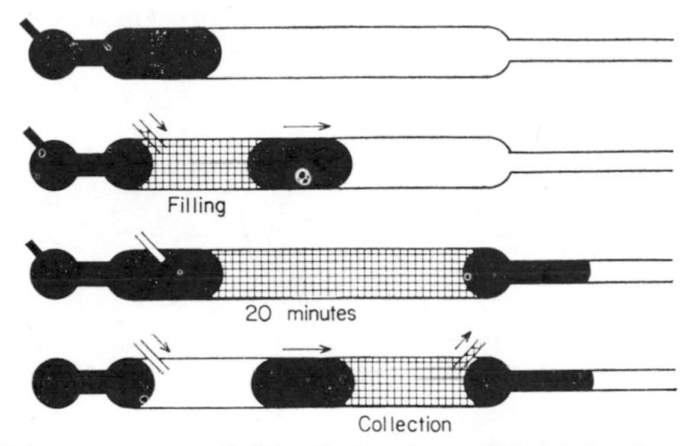

Fig. 10.12. Measurement of the absorption of fluid in the renal tubule. A droplet of oil is injected into the tubule and this droplet is split in half by the injection of a fluid of known composition. The fluid drop bounded by two oil droplets, is withdrawn after about 20 min and the change in composition studied. (Solomon, "Method of Isotopic Tracers", Pergamon, 1959.)

Mannitol Diuresis. Because the reabsorptive system is unable to establish significant gradients of osmolality across the proximal tubule, the presence of an impermeant solute, such as mannitol, tends to restrict the reabsorption of salt and water, with the result that less than the usual 85 per cent of the filtrate is reabsorbed and therefore the distal nephron is presented with a larger load of water to deal with; if it is incapable of handling this extra load, more water is excreted and we have the phenomenon of *mannitol diuresis.*

Ouabain. The mechanism of the active transport of Na^+ is presumably similar to that involved in many other processes in which an

*There has been some disagreement as to the existence of a measurable difference of potential across the proximal tubule; it is certainly less than 20 mV and the most recent figure, obtained by Burg and Orloff on isolated segments, was 4 mV, the inside being negative.

iso-osmolar solution is caused to flow across an epithelial layer (p. 450); it may be inhibited by cardiac glycosides and thus depends on a Na^{+}-K^{+}-activated ATPase, and it would seem that this enzyme is concentrated on the epithelial cell membrane in the region of the basal infoldings.

Passive Reabsorption

The iso-osmotic reabsorption of some 85 per cent. of the fluid in the proximal tubule will tend to raise the concentrations of many solutes above those in the blood plasma and thus concentration gradients will be established favouring reabsorption into the blood. This passive reabsorptive process is a significant factor with urea, and we see from Table I that there is absorption from the proximal tubule, since the late proximal urea/inulin clearance-ratio is 0·5. In the urine the ratio

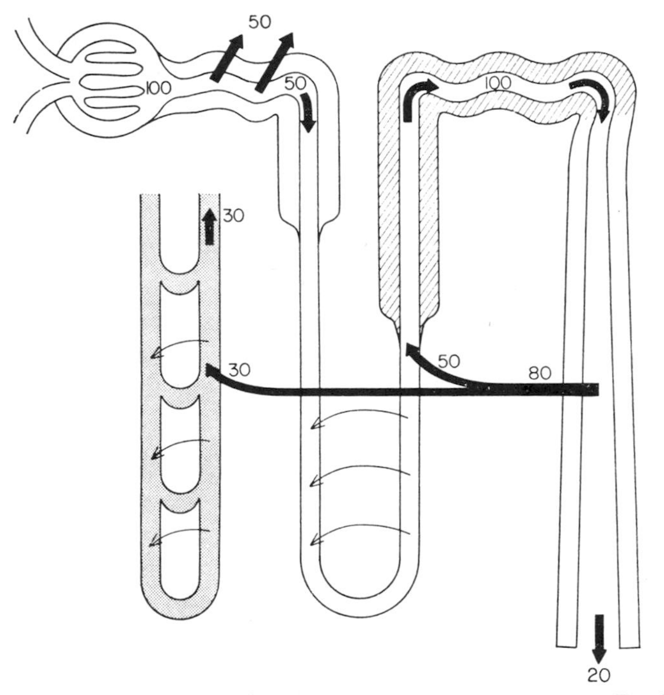

Fig. 10.13. Urea movements in the countercurrent system. During antidiuresis half of the filtered urea is absorbed proximally. A large part (5/8) of the amount leaving the collecting ducts by diffusion is recirculated by way of the ascending limb and distal tubule which are relatively impermeable to urea (hatched area). The rest (3/8) is lost via the blood vessels. Thin arrows—countercurrent diffusion of urea. Thick arrows—net flux of urea. (Ullrich *et al.*, *Progr. Cardiovasc. Dis.* 1961, **3,** 315.)

is much less, 0·13, indicating some 87 per cent absorption, but the situation is complicated since in the early distal tubule, i.e. after the fluid has passed through Henle's loop, the ratio has returned to about unity. The probable explanation for this is that urea has returned to the fluid by diffusion out of the collecting duct, where it has been highly concentrated through absorption of water, across the interstitial fluid into the ascending loop of Henle, as illustrated in Fig. 10.13.

Active Transport of Sugars and Amino Acids

The reabsorption of urea is accounted for, at any rate in many species, by passive diffusion down gradients of concentration created by the loss of water from the tubule. Clearly, similar forces will operate on all the constituents of the glomerular filtrate, but with many substances, such as glucose and amino acids, passive diffusion cannot account for the complete removal of the solutes; and we must invoke active transport.

Sugars

When different sugars are compared, it is found, as with absorption from the intestine, that the rates of reabsorption are different the specificity being so great that the reabsorptive mechanism differentiates easily between optical isomers—the absorption of L-glucose being much less than that of D-glucose. Similarly, when a racemic mixture of an amino acid is presented to the tubules, the L-isomer is preferentially reabsorbed. As with intestinal absorption, the reabsorption of these compounds shows competition, so that the more actively reabsorbed D-glucose tends to depress the reabsorption of D-fructose or xylose, and in accordance with the degrees of mutual inhibition the sugars and amino acids may be arranged in orders of decreasing affinity for a hypothetical carrier. Thus the order for monosaccharides is:

Glucose → Galactose → Mannose → Fructose → Xylose → Arabinose

The rat kidney can convert fructose to glucose, and this seems to be the mode in which it is reabsorbed as with the intestine.

Amino Acids

The amino acids do not all show mutual interference, so that we may divide them into classes and assign a theoretical carrier on which a class is carried. Thus the basic amino acids, such as arginine and histidine do not interfere with the reabsorption of acid amino acids, such as aspartic and glutamic acids; neutral amino acids likewise seem to have a separate mechanism which may, however, be shared to some extent with basic amino acids.

Maximal Reabsorptive Capacity

When the concentration of glucose in the blood is raised, either experimentally or in diabetes mellitus, a point is reached when glucose appears in the urine; this occurs at a fairly sharply defined concentration, and the phenomenon gives rise to the concept of a maximal reabsorptive capacity for the tubules, such that when the filtrate contains more than a certain quantity of the solute there is a failure

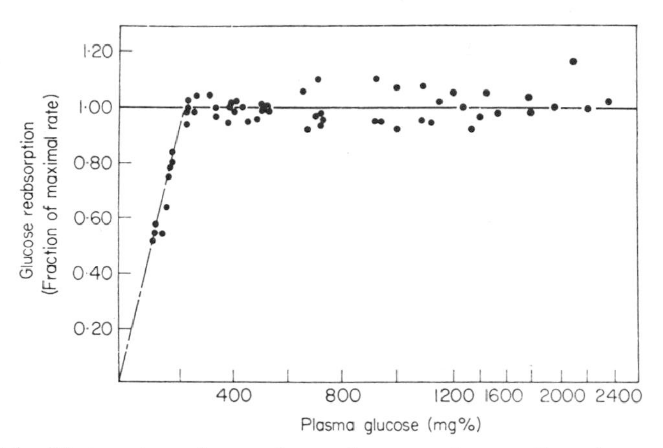

Fig. 10.14. Illustrating the reabsorption maximum for glucose. As the concentration of glucose in the plasma is raised, the amount reabsorbed increases, but only up to a certain limit when the maximal reabsorptive capacity T_m is reached. (Shannon and Fisher, *Am. J. Physiol.* 1938, **122,** 765.)

to reabsorb it completely. When the measured reabsorption was plotted against the plasma sugar concentration, this reached a plateau value at a sharply defined point, and the reabsorption at this point was called the *maximal reabsorptive capacity*, indicated by the symbol T_m. The existence of such a sharply defined inflexion point on the graph (Fig. 10.14) suggests that all the tubules reach their maximum rates at the same level of plasma concentration, a surprising fact when the variations in tubular length are considered, so that a matching of the glomerulus to the proximal tubule has been postulated such that a long tubule has a larger volume of filtrate presented to it than a short tubule.

Glomerular Filtration Rate and T_m. Furthermore, the concept of a T_m implies that the reabsorption is independent of the rate of filtration, otherwise we should not expect to find excretion of glucose to occur at a sharply defined plasma level. Subsequent work has shown that it is only because, in most studies, filtration rate is reasonably constant that this sharply defined plateau-level is obtained. If, however, reabsorption divided by filtration rate is plotted against plasma glucose, a fairly sharply defined plateau is obtained (Fig. 10.15).

Threshold. This phenomenon of the T_m illustrates the basic mechanism for maintaining a certain degree of homeostasis in the

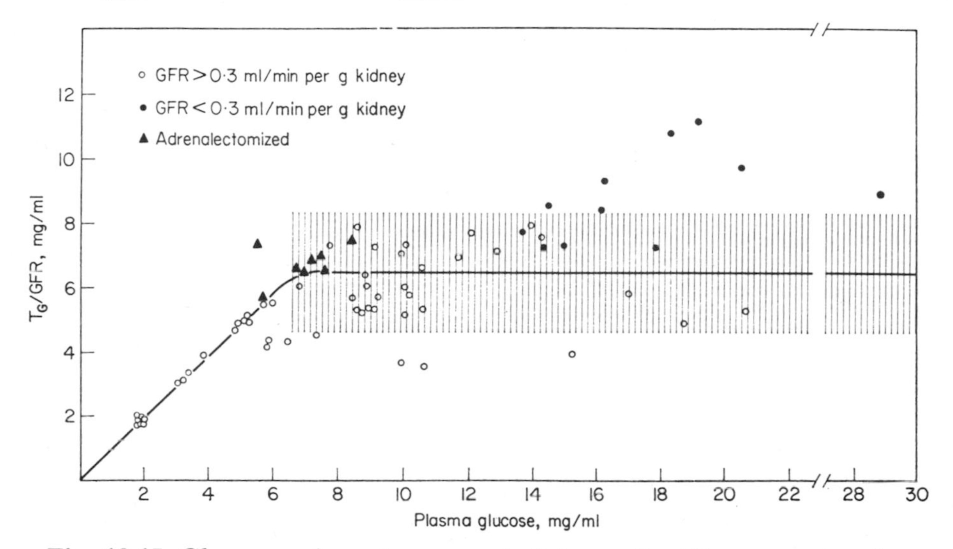

Fig. 10.15. Glucose reabsorption per ml of glomerular filtrate as a function of plasma glucose concentration. The horizontal portion of the curve represents the average of all points with a plasma glucose concentration above threshold. Shaded area indicates standard deviation of points. (Van Liew *et al.*, *Pflug. Arch.* 1967, **295,** 232.)

blood composition. The plasma concentrations of important metabolites are allowed to rise to a certain point, and over this range the kidney makes no effort to reduce them; beyond this point the metabolite is allowed to "spill over" into the urine. Substances of this type are called "*threshold substances*", a certain threshold plasma level being attained before they appear in the urine.

Inhibition of Reabsorption

Because reabsorption depends on active transport mechanisms we may expect that metabolic inhibitors will affect the process; alternatively substances that compete for the hypothetical carrier may inhibit reabsorption. The study of metabolic inhibitors in the intact animal is difficult because of their general toxic effects, but modern studies, in which tubules may be isolated and perfused with solutions through fine tubes (Fig. 10.16), have shown that inhibitors such as dinitrophenol also abolish the active removal of sugars and amino acids. A substance that blocks access to the carrier for sugars is *phlorizin*, and this may be given to the intact animal, when it causes a phlorizin diabetes, reabsorption of glucose in the proximal tubule being abolished.

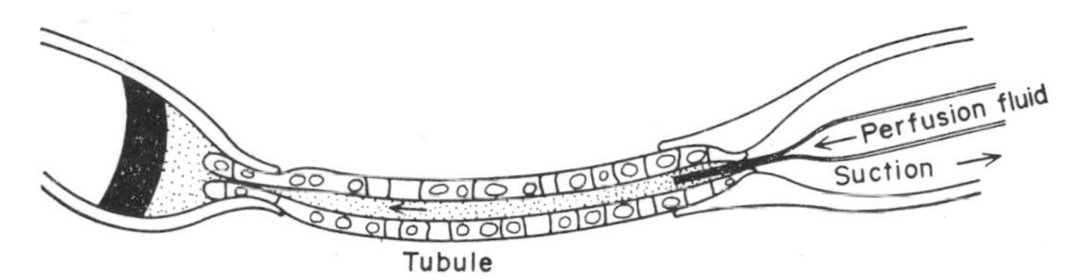

Fig. 10.16. The arrangement for the perfusion of single kidney tubules *in vitro*. The tubules are isolated and a fine cannula introduced into one end by an outer suction jacket. Fluid of varying composition may be passed through the inner tube and collected at the other end by another tube into which the tubule has been introduced.
(Burg and Orloff, *J. Clin. Invest.* 1968, **47,** 2016.)

Dependence on Na^+

We have seen that the absorption of sugars and amino acids by the intestine is linked with active transport of Na^+; the same is true of tubular reabsorption; hence ouabain, which inhibits the Na^+-K^+-activated ATPase concerned in active transport of Na^+, blocks reabsorption.

Renal Handling of Protein

The glomerular filter is not perfect, so that the proximal fluid contains some 40 mg/100 ml; if this were not reabsorbed this would lead to a very considerable depletion of the blood protein. The reabsorption of such large molecules through the tubular epithelial layer may be expected to occur by a process of endocytosis, the protein being engulfed in small vesicles into the cytoplasm of the epithelial cell. Histologically, when haemoglobin is in the filtrate this absorption can be demonstrated, the cells becoming filled with haemoglobin-containing droplets. In the electron microscope the small protein, horseradish peroxidase, which passes the glomerular filter easily, may be seen accumulated in vacuoles in the apex of the tubular cell within 90 seconds after the intravenous injection. Later the protein can be seen accumulated in larger droplets. As to the ultimate fate of the absorbed protein, it seems likely that it is broken down to amino acids and returned to the blood, so that the *absorption droplets* are probably to be viewed as phagosomes, etc. (p. 49).

Tubular Secretion

Aglomerular Excretion

The glomerular kidney, providing a preliminary filtration of the blood, is a vertebrate characteristic, so that renal excretion in non-vertebrates is brought about by a direct transfer of unwanted solutes and water into the tubular system. In certain fish, moreover, the glomerulus has become vestigial, and the goose-fish, for example, relies on a process of tubular secretion for the elimination of its solutes and water. In the vertebrate kidney the filtration–reabsorption mechanism is supplemented by direct transfer into the tubule.

PAH Clearance

The process is illustrated in its most striking form when the clearances of certain foreign substances such as p-aminohippurate (PAH), phenol red, Diodrast, etc., are measured; these greatly exceed the inulin clearance and thus indicate that excretion is not only through filtration but also through direct transfer into the tubule. The site of this secretion is the proximal tubule; this is easily demonstrated either by micropuncture techniques or by the stop-flow technique.

Stop-flow. With this latter technique the flow of urine is abruptly stopped by ligating the ureter for a short time; during this period the various reabsorptive and secretory processes continue, i.e. glucose is removed from the proximal tubule, inulin is concentrated and there is reabsorption of water, and so on. Flow is then allowed to begin, and successive samples of urine are collected; the early samples are those from the collecting ducts and later samples from the ever more proximal parts of the nephrons. It is found that the peak of PAH concentration occurs at the trough of glucose concentration, indicating a common site in the proximal tubules (Fig. 10.17). It will be seen from Fig. 10.17 that inulin appears in the 12th–13th sample; this is because the inulin was injected into the blood during the period of stopped flow, so that it only entered the tubules on restarting flow and collection of the samples; thus the 12th–13th sample represents the point where new filtrate has entered the system, and corresponds with the early proximal samples. Creatinine, another substance excreted by filtration without reabsorption, and thus similar to inulin, was present in the blood before the stopped flow, and appears in all samples. In general, experience has shown that the sample with maximal PAH concentration is representative of almost pure proximal fluid whilst pure distal fluid is given by the earlier sample containing minimal Na^+; fluid from the collecting duct is considered to be represented by a still earlier sample with maximal concentration of K^+.

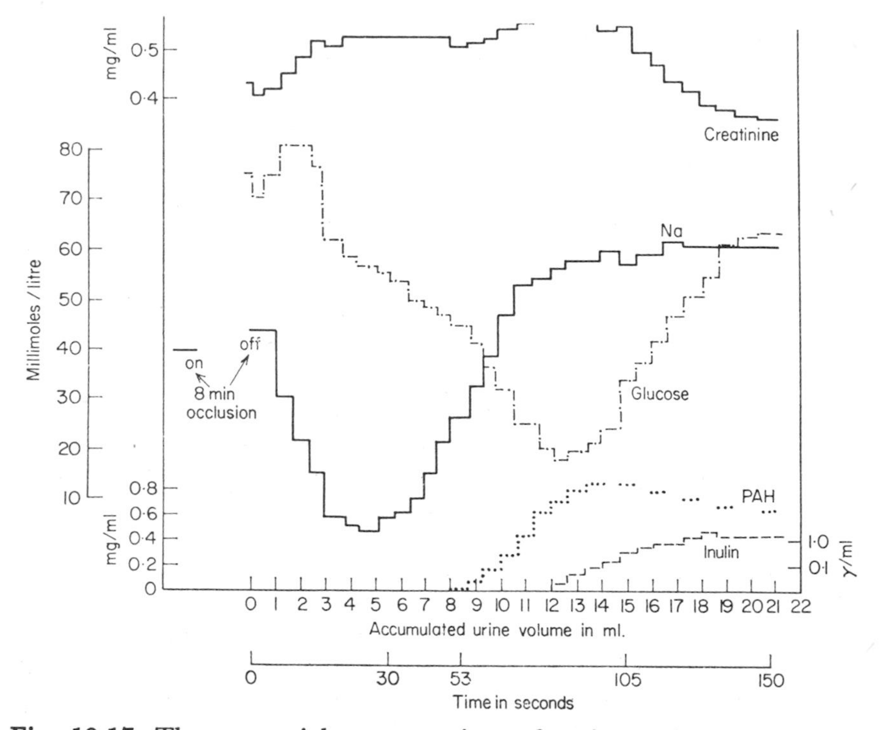

Fig. 10.17. The sequential concentrations of various substances in urine released after a clamp has been placed on the ureter—the stopped-flow technique. The concentration of each of the samples is taken to be the same as that in the particular zone of the kidney tubule from which the sample originated. Inulin only enters the 12th sample, as it was injected during the period of stopped flow and represents the first of the newly filtered fluid. The early samples are from the collecting ducts and later samples from the tubular parts of the nephron. Note that the peak of PAH concentration occurs at the trough of glucose concentration, indicating that the secretion of PAH and absorption of glucose occur at the same site in the proximal tubule. (Malvin *et al.*, *Am. J. Physiol.* 1958, **194,** 135.)

Measurement of Renal Blood-flow. So efficient is the tubular secretion of PAH that when the concentration is low, the blood passing through the kidney is very nearly cleared of all its PAH in a single circuit, and the fact that the extraction-ratio is not unity, but about

0·9, is probably due to the short-circuiting of some blood, which results in its failure to pass through the tubular blood capillaries. In general, however, the PAH-clearance is used as a measure of the flow of blood through the kidney in clinical studies.

Maximal Secretory Capacity

As with absorption, there is a maximum capacity, this time called the *maximal secretory capacity*, T_m; it must be appreciated, however, that, because PAH may pass in the filtrate, there is no maximum to the

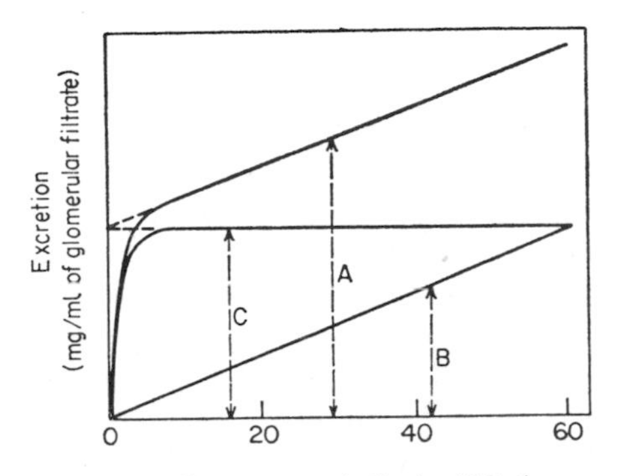

Fig. 10.18. The maximal secretory capacity, T_m. The curves describe the excretion of phenol red. Curve A represents the excretion of phenol red as a function of plasma concentration. The excretion rises steeply then there is a linear increase with plasma concentration about 5 mg/100 ml. Curve B, derived from inulin clearance, gives the amount filtered and by subtraction from A gives curve C which is a measure of the secretory activity. As can be seen from curve C there is a definite limit to the secretory capacity of the tubules, and above 5 mg/100 ml plasma concentration this capacity is exceeded. (Pitts, *J. Cell. Comp. Physiol.* 1938, **11,** 99.)

excretory capacity, the higher the blood concentration the more will be eliminated in unit time. Thus in Fig. 10.18, Curve A represents the excretion of phenol red as a function of plasma concentration; there is a steep rise, and then a linear increase above a plasma concentration in the region of 5 mg/100 ml. The amount of the excretion due to filtration is easily calculated from the inulin clearance which, we have seen, measures the rate of filtration. This is given by the straight line, B. When the excretion due to filtration is subtracted from the total, we obtain Curve C indicating that, beyond about 5 mg/100 ml, the secretory capacity of the tubules has been exceeded.

Competition

As with reabsorption, there is competition for tubular secretion but this does not extend to all secreted substances which, as with the amino acids, may be grouped in such a way that members of a group compete with each other and not with those in other groups.

Significance of Tubular Secretion

PAH is classified in a group of organic acids in which are included sulphonamides, uric acid and probenecid. As so many of these substances are foreign to the body, the physiological significance of their tubular secretion is not obvious; however, many toxic substances are highly lipid-soluble and this lipid-solubility would impair renal excretion since they would tend to diffuse back into the blood during their passage through the tubule; such back-diffusion is easy to demonstrate with lipid-soluble substances, e.g. the thioureas. The means adopted by the body to eliminate them is for the liver to conjugate them with an acid, such as glucuronic acid, thereby giving them water-solubility. In this "detoxicated" condition they remain in the filtrate, whilst the presence of the conjugating molecule gives them sufficient structural specificity to allow them to employ the tubular secretory mechanism. The corticosteroid hormones are highly lipid-soluble, and their elimination is probably brought about because of a preliminary conjugation, e.g. 17-H-corticosteroid is conjugated with glucuronic acid and this gives it sufficient water-solubility, and a characteristic atomic grouping, to permit rapid excretion.

Uric Acid

This is an unwanted product of purine metabolism and is excreted by the kidneys; its clearance is about a third of the filtration rate in man and most mammals, indicating partial reabsorption. More careful examination of the process indicates, however, a process of tubular secretion taking place in addition to the reabsorption. In birds and reptiles the excretion is entirely through combined filtration and secretion, without any reabsorption, and the competition with such substances as probenecid indicates that the organic acid secretory mechanism is employed. In other species, including the Dalmatian coach hound, there is a large amount of tubular secretion, but if this is inhibited by probenecid, a tubular reabsorption becomes manifest. As indicated above, in man there is a net reabsorption, so that the clearance is less than the filtration rate, but this is a combination of tubular secretion, tending to make the clearance greater than that of inulin, and tubular reabsorption having the opposite effect. Under special conditions of uric acid loading, clearances greater than the filtration rate can be induced.

Organic Bases

In addition to the organic acid group there is a group of organic bases that are secreted by the tubules; these include tetraethylammonium (TEA), choline and N′-methylnicotinamide, and their secretion may be competitively inhibited by Darstine and quinine. Strong bases of this type are closely related chemically to neurotransmitters, and it may well be that such excretory mechanisms are developed to enable the blood to be cleared of such pharmacologically active substances should they appear in the blood adventitiously.

HOMEOSTASIS OF THE BLOOD

Glucose

The basic mechanisms of the kidney, described so far, enable us to picture, in general terms, the manner in which it contributes to maintaining the composition of the blood within certain limits. Thus a large rise in the concentration of glucose in the blood, due, say, to a carbohydrate meal, will increase the "filtered load" of glucose presented to the proximal tubule, and if this load exceeds the T_m, i.e. the maximal reabsorptive capacity, some of the glucose will "spill over" into the urine. A similar mechanism operates with amino acids and several other solutes, but it must be emphasized that such a T_m-type of control is not likely to be very precise if left to itself, since by its nature it must deal with the relatively large fluctuations required to exceed the T_m. Thus the blood glucose can rise from an average resting value of about 100 mg/100 ml to over 150 mg/100 ml without any glucose appearing in the urine, the primary role in the homeostasis being, in fact, uptake by the liver.

Urea

With urea, which is continually being formed and must be as continually eliminated from the body, fairly large fluctuations can occur normally, the elimination relying, in man and many other mammals, on a passive process of filtration with only partial reabsorption; the greater the concentration in the blood the greater the concentration in the filtrate and, if the fraction reabsorbed is tolerably constant, the greater the amount that will be eliminated. In Vol. 3 we shall discuss in some detail the finer points in the homeostasis of the blood, showing the co-operation of the hepatic, renal and other mechanisms in achieving this; for the present we may confine ourselves to one very important aspect, namely the mechanism whereby changes in osmolality of the blood may be resisted, since it is here that the kidney exerts a dominant role.

Osmolality of the Blood

Control over the osmolality of the blood involves control over the concentration of its dominant solute, namely Na^+, so that the renal handling of this ion is a first consideration. If the diet contains appreciable quantities of Na^+ ion, as indeed it usually does, this is absorbed from the intestine by an active process, so that its continual elimination from the body is a usual requirement. This does in fact occur in the urine, the Na^+ in the filtrate being incompletely reabsorbed during its passage through the nephron. If the subject drinks plenty of water, i.e. more than enough to compensate for the losses due to evaporation, sweat, etc., the eliminated salt will be dissolved in the extra water taken in; other solutes such as urea will also be dissolved in the urine as finally excreted, and it will depend primarily on the relative amounts of salt and water in the diet as to whether the urine will be hypo-, iso- or hyperosmolal to the blood.

Selective Reabsorption of Na^+ and H_2O

In general it is found, with normally functioning kidneys, that the osmolality of the urine reflects faithfully the salt and water balance of the animal. This is achieved partly by the selective control over the reabsorption of salt, principally NaCl and $NaHCO_3$, from the glomerular filtrate on its passage through the nephron, but also through selective control over the reabsorption of water. Such a combined control is necessary, since there are occasions where the needs of the body to conserve salt and eliminate water, or to conserve both salt and water, conflict with each other, so that the final response of the kidney to a given situation may represent a compromise between the two requirements. Thus during severe water-lack the blood becomes hyperosmolal and strong control is exerted over the reabsorptive process for water, the ideal condition, of course, being complete cessation of water-loss through this source. Elimination of extra salt, by less complete reabsorption, will tend to reduce the osmolality of the blood, and so co-operate with the water-conserving mechanism to maintain osmolality constant. The increased elimination of salt requires, however, elimination of water, since the salt cannot be excreted as a solid, and in fact the concentration at which it can be eliminated is strictly limited, as we shall see. Thus there is a limit to which the salt and water mechanisms can co-operate in holding down osmolality. A further consideration that makes separate control mechanisms necessary is the requirement to maintain the volume of the blood tolerably constant; elimination of salt, requiring as it does elimination of water, may lead to too great a diminution in blood-volume and thus to failure of the heart's pumping mechanism.

Free-water and Osmolal Clearances

To return to the control mechanisms, we may express the effects of the renal adaptation to osmolality stress, be it absorption of excess water or water-deficiency, as the excretion or otherwise of "solute-free water", or "free-water". Thus, after drinking water in excess of requirements, the urine, if it is to compensate for the dilution of the blood, must be hypo-osmolal, and the amount of water that would be required to bring this volume of excreted urine up to iso-osmolality with the blood is the solute-free water excreted. Alternatively, in water-deficiency, there will be a negative excretion of solute-free water, or an excretion of solute in excess of that required to keep the urine iso-osmolal. It is customary to employ, in these respects, the concepts of osmolal clearance, and free-water clearance.

Osmolal Clearance. The osmolal clearance is defined, like the clearance of any solute, by:

$$C_{osm} = \frac{U_{osm}}{P_{osm}} \times V \tag{1}$$

and, expressed in words, it is the volume of fluid, with the same osmolality as that of plasma, that would be required to excrete the osmolal contents of the urine. If the osmolal clearance is greater than the actual volume of the urine there will have been a net excretion of water, i.e. a positive free-water clearance; if the osmolal clearance is less than the volume of the urine, then there will be net excretion of solute or a negative free-water clearance.

Free-water Clearance. If V is the volume-flow of urine, then we may define the free-water clearance, C_{H_2O}, by:

$$V = C_{osm} + C_{H_2O} \tag{2}$$

and by substituting the value of C_{osm} from equation (1) in equation (2) we obtain:

$$C_{H_2O} = V\left(1 - \frac{U_{osm}}{P_{osm}}\right)$$

Thus, when urine and plasma osmolalities are equal, the free-water clearance is zero; when the osmolality of urine is greater than that of plasma the free-water clearance is negative, and when the osmolality of urine is less than that of the plasma, the free-water clearance is positive.

Urine Dilution

In general the clearance or excretion of solute-free water presents no special problems to the physiologist. It requires the ability of the nephron to reabsorb the solutes of the filtrate to the point that, by the

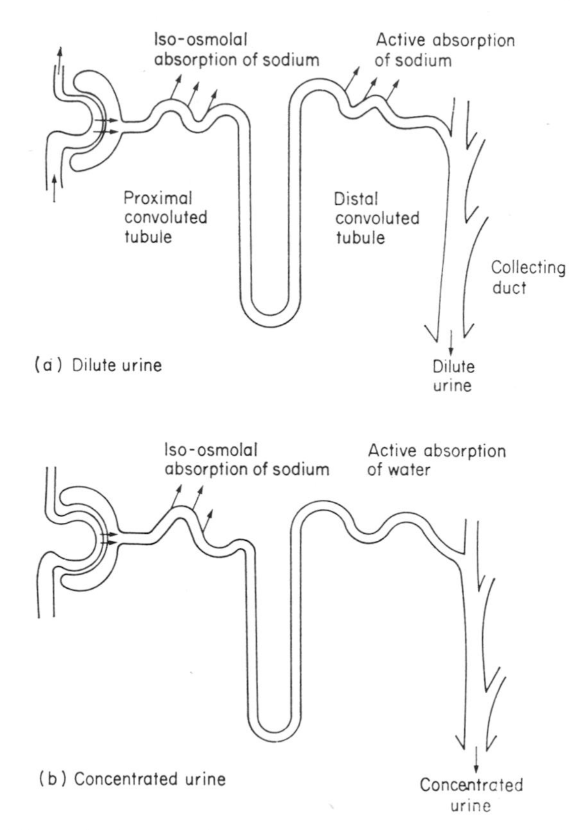

Fig. 10.19. The dilution and concentration of urine. (*a*) The production of a dilute urine by the kidney can be explained by an iso-osmolal absorption of sodium and water in the proximal convoluted tubule and the reabsorption of salt from the distal tube. (*b*) The production of a concentrated urine could be explained by iso-osmolal absorption of sodium and water in the proximal convoluted tubule and the active absorption of water in the distal convoluted tubule.

time it has left the collecting ducts and passed into the ureter, it is hypo-osmolal (Fig. 10.19a). This can be, and is, achieved by selective reabsorption of Na^+; thus when large quantities of water have been drunk, the urine formed during the consequent water-diuresis, is very "dilute", the concentrations of all solutes being low and that of Na^+ very low indeed by comparison with that in plasma. In the proximal tubule we have seen that the reabsorption of solute is *iso-osmolal*, the reabsorption of Na^+ leading to an osmotic flow of water that results in no net change in osmolality of the filtrate left behind in the tubule. Thus the osmotic adjustment required to eliminate excess water (free-water clearance) must take place distally to the proximal convoluted tubule, and micropuncture studies have shown that during diuresis, or free-water clearance, the hypo-osmolality is brought about by reabsorption of salt from the *distal tubule*.

Mechanism of Urine Concentration

In the response to water-deficit we must have a negative free-water clearance leading to the establishment of a hyperosmolal urine. For the establishment of this hyperosmolality it was postulated for a long time that there was an active transport of water from the more distal parts of the nephron, as illustrated by Fig. 10.19b.

Tissue Osmolality

The loop of Henle was implicated since it was observed that animals with long loops were able to form more concentrated urines than those with short loops, such as birds. However, the active transport of water has never been demonstrated unequivocally in vertebrate systems. Moreover, if this were occurring in, say, the loop of Henle or in the collecting ducts, we should expect the osmolality of the tissue of the kidney to be lower than that of the fluid in the ducts and tubules. In fact, it was shown by Wirtz that the osmolality of the kidney tissue progressively increased from the outermost to the innermost zones when the animal was in antidiuresis, i.e. compensating for water-deficiency, whilst during diuresis the gradient of osmolality disappeared, the osmolality being equal to that of the blood.

Collecting Tubule. When the fluid emerging from the loop of Henle into the distal convoluted tubule was examined, it was found to be actually hypo-osmolal to blood, so that there had been no concentration during passage through the loop, whereas examination of the fluid at successive stages along the collecting tubules showed that, as the fluid passed deeper and deeper into the medulla of the kidney, its osmolality increased. Thus the process of concentration, however it was brought about, took place in the collecting tubules, running radially in the medulla in company with the loops of Henle.

Countercurrent Exchange Theory

The countercurrent exchange theory of urine concentration rejects the notion of an active transport of water; the mechanism is considered to consist primarily of an osmotic removal of water from the

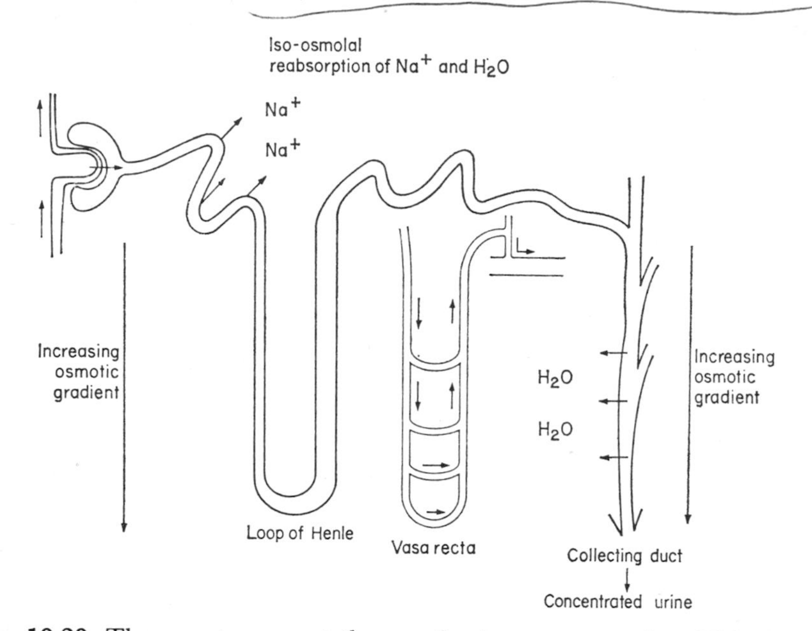

Fig. 10.20. The countercurrent theory of urine concentration. This theory rejects the active transport of water and moves water out of the collecting duct by a zone of raised osmolality in the medullary region.

fluid in the collecting ducts due to the greater osmolality of the tissue-fluid in the medullary region, as illustrated in Fig. 10.20; the water so removed is carried away by the blood in the vessels running parallel with the collecting ducts, namely the vasa recta. Such a mechanism poses two main questions. First, how is the gradient of osmolality of the medullary tissue, established during antidiuresis, to be lost when the animal no longer conserves water, i.e. in diuresis. Second, how is the water, removed osmotically, carried out of the kidney? And why does not osmosis level out the difference of osmolality.

Active Transport of Na^+

The first is the main problem, and the explanation is based on what has been called a *countercurrent mechanism*. According to this, Na^+ is actively transported out of the ascending limb of the loop of Henle, so that the fluid leaving the loop is hypotonic, as indeed is found. The salt transferred diffuses in the interstitital space between the limbs of the loop and passes into the incoming tubular fluid in the descending

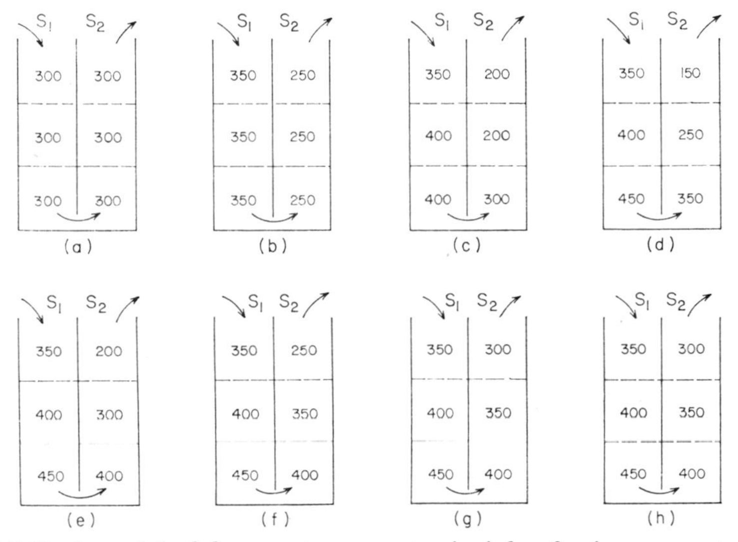

Fig. 10.21. A model of the countercurrent principle of urine concentration. In (*a*) a "U" tube with a solution flowing through it with concentrations as shown, no active transport has occurred. In (*b*) active transport starts in S_2 and some 50 mEq Na^+/litre has been transported to S_1. In (*c*) the active transport continues, but because of the flow the fluid S_2 now receives some of the salt it has transported. (*d*) to (*h*) show successive stages of the process leading ultimately to a steady state. (Davson, "A Textbook of General Physiology", Churchill. 1970.)

loop. At the top of this loop, the fluid is isotonic, having undergone iso-osmotic reabsorption in the proximal tubule; as it passes down the descending limb it acquires more and more salt and thus, at the turn of the loop, reaches a maximum of hypertonicity; as it ascends the loop, salt leaves because of the active transport process, and a stable steady state is achieved with a steadily increasing concentration in both descending and ascending limbs and the adjacent interstitial fluid.

Steady-state Concentrations. The establishment of this steady state is illustrated in Fig. 10.21 where in (a) no active transport of

Na^+ has happened; in (b) some 50 mEq Na^+/litre has been transported to the descending limb; in (c) the active transport has continued but, because of flow, the fluid at the bottom of the ascending loop has received some of the salt it transported; (d) to (h) shows successive

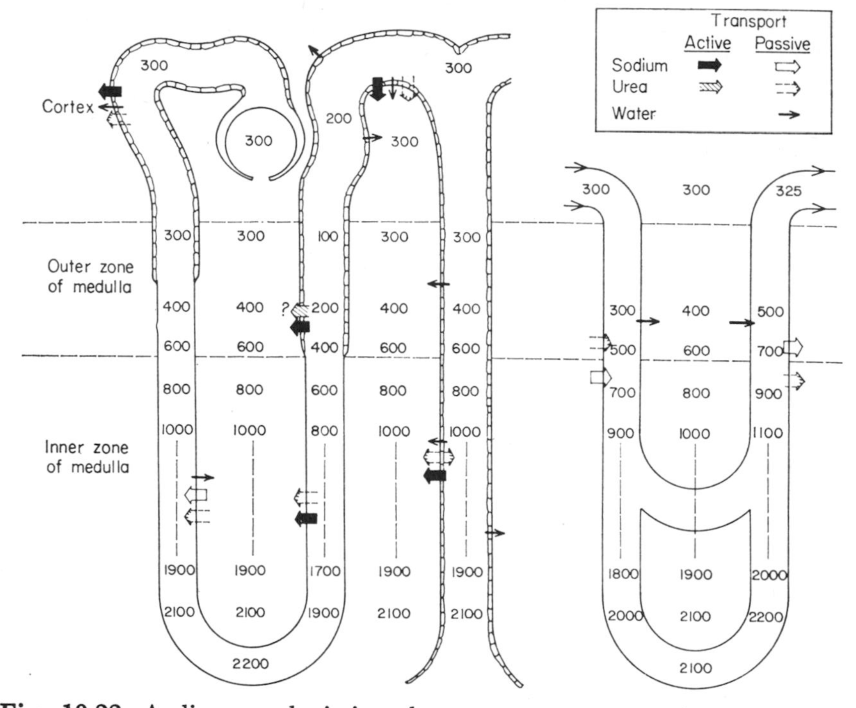

Fig. 10.22. A diagram depicting the countercurrent mechanism as it is believed to operate in a nephron, with a long loop, and in the vasa recta. Not all loops or vasa recta reach the tip of the papilla and hence the fluid in them does not become as concentrated as that of the final urine but only as concentrated as that at medullary level. This model requires that the ascending limb of the loop of Henle be relatively impermeable to water yet actively transport sodium out of the lumen. (Gottschalk and Mylle, *Am. J. Physiol.* 1959, **196,** 927.)

hypothetical stages leading ultimately to a steady state. In the illustration, gross steps have been envisaged and, as a result, the concentrations of solute in opposite segments of the loops are not equal. However, if the steps were made more gradually the discrepancies would be less so that, in effect, there would be approximately equal osmolality in a horizontal direction.

Application to the Nephron. The application of this system to the nephron with its associated straight collecting tubule is shown in Fig. 10.22, where the approximate osmolalities are indicated; it will be seen that the fluid flowing down the collecting duct becomes more and more concentrated as the result of osmosis into the tissue. An important assumption in this theoretical treatment is that the ascending limb has a low permeability to water, since it is only by virtue of this that we can envisage an active transport of solute out of the limb creating a large difference of osmolality; the ascending limb is thus similar to the distal tubule of the frog in that it can establish gradients of osmolality, but the countercurrent arrangement of the loop permits this difference of osmolality to be applied to an ultimate concentrating mechanism.

Changes in Distal Nephron. When the fluid reaches the distal tubule it is hypotonic, as experimentally found. During passage along the distal tubule there is considerable reabsorption of salt and water, but this occurs approximately iso-osmotically and, by the time the fluid has reached the beginning of the collecting tubule, most of the fluid will have been reabsorbed. The osmolality will be about equal to that of plasma, 300, but the concentration of Na^+ will be low since the osmolality is largely made up of urea. As osmosis proceeds in the descending route along the collecting tubule, the concentrations of all solutes build up; if salt must be retained as well as water, active transport out of the collecting tubule will also occur, as indicated by the broad arrow of Fig. 10.22. In fact as Table I shows, there is considerable removal of Na^+ from the fluid in the collecting duct, as revealed by the fall in the F/P ratio on passing from the bend of the loop to the duct, in micropuncture experiments on the hamster where the papilla is easily accessible.

Vasa Recta

It is not difficult to envisage the setting up of concentration gradients in a stagnant tissue; it can be argued, however, that the passage of blood through the system, with its osmolality of 300 mosm/litre, would quickly level out gradients. Nevertheless, the special feature of the blood circulation in the medulla is the hair-pin character of the vessels, the *vasa recta*, so that the blood operates its own countercurrent mechanism actually tending to stabilize the concentration gradients in the tissue rather than to break them down. Thus the blood entering the descending limb of a vasum rectum has an osmolality of 300; as it passes down it acquires higher and higher osmolalites by diffusion of solutes into it; thus on its way down it has tended to remove solutes, and the deeper it has gone the more solutes it has taken up so the flow has, indeed, tended to equalize the gradient of concentration. On the return journey, however, it removes less at the bottom of the ascending limb and will actually tend to restore what it has taken at successive levels, the more it has taken from a given level on its way down the more it will tend to return at the same level on the way up, and it is this reciprocity that constitutes the basis of a countercurrent system, a system that stabilizes the gradients. As Fig. 10.22 shows, the net effect of the flow of blood is to remove a little solute, but far less than if there had been no countercurrent system. In general, the osmosis of water into the interstitium during flow down the collecting tubule will increase the hydrostatic pressure in the tissue and favour bulk absorption in the capillaries of the vasa recta, and it is in this way, as well as in the lymphatic system, that the fluid removed from the tubules and collecting ducts is returned to the main vascular system.

Vasopressin (ADH)

The kidney may behave as a urine concentrating or diluting mechanism according to the needs of the organism. The switch-over from diuresis, when the urine is diluted, to antidiuresis, when the urine is concentrated, is achieved mainly by the secretion of the *antidiuretic hormone* (ADH), or *vasopressin*, into the blood by the posterior pituitary. As a result of the presence of the hormone in the blood, the permeability of the collecting tubule to water is increased, and this favours osmotic reabsorption of water.

Tubular Permeability. Experimentally the permeability of the distal tubule may be measured by injecting, say, a hyperosmolal solution and measuring the influx of water, and this is found to be markedly increased during antidiuresis in the rat. Again, isolated fragments of collecting tubule may be perfused (Fig. 10.16) and the permeability to water measured; this was increased by treatment with ADH. Finally, we may measure the permeability of the ascending and descending limbs of Henle's loop; as the theory demands, the permeability of the ascending loop to water is small and that of the descending loop large. Some results obtained by Morgan and Berliner, by perfusing the ascending and descending loops *in vivo*, are as follows:

Loop of Henle		*Collecting duct*	
Descending	Ascending	No ADH $+ 2$	ADH $+ 2$
$58{\cdot}6 \pm 6$	$4{\cdot}4 + 1$	$4{\cdot}2$	30

the figures representing hydraulic coefficients in nanolitres cm^{-2} milliosmole^{-1} min^{-1}).

Hyaluronidase. As to the mechanism by which ADH increases water-permeability the position is not clear; it has been suggested that the hormone causes the amount of intercellular cement of the collecting duct to be reduced in amount, and this is thought to favour increased permeability to water. When rats in diuresis were given water, the amount of interstitial cement increased markedly, and since this cement substance may well be hyaluronic acid the possible basis for ADH action might be to cause the liberation by the tubular cells of the enzyme, hyaluronidase, which attacks hyaluronic acid. Urine certainly contains an enzyme that depolymerizes hyaluronic acid. Failure of the organism to secrete ADH leads to a condition in which copious quantities of urine are excreted—*diabetes insipidus*—a condition that is remedied by injections of ADH. In the condition of nephrogenic diabetes insipidus, however, ADH has no effect so that there is some other cause, and it is interesting that ADH had no effect on the excretion of hyaluronidase in the urine although in the normal subject antidiuresis, stimulated in this way, causes an increased excretion of the enzyme.

Vascular Changes in Antidiuresis

The antidiuretic hormone, as its name vasopressin indicates, has a pronounced vascular effect, raising arterial blood pressure by virtue of a constriction of resistance vessels. It is natural to enquire whether the vascular effects are extended to the kidney and if so whether they contribute to antidiuresis; thus an increased flow through the cortical circuits and diminished flow through the vasa recta might favour the maintenance of the osmolality gradient in the medulla. In fact, when the blood-flow through the kidneys was examined, it was found that immediately after induction of a diabetes insipidus through a section of the hypothalamic-hypophysial tract, there was an increased total renal blood-flow that was suppressed by administering ADH. This could be shown, by autoradiography, to be due to a marked increase in cortical flow, while that through the juxtamedullary and outer medullary regions is actually decreased. The increased flow was due to a decrease in glomerular and efferent arteriolar resistance, and since these vessels are not apparently under nervous control (whilst the afferent arteriole is), it may be that hormonal control through ADH is exerted exclusively on the glomerular and postglomerular regions of the circulation. There seems little doubt from the experiments of Fisher, Grunfeld and Berger, that the diuresis provoked by absence of ADH is accompanied by vascular changes, but their importance in diuresis remains to be established.

Diuresis

Osmotic Diuresis

By diuresis is meant the excretion of a copious flow of urine; in diabetes mellitus this is due to the overloading of the proximal tubule osmotically so that the iso-osmotic reabsorption of NaCl is impaired (Fig. 10.23). A similar diuresis may be provoked experimentally by infusion of mannitol intravenously; this appears in the glomerular filtrate but is reabsorbed very slowly by the proximal tubule. Hence

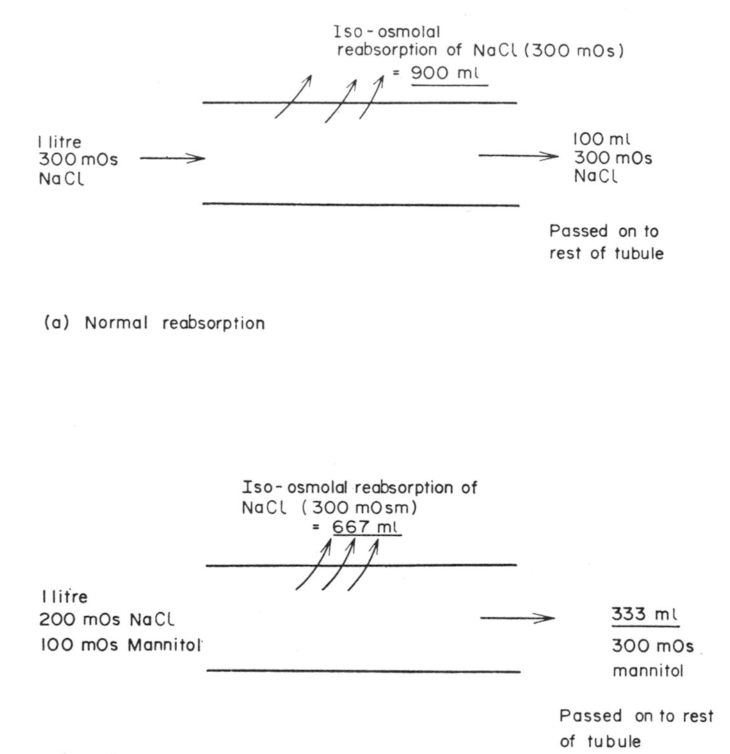

Fig. 10.23. Osmotic diuresis. The absorption of water in the proximal tubule can only occur in iso-osmolal amounts with sodium. As shown in (*a*) some 90 per cent of the filtrate is reabsorbed by this process, leaving only 10 per cent to be handled by the rest of the tubule. If mannitol is given intravenously this appears in the filtrate and contributes to the total osmolality as shown in (*b*). The iso-osmolal reabsorption in the proximal tubule can now only proceed in a restricted fashion since the osmotic effect of the mannitol, which is only reabsorbed slowly, holds back some of the water in the tubule. An increased water load is passed on to the distal nephron and this exceeds its reabsorptive capacity and diuresis occurs.

removal of NaCl iso-osmotically tends to build up the osmotic pressure of the remaining fluid and brings the process to a halt because the proximal tubule's power to establish osmotic gradients is small or non-existent.

Mercurial Diuresis

Another means of causing a diuresis is by poisoning the reabsorption of important solutes, such as Na^+; this can be done by a direct attack on the active transport of Na^+, and this is probably the basis of the action of mercurial diuretics, which presumably poison SH-containing enzymes concerned with some step in the supply of metabolic energy to the active reabsorption of Na^+. This has been generally located in the proximal tubule, but more recent work suggests that the main attack is on the Na^+-reabsorbing mechanisms in the distal nephron. The loss of body fluid resulting from the diuresis apparently initiates a defence reaction that leads to a reduction in proximal reabsorption, which is thus secondary to the primary attack on the distal nephron. As we shall see, loss of body fluid as such reduces proximal tubular reabsorption of water and salt (Vol. 3).

Diamox

The reabsorption of Na^+ can be reduced if the reabsorption of its accompanying anions is interfered with; this is probably the basis of the diuretic action of the carbonic anhydrase inhibitor acetazoleamide or Diamox. This interferes with the reabsorption of bicarbonate, presumably because this ion is mainly absorbed first as CO_2 which becomes converted into HCO_3^- outside the tubule. At any rate, the diuresis is accompanied by a large excretion of bicarbonate in the urine.

Salt-retention

Aldosterone

As indicated earlier, the control over the body's water-balance would be crude if it were based only on the osmolality of the plasma and exerted only through selective excretion of water. An additional control is achieved through the adrenal gland which secretes a hormone, *aldosterone*, that controls the excretion of salt; it is a salt-retaining hormone so that, when injected into the blood, the excretion of Na^+ is reduced. In Addison's disease, when the adrenal cortex is defective, there is excessive excretion of salt with lowered blood-Na^+ and raised blood-K^+, due to inhibited excretion of this ion. Under conditions of sodium depletion, the amount of aldosterone in the blood increases; on feeding excessive NaCl to human subjects the concentration of aldosterone in the blood is negligible. Examination of the excretory process by stop-flow analysis showed that the main action was on the distal tubule's reabsorption, so that the Na^+-trough (Fig. 10.17) disappears in the adrenalectomized animal.* Puncture studies have been conflicting, but on balance an impairment of both proximal and distal reabsorption seems to occur in adrenalectomy.

Aldosterone and antidiuretic hormone co-operate in so far as the increased active transport of Na^+ in the distal nephron, promoted by aldosterone, assists the urine concentrating mechanism of the countercurrent system. In fact when a human subject is short of water (hydropenia) he can excrete a more concentrated urine than normally hydrated subjects given a maximal dose of ADH; furthermore, in spite of the greater osmolality of the urine of the hydropenic subjects, the concentration of salt was diminished. Thus in the hydropenic state both hormones are acting, and experimentally it is found that in normally hydrated humans combined treatment with ADH and aldosterone gives a higher urine osmolality and lower total solute output than ADH alone. In the same way, when, as a result of haemorrhage, there is a necessity to conserve water and salt we find that not only is there an increased secretion of aldosterone but also of ADH.

* It must be emphasized that the study of the adrenalectomized animal is not the study of the animal simply without its aldosterone-secreting mechanism; the adrenal gland also secretes a "glucocorticoid-hormone"—hydrocortisone—which influences sugar metabolism and also affects renal haemodynamics, so that it is difficult to separate the effects of the two deficiences.

BIBLIOGRAPHY

The references given below are intended primarily as guides permitting the interested student to pursue any of the subjects in greater depth. To some extent this can be achieved by consultation of one of the large academically orientated texts, and to avoid repetition these may be mentioned now:

As further reading to Chapters 1, 2, 3, 6, 9, 10:

Davson, H. *A Textbook of General Physiology*. Churchill, London.

As further reading to Chapters 4, 5, 7, 9, 10:

Davson, H. and Eggleton, G. (Eds.) *Principles of Human Physiology*. Churchill, London.

Keele, C. A. and Neil, F. *Samson Wright's Applied Physiology*. O.U.P., London.

Mountcastle, V. B. (Ed.) *Medical Physiology*. Mosby, St. Louis.

Ruch, T. C. and Patton, H. D. (Eds.) *Physiology and Biophysics*. Saunders, Philadelphia.

Chapter 1

Abercrombie, M. (1961). The locomotory behaviour of cells. In *Cells and Tissues in Culture*, Vol 1 (Ed. E. N. Willmer), pp. 177–202. Academic Press, London, New York, San Francisco.

Bajer, A. S. and Molè-Bajer, J. (1972). *Spindle Dynamics and Chromosome Movements*. Academic Press, New York, London, San Francisco.

Bangham, A. D. (1964). The adhesiveness of leukocytes with special reference to Zeta potential. *Ann. N.Y. Acad. Sci.*, **116,** 945–949.

Cairns, J. (1973). DNA synthesis. *Brit. Med. Bull.*, **29,** 188–191.

Callan, H. G. (1973). Replication of DNA in eukaryotic chromosomes. *Brit. Med. Bull.*, **29,** 192–195.

Cambell, P. N. (1965). The biosynthesis of proteins. *Progr. Biophys.*, **15,** 3–38

Cove, D. J. (1972). *Genetics*. C.U.P.

Curtis, A. S. G. (1966). Cell adhesion. *Science Progr.*, **54,** 61–86.

De Bruyn, P. P. H. (1947). Theories of amoeboid movement. *Q. Rev. Biol.*, **22,** 1–24.

Franke, W. W. (1970). On the universality of nuclear pore complex structure. *Z. Zellforsch.*, **105,** 405–429.

Friend, D. S. and Gilula, N. B. (1972). Variations in tight and gap junctions in mammalian tissues. *J. Cell. Biol.*, **53,** 758–776.

Hendler, R. W. (1971). Biological membrane ultrastructure. *Physiol. Rev.*, **51,** 66–97.

Kaplan, D. M. and Criddle, R. S. (1971). Membrane structural proteins. *Physiol. Rev.*, **51,** 249–272.

Mehrishi, J. N. (1972). Molecular aspects of the mammalian cell surface. *Progr. Biophys.*, **25,** 3–70.

Schroeder, T. E. (1972). The contractile ring. II. *J. Cell Biol.*, **53,** 419–434.
Tilney, L. G. (1971). How microtubule patterns are generated. *J. Cell Biol.*, **51,** 837–854.
Wessels, N. K. *et al.* (1971). Microfilaments in cellular and developmental processes. *Science, N.Y.*, **171,** 135–143.
Willmer, E. N. (1961). Morphological problems of cell type, shape and identification. In *Cells and Tissues in Culture*, Vol. 1 (Ed. E. N. Willmer), pp. 143–176. Academic Press, London, New York, San Francisco.

Chapter 2

Davson, H. and Danielli, J. F. (1943). *The Permeability of Natural Membranes.* C.U.P. (Reprinted, 1970; Hafner Publ. Co., Darien, Conn.).
Diamond, J. M. and Bossert, W. H. (1967). Standing gradient osmotic flow. A mechanism for coupling of water and solute transport in epithelia. *J. gen. Physiol.*, **50,** 2061–2083.
Diamond, J. M. and Bossert,, W. H. (1968). Functional consequences of ultrastructural geometry in "backwards" fluid-transporting epithelia. *J. Cell Biol.*, **37,** 694–702.
Hendler, R. W. (1971). Biological membrane ultrastructure. *Physiol. Rev.*, **51,** 66–97.
House, C. R. (1974). *Water Transport in Cells and Tissues.* Arnold, London.
Keynes, R. D. (1969). From frog skin to sheep rumen: a survey of active transport of salts and water across multicellular membranes. *Q. Rev. Biophys.*, **2,** 177–281.
Northcote, D. H. (Ed.) (1968). Structure and function of membranes. *Brit. Med. Bull.*, **24,** 99–184.
Stein, W. D. (1967). *The Movement of Molecules across Cell Membranes.* Academic Press, New York, London, San Francisco.
Ussing, H. H. (1960). Active and passive transport across epithelial membranes. In *Methods of Isotopic Tracers Applied to the Study of Active Ion Transport* (Ed. J. Coursaget), pp. 139–154. Pergamon Press, London.
Whittam, R. (1964). *Transport and Diffusion in Red Blood Cells.* Arnold, London.

Chapter 3

Benzinger, T. H. (1969). Heat regulation: homeostasis of central temperature in man. *Physiol. Rev.*, **49,** 671–759.
Kerslake, D. McK. (1972) *The Stress of Hot Environments.* C.U.P.
Taylor, C. R. (1970). Strategies of temperature regulation: effect on evaporation in East African ungulates. *Am. J. Physiol.*, **219,** 1131–1135.
Smith, R. E. and Horwitz, B. A. (1969). Brown fat and thermogenesis. *Physiol. Rev.*, **49,** 330–425.

Chapter 4

Bull, H. B. (1964). *An Introduction to Physical Biochemistry.* F. A. Davis Co., Philadelphia.
Dixon, M. and Webb, E. C. (1964). *Enzymes.* 2nd Ed. Longmans, London.
Doonan, S., Vernon, C. A. and Banks, B. E. C. (1970). Mechanisms of enzyme action. *Progr. Biophys. Mol. Biol.*, **20,** 247–327.
Martin, R. B. (1964). *Introduction to Biophysical Chemistry.* McGraw-Hill, New York.
Phillips, D. C. (1970, 1971). On the stereochemical basis of enzyme action: Lessons from lysozyme. *Harvey Lectures*, Series **66.** Academic Press, New York, London, San Francisco.

Chapter 5

Bloch, E. H. (1962). A quantitative study of the hemodynamics in the living microvascular system. *Am. J. Anat.*, **110,** 125–145.

Burton, A. C. (1954). Relation of structure to function of the tissues of the wall of blood vessels. *Physiol. Rev.*, **34,** 619–642.

Burton, A. C. (1970). *Biophysics of the Circulation.* Year Book Med. Publ. Co., Chicago.

Folkow, B. and Neil, E. (1971). *Circulation.* O.U.P., New York.

Goldsmith, H. L. (1968). The microrheology of red blood cell suspensions. *J. gen. Physiol.*, **52,** 5–28 s.

Henry, J. P. and Meehan, J. P. (1971). *The Circulation: an Integrated and Physiological Study.* Year Book Med. Publ. Co., Chicago.

McDonald, D. A. (1960). *Blood Flow in Arteries.* Arnold, London.

Schmid-Schönbein, H. and Wells, R. E. (1971). Rheological properties of human erythrocytes and their influence upon the "anomalous viscosity" of blood. *Ergebn. Physiol.*, **63,** 146–219.

Whittaker, S. R. F. and Winton, F. R. (1932). The apparent viscosity of blood flowing in the isolated hindlimb of the dog and its variation with corpuscular concentration. *J. Physiol.*, **78,** 339–369.

Chapter 6

Garlick, D. G. and Renkin, E. M. (1970). Transport of large molecules from plasma to interstitial fluid and lymph in dogs. *Am. J. Physiol.*, **219,** 1595–1605.

Guyton, A. C., Granger, H. J. and Taylor, A. E. (1971). Interstitial fluid pressure. *Physiol. Rev.*, **51,** 527–563.

Kirsch, K., Rafflenbeul, W. and Roedel, H. (1971). Untersuchungen zur Ursache des negativen interstitiellen Gewebsdruckes (Guyton-Kapsel). *Pflüg. Arch. ges. Physiol.*, **328,** 193–204.

Reynolds, S. R. M. and Zweifach, B. W. (1959). *The Microcirculation.* Univ. Illinois Press, Urbana.

Yoffey, J. M. and Courtice, F. C. (1970). *Lymphatics, Lymph, and the Lymphomyeloid Complex.* Academic Press, New York, London, San Francisco.

Chapter 7

Blake, C. C. F. (1972). X-Ray studies of crystalline proteins. *Progr. Biophys.*, **25,** 85–130.

Comroe, J. H. *et al.* (1962). *The Lung: Clinical Physiology and Pulmonary Function Tests.* 2nd Ed. Year Book Med. Publ. Co. Chicago.

Davenport, H. W. (1969). *The ABC of Acid-Base Chemistry.* Univ. Press, Chicago.

Hemmingsen, E. A. (1965). Accelerated transfer of oxygen through solutions of heme pigments. *Acta physiol. scand.*, Suppl., **246.**

Lehmann, H. and Huntsman, R. G. (1970). *Man's Haemoglobins.* North Holland Publ. Co., Amsterdam.

Perutz, M. F. (1970). Stereochemistry of cooperative effects in haemoglobin. *Nature, Lond.*, **228,** 726–739.

Stein, T. R., Martin, J. C. and Keller, K. H. (1971). Steady-state oxygen transport through red blood cell suspensions. *J. appl. Physiol.*, **31,** 397–404.

Wittenberg, J. B. (1970). Myoglobin-facilitated oxygen diffusion: role of myoglobin in oxygen entry into muscle. *Physiol. Rev.*, **50,** 559–636.

Zander, R. and Schmid-Schönbein, H. (1972). Influence of intracellular convection on the oxygen release by human erythrocytes. *Pflüg. Arch. ges. Physiol.*, **335,** 58–73.

Chapter 8

Ciba Symposium. (1962). *Pulmonary Structure and Function.* Churchill, London.
Clements, J. A. (1962). Surface phenomena in relation to pulmonary function. *Physiologist*, **5,** 11–28.
Comroe, J. H. *et al.* (1962). *The Lung: Clinical Physiology and Pulmonary Function Tests.* 2nd Ed. Year Book Med. Publ. Co., Chicago.
Hatasa, K. and Nakamura, T. (1965). Electron microscopic observations of lung alveolar epithelium cells of normal young mice. *Z. Zellforsch.*, **68,** 266.
Slonim, N. and Hamilton, L. H. (1971). *Respiratory Physiology.* 2nd Ed. C. V. Mosby, St. Louis.
West, J. B. (1967). *Ventilation/Blood Flow and Gas Exchange.* 2nd Ed. Blackwell, Oxford.

Chapter 9

Burgen, A. S. V. and Emmelin, N. G. (1961). *Physiology of the Salivary Gland.* Arnold, London.
Case, R. M., Scratcherd, T. and Wynne, R. D. A. (1970). The origin and secretion of pancreatic juice bicarbonate. *J. Physiol.*, **210,** 1–15.
Castle, J. D., Jamieson, J. D. and Palade, G. E. (1972). Radioautographic analysis of the secretory process in the parotid acinar cell of the rabbit. *J. Cell Biol.*, **53,** 290–311.
Davenport, H. W. (1966). *Physiology of the Digestive Tract.* Year Book Med. Publ. Co., Chicago.
Diamond, J. M. and Bossert, W. H. (1967). Standing gradient osmotic flow. A mechanism for coupling of water and solute tranpsort in epithelia. *J. gen. Physiol.*, **50,** 2061–2083.
Diamond, J. M. and Bossert, W. H. (1968). Functional consequences of ultrastructural geometry in "backwards" fluid-transporting epithelia. *J. Cell. Biol.*, **37,** 694–702.
Edmonds, C. J. (1967). The gradient of electrical potential difference and of sodium and potassium in the gut contents along the caecum and colon of normal and sodium-depleted rats. *J. Physiol.*, **193,** 571–588.
Gibson, Q. H. and Wiseman, G. (1951). Selective absorption of stereo-isomers of amino acids from loops of the small intestine of the rat. *Biochem. J.*, **48,** 426–429.
Gregory, R. A. (1961). *Secretory Mechanisms of the Gastrointestinal Tract.* Arnold, London.
Hargreaves, T. (1968). *The Liver and Bile Metabolism.* North Holland Publ. Co., Amsterdam.
Kurosumi, K. (1961). Electron microscopic analysis of the secretion mechanism. *Int. Rev. Cytol.*, **11,** 1–124.
Mangos, J. A. and Braun, G. (1966). Excretion of total solute, sodium and potassium in the saliva of the rat parotid gland. *Pflüg. Arch. ges. Physiol.*, **290,** 184–192.
Mangos, J. A., Braun, G. and Hamann, K. F. (1966). Micropuncture study of sodium and potassium excretion in the rat parotid saliva. *Pflüg. Arch. ges. Physiol.*, **291,** 99–106.

Preisig, R., Cooper, H. L. and Wheeler, H. O. (1962). The relationship between taurocholate secretion rate and bile production in the unanaesthetized dog during cholinergic blockade and during secretin administration. *J. clin. Invest.*, **41,** 1152–1162.

Ritchie, J. A. (1968). Colonic motor activity and bowel function. *Gut*, **9,** 442–456.

Schultz, S. G. and Zalusky, R. (1964). Ion transport in isolated rabbit ileum. I. Short-circuit current and Na fluxes. *J. gen. Physiol.*, **47,** 567–584.

Scratcherd, T. (1965). Electrolyte composition and control of biliary secretion in the cat and rabbit. In *The Biliary System* (Ed. W. Taylor). Blackwell, Oxford.

Sperber, I. (1959). Secretion of organic anions in the formation of urine and bile. *Pharmacol. Rev.*, **18,** 109–134.

Wiseman, G. (1964). *Absorption from the Intestine*. Academic Press, New York, London, San Francisco.

Yoshida, Y., Sprecker, R. L., Schneyer, C. A. and Schneyer, L. H. (1967). Role of β-receptors in sympathetic regulation of electrolytes in rat submaxillary saliva. *Proc. Soc. exp. Biol. N.Y.*, **126,** 912–916.

Young, J. A., Frömter, E., Schögel, F. and Hamann, K. F. (1967). A microperfusion investigation of sodium resorption and potassium secretion by the main excretory duct of the rat submaxillary gland. *Pflüg. Arch ges. Physiol.*, **295,** 157–172.

Young, J. A., Martin, C. J. and Weber, F. D. (1971). The effect of a sympatho- and a parasympathomimetic drug on the electrolyte concentrations of primary and final saliva of the rat submaxilary gland. *Pflüg. Arch. ges. Physiol.*, **327,** 285–302.

Young, J. A. and Schlögel, E. (1966). Micropuncture investigation of sodium and potassium excretion in rat submaxillary saliva. *Pflüg. Arch. ges. Physiol.*, **291,** 85–98.

Chapter 10

De Wardener, H. E. (1967). *The Kidney. An Outline of Normal and Abnormal Structure and Function*. Churchill, London.

Dicker, S. E. (1970). *Mechanism of Urine Concentration and Dilution in Mammals*. Arnold, London.

Fisher, R. D., Grünefeld, J.-P. and Barger, A. C. (1970). Intrarenal distribution of blood flow in diabetes insipidus: role of ADH. *Am. J. Physiol.*, **219,** 1348–1358.

Gottschalk, C. W. and Mylle, M. (1959). Micropuncture study of the mammalian urinary concentrating mechanism. *Am. J. Physiol.*, **196,** 927–936.

Graham, R. C. and Karnovsky, M. J. (1966). Glomerular permeability. *J. exp. Med.*, **124,** 1123–1134.

Lassiter, W. E., Gottschalk, C. W. and Mylle, M. (1961). Micropuncture study of net transtubular movement of water and urea in nondiuretic mammalian kidney. *Am. J. Physiol.*, **200,** 1139–1147.

Morgan, T. and Berliner, R. W. (1968). Permeability of the loop of Henle, vasa recta, and collecting duct to water, urea, and sodium. *Am. J. Physiol.*, **215,** 108–115.

Navar, L. G. (1970). Minimal preglomerular resistance and calculation of normal glomerular pressure. *Am. J. Physiol.*, **219,** 1658–1664.

Pitts, R. F. (1963). *Physiology of Kidney and Body Fluids*. Year Book Med. Publ. Co., Chicago.

Smith, H. (1951). *The Kidney*. O.U.P., New York.

Solomon, A. K. (1959). Ion and water transport in single proximal tubules of the *Necturus* kidney. In *The Method of Isotopic Tracers Applied in the Study of Active Ion Transport* (Ed. J. Coursaget), pp. 106–124. Pergamon, London.

Ullrich, K. H., Kramer, K. and Boylan, J. W. (1961). Present knowledge of the counter-current system in the mammalian kidney. *Progr. Cardiovasc. Dis.*, **3**, 395–431.

SUBJECT INDEX

A

I

L

Q

R

S